# TRIGONOMETRY
## A GRAPHING APPROACH

# TRIGONOMETRY

## A GRAPHING APPROACH

**FRANKLIN DEMANA • BERT K. WAITS**

with the assistance of
ALAN OSBORNE • GREGORY D. FOLEY

The Ohio State University

ADDISON-WESLEY PUBLISHING COMPANY

Reading, Massachusetts • Menlo Park, California • New York
Don Mills, Ontario • Wokingham, England • Amsterdam • Bonn
Sydney • Singapore • Tokyo • Madrid • San Juan

| | |
|---|---|
| Executive Editor | David F. Pallai |
| Associate Editor | Stephanie Botvin |
| Production Administrator | Catherine Felgar |
| Senior Electronic Production Supervisor | Mona Zeftel |
| Electronic Production Consultant | Laurie Petrycki |
| Electronic Production Assistant | Sally Simpson |
| TeX Consultant | Frederick H. Bartlett of The Bartlett Press, Inc. |
| Text Designer | Geri Davis of Quadrata, Inc. |
| Copyeditor | Lorraine Ferrier |
| Manufacturing Supervisor | Roy Logan |
| Cover Designer | Marshall Henrichs |

This book was produced with TeXtures.

Library of Congress Cataloging-in-Publication Data

Demana, Franklin D., 1938–
    Trigonometry, a graphing approach/ by Franklin Demana and Bert K. Waits; with the assistance of Alan Osborne and Gregory D. Foley.

    p. cm.
    Includes index.
    ISBN 0-201-52812-6
    1. Trigonometry–Graphic methods. I. Waits, Bert K. II. Title.
QA538.D45 1989
516.24'0285–dc20                                                                                                                89-6790

Copyright © 1990 by Addison-Wesley Publishing Company, Inc. All rights reserved. No part of this publication may be reproduced, stored in a retrieval system, or transmitted, in any form or by any means, electronic, mechanical, photocopying, recording, or otherwise, without the prior written permission of the publisher. Printed in the United States of America.

ABCDEFGHIJ-MU-954321089

The authors dedicate this book to their wives, Christine Demana and Barbara Waits, without whose patience, love, and understanding this book would not have been possible.

# Preface

This text is designed for a one-quarter or one-semester college trigonometry course. It was written because the authors believe that conventional texts that do not incorporate technology do not prepare students well for further study in mathematics and science in the 1990s. For example, only about 17% of the entering freshmen at The Ohio State University with four or more years of high school college-preparatory mathematics, including a precalculus course, are ready to begin their collegiate study of mathematics with calculus. This is consistent with national evidence that performance in college calculus courses is dismal at best.

Calculator- and computer-based graphing technology is incorporated in this text to enhance the teaching and learning of precalculus mathematics. Students are expected to have *regular and frequent* access to graphing calculators or computers with appropriate software for homework outside of class and perhaps for occasional classroom laboratory activities as well. Use of computer graphing or graphing calculator technology in this text is *not* optional.

Modern technology has evolved to the stage that it should be routinely used by mathematics students at all levels. Computer- or calculator-based graphing removes the need for contrived problems and opens the door for students to explore and solve realistic and interesting applications. The teaching and learning of traditional topics can be improved with the full use of these new tools. Computer- and calculator-based technology can turn the mathematics classroom into a mathematics laboratory. Technology gives rise to interactive instructional approaches that permit a focus on problem solving and encourage generalizations based on strong geometric evidence. The new instructional approaches that are possible with today's technology make you and your students active partners in an exciting, rewarding, enjoyable, and intensive educational experience. It is in this spirit of exploration and experimentation that this text is written.

## Content Features

- **Applications**  Problems are used to introduce and develop much of the mathematics in this text. Students using this approach become flexible problem solvers. Using problems as a basis for discussion makes the mathematics understandable to students, and students come to value mathematics because they appreciate its power. More realistic applications are possible because of the speed and power of technology (see Example 5 of Section 5.6 and Example 6 of Section 6.2). You can select exercises with a business and economics theme, a science theme, or other themes.

- **Graphs**  In the past careful, time consuming numerical and analytical techniques were used to produce accurate graphs that were rarely used. Now accurate graphs are obtained quickly and used to study the numerical and analytical properties of functions (see Example 5 of Section 4.3). Important questions are often generated by students viewing graphs. Sometimes these questions are answered using algebraic manipulation, which is then often well received by students. Other times they are answered by student exploration, taking advantage of the speed and power of graphing technology (see Examples 1 and 8 of Section 3.1).

- **Scale**  Scale is a crucial issue. The shape of a graph depends on the viewing window or rectangle in which the graph is viewed (see Example 6 of Section 1.2). Care in selecting viewing rectangles must be exercised because familiar graphs may not be recognized when they are distorted.

- **Multiple Representations**  Algebraic representations (equations, inequalities, etc.) and geometric representations (graphs of algebraic representations) are established for a given problem situation (see Example 1 of Section 1.1). Then connections among algebraic representations, geometric representations, and the problem situation are exploited to provide understanding about mathematical concepts and to give a geometric base for the algebraic ideas. Modeling receives special attention in this text (see Example 2 of Section 1.6).

- **Foreshadowing Calculus**  Maxima and minima of functions are found in this text by using graphs (see Example 5 of Section 2.7). Intervals where functions are increasing or decreasing (see Example 5 of Section 1.7) and limiting behavior of functions are determined graphically. We do *not* borrow the techniques of calculus—rather we lay the foundation for the later study of calculus by providing students with rich intuitions and understandings about functions and graphs.

- **Answers**  Emphasis on exact answers is reduced, and approximate answers are emphasized. Technology provides a proper balance between exact answers that are rarely needed in the real world and approximate answers. What is usually needed is an answer with prescribed accuracy (see Example 7 of Section 1.4). Graphing techniques, such as zoom-in, provide an excellent geometric vehicle for discussing error in answers. Students can read answers from graphs with accuracy up to the limits of machine precision.

- **Algebraic Manipulation**  You will find a good deal of ordinary algebraic manipulation in this text. However, the algebraic techniques often arise from problem situations or are used to answer questions generated by graphs (see Example 2 of Section 3.4). Students are more willing to perform algebraic manipulations when they are developed in context but are *not* the focus of a lesson.

- **Geometric Transformations**   The exploratory nature of graphing technology helps students learn how a given graph may be obtained from a basic graph using the following geometric transformations: horizontal or vertical shifting, horizontal stretching or shrinking, vertical stretching or shrinking, reflection with respect to a coordinate axis. This technology-enhanced approach develops students' abilities so that they can sketch graphs of complicated functions (see Example 3 of Section 4.3 and Example 4 of Section 3.3).

- **Trigonometric Identities**   We deliberately spend less time in this textbook verifying identities algebraically and believe this to be appropriate. We do use graphs to investigate the plausibility of equations that are possible identities.

## Pedagogical Features

Each chapter begins with an introductory overview of the material to be covered. The first paragraph of a section gives an overview of the section. The text requires considerably more reading than conventional texts. Students are expected to read the text and work through the examples.

- **Exercises**   Directions for some exercises state *not* to use graphing technology. Whenever possible, it is good practice to have students verify responses to these exercises with graphing technology. Many exercises will require the use of technology. We have *not* written an example for each type of exercise given. Some exercises that are more difficult or extend the ideas of the section are marked with an asterisk. There are more exercises than can be reasonably assigned to any one student. The exercises were designed to provide you with flexibility to accommodate students with different backgrounds and interests.

- **Intensity**   The extensive examples found within a section, coupled with the interrelated explanations, provide for an intensity of topic coverage. Thus, it is usually not possible to cover a section in one class meeting. Some examples should be left as reading exercises for students. You may find it helpful sometimes to work one problem through completely even if it takes the entire class, especially early in the course. However, you may find it necessary to omit some topics.

- **The Role of the Instructor Using Technology**   The role of an instructor changes from a lecturer to a facilitator of learning. Students become active partners in the learning process as a consequence of technology. They learn to explore and experiment with mathematical concepts because of the speed and power of technology. You can use a single computer or an overhead graphing calculator in the classroom to provide an interactive lecture-demonstration, or use a guided-discovery approach in a computer lab or in a classroom where each student has a graphing calculator. You can use realistic problems to motivate and teach mathematical concepts because technology makes such problems accessible. Technology provides a much richer classroom environment that fosters student involvement in the educational experience. You should encourage students to explore and experiment with the technology.

- **Visualization** Graphing helps students learn how to see and describe graphs and their characteristics. You use the power of visualization by carefully selecting a sequence of visual experiences to help students understand or discover mathematical concepts. Graphing technology makes the addition of geometric representations to the usual numeric and algebraic representations very natural. Exploring the connections between representations and problems deepens student understanding about mathematics and helps students value mathematics.

- **Review** Each chapter ends with a list of key terms introduced in that chapter and an extensive set of exercises. These exercises can be used to prepare students for exams or used for quizzes.

## Supplements for the Instructor

- **Instructor's Manual** The Instructor's Manual contains an introductory chapter that gives an extensive overview and advice about using a technology-enhanced approach. There is a chapter-by-chapter commentary including remarks about selected exercises. Two versions are given for each chapter test. The Instructor's Manual contains answers to all exercises. The text contains the answers to approximately half of the exercises—usually the odd ones.

- **Tests** The Instructor's Manual that accompanies this text contains two versions of each chapter test.

- **Software** Instructors of class-size adoptions of this text receive *Master Grapher*, a powerful, interactive computer software package designed by the authors that graphs functions, conic equations, parametric equations, polar equations, and functions of two variables. *Master Grapher* comes in versions that work on Apple II, Macintosh, IBM, and most IBM-compatible computers.

- **Graphing Calculator and Computer Graphing Laboratory Manual** This laboratory manual gives details about the use of graphing calculators and computers to support the text. Students also receive a copy of the manual. It contains a user's guide to the *Master Grapher* graphing software as well as guides to the use of a Casio or Sharp graphing calculator.

## A Request

For many of you using this text, incorporating technology into the teaching of mathematics is a new venture. You are a pathfinder in the true sense of the word and, consequently, will enjoy some highs when instruction goes well but some lows when your experiences do not match your expectations.

We would like to hear from you concerning your successes and frustrations. This is partly to guide the design of a revision of this text and partly to gain understanding of what it means for instructors and institutions to shift to using technology for mathematics teaching.

## Acknowledgements

The authors are indebted to a great many people who participated in the development of this text. We very much appreciate the important and constructive suggestions of our colleagues Alan Osborne and Greg Foley. Special thanks are due to Janice McDonald and Barbara Waldron, our skillful TEX typists, and to Amy Edwards, who prepared the artwork for the manuscript and worked tirelessly to help the authors meet important deadlines. We also appreciate the efforts of David New, our resident TEXpert, who facilitated production of the book; and Jill Baumer, Cindy Bernlohr, Christine Browning, Sandy Davey, Tony DeGennaro, Penny Dunham, Ann Farrell, Bishnu Naraine, Jeri Nichols, Anne Sadeghipour, Linda Taylor, and Laurie Wern, who typed drafts, prepared answers, proofread, and made numerous helpful suggestions. The staff at Addison-Wesley, especially Stephanie Botvin and David Pallai, have provided enthusiastic support and expert guidance throughout the production process, making us proud to be part of the Addison-Wesley family.

We sincerely thank our colleagues who participated in the 1988–89 field test of a preliminary version of the textbook and who made valuable contributions that helped shape this book.

## 1988–89 College Field Test Instructors

Chris Allgyer, *Mountain Empire Community College*
George R. Barnes, *University of Louisville*
Daniel Buchanan, *Henry Ford Community College*
James G. Carr, *Iona College*
Gloria Child, *Rollins College*
Carolyn Crandell, *Muskingum College*
Professor Philip A. DeMarois, *National College of Education*
Ann Dinkheller, *Xavier University*
Gloria Dion, *Pennsylvania State University-Ogontz*
Eunice Everett, *Seminole Community College*
Max O. Gerling, *Eastern Illinois University*
Margaret J. Greene, *Florida Community College at Jacksonville*
Thomas Gregory, *The Ohio State University-Mansfield*
William L. Grimes, *Central Missouri State University*
John G. Harvey, *University of Wisconsin-Madison*
Robert Hathway, *Illinois State University*
Ingrid Holzner, *University of Wisconsin-Milwaukee*
Spencer P. Hurd, *The Citadel*
Lynne K. Ipiña, *University of Wyoming*
Bill Jordan, *Seminole Community College*
Rose Kaplan, *The Ohio State University-Newark*
Thomas J. Kearns, *Northern Kentucky University*
Stephen C. King, *The University of South Carolina at Aiken*
David E. Kullman, *Miami University*
Larry Lance, *Columbus State Community College*
Edward Laughbaum, *Columbus State Community College*

Robert Lavelle, *Iona College*
Tommy Leavelle, *John Brown University*
Millianne Lehmann, *University of San Francisco*
Gerald Leibowitz, *The University of Connecticut*
John Long, *University of Rhode Island*
Mary E. Maxwell, *The University of Akron*
Roger B. Nelsen, *Lewis and Clark College*
Henry Nixt, *Shawnee State University*
Susan D. Penko, *Baldwin Wallace College*
Anthony Perrescini, *University of Illinois–Urbana-Champaign*
Ruth A. Pruitt, *Fort Hays State University*
William Rettig, *Indiana University of Pennsylvania*
John Savige, *St. Petersburg Junior College*
Lawrence Sher, *Manhattan Community College (CUNY)*
Donald Shriner, *Frostburg State College*
Al Stickney, *Wittenberg University*
Karen Sutherland, *The College of St. Catharine*
Frederic Tufte, *University of Wisconsin-Platteville*
Kathy Underdown, *Rollins College*
Marjie Vittum-Jones, *South Seattle Community College*
Chuck Vonder Embse, *Central Michigan University*
Ron Waite, *Blue Mountain Community College*
Suzanne Welsch, *Sierra Nevada College*
Howard L. Wilson, *Oregon State University*

*Columbus, Ohio*

F.D.
B.K.W.

# Contents

## 1 Functions — 1

- 1.1 Cartesian Coordinate System and Complete Graphs — 3
- 1.2 Functions and Graphing Utilities — 19
- 1.3 More on Functions — 31
- 1.4 Solving Equations — 41
- 1.5 Solving Systems of Equations — 57
- 1.6 Solving Inequalities Algebraically and Graphically — 70
- 1.7 Maximum and Minimum Values — 83
- 1.8 Increasing and Decreasing Functions — 91
- 1.9 Continuity and End Behavior — 98
- Key Terms — 113
- Review Exercises — 115

## 2 Trigonometric Functions — 125

- 2.1 Angle — 126
- 2.2 Right Triangle Trigonometry — 133
- 2.3 Applications — 141
- 2.4 Trigonometric Functions of Any Angle — 148
- 2.5 Radian Measure and Graphs of $\sin x$ and $\cos x$ — 162
- 2.6 Graphs of $\tan x$, $\cot x$, $\sec x$, and $\csc x$ — 176
- 2.7 Trigonometric Graphs — 186
- Key Terms — 195
- Review Exercises — 195

## 3 Analytic Trigonometry — 199

- 3.1 More on Trigonometric Graphs — 199
- 3.2 Trigonometric Identities — 209
- 3.3 Inverse Relations — 222
- 3.4 Inverse Trigonometric Functions — 232
- 3.5 Solving Equations and Inequalities — 242
- 3.6 The Law of Sines — 253
- 3.7 The Law of Cosines and Area of Triangles — 264
- Key Terms — 273
- Review Exercises — 273

## 4 Logarithmic and Exponential Functions — 279

- 4.1 Operations on Functions — 280
- 4.2 Exponential Functions — 291
- 4.3 Economic Applications — 302
- 4.4 Logarithmic Functions — 314
- 4.5 More on Logarithms — 324
- 4.6 Equations, Inequalities, and Extreme-Value Problems — 335
- Key Terms — 344
- Review Exercises — 344

## 5 Trigonometry of Complex Numbers, Vectors, Polar Equations, and Parametric Equations — 349

- 5.1 Trigonometry and the Geometry of Complex Numbers — 350
- 5.2 De Moivre's Theorem, Roots of Complex Numbers, and Binomial Theorem — 360
- 5.3 Polar Coordinates and Graphs — 369
- 5.4 Vectors — 380
- 5.5 Parametric Equations — 393
- 5.6 Motion Problems and Parametric Equations — 399
- Key Terms — 408
- Review Exercises — 409

## 6 Conics — 413

- 6.1 Parabolas — 414
- 6.2 Ellipses — 427
- 6.3 Hyperbolas — 438

## Contents

| | | | |
|---|---|---|---|
| | 6.4 | Quadratic Forms and Conics | 450 |
| | 6.5 | Nonlinear Systems of Equations and Inequalities | 459 |
| | 6.6 | Polar Equations of Conics | 467 |
| | | Key Terms | 478 |
| | | Review Exercises | 478 |

## Answers                                                                   A1

## Index                                                                      I1

# TRIGONOMETRY
A GRAPHING APPROACH

# 1  Functions

## • Introduction

In this chapter we introduce models, algebraic representations, geometric representations, and complete graphs—important concepts that will be used throughout the textbook. The definitions of these terms will not be precise, but their meanings will become clear as examples are given throughout the textbook. Roughly speaking, algebraic representations are equations, inequalities, or systems of equations or inequalities, and geometric representations are graphs. We use algebraic representations and geometric representations as models of problem situations. We introduce functions and relations and draw graphs in the Cartesian coordinate system in this chapter. We will explain what is meant by the domain, range, and graph of a relation, function, equation, algebraic representation, and problem situation. Functions, relations, equations, inequalities, systems of equations and inequalities, and graphs are examples of models. Models can be very complex, and generally there is no single, correct model for a given problem situation. The person analyzing the problem situation usually chooses a model to represent the problem situation. Ideally a model should contain all the pertinent information about the problem situation and not be unduly complicated.

In this textbook we pay a good bit of attention to the way the following topics are interrelated: problem situations, algebraic representations of problem situations, and their graphs. Textbooks often focus most of their attention on working with various algebraic representations that are common in mathematics and give too little attention to geometric representations and their connections with algebraic representations and problem situations. Applications and graphical problem solving are important themes in this textbook.

Functions are important models in mathematics. We use graphing calculators or computer graphing software to obtain graphs of functions and certain relations. The vertical line test is used to decide if the relation determined by a graph is a function. We will see that graphing is an extremely important tool in mathematics. The absolute value function, the greatest integer function, and piecewise-defined functions will be introduced in this chapter.

We find real number solutions to equations, inequalities, and systems of equations using both algebraic and graphical techniques. We introduce the important graphical problem-solving technique called *zoom-in*. Computer based graphing techniques are more general and more powerful than algebraic techniques. In this chapter we concentrate on equations, inequalities, and systems of equations that involve *polynomials*.

In the following definition we use the symbol $a_i$, where $i$ is a positive integer. The $i$ in $a_i$ is called a **subscript**, and $a_i$ is read "$a$ sub $i$." There are 26 letters in the alphabet. Subscripting letters of the alphabet is a common way to expand the list of letters or variables for use in mathematics.

• **DEFINITION**   Any expression of the form

$$a_n x^n + a_{n-1} x^{n-1} + \cdots + a_1 x + a_0,$$

where $n$ is a nonnegative integer and $a_n$, $a_{n-1}$, ..., $a_0$ are real numbers is called a **polynomial in $x$**. If $a_n \neq 0$, then $n$ is called the **degree** of the polynomial. The $a_i$'s are called **coefficients**, $a_n$ is often called the **leading coefficient** of the polynomial and $a_0$ the **constant term**. If a function $f$ is a polynomial in $x$, then $f$ is called a **polynomial function**.   •

A polynomial of degree 1 is called a **linear polynomial**, and a polynomial of degree 2 is called a **quadratic polynomial**. For example, $\frac{1}{2}x - 3$ is a linear polynomial and $2x^2 - x + 1$ is a quadratic polynomial. The expression $3x^4 - 2x^2 - x + 15$ is a polynomial of degree 4. Be sure you understand why $\frac{2x-1}{x^2-1}$ and $3\sqrt{x} + 5$ are *not* polynomials.

• **DEFINITION**   Any equation of the form

$$a_n x^n + a_{n-1} x^{n-1} + \cdots + a_1 x + a_0 = 0,$$

is called a **polynomial equation**. If $a_n \neq 0$, the equation is called a **polynomial equation of degree $n$**.   •

The equation $\frac{1}{2}x - 3 = 0$ is a polynomial equation of degree 1, and the equation $2x^2 - x + 1 = 0$ is a polynomial equation of degree 2. The equations $\frac{2x-1}{x^2-1} = 0$ and $3\sqrt{x} + 5 = 0$ are *not* polynomial equations because $\frac{2x-1}{x^2-1}$ and $3\sqrt{x} + 5$ are *not* polynomials.

You have used algebraic methods and formulas in previous mathematics courses to solve equations. Solving polynomial equations graphically is a general solution method that does not depend on the degree. A graphical solution is quick, accurate, and leads to important geometric intuition about the meaning of solutions to an equation, inequality, or system of equations.

We establish the connection between finding solutions algebraically and geometrically. Solving equations graphically is equivalent to finding the *x-intercepts* of a related graph or points *where two graphs intersect*. Solving inequalities graphically involves determining when a graph is above or below the *x*-axis or when one graph is above or below another. When solving problems graphically, we must concern ourselves with the *accuracy* of the solutions found. A discussion about error is included in this chapter.

We find the intervals on which a function is increasing or decreasing, and we determine the coordinates of local maximum and local minimum values of functions. Graphs of inequalities of the form $y > f(x)$, $y \geq f(x)$, $y < f(x)$, and $y \leq f(x)$, where $f$ is a polynomial, are drawn. Continuity, end behavior, and the Intermediate Value Property are introduced in this chapter.

## 1.1 Cartesian Coordinate System and Complete Graphs

In this section we introduce the Cartesian coordinate system and explain what is meant by domain, range, and the graph of a relation, equation, and problem situation. The meaning of *complete graph* will be introduced. We also explain what is meant by an algebraic representation of a problem situation and a geometric representation of a problem situation. A geometric representation will be given for a problem situation and compared with the graph of an algebraic representation of the problem situation.

Data that arise in problem situations can be graphed. A branch of mathematics called *data analysis* is devoted to analyzing data and determining appropriate algebraic representations of data. In this textbook, data will generally be obtained from well defined algebraic relationships. In the first example we use a familiar problem situation to give an example of an algebraic representation and a geometric representation of a problem situation. In this case the algebraic representation will be an equation and the geometric representation a portion of a straight line.

- **EXAMPLE 1:** Quality Rent-a-Car charges $15 plus $0.20 a mile to rent a car. Give an algebraic representation and a geometric representation that shows the relationship between the number of miles driven and the charges.

SOLUTION: If 50 miles are driven, the charges are $0.20(50) + 15$ or $25. If $x$ represents the number of miles driven and $y$ the corresponding charges in dollars, then the equation $y = 0.2x + 15$ can be used to compute additional possibilities.

| $x$ (miles driven) | 50 | 75 | 100 | 200 |
|---|---|---|---|---|
| $y$ (rental charges) | 25 | 30 | 35 | 55 |

This equation is an algebraic representation of the problem situation. We can represent each of the possible miles driven and corresponding charges as a point on a graph. For

Figure 1.1.1

example, the point representing 50 miles driven ($x = 50$) and charges of \$25 ($y = 25$) is located above 50 on the horizontal axis and across from 25 on the vertical axis (Figure 1.1.1). Can any real number greater than or equal to zero be the number of miles driven? Many would say that the answer is no. They would claim that miles driven should be given as positive integers, or maybe as positive numbers in tenths. Regardless of your point of view, we cannot graph all the possible points because there are infinitely many. However, if you continue to plot additional points, you will be convinced that the graph of the relationship between the number of miles driven and the charges appears to be the ray indicated in Figure 1.1.1. The ray is a geometric representation of the problem situation. The arrow on the graph indicates that the graph continues in the direction suggested by the arrow, even though we can only draw a part of it on the page. •

The collection of points on the ray indicated in Figure 1.1.1 that corresponds only to possible miles driven is a *graph of the Quality Rent-a-Car problem situation* because the graph geometrically describes the complete relationship between the number of miles driven and the charges. Because of the scale in Figure 1.1.1 the graph will appear to be the same whether you use all positive real numbers, only positive numbers in tenths, or only positive integers as possible number of miles driven.

## Scale

In Figure 1.1.1 each scale mark on the horizontal axis represents 50 (miles) and each scale mark on the vertical axis represents 5 (dollars). If we had chosen the scale marks on the $y$-axis and $x$-axis to represent one dollar and one mile, respectively, then we would need a very large piece of paper to display the same information that is shown in Figure 1.1.1. The scale will often have to be different on the two axes, and should be based on the size of the numbers involved in the problem situation.

We will see that we do *not* always need a graph of a problem situation to answer questions about the problem. Often a graph that contains the graph of the problem situation will be used to answer questions about the problem situation. For example, we can use the graph in Figure 1.1.1 to answer questions about renting cars from Quality Rent-a-Car.

- **EXAMPLE 2**: Sarah is charged $50 for renting a car at Quality Rent-a-Car. How many miles did she drive the car?

  **SOLUTION**: To solve this problem graphically, we draw a horizontal line from the point marked 50 ($50) on the vertical axis (rental charges) to the graph in Figure 1.1.1. Then, we draw a vertical line down to the horizontal axis (miles driven) to read the answer to this question from the graph. The result is shown in Figure 1.1.2. We estimate from the graph that Sarah drove 175 miles. In this case, 175 miles is the exact solution. This can be confirmed by direct substitution, that is, $0.2(175) + 15 = 50$, or algebraically by solving the equation $0.2x + 15 = 50$ for $x$. •

Next we give some of the standard terminology used in graphing.

## The Real Number Line

We call $1, 2, 3, \ldots$ the **positive integers**, and $-1, -2, -3, \ldots$ the **negative integers**. The positive and negative integers together with zero form the **integers**. The positive integers together with 0 are called the **whole numbers**. A **rational number** is any number of the form $\frac{a}{b}$, where $a$ and $b$ are integers with $b \neq 0$. The decimal form of a rational number either terminates or is nonterminating *and* repeating. For example, $\frac{3}{4} = 0.75$ terminates and $\frac{13}{22} = 0.5909090\ldots$ is nonterminating (the digits 9 and 0 repeat forever as suggested by the pattern). A real number that is not rational is called an **irrational number**. Some examples of irrational numbers are $\sqrt{2}$, $-\sqrt{3}$, $\pi$, and $\frac{1}{\pi+1}$. Irrational numbers have infinite decimal representations that do *not* repeat; for example, $0.4404004000400004\ldots$. Notice that the number of 0's between a pair of consecutive 4's increases by one as we move from left to right through the decimal representation.

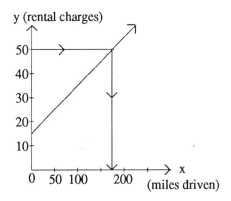

Figure 1.1.2

The **real number line** provides a geometric representation of the real numbers. Each real number corresponds to one and only one point on the number line, and each point on the number line corresponds to one and only one real number. We first choose a point on the number line to be labeled zero (0). Then we mark equally spaced points on each side of 0. The points to the right of 0 are labeled $1, 2, 3, \ldots$, and the points to the left of 0 are labeled $-1, -2, -3, \ldots$ (Figure 1.1.3). Every real number, including irrational numbers, corresponds to one and only one point on the number line. The small vertical line segments used to mark the points are called **scale marks**.

The points corresponding to rational numbers are obtained by subdividing the unit line segment. For example, if the interval between 1 and 2 is divided into three equal segments, the right-hand end point of the second segment is labeled $\frac{5}{3}$ (Figure 1.1.3).

We use the notation $a$ to represent both the real number $a$ and the name of the point on the number line corresponding to $a$. We call $a$ the **coordinate** of the point on the real number line corresponding to the real number $a$.

## The Cartesian Coordinate System

A **Cartesian coordinate system** can be introduced in a plane by placing two real number lines (coordinate lines) in the plane so that they intersect at the zero on each line and are perpendicular. Usually one of the lines is horizontal with positive direction to the right, and the other line is vertical with positive direction upward as illustrated in Figure 1.1.4. The point $O$ of intersection of the two lines is called the **origin**. The two lines are called **coordinate axes**, the horizontal line the ***x*-axis**, the vertical line the ***y*-axis**, and the plane a **coordinate plane**. The coordinate axes divide the plane into four parts called the **first, second, third,** and **fourth quadrants** as shown in Figure 1.1.4. The points on the coordinate axes are *not* in any quadrant.

Each point $P$ in the coordinate plane determines a unique ordered pair of real numbers. If the vertical line through $P$ (Figure 1.1.4) intersects the *x*-axis at $a$ and the horizontal line through $P$ intersects the *y*-axis at $b$, then $P$ determines the **ordered pair** $(a, b)$. We call these pairs *ordered pairs* because in general $(a, b) \neq (b, a)$; in fact $(a, b) = (b, a)$ if and only if $a = b$. In general, $(a, b) = (c, d)$ if and only if $a = c$ and $b = d$. We call $a$ the

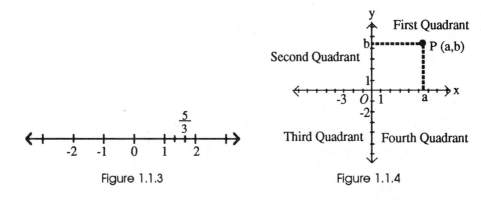

Figure 1.1.3         Figure 1.1.4

$x$-coordinate of $P$ and $b$ the $y$-coordinate of $P$. In addition, we say $P$ has **coordinates** $(a, b)$ and write $P(a, b)$. The origin $O$ has coordinates $(0, 0)$.

Moreover, each ordered pair of real numbers $(a, b)$ determines a unique point $P$ in the coordinate plane. $P$ is the point of intersection of the line perpendicular to the $x$-axis through $a$ and the line perpendicular to the $y$-axis through $b$. In this way we have established a one-to-one correspondence between the set of all points in the coordinate plane and the set of all ordered pairs of real numbers. For brevity, we will refer to the point with coordinates $(a, b)$ as the point $(a, b)$ or simply $(a, b)$.

## Relations and Graphs

- **DEFINITION** Any set of ordered pairs of real numbers is called **a relation**, and the set of points in the coordinate plane corresponding to the ordered pairs is called the **graph of the relation**. The set of all first entries $a$ of the ordered pairs $(a, b)$ is called the **domain of the relation**, and the set of all second entries $b$ is called the **range of the relation**. •

Example 1 determines a relation with miles driven as the first entry of each ordered pair and the rental charges as the second entry. The portion of the graph in Figure 1.1.1 that corresponds to the number of miles driven and the charges is a graph of this relation. The domain of this relation depends on your point of view. The range consists of all real numbers determined by $0.2x + 15$ as $x$ takes all values in the domain. Now $0.2x + 15$ takes on all real numbers greater than or equal to 15 as $x$ takes on all real numbers greater than or equal to zero. Thus, if we assume the domain of this relation (miles driven) is the set of all real numbers greater than or equal to zero then the range (rental charges) is the set of all real numbers greater than or equal to 15. If we assume that the domain is the set of positive integers, then the range is $0.2(1) + 15 = 15.2$, $0.2(2) + 15 = 15.4$, and so forth.

The ray in Figure 1.1.1 contains a graph of the Quality Rent-a-Car problem regardless of your point of view. We make the following definition.

- **DEFINITION** Any graph that represents *all* and *only* the data from a given problem situation is called **a graph of the problem situation**. •

- **DEFINITION** Any graph that contains a graph of a problem situation is called **a geometric representation** of the problem situation. •

Thus, the graph in Figure 1.1.1 is a *geometric representation* of the Quality Rent-a-Car problem. The collection of points on this graph that represent miles driven and rental charges is a *graph of the problem situation*.

Next we define what we mean by a *solution* to an equation in two variables.

- **DEFINITION**  The pair of real numbers $x = a$ and $y = b$ is called a **solution** to an equation in the two variables $x$ and $y$ if we get a true statement when $x$ and $y$ are replaced by $a$ and $b$, respectively, everywhere in the equation.  •

The pair of numbers $x = 2$ and $y = 9$ is a solution to the equation $y = 2x^2 + 1$ because $9 = 2(2)^2 + 1$. The pair $x = -1$ and $y = 3$ is also a solution because $3 = 2(-1)^2 + 1$. The equation $y = 2x^2 + 1$ has *infinitely* many solutions. The primary focus in this textbook is on real number solutions.

Each pair $x = a$ and $y = b$ of real numbers that is a solution to $y = 2x^2 + 1$ determines a point $(a, b)$ in the plane. The set of all such ordered pairs of real numbers is a *relation*, and the corresponding set of points in the plane is the *graph of this relation*.

- **DEFINITION**  The set of all points in the coordinate plane determined by the solutions to an equation in two variables is called the **graph of the equation**. The set of all first coordinates of the points on the graph of an equation is called the **domain of the equation**, and the set of all second coordinates of the points on the graph is called the **range of the equation**. The domain variable is called the **independent variable**, and the range variable the **dependent variable**.  •

The domain and range of an equation are, respectively, the domain and range of the relation determined by the set of all solutions to the equation. In Example 1 we used the algebraic equation $y = 0.2x + 15$ to compute charges in the Quality Rent-a-Car problem. This equation is an example of an algebraic representation of the Quality Rent-a-Car problem. Solutions to this equation with $x$ greater than zero represent the Rent-a-Car problem, but solutions with $x$ less than or equal to zero do not. (Why?) We use the following *informal* definition.

- **DEFINITION**  An **algebraic representation** of a problem situation is an equation, inequality, system of equations or inequalities, or other symbolic expression that represents the problem situation. The domain of the algebraic representation must contain the domain of the problem situation. If $a$ is in the domain of the problem situation, then the algebraic representation must give a good approximation to the element that corresponds to $a$ in the problem situation.  •

The process of **modeling** is important in this textbook. The process involves creating an appropriate algebraic or geometric representation of a problem situation and then determining the portion of the domain of the representation that is the domain of the problem situation.

In the next example we see that a graph of an algebraic representation contains the graph of the problem situation. Thus, a graph of an algebraic representation of a problem situation *is* a geometric representation of the problem situation. However, a graph of the algebraic representation may also contain information that is not meaningful in the problem situation. Next to the actual graph of a problem situation, the graph of an algebraic representation of a problem situation is the most important and most frequently used geometric representation of a problem situation.

- **EXAMPLE 3**: Draw a graph of the equation $y = 0.2x + 15$ and specify the domain and range of the equation. What part of the graph represents the Quality Rent-a-Car problem?

    SOLUTION: Every point on the graph in Figure 1.1.1 is a point on the graph of the equation $y = 0.2x + 15$. However, there are points on the graph of this equation with negative $x$-coordinates. For example, if $x = -50$, then $y = 5$. The graph of this equation consists of all points on the straight line indicated in Figure 1.1.5. Both the domain and range of the equation are the set of all real numbers.

    The portion of the graph in Figure 1.1.5 to the right of the $y$-axis with $x$-coordinates that are possible miles driven is the part of the graph that represents the Quality Rent-a-Car problem. Regardless of the agreement about what numbers represent miles driven, the graph of the problem situation will appear to be all of the graph in Figure 1.1.5 to the right of the $y$-axis.

We say that the graph drawn in Figure 1.1.5 is *complete* because it suggests all of the behavior of the relation determined by the equation $y = 0.2x + 15$. Loosely speaking, we say that a graph of a relation is *complete* if it suggests all of the points of the graph of the relation and all of the important features of the graph. You will need to do many examples before this idea becomes clear to you. However, determining complete graphs is an activity of most sections of this textbook. In this textbook, we use the following *informal* definition.

- **DEFINITION**    A graph of a relation or equation is said to be **complete** if it suggests all points of the graph and all of the important features of the graph of the relation or equation.

We say that the graph in Figure 1.1.5 is *complete* because it conveys the fact that the graph of the equation $y = 0.2x + 15$ is a straight line.

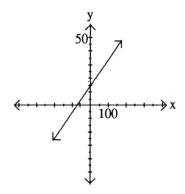

Figure 1.1.5

**Inverse Variations.** Inverse variations, or reciprocal relationships, are frequently encountered in real world situations. For example, the total time required for a trip is inversely related to the speed of travel of the trip. Pressure and volume of a gas are inversely related according to Boyle's law. Intensity of light and the square of the distance from the source of the light are inversely related. In general, $y$ is inversely related to $x$ if $y = \frac{K}{x}$, where $K$ is a real number. In the next example, we investigate the most basic of all inverse relationships.

- **EXAMPLE 4:** The second of two numbers is the reciprocal of the first. Give a geometric and an algebraic representation of this problem situation.

**SOLUTION:** Let $x$ represent the first number and $y$ the second number. Then $y = \frac{1}{x}$. This equation is an *algebraic representation* of the problem situation because every solution to the problem is also a solution to the equation. A graph of the algebraic representation $y = \frac{1}{x}$ is a *geometric representation* of the problem situation. In this case, every solution to the equation is also a solution to the problem situation. Thus, a *complete graph* of the algebraic representation $y = \frac{1}{x}$ is also a *complete graph* of the problem situation.

Any real number different from zero has a reciprocal. Zero does *not* have a reciprocal. Thus, there is no second number with first number zero. Table 1 gives values of the second number $y$ for some values of the first number $x$. Each value of $y$ is rounded to hundredths. The graph of the 12 points determined by Table 1 is given in Figure 1.1.6.

You might be tempted to connect consecutive points from left to right. This gives an *incorrect* graph. We need more information about this relation in order to complete its graph. If $x$ is 0.1, 0.01, or 0.001, then $y$ is 10, 100, 1000, respectively. (Why?) If $x$ is $-0.1$, $-0.01$, or $-0.001$, then $y$ is $-10, -100, -1000$, respectively. We can see that if $x$ is very close to zero, then $y$ is very far from zero. Similar computations show that if $x$ is very far from zero, then $y$ is very close to zero.

| $x$ | $y$ |
|---|---|
| $-5$ | $-0.2$ |
| $-4$ | $-0.25$ |
| $-3$ | $-0.33$ |
| $-2$ | $-0.5$ |
| $-1$ | $-1$ |
| $-0.5$ | $-2$ |
| 0.5 | 2 |
| 1 | 1 |
| 2 | 0.5 |
| 3 | 0.33 |
| 4 | 0.25 |
| 5 | 0.2 |

Table 1

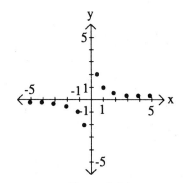

Figure 1.1.6

## 1.1 Cartesian Coordinate System and Complete Graphs • 11

There is a point on the graph with first coordinate $a$ for every real number $a$ different from zero. If we were able to plot all these points we would get the two solid curves indicated in Figure 1.1.7. Once again we use arrows on the graph to indicate that the graph continues as suggested by the arrows. The graph in Figure 1.1.7 is a *complete graph* of the equation $y = \frac{1}{x}$ because this graph suggests the complete behavior of the relation determined by the equation. It is also a geometric representation of the problem situation. Notice that the points on the graph with $x$-coordinates very far from zero are very close to the horizontal axis, and the points with $x$-coordinates very close to zero are very close to the vertical axis. •

Calculators and computers can be used to define relations. The first entry of an ordered pair is the real number that we enter into a calculator or computer, and the second entry is the corresponding real number the calculator or computer gives as an output. For example, the square key $\boxed{x^2}$ on a calculator defines a specific relation. We use the symbol $\boxed{\phantom{x}}$ to represent a key on a calculator. If we input the number 3, then the calculator gives 9 as the output.

We agree that, when not explicitly stated, the domain of a relation defined by a calculator key is the set of all real numbers that would produce real numbers as output using the relation, and the range is the corresponding set of all output numbers obtained using the relation. For the calculator key $\boxed{\sqrt{\phantom{x}}}$, the domain and range are the set of all **nonnegative numbers** (all real numbers greater than or equal to zero). For the calculator key $\boxed{x^2}$, the domain is the set of all real numbers and the range is the set of all nonnegative real numbers.

In the next two examples we explore the meaning of *complete graph*. Specifically, we will see that if a few points of a graph are known, there are infinitely many possibilities for the *complete graph* unless we have an explicit rule for the points of the graph.

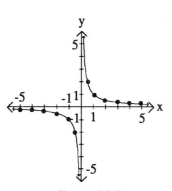

Figure 1.1.7

| Input | Output |
|-------|--------|
| $-3$ | $-15$ |
| $-2.5$ | $-5.625$ |
| $-2$ | $0$ |
| $-1.5$ | $2.625$ |
| $-1$ | $3$ |
| $-0.5$ | $1.875$ |
| $0$ | $0$ |
| $0.5$ | $-1.875$ |
| $1$ | $-3$ |
| $1.5$ | $-2.625$ |
| $2$ | $0$ |
| $2.5$ | $5.625$ |
| $3$ | $15$ |

Table 2

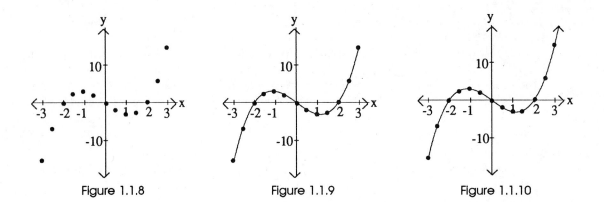

Figure 1.1.8         Figure 1.1.9         Figure 1.1.10

- **EXAMPLE 5:** Table 2 gives some values of a certain relation. Graph the points determined by the table. Then give three possibilities for a complete graph of the relation, and specify the domain and range of each.

  **SOLUTION:** The graph of the points determined by Table 2 is given in Figure 1.1.8. Here the first entries ($x$ values) are called inputs, and the second entries ($y$ values) corresponding to $x$ are called outputs. There are many possibilities for a complete graph of this relation. One possibility is that these are all of the values of the relation so that the graph in Figure 1.1.8 is one possible *complete* graph. In this case the domain is the set of all numbers in the input column of the table, and the range is the set of all numbers in the output column. Two other possible complete graphs of the relation are given in Figures 1.1.9 and 1.1.10.

  The solid dots in Figure 1.1.9 are used to indicate that the graph starts at the point $(-3, -15)$ and stops at the point $(3, 15)$. The domain of this relation is the set of all real numbers between $-3$ and $3$, including $-3$ and $3$, which can be denoted by $-3 \leq$ Input $\leq 3$; the range is the set of real numbers between $-15$ and $15$, including $-15$ and $15$, or $-15 \leq$ Output $\leq 15$.

  Figure 1.1.10 indicates that the graph starts at the point $(-3, -15)$ and continues in the direction suggested by the *arrow* above $(3, 15)$. The domain of this relation is the set of all real numbers greater than or equal to $-3$, or Input $\geq -3$; the range is the set of all real numbers greater than or equal to $-15$, or Output $\geq -15$. •

  Generally it is difficult to graph accurately points such as the ones determined by Table 2. For example, to display carefully both 1.875 and 15 would require a very large piece of graph paper. The reason we used these numbers will become clear in the next example.

- **EXAMPLE 6:** Complete the graph of the relation started in Figure 1.1.8 assuming that the output $y$ is obtained from the input $x$ using the equation $y = x^3 - 4x$.

  **SOLUTION:** You can verify that the entries in Table 2 can be obtained using the equation $y = x^3 - 4x$ with $x$ as input and $y$ as output. For example, $(2.5)^3 - 4(2.5) = 5.625$. Now

## 1.1 Cartesian Coordinate System and Complete Graphs

there is only *one* possible *complete graph*, and it is indicated in Figure 1.1.11. Arrows are drawn on the graph to indicate that it continues in both directions.

If we are given a complete graph of a relation, then we can use that graph to estimate the domain and range of the corresponding relation.

- **EXAMPLE 7:** Assume the graph given in Figure 1.1.12 is complete.

    (a) Complete the table for the relation determined by this graph.

    | $x$ | $y$ |
    |---|---|
    | 0 | |
    | 1 | |
    |  | 5 |

    (b) What are the domain and range of this relation?

**SOLUTION:**

(a) We can read from the graph in Figure 1.1.12 that $y$ is 1 when $x$ is 0, and $y$ is 2 when $x$ is 1. If we draw a horizontal line segment from 5 on the $y$-axis to the graph and then a vertical line segment down to the $x$-axis, then we can read that $x$ is 2 when $y$ is 5. The completed table is as follows.

| $x$ | $y$ |
|---|---|
| 0 | 1 |
| 1 | 2 |
| 2 | 5 |

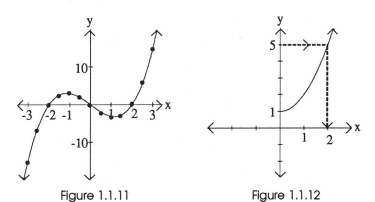

Figure 1.1.11              Figure 1.1.12

(b) The domain of this relation is the set of all real numbers $x$ greater than or equal to 0 ($x \geq 0$), and the range is the set of all real numbers $y$ greater than or equal to 1 ($y \geq 1$).

## EXERCISES 1-1

Exercises in this textbook that are marked with an asterisk (*) either are unusually challenging or have not received attention in the narrative.

1. Estimate the coordinates of the seven points in Figure 1.1.13.

2. Graph the points in (a) – (h) on the same coordinate plane. Identify the quadrant containing each point.
   (a) $(2,4)$     (b) $(0,3)$     (c) $(3,0)$     (d) $(-1,2)$
   (e) $(-2,3)$    (f) $(2,-\frac{5}{2})$    (g) $(3,-2)$    (h) $(-\frac{1}{2},-5)$

Graph the three points on the same coordinate system. Choose an *appropriate scale*.

3. $(5, 200), (3, 100), (7, 160)$   4. $(2.01, 5), (2.03, 8), (1.98, -4)$

5. Draw and identify the polygons determined by each set of vertices. How many different polygons can be determined in each case?
   (a) $(-1,2), (3,5), (2,-4)$          (b) $(0,0), (2,3), (-1,3), (-1,0)$
   (c) $(-1,-2), (6,-4), (10,1), (3,3)$  (d) $(-3,0), (-2,3), (0,4), (3,0), (2,-3)$

6. The point $(a, 6)$ is on the graph of $y = 2x$. What is $a$?

7. The point $(-2, T)$ is on the graph of $y = 5x - 3$. What is $T$?

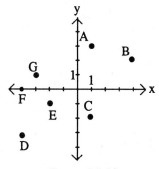

Figure 1.1.13

# 1.1 Cartesian Coordinate System and Complete Graphs

8. Draw complete graphs of three different relations that include the data given in the table.

| First number | Second number |
|---|---|
| −4 | −2 |
| −1 | 0 |
| 1 | 3 |
| 4 | 6 |
| 9 | 11 |

9. Draw complete graphs of three different relations that include the data given in the table.

| $x$ | $y$ |
|---|---|
| 1 | 1 |
| 2 | 4 |
| 3 | 9 |
| 4 | 16 |
| 5 | 25 |

10. (a) Complete the table for the relation given by the complete graph in Figure 1.1.14.
    (b) What are the domain and range of this relation?

| First number | Second number |
|---|---|
| 0 |  |
|  | 0 |
| 2 |  |
|  | −2 |

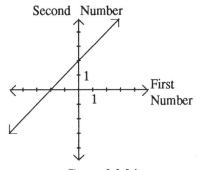

Figure 1.1.14

11. (a) Complete the table for the relation given by the complete graph in Figure 1.1.15.

| Height (ft) | Weight (lbs) |
|---|---|
| 2 | |
| | 100 |
| 4 | |
| | 200 |

(b) What are the domain and range of this relation?

12. (a) Complete the table for the relation given by the complete graph in Figure 1.1.16.

| Input | Output |
|---|---|
| 0 | |
| | 2 |
| 1 | |
| | 0 |
| −3 | |

(b) What are the domain and range of this relation?

Draw a complete graph of each equation. Choose an appropriate scale for each axis.

13. $x + y = 6$
14. $y = 200x - 10$
15. $y = 2x^2 + 1$
16. $y = 10x^2 + 500$
17. $y = -\frac{2}{x}$
18. $y = x^3 + x$

Figure 1.1.15

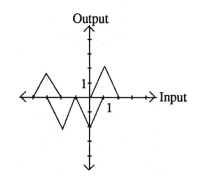

Figure 1.1.16

# 1.1 Cartesian Coordinate System and Complete Graphs

19. (a) Use the square root key $\sqrt{\phantom{x}}$ on a calculator to complete the table. Give your answers accurate to hundredths.

| Input | Output (Square Root) |
|---|---|
| −3 | |
| −2 | |
| 0 | |
| 1 | |
| 2 | |
| 3 | |
| 4 | |
| 5 | |
| 6 | |

(b) Graph the points determined by the table in (a).

(c) Draw a complete graph of the relation determined by the calculator square root key $\sqrt{\phantom{x}}$. (The output is the square root of the input.)

(d) What are the domain and range of the square root relation?

20. (a) Use the natural logarithm key $\boxed{\ln}$ on a calculator to complete the table. Give your answers accurate to hundredths.

| Input | Output (ln) |
|---|---|
| 0.25 | |
| 0.5 | |
| 1 | |
| 1.5 | |
| 2 | |
| 3 | |
| 4 | |
| 5 | |

(b) Graph the points determined by the table in (a).

(c) Draw a complete graph of the relation determined by the calculator key $\boxed{\ln}$. (The output is the natural logarithm of the input.)

(d) What are the domain and range of the natural logarithm relation?

21. (a) Use the sine key [sin] on a calculator to complete the tables. Use radian mode. Give your answers accurate to hundredths.

| Input | Output (sin) |
|---|---|
| 0 | |
| 0.52 | |
| 1.05 | |
| 1.57 | |
| 2.09 | |
| 2.62 | |
| 3.14 | |
| 3.67 | |
| 4.19 | |
| 4.71 | |

| Input | Output (sin) |
|---|---|
| 5.24 | |
| 5.76 | |
| 6.28 | |
| 6.81 | |
| 7.33 | |
| 7.85 | |
| 8.38 | |
| 8.90 | |
| 9.42 | |
| 9.95 | |

(b) Graph the points determined by the table in (a).
(c) Assuming the pattern in the table continues (repeats itself), draw a complete graph of the relation determined by the [sin] key.
(d) What are the domain and range of the sine relation?

22. A family takes a 100-mile trip. Let $R$ be their average speed in miles per hour and $T$ be the total time of the trip in hours.
 (a) Write an equation (algebraic representation) that shows how $T$ (output) depends on $R$ (input).
 (b) Draw a complete graph of the algebraic representation (equation) in (a).
 (c) Which part of the graph in (b) represents the problem situation, that is, the family trip?
 (d) Use the graph in (b) to determine the total time $T$ if the average speed is 40 mph.

23. If the point $(a, b)$ is on the graph of $T = \frac{100}{R}$, then find $ab$.

24. A school club pays \$35.75 to rent a video cassette player to show a movie for a fund raising project. The club charges \$0.25 per ticket for the movie.
 (a) Write an equation (algebraic representation) that shows how the club's profit (or loss) $P$ (output) depends on the number $n$ (input) of tickets sold.
 (b) Draw a complete graph of the equation in (a).
 (c) What are the domain and range of the algebraic representation (equation)?
 (d) Which part of the graph in (b) represents the problem situation? *Hint:* You cannot sell half a ticket!
 (e) Use the graph in (b) to determine the minimum number of tickets that must be sold for the club to realize a profit.

25. Let the point $(z, w)$ be on the graph of $P = 0.25n - 35.75$. Determine a formula for $w$ in terms of $z$.

26. Does the graph of the relation given in Figure 1.1.7 ever touch the horizontal axis or vertical axis? (Why?)

*27. Consider the relation *the second number is less than or equal to the first number.*
   (a) Give an algebraic representation of this relation.
   (b) Draw a complete graph of this relation.
   (c) What are the domain and range of this relation?

## 1.2 Functions and Graphing Utilities

In this section we begin the study of functions and use computers or graphing calculators to obtain their graphs. A function is a special type of relation. Recall that a relation is any set of ordered pairs of real numbers. A vertical line test is introduced and used to identify graphically relations that are functions. We investigate an application that has a function for an algebraic representation.

Functions are encountered routinely in our daily lives in tabular form: height–weight charts, time–temperature charts, price-sales-tax charts, income tax tables, and so forth. Functions play a very crucial role in the study of mathematics and science. They provide important models for problem situations. For example, functions can be used to model the growth of a population, the growth of money in a savings account, the force of attraction between objects, the illumination provided by a source of light, the cost of producing goods, and so forth.

Calculators can be used to give examples of functions. In fact, calculators are sometimes referred to as function machines. If a calculator computes unique output for given input, then the calculator is producing values of a function.

- **DEFINITION** A **function** is a relation with the property: If $(a,b)$ and $(a,c)$ belong to the relation, then $b = c$. The set of all first entries of the ordered pairs is called the **domain of the function**, and the set of all second entries is called the **range of the function**. •

Using an input–output analogy, a relation is a function if each input is paired with no more than one output. Suppose that $(2,b)$ and $(2,c)$ are two ordered pairs of a given relation. If the relation is a function, then $b$ must be equal to $c$. That is, *only one* output is possible for the input 2. This means each second entry of an ordered pair of a function is *uniquely determined* by the corresponding first entry.

- **EXAMPLE 1:** Which of the following relations are functions?
   (a) $R = \{(0,0), (-1,1), (1,1), (2,4)\}$    (b) $T = \{(0,0), (1,-1), (1,1), (4,2)\}$

SOLUTION: In order for a relation to be a function, there cannot be *two* different ordered pairs with the same first entry. The relation $R$ is a function because there is one and only one pair with a given first entry: 0, −1, 1, or 2. The relation $T$ is *not* a function because there are two ordered pairs with first entry 1, namely $(1, −1)$ and $(1, 1)$. Alternatively, if we think of the first entry as input and the second entry as output, then the input 1 produces two distinct outputs: $−1$ and $1$. Thus, the input–output relation $T$ is not a function. •

In the next example we consider relations defined by calculator keys.

- **EXAMPLE 2:** Show that the relations determined by the following calculator keys are functions: $\boxed{\sqrt{\phantom{x}}}$, $\boxed{x^2}$, $\boxed{\sin}$, $\boxed{\log}$.

SOLUTION: The relation determined by the calculator key $\boxed{\sqrt{\phantom{x}}}$ consists of ordered pairs $(x, \sqrt{x})$ where $\sqrt{x}$ stands for the nonnegative square root of $x$. This relation is a function because $\sqrt{x}$ is uniquely determined by $x$, that is, if $a = b$ then $\sqrt{a} = \sqrt{b}$. Similarly, the key $\boxed{x^2}$ determines a function. Later in this textbook we study the functions $\sin x$ and $\log x$ in greater detail. For now, the fact that the calculator keys $\boxed{\sin}$ and $\boxed{\log}$ produce a unique output for a given input means that the relation determined by each of these keys is a function. •

- **EXAMPLE 3:** Show that the relation $R = \{(x, y) \mid y^2 = x\}$ determined by the equation $y^2 = x$ is not a function.

SOLUTION: The relation determined by $y^2 = x$ is the set of all solutions to $y^2 = x$ considered as ordered pairs of real numbers. The ordered pair $(1, 1)$ belongs to the relation $R$ because the pair $x = 1$ and $y = 1$ is a solution to the equation $y^2 = x$. Similarly, the ordered pair $(1, −1)$ belongs to $R$ because the pair $x = 1$ and $y = −1$ is also a solution to the equation $y^2 = x$. $R$ is *not* a function because the two different ordered pairs $(1, 1)$ and $(1, −1)$ have the same first entry. In other words, the first entry 1 does not uniquely determine the second entry. •

In Section 6 we show that the graph in Figure 1.2.1 is a complete graph of the equation $y^2 = x$. Notice that the two points $(1, 1)$ and $(1, −1)$ lie on the vertical line $x = 1$.

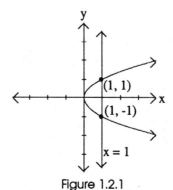

Figure 1.2.1

# 1.2 Functions and Graphing Utilities

If a relation is *not* a function, then the relation must contain at least two ordered pairs $(a, b)$ and $(a, c)$ with $b \neq c$. Because both of these points lie on the vertical line $x = a$, the vertical line $x = a$ must intersect the graph of the relation in at least two points. Thus, a complete graph of a relation can be used to decide if the relation is a function.

- **Vertical Line Test for Functions.** If every vertical line intersects the graph of a relation in at most one point, then the relation is a function.

- **EXAMPLE 4:** Which of the relations represented by the graphs in Figure 1.2.2 are functions?

   SOLUTION:

   (a) The vertical lines $x = a$ with $a$ greater than 2 or less than $-2$ do not intersect this graph. The vertical lines $x = a$ with $a$ between $-2$ and 2, including 2 and $-2$, intersect the graph in exactly one point. Thus, the relation represented by this graph is a function.
   (b) The vertical lines $x = a$ with $a$ negative do not intersect this graph. The line $x = 0$ ($y$-axis) intersects the graph exactly once, but each line $x = a$ with $a$ positive

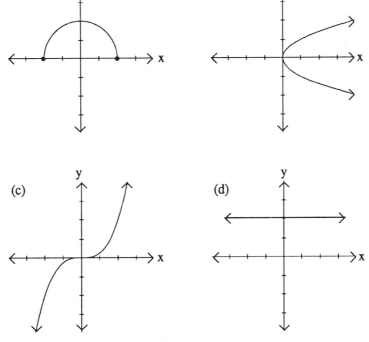

Figure 1.2.2

intersects the graph in two points (Figure 1.2.3). Thus, the relation represented by this graph is not a function.
(c) Every vertical line intersects this graph exactly once, so this graph is the graph of a function.
(d) Every vertical line intersects this horizontal line in exactly one point, so this graph is a graph of a function.  •

If $f$ is a function and $x$ a first entry of an ordered pair in $f$, we denote the second entry by $f(x)$. The notation $f(x)$ for the number paired with $x$ by the function $f$ is read "**$f$ of $x$**" or "**$f$ at $x$.**" Often we call $f(x)$ the **value** of $f$ at $x$ or the **output** associated with **input** $x$. If $(x, y)$ is an ordered pair in $f$, then $y = f(x)$.

• **EXAMPLE 5:** Let $f$ be the relation defined by the equation $y = x^2 + 1$. Show that $f$ is a function, and find $f(-1)$, $f(3)$, $f(a)$, $f\left(\frac{1}{a}\right)$, $f(a+h)$, and $f(x)$.

**SOLUTION:** First, we must show that $f$ is a function. If $f$ contains two ordered pairs with the same first entry, we must show that the second entries are also the same; in other words, the two ordered pairs are really the same ordered pair. The number paired with 2 by $f$ is $2^2 + 1$ or 5. Notice that 5 is the only possible number that can be paired with 2 by $f$. More precisely, if $(a, b)$ and $(a, c)$ are ordered pairs in $f$, we must show that $b = c$. If $(a, b)$ and $(a, c)$ are ordered pairs in $f$, then, by the definition of $f$, $b = a^2 + 1$ and $c = a^2 + 1$. Thus, $b = c$, and $f$ is a function by definition. Usually we will not take the time to verify algebraically that a given relation is a function.

The notation $f(-1)$ stands for the number paired with $-1$ by $f$, or $f(-1)$ is the output produced by $f$ with input $-1$. Thus, $f(-1) = (-1)^2 + 1 = 2$. Similarly, $f(3) = 3^2 + 1 = 10$, $f(a) = a^2 + 1$, and $f\left(\frac{1}{a}\right) = \left(\frac{1}{a}\right)^2 + 1 = \frac{1}{a^2} + 1$. Next, $f(a+h) = (a+h)^2 + 1 = a^2 + 2ah + h^2 + 1$. Finally, $f(x) = x^2 + 1$.  •

Now that we have had some practice with drawing graphs by computing values and plotting points, we will use graphing utilities to draw graphs of functions. Because a function is a relation, a **graph of a function** is the graph of the function considered as a relation.

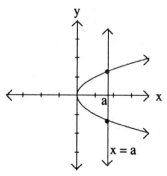

Figure 1.2.3

## 1.2 Functions and Graphing Utilities

We use the term **function graphing utility** to refer to any device that draws a graph of a function. A computer with appropriate software and a graphing calculator are graphing utilities. We take advantage of the speed and power of these utilities to draw graphs of many more functions than would be reasonably possible using only pencil and paper.

We think of the display screen of a graphing utility as representing a portion of the coordinate plane. In fact, if the four corners of the screen have coordinates as illustrated in Figure 1.2.4, we are looking at a graph in the **viewing rectangle** $[L, R]$ by $[B, T]$ where $L \leq x \leq R$ and $B \leq y \leq T$. We use the notations $L \leq x \leq R$ to represent all the real numbers between $L$ and $R$, including $L$ and $R$. Similarly, the notation $B \leq y \leq T$ represents all the real numbers between $B$ and $T$. It is easy to remember this choice of notation if we think of $L$, $R$, $B$, and $T$ as meaning *left*, *right*, *bottom*, and *top*, respectively. Sometimes $L, R, B,$ and $T$ are referred to as "$x$ min," "$x$ max," "$y$ min," and "$y$ max," respectively. It is important to understand that the viewing rectangle frames a portion, usually not all, of a graph.

You are expected to have access to a graphing calculator or computer on a regular basis for homework. This utility should allow you to

- enter and obtain the graph of any function;
- enter any number between $10^{-6}$ and $10^6$ for the value of $L$, $R$, $B$, or $T$;
- overlay the graph of a second function on the same screen with the graph of the first function.

Consult the accompanying *Graphing Calculator and Computer Graphing Laboratory Manual* to see how to use a graphing utility. All graphs in this textbook have been plotted using the Apple Macintosh$^{TM}$ version of a graphing utility called *Master Grapher*.

Most figures that appear in this text specify an equation for the function or relation whose graph appears in the figure together with the viewing rectangle used. In addition, the graph will be framed by a border to suggest what actually appears on the display screen of a graphing utility in the specified viewing rectangle. Scale marks that appear on the coordinate axes are equally spaced. The distance represented by the interval between consecutive scale marks on the $x$-axis often is different from the distance represented by the interval between consecutive scale marks on the $y$-axis. The shape of a given graph depends on the viewing rectangle in which it is viewed.

- **DEFINITION**     A graph of a function is said to be **complete** if it suggests all of the points of the graph and all of the important features of the graph. •

Figure 1.2.4

We need to choose viewing rectangles carefully in order to investigate critical portions of a graph and to determine a complete graph of a function. Sometimes we need to display a graph in several different viewing rectangles to describe a complete graph of a given function.

- **EXAMPLE 6:** Draw a complete graph of the function $f(x) = x^2 + 1$ and determine its domain and range.

**SOLUTION:** First, we choose a viewing rectangle to frame a portion of the graph of $f$. Suppose we choose the viewing rectangle $[-1, 1]$ by $[-1, 1]$, that is, $-1 \leq x \leq 1$ and $-1 \leq y \leq 1$. Because $(0, 1)$ is the only point of the graph of $f$ in this viewing rectangle, the display screen will appear to be blank except for the single point and the coordinate axes. In fact, the point will not be visible because it is covered up by the $y$-axis. We have positioned our viewing rectangle (framed our picture) so that only one point of the graph appears on the display screen.

If we now choose $[-1, 1]$ by $[-2, 2]$ for the viewing rectangle we get the graph in Figure 1.2.5. This time more than one point on the graph of $f$ appears in the viewing rectangle. Notice that the equation $f(x) = x^2 + 1$ and the viewing rectangle $[-1, 1]$ by $[-2, 2]$ are displayed with the graph in Figure 1.2.5.

The graph in Figure 1.2.5 gives a clue to the complete behavior of $f$. However, we must be careful not to jump to conclusions too quickly. The graph of $f$ in the **standard viewing rectangle** $[-10, 10]$ by $[-10, 10]$ is given in Figure 1.2.6. We can convince ourselves that the graph in Figure 1.2.6 is *complete* by viewing the graph in a large viewing rectangle (Figure 1.2.7). Thus, as we move from left to right through $x$ the graph of $f$ steadily increases. The lowest point of the graph of $f$ is $(0, 1)$. Notice the shape of the graph of $f$ is different in these three figures.

In this case, the graph of $f$ in the standard viewing rectangle is a *complete graph*. Example 7 will show that this does not always happen, especially when drawing the graph of a problem situation.

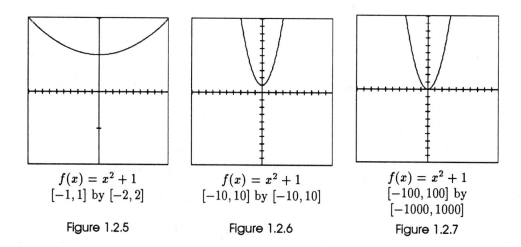

$f(x) = x^2 + 1$
$[-1, 1]$ by $[-2, 2]$

Figure 1.2.5

$f(x) = x^2 + 1$
$[-10, 10]$ by $[-10, 10]$

Figure 1.2.6

$f(x) = x^2 + 1$
$[-100, 100]$ by $[-1000, 1000]$

Figure 1.2.7

Notice that the scale marks on the coordinate axes in Figure 1.2.6 are one unit apart. In Figure 1.2.7 the scale marks on the $x$-axis are 10 units apart and the scale marks on the $y$-axis are 100 units apart. These scale marks can be used to estimate coordinates of points on the graph. We can see from Figure 1.2.6 that the domain of $f$ is the set of all real numbers, and the range of $f$ the set of all real numbers greater than or equal to 1, or $y \geq 1$. •

It is important to understand how graphing utilities draw graphs of functions in viewing rectangles $[L, R]$ by $[B, T]$. Generally, the values of the functions are computed for a specified number of equally spaced values of $x$ between $L$ and $R$, and then the corresponding points are graphed if they fall within the rectangle $[L, R]$ by $[B, T]$. Usually, these points are joined with straight line segments to produce the graph.

A complete graph of $f(x) = ax^2 + bx + c$ with $a \neq 0$ looks like one of the two parabolas in Figure 1.2.8. The domain of any such function is the set of all real numbers. If $a$ is positive, the graph has a lowest point and the range is the set of all real numbers greater than or equal to the smallest value of the function. If $a$ is negative, the graph has a highest point and the range is the set of all real numbers less than or equal to the largest value of the function. The highest or lowest point is called the vertex of the parabola. The coordinates of the vertex are $\left(-\frac{b}{2a}, f\left(-\frac{b}{2a}\right)\right)$.

Next, the area of a sidewalk that surrounds a swimming pool will be found. A careful choice of viewing rectangles is required to obtain a complete graph of an algebraic representation of this area.

A rectangular swimming pool with dimensions 30 feet by 50 feet is surrounded by a walk of uniform width $x$ as illustrated in Figure 1.2.9. The area of the sidewalk is the area of the outer rectangle minus the area of the inner rectangle. If $A$ stands for the area of the sidewalk, then the following gives $A$ as a function of $x$.

$$A(x) = (30 + 2x)(50 + 2x) - (30)(50)$$
$$= 4x^2 + 160x$$

The function $A$ is an algebraic representation of the sidewalk problem. In the next example, we draw a *complete* graph of this model.

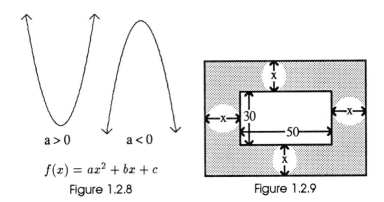

a > 0    a < 0

$f(x) = ax^2 + bx + c$

Figure 1.2.8

Figure 1.2.9

- **EXAMPLE 7:** Let $A(x) = 4x^2 + 160x$.

    (a) Draw a complete graph of the function $A$.
    (b) What portion of the graph in (a) represents the problem situation?
    (c) What are the domain and range of $A$?

    SOLUTION:

    (a) A complete graph of $A$ must look like one of the two graphs in Figure 1.2.8. We know that $(0,0)$ is one point of the graph of $A$ since $A(0) = 0$. The graph of $A$ in the standard viewing rectangle $[-10, 10]$ by $[-10, 10]$ gives very little information about $A$ because most graphing utilities plot very few, if any, points of the graph of $A$ in this viewing rectangle. If we look at some values of $A$ for $x$ near zero (Table 1), we see why this is so. Notice that the values of $A(x)$ in Table 1 are greater than 10 for $x \geq 0.5$, and the values of $A(x)$ are less than $-10$ for $x \leq -0.5$. Thus, most of the points of the graph of $A$ that a graphing utility would try to produce lie outside the standard viewing rectangle $[-10, 10]$ by $[-10, 10]$. To see the graph of $A$ near $(0,0)$ we have to either choose $L$ and $R$ very close together or make $B$ and $T$ further apart. For example, the graph in $[-1, 1]$ by $[-10, 10]$ or in $[-10, 10]$ by $[-100, 100]$ gives more information about the behavior of $A$ for $x$ near $0$ than does the standard viewing rectangle.

    We obtain a complete graph of $A$ by viewing the graph of $A$ in the $[-80, 80]$ by $[-2000, 2000]$ viewing rectangle (Figure 1.2.10). We know this is a complete graph because it looks like the graph on the left in Figure 1.2.8. Notice that this graph has the same general shape as the graph in Example 6.

    (b) If we are interested only in the values of $x$ that could be possible widths of the sidewalk, then $x$ must be positive. Thus, only the portion of the graph of $A$ in Figure 1.2.10 to the right of the $y$-axis represents the sidewalk problem.

    (c) The domain of $A$ is the set of all real numbers. To find the range of $A$ we need to determine the coordinates of the *vertex* (lowest point) of the graph of $A$. We can use the graph in Figure 1.2.10 to estimate that the coordinates of the vertex are $(-20, -1600)$. This estimate can be confirmed using the formula for the vertex given earlier in this section. The $x$-coordinate of the vertex is $-\frac{b}{2a}$, where $a$ is the coefficient of $x^2$ and $b$ is the coefficient of $x$. Thus, $-\frac{160}{8} = -20$ is the $x$-coordinate

| $x$ | $A(x)$ |
|---|---|
| 0 | 0 |
| 0.5 | 81 |
| −0.5 | −79 |
| 1 | 164 |
| −1 | −156 |

Table 1

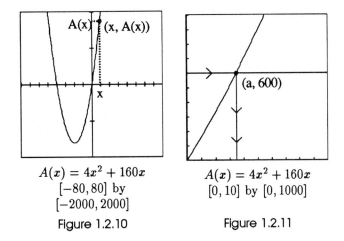

$A(x) = 4x^2 + 160x$
$[-80, 80]$ by
$[-2000, 2000]$
Figure 1.2.10

$A(x) = 4x^2 + 160x$
$[0, 10]$ by $[0, 1000]$
Figure 1.2.11

of the vertex. The $y$-coordinate of the vertex is $A(-20) = -1600$. Thus, the range is the set of all real numbers greater than or equal to $-1600$, or $y \geq -1600$. •

• **EXAMPLE 8**: Use a graph to approximate the width of the sidewalk in Figure 1.2.9 if its area is 600 square feet.

**SOLUTION**: If $x$ is positive and $(x, A(x))$ a point on the graph in Figure 1.2.10, then $x$ is a possible width of the sidewalk and $A(x)$ the corresponding area of the walk. Thus, Figure 1.2.11 is a graph of the problem situation. Because we want the area of the walk to be 600, we need to find a point on the graph of $A$ with second coordinate 600. The intersection of the horizontal line $y = 600$ with the graph of $A$ in Figure 1.2.11 gives a point with second coordinate 600. Let the first coordinate of this point be $a$. We can read from the graph in Figure 1.2.11 that the width of the sidewalk (the value of $a$) is approximately 3.5 feet. You can check that $A(3.5) = 609$. In Chapter 2 we review how to solve the quadratic equation $A(x) = 600$ algebraically and determine the exact solution. •

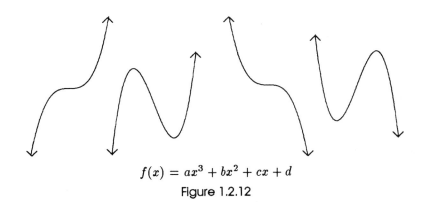

$f(x) = ax^3 + bx^2 + cx + d$
Figure 1.2.12

In the exercises you will be asked to find complete graphs of functions of the form $f(x) = ax^3 + bx^2 + cx + d$ $(a \neq 0)$. You may assume that a complete graph of such a function looks like one of the graphs in Figure 1.2.12. Each such function has both its domain and its range equal to the set of all real numbers.

## EXERCISES 1-2

1. Which of the graphs in Figure 1.2.13 are graphs of functions?

Use a graph to determine whether the relation is a function.

2. $y = x + 1$
3. $y = 2x - 4$
4. $y = \frac{2}{x}$
5. $y = x^2 - 2$
6. $y^2 = x$
7. $x = 3$
8. $y = -2$
9. $x + y = 0$
10. Let $f(x) = x^2 - 1$. Find $f(0)$, $f(1)$, $f(3)$, $f(-5)$, $f(t)$, $f(-t)$, $f\left(\frac{2}{t}\right)$, $f(a+h)$, and $\frac{f(a+h)-f(a)}{h}$.
11. Let $g(x) = \frac{1}{x+1}$. Find $g(0)$, $g(1)$, $g(3)$, $g(-5)$, $g(t)$, $g(-t)$, $g\left(\frac{2}{t}\right)$, $g(a+h)$, and $\frac{g(a+h)-g(a)}{h}$.
12. Consider the graph of the function $y = f(x)$ given in Figure 1.2.14. Estimate the indicated values.
    (a) $f(0)$, $f(-1)$, $f(-2)$, $f(4)$
    (b) $x$ if $f(x) = 0$
    (c) $x$ if $f(x) = 1$
    (d) $x$ if $f(x) = -8$
    (e) $x$ if $f(x) = 8$

Figure 1.2.13

## 1.2 Functions and Graphing Utilities

13. Consider the graph of the function $y = V(x)$ given in Figure 1.2.15. Estimate the indicated values.
    (a) $V(0), V(-5), V(4), V(11)$ (b) $x$ if $V(x) = 0$     (c) $x$ if $V(x) = 100$
    (d) $x$ if $V(x) = 1000$     (e) $x$ if $V(x) = -100$

Draw the graph of each pair of functions in the same $[-10, 10]$ by $[-10, 10]$ viewing rectangle. Compare the graphs and explain why they are different. The caret symbol, $\wedge$, stands for exponentiation and the asterisk, $*$, for multiplication. If necessary, replace these symbols with your graphing utility's symbol for these operations.

14. (a) $y = x \wedge (1/2)$     (b) $y = x \wedge 1/2$
15. (a) $y = 1/(x - 2)$     (b) $y = 1/x - 2$
16. (a) $y = 3 * x \wedge 2$     (b) $y = (3 * x) \wedge 2$
17. Which of the following points lie in the viewing rectangle $[-2, 3]$ by $[1, 10]$?
    (a) $(0, 0)$     (b) $(0, 5)$     (c) $(5, 0)$
    (d) $(3, 2)$     (e) $(1, 6)$     (f) $(1, -1)$
18. Choose a viewing rectangle $[L, R]$ by $[B, T]$ that includes all of the indicated points.
    (a) $(-9, 10), (2, 8)$ and $(3, 12)$
    (b) $(20, -3), (13, 56)$ and $(-11, 2)$
19. Which of the following viewing rectangles gives the best complete graph of $f(x) = x^2 - 30x + 225$?
    (a) $[-5, 5]$ by $[-5, 5]$     (b) $[-10, 10]$ by $[-10, 10]$
    (c) $[15, 30]$ by $[-50, 50]$     (d) $[5, 25]$ by $[-100, 100]$
    (e) $[10, 20]$ by $[-1000, 1000]$
20. Which of the viewing rectangles gives the best complete graph of $f(x) = 20 - 30x + x^3$?
    (a) $[-10, 10]$ by $[-10, 10]$     (b) $[-50, 50]$ by $[-10, 10]$
    (c) $[-10, 10]$ by $[-50, 50]$     (d) $[0, 10]$ by $[-100, 100]$
    (e) $[-10, 10]$ by $[-50, 100]$

 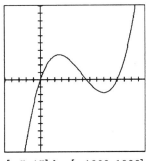

$[-5, 15]$ by $[-10, 10]$     $[-5, 15]$ by $[-1000, 1000]$

Figure 1.2.14     Figure 1.2.15

Draw a graph of the relation in several different viewing rectangles. Use this information to determine a complete graph. Determine whether the relation is a function.

21. $y = 1 + \log x$

22. $y = -\cos x$

23. $y = 3 \sin x$

24. $y = 1 - \log x$

Draw a complete graph of each function. Explain why the graph is a complete graph.

25. $g(x) = \dfrac{x-4}{5}$

26. $f(x) = x^2 - 4$

27. $t(x) = 2x^2 - 3x + 5$

28. $f(x) = \left(\dfrac{x+3}{5}\right)^2 + 2x + 4$

29. $f(x) = x^3 - x + 1$

30. $h(x) = 4 - x - x^3$

31. Determine the domain and range of the function in Exercise 25.

32. Determine the domain and range of the function in Exercise 27. *Hint:* You need to find the coordinates of the vertex of the graph of $t$ in order to determine the range of $t$.

33. Determine the domain and range of the function in Exercise 28.

34. Jerry runs at a constant speed of 4.25 mph.
   (a) Express the distance $d$ that Jerry runs as a function of time $t$.
   (b) Draw a complete graph of the function $y = d(t)$.
   (c) Use the graph in (b) to estimate the time that it takes Jerry to run a 26-mile marathon.

35. A 4-inch-by-6.5-inch picture pasted on a cardboard sheet is surrounded by a region (border) of uniform width $x$.
   (a) Determine the area $A$ of the border region as a function of $x$.
   (b) Draw a complete graph of the function $y = A(x)$ found in (a).
   (c) What are the domain and range of the function $A$?
   (d) What values in the domain are possible widths?
   (e) Use a graph to determine the width of the border if its area is 20 square inches.

36. Eighty feet of fencing is cut into two unequal lengths, and each piece is used to make a square enclosure. Let $x$ be the perimeter of the smaller enclosure.
   (a) Express the total area $A$ of both enclosures as a function of $x$.
   (b) Draw a complete graph of the function in (a).
   (c) What are the domain and range of the function $A$? What numbers in the domain represent possible lengths of the smaller piece of fence?
   (d) If $A = 300$ square feet use the graph in (b) to approximate the smaller length of fencing.

Draw a complete graph of each function. Record each series of graphs on the *same* coordinate system.

37. $y = x^2$, $y = (x-1)^2$, $y = (x-2)^2$, $y = (x+3)^2$.

38. $y = -1 + x^2$, $y = 2 + x^2$, $y = 4 - x^2$, $y = x^2 - 4$.

39. Consider the two relations in Example 1. Determine an equation for each with domain that contains the domain of the given relation.

40. Consider the relation $R = \{(-2,-1), (-1,-4), (0,-5), (1,-4)\}$.
    (a) Graph this relation.
    (b) Determine an equation with domain that contains the domain of this relation.
41. Use the method in Example 5 to show that the relation $y = 3x - 1$ is a function.
42. Use the method in Example 5 to show that the relation $y = 2x^2 - 1$ is a function.

## 1.3 More on Functions

Another way to define a function is to say that it is a *domain* together with a *rule* that assigns an element of the *range* to each element of the domain. Sometimes a separate rule is used for various portions of the domain. Such functions are said to be **piecewise defined** and are introduced in this section. We find the domain of a function by analyzing the rule that defines the function. The graphs of some of the functions that are built-in on most computers and graphing calculators are studied in this section.

Finding the domain and range of a relation is often a difficult task. We have found the domain and range of a relation given by an equation by reading this information from a complete graph of the relation. To obtain this complete graph, we first found all the real solutions to the equation. Now we want to turn our attention to a method of finding the domain of a function by analyzing the equation or rule that defines the function.

### Domain of a Function

- **Understood Domain of a Function.** Let $y = f(x)$ be a function. The **understood domain** of $f$ is the largest subset of the real numbers for which the equation $y = f(x)$ makes sense. That is, the real number $a$ is in the *understood domain* of $f$ if and only if $f(a)$ is a real number.

Unless otherwise specified, the *domain* of a function is its *understood domain*.

- **EXAMPLE 1:** Find the domain of the function $f(x) = x^2$.

  SOLUTION: For every real number $a$, $f(a) = a^2$ is also a real number. Thus, the domain of $f$ is the set of all real numbers. ●

- **EXAMPLE 2:** Find the domain and range of the function $f(x) = \sqrt{x-3}$.

  SOLUTION: Notice that $f(2) = \sqrt{2-3} = \sqrt{-1}$ is not a real number. Thus, 2 is not in the domain of $f$. If $a$ is any real number less than 3, then $f(a)$ is *not* a real number. If $a$ is greater than or equal to 3, then $f(a)$ is a real number. Therefore, the domain of $f$ is the set of all real numbers greater than or equal to 3, or $x \geq 3$. The computer drawn graph of $f$ in Figure 1.3.1 confirms the domain determined algebraically. We can read from this graph that the range is $y \geq 0$. ●

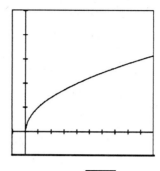

$y = \sqrt{x-3}$
$[-100, 1000]$ by $[-10, 50]$
Figure 1.3.1

We can algebraically find the domain of a function by analyzing its rule. Finding the range of the equation by analyzing the rule algebraically will often be difficult. Most of the time we will still read the range from a complete graph.

## Built-In Functions

Computers have special built-in functions that can be accessed by typing a special code. For example, typing SQR($x$) (or SQRT($x$) on some computers) will usually produce the **square root function** $f(x) = \sqrt{x}$. This function can also be accessed by entering it in exponential form: $x \wedge (1/2)$ or $x \wedge 0.5$. The parentheses around the 1/2 are needed because computers follow the conventional rules for order of operations. Entering $x \wedge 1/2$ produces the function $\frac{x^1}{2}$, that is, $\frac{x}{2}$.

Inequalities will be studied in Section 1.6. For the time being we use the following notation.

| | |
|---|---|
| $x \geq a$ | The set of all real numbers greater than or equal to $a$. |
| $x > a$ | The set of all real numbers greater than $a$. |
| $x \leq a$ | The set of all real numbers less than or equal to $a$. |
| $x < a$ | The set of all real numbers less than $a$. |
| $a \leq x \leq b$ | The set of all real numbers between $a$ and $b$, including $a$ and $b$. |
| $a \leq x < b$ | The set of all real numbers between $a$ and $b$, including $a$ but *not* $b$. |
| $a < x \leq b$ | The set of all real numbers between $a$ and $b$, including $b$ but *not* $a$. |
| $a < x < b$ | The set of all real numbers between $a$ and $b$, excluding both $a$ and $b$. |

- **EXAMPLE 3**: Find the domain and range of the function $f(x) = \frac{\sqrt{x}}{2x-4}$.

**SOLUTION:** The denominator $2x - 4$ of the rule for $f$ is zero if $x$ is 2. Division by zero does *not* produce a real number. Thus, 2 is *not* in the domain of $f$. The numerator $\sqrt{x}$ of the rule for $f$ is real if and only if $x$ is nonnegative. Thus, the domain of $f$ consists of all nonnegative real numbers different from 2, or $0 \leq x < 2$ together with $x > 2$. The computer drawn graph

# 1.3 More on Functions

of $f$ in Figure 1.3.2 confirms the domain found algebraically. This figure suggests that the range of $f$ is the set of all real numbers. Notice that $f(0) = 0$ so that 0 is in the range of $f$. •

Another function that is built-in on most computers is the greatest integer function.

- **DEFINITION**  The **greatest integer function** pairs a number with the largest integer that is less than or equal to the number. This function is denoted by INT($x$). Another notation often used for this function is $[\![x]\!]$. •

- **EXAMPLE 4**: Compute:
    (a) INT(2), INT(2.01), and INT(2.999).
    (b) INT($-2.01$), INT($-2.999$), and INT($-3$).

    SOLUTION:

    (a) Notice that 2 is the largest integer that is less than or equal to 2, 2.01, and 2.999. Thus,
    $$\text{INT}(2) = \text{INT}(2.01) = \text{INT}(2.999) = 2.$$

    (b) The largest integer that is less than or equal to $-2.01$, $-2.999$, and $-3$ is $-3$. (Why?) Thus,
    $$\text{INT}(-2.01) = \text{INT}(-2.999) = \text{INT}(-3) = -3.$$ •

- **EXAMPLE 5**: Draw a complete graph of $f(x) = \text{INT}(x)$ and determine its domain and range.

    SOLUTION: There is a unique largest integer less than or equal to any number $x$. Therefore, the domain of this function is the set of all real numbers. The graph of this function in the

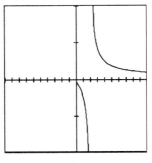

$y = \frac{\sqrt{x}}{2x-4}$
$[-10, 10]$ by $[-2, 2]$
Figure 1.3.2

$[-10, 10]$ by $[-10, 10]$ viewing rectangle is shown in Figure 1.3.3. The scale marks on the coordinate axes are one unit apart.

If $x$ is between 0 and 1, then $\text{INT}(x) = 0$. Therefore, the graph of $f(x) = \text{INT}(x)$ between $x = 0$ and $x = 1$ coincides with the $x$-axis. This is not apparent from the computer drawn graph in Figure 1.3.3.

If $x$ is an integer, then $\text{INT}(x) = x$. This is also not clear from the computer drawn graph in Figure 1.3.3. For example, from the graph in Figure 1.3.3 it appears that $\text{INT}(3)$ is either 2 or 3. To clarify that $\text{INT}(3)$ is 3 and not 2, we put a solid circle at $(3, 3)$ and an open circle at $(3, 2)$ as illustrated in Figure 1.3.4. We can, of course, do this at each integer value of $x$. We sometimes call this function a **step function** because its completed graph looks like a set of stair steps.

Finally, we can see that the range of $f(x)$ is the set of integers. By definition, no real number that is not an integer can be a value of this function. •

- **Warning.** There are some graphing calculators that use $\text{INT}(x)$ for a function different from the greatest integer function. Their graphs agree for $x \geq 0$ but disagree for $x < 0$.

Another function that is built-in on a computer is the **absolute value function**.

- **EXAMPLE 6:** Compare the graphs of $f(x) = x^2 - 4$ and $g(x) = |x^2 - 4|$. Determine the domain and range of each function.

**SOLUTION:** A complete graph of $f(x) = x^2 - 4$ is given in Figure 1.3.5. You may need to rewrite $g$ in the form $g(x) = \text{ABS}(x^2 - 4)$ in order to enter it on your graphing utility. Figure 1.3.6 gives the graph of $g$ in the $[-10, 10]$ by $[-10, 10]$ viewing rectangle. The graphs of $f$ and $g$ coincide for $x \geq 2$ and for $x \leq -2$. If the coordinate plane is folded along the $x$-axis, then the portion of the graph of $f$ below the $x$-axis coincides with the graph of $g$ for $x$ between $-2$ and $2$. The graph of $g$ can never be below the $x$-axis. (Why?) The effect of the absolute value in $g$ is to bring the part of the graph of $f$ that is below the $x$-axis above

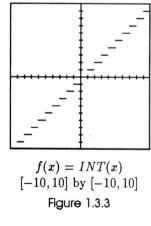

$f(x) = \text{INT}(x)$
$[-10, 10]$ by $[-10, 10]$

Figure 1.3.3

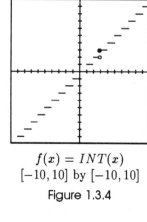

$f(x) = \text{INT}(x)$
$[-10, 10]$ by $[-10, 10]$

Figure 1.3.4

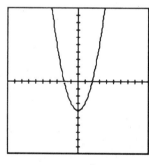

$f(x) = x^2 - 4$
$[-10, 10]$ by $[-10, 10]$

Figure 1.3.5

the $x$-axis to produce the graph of $g$. This happens because the values of $f$ are negative if the graph of $f$ lies below the $x$-axis, and the absolute value is always positive.

The domain of $f$ is the set of all real numbers and the range of $f$ is $y \geq -4$. The domain of $g$ is also the set of all real numbers, but its range is $y \geq 0$. •

Example 6 illustrates how to find a complete graph of $y = |f(x)|$ whenever we have a complete graph of $y = f(x)$. The portions of the graph of $y = f(x)$ that are below the $x$-axis must be reflected through the $x$-axis using folding symmetry with respect to the $x$-axis. The resulting graph is a complete graph of $y = |f(x)|$. The same general strategy can be applied to determine complete graphs of $y = \sqrt{f(x)}$ and $y = \sqrt{|f(x)|}$ from a complete graph of $y = f(x)$. We will see other applications of reflections in subsequent chapters.

• **EXAMPLE 7:** Draw a complete graph of the function $f(x) = 16x - x^3$. Find the domain and range of $f$.

**SOLUTION:** The graph of $f$ in the default viewing rectangle $[-10, 10]$ by $[-10, 10]$ is given in Figure 1.3.7. Careful observation while a graphing utility draws the graph in Figure 1.3.7 indicates that the graph goes off the screen and then comes back on the screen twice during the plotting process. This means that we need to see more of this graph in the vertical direction. Figure 1.3.8 shows the graph of $f$ in the $[-10, 10]$ by $[-30, 30]$ viewing rectangle. In Section 1.2 we indicated that there are four possibilities for the complete graph of a function of the form $f(x) = ax^3 + bx^2 + cx + d$ with $a \neq 0$. We know the graph in Figure 1.3.8 is a complete graph because it looks like the second of the four graphs in Figure 1.2.10.

We can use the rule of $f$ or the graph in Figure 1.3.8 to see that the domain of $f$ is the set of all real numbers. It is clear from the graph that the range of $f$ is also the set of all real numbers. •

Functions can be used to model many problem situations. Computer generated graphing can help us better understand the problem situation. One classic problem concerns the behavior of an object thrown straight up into the air with some initial velocity of $v_0$ feet

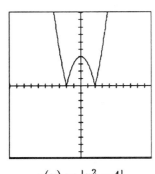

$g(x) = |x^2 - 4|$
$[-10, 10]$ by $[-10, 10]$
Figure 1.3.6

$f(x) = 16x - x^3$
$[-10, 10]$ by $[-10, 10]$
Figure 1.3.7

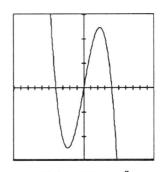

$f(x) = 16x - x^3$
$[-10, 10]$ by $[-30, 30]$
Figure 1.3.8

per second. The height $s$ of the object above ground at any time $t$ is given by $s = -16t^2 + v_0 t$. Notice that the initial velocity is the coefficient of $t$. If the object is thrown from a point $s_0$ feet above ground with the same initial velocity, then the height $s$ is given by $s = -16t^2 + v_0 t + s_0$ (Figure 1.3.9). The number 16 is half the acceleration due to gravity.

- **EXAMPLE 8**: An object is shot straight up from level ground with initial velocity of 64 feet per second.

    (a) Write the height $s$ of the object above ground as a function of time $t$.
    (b) Draw a complete graph of the function in (a) indicating what portion of the graph represents the problem situation.
    (c) Find the maximum height reached by the object and the time required to reach that height.

    SOLUTION:

    (a) The height above ground $t$ seconds after launch is given by $s = -16t^2 + 64t$.
    (b) You may need to view the graph of $s$ in several viewing rectangles to find the complete graph given in Figure 1.3.10. Because distance above ground must be nonnegative, only the *first* quadrant portion of the graph in Figure 1.3.10 represents the problem situation.
    (c) We can estimate from the graph in Figure 1.3.10 that the coordinates of the highest point are about $(2, 65)$. A formula for the exact coordinates of the highest point was given in Section 1.2. We know from Section 1.2 that the $t$ coordinate of the highest point in Figure 1.3.10 is $-\frac{64}{2(-16)} = 2$. The corresponding value of $s$ is 64. Thus, the object reaches a maximum height of 64 feet exactly 2 seconds after launch. •

- **EXAMPLE 9**: At what time will the object in Example 8 be 50 feet above ground?

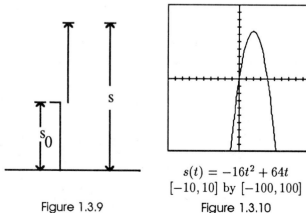

$s(t) = -16t^2 + 64t$
$[-10, 10]$ by $[-100, 100]$

Figure 1.3.9  Figure 1.3.10

# 1.3 More on Functions

SOLUTION: We know that only the first quadrant portion of the graph in Figure 1.3.10 represents the problem situation. Notice that $s$ is 0 when $t$ is 4. Figure 1.3.11 gives a graph of the problem situation and the horizontal line $s = 50$. The *two* points of intersection in Figure 1.3.11 have second coordinate 50. The first coordinates give values of $t$ for which $s(t) = -16t^2 + 64t = 50$. These two values of $t$ give the times when the object is 50 feet above ground; once on the way up and once on the way down. We can estimate from Figure 1.3.11 that the two times are 1.1 and 2.9 seconds after launch. •

The graph in Figure 1.3.11 represents the height above ground of the object at any time $t$. It does *not* represent the path of the object. The object goes straight up and comes straight down. The actual path of the object lies on a vertical line. In the next chapter, we review algebraic methods for solving equations and introduce a powerful geometric method for solving equations.

## Piecewise-Defined Functions

Piecewise-defined functions are often used to provide examples of functions with specified properties or to give counterexamples to certain conjectures. We can use a graphing utility to help us draw a complete graph of a piecewise-defined function.

• EXAMPLE 10: Draw a complete graph of the function

$$f(x) = \begin{cases} 3 - x^2, & \text{for } x < 1 \\ x^3 - 4x, & \text{for } x \geq 1 \end{cases}.$$

Determine the domain and range of $f$.

SOLUTION: The computer drawn graph of $y = 3 - x^2$ gives the portion of the graph of $f$ to the left of the line $x = 1$ (Figure 1.3.12). The computer drawn graph of $y = x^3 - 4x$ gives the portion of the graph of $f$ to the right of the line $x = 1$ (Figure 1.3.13). Notice that $f(1) = -3$. Combining the unshaded portions of the graphs in Figures 1.3.12 and 1.3.13

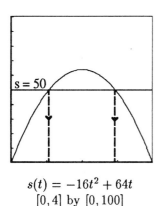

$s(t) = -16t^2 + 64t$
$[0, 4]$ by $[0, 100]$

Figure 1.3.11

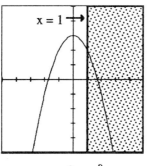

$y = 3 - x^2$
$[-5, 5]$ by $[-5, 5]$

Figure 1.3.12

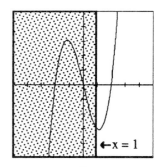

$y = x^3 - 4x$
$[-5, 5]$ by $[-5, 5]$

Figure 1.3.13

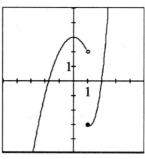

$$y = \begin{cases} 3 - x^2, & \text{for } x < 1 \\ x^3 - 4x, & \text{for } x \geq 1 \end{cases}$$

Figure 1.3.14

and putting an open circle at $(1, 2)$ and a closed circle at $(1, -3)$ gives the complete graph of $f$ shown in Figure 1.3.14. The domain and range of $f$ are the set of all real numbers. •

## • EXERCISES 1-3

Draw a complete graph of each function.

1. $y = \sqrt{x + 1}$
2. $y = -1 - \sqrt{x - 1}$
3. $y = 1 + |x - 2|$
4. $y = -|2x - 3|$
5. $y = \text{INT}(x + 2)$
6. $y = x^2 - 5x - 10$
7. $y = (x - 30)(x + 20)$
8. $y = \left(\dfrac{x - 20}{2}\right)^2 - 50x$
9. $y = |x^2 - 6x - 12|$
10. $y = 2x^3 - x + 3$
11. $y = x^3 - 8x$
12. $y = 10x^3 - 20x^2 + 5x - 30$
13. What are the domain and range of the function in Exercise 1?
14. What are the domain and range of the function in Exercise 6?

Algebraically determine the domain of the function. Check with a graphing utility. Resolve conflicting results.

15. $f(x) = \dfrac{x^2 - 5}{3}$
16. $g(x) = \sqrt{8 - x}$
17. $f(x) = \dfrac{x - 2}{x^2 + 5}$
18. $f(x) = \sqrt{\dfrac{1}{x + 1}}$
19. $g(x) = \dfrac{\sqrt{x - 5}}{x^2 - 5}$
20. $T(x) = \sqrt{x^3 - 8x}$

21. Compare complete graphs of $f(x) = \dfrac{|x|}{x}$ and $g(x) = \dfrac{x}{|x|}$.

22. Which of the computer expressions are equal to $3x^{1/2}$?
    (a) $3 * x \wedge (1/2)$
    (b) $3 * (x \wedge (1/2))$
    (c) $(3 * x) \wedge (1/2)$
    (d) $3 * x \wedge 1/2$

## 1.3 More on Functions

23. Which of the viewing rectangles gives the best complete graph of $f(x) = x^3 + 10x + 100$?
    (a) $[-10, 10]$ by $[-10, 10]$
    (b) $[-10, 10]$ by $[-100, 100]$
    (c) $[-10, 10]$ by $[0, 100]$
    (d) $[-10, 10]$ by $[-100, 0]$
    (e) $[-10, 10]$ by $[-1000, 1000]$

24. Which of the viewing rectangles gives the best complete graph of $y = 12x^2 - 5x + 19$?
    (a) $[-10, 10]$ by $[-10, 10]$
    (b) $[-10, 10]$ by $[-100, 100]$
    (c) $[-50, 50]$ by $[-5, 5]$
    (d) $[-5, 5]$ by $[-50, 50]$
    (e) $[-10, 0]$ by $[0, 50]$

25. Which of the viewing rectangles gives the best complete graph of $f(x) = 0.005x^3 - 5x^2 - 60$?
    (a) $[-10, 10]$ by $[-100, 100]$
    (b) $[-100, 100]$ by $[-100,000, 100,000]$
    (c) $[-1000, 2000]$ by $[-1,000,000, 1,000,000]$
    (d) $[0, 2000]$ by $[0, 1,000,000]$
    (e) $[-1000, 1000]$ by $[-1000, 1000]$

26. Determine a viewing rectangle that gives a complete graph of $f(x) = 50 - 25x + 100x^2 - x^3$. Sketch a complete graph of the function. Determine the domain and range.

27. $f(x) = \begin{cases} x^2, & x < 3 \\ x - 2, & x \geq 3 \end{cases}$

28. $g(x) = \begin{cases} \frac{1}{x}, & x < -2 \\ 2x, & x \geq -2 \end{cases}$

29. $f(x) = \begin{cases} 4 - x^2, & |x| \leq 5 \\ 6, & |x| > 5 \end{cases}$

30. $g(x) = \begin{cases} -x, & 0 \leq x < 4 \\ \sqrt{x - 3}, & x \geq 4 \end{cases}$

31. $f(x) = \begin{cases} 4 - x + x^2, & x < 2 \\ 1 - 2x + 3x^2, & x \geq 2 \end{cases}$

32. $g(x) = \begin{cases} 2x^3 - 4x^2 + x - 6, & x < 0 \\ \sqrt{2x + 3}, & x \geq 0 \end{cases}$

33. Sue charges $3 per hour for babysitting plus $5 to cover her transportation costs.
    (a) Express Sue's babysitting fee $F$ as a function of time $t$.
    (b) Draw a complete graph of the function determined in (a).
    (c) What are the domain and range of the function in (a)?
    (d) What part of the domain of the function represents the problem situation?
    (e) Use the graph in (b) to estimate the time Sue spent babysitting if her charges amounted to $14.50.

34. A trucker averages 42 mph on a cross country trip from New York City to San Francisco, California. Let $t$ be the time elapsed since the trucker left New York.
    (a) Express the total distance the trucker travels as a function of time $t$.
    (b) Draw a complete graph of the function determined in (a).
    (c) What are the domain and range of the function in (a)?
    (d) What part of the domain of the function represents the problem situation?
    (e) Use the graph in (b) to determine how many hours have elapsed since the trucker left New York if the distance traveled is 850 miles.

35. The annual profit $P$ of a candy manufacturer is determined by the formula $P = R - C$, where $R$ is the total revenue generated from selling $x$ pounds of candy and $C$ is the total cost of

making and selling $x$ pounds of candy. Each pound of candy sells for $1.50 and costs $1.25 to make. The fixed costs of making and selling the candy are $20,000 annually.

(a) Express the company's annual profit as a function of $x$.
(b) Draw a complete graph of the function determined in (a).
(c) How many pounds of candy must be sold for the company to "break even?"
(d) How many pounds of candy must be sold for the company to make a profit of $10,000?
(e) Suppose the company gives some buyers a reduced price of $1.40 a pound when large quantities of candy are purchased. If exactly 300 pounds of the candy are sold at the reduced price, what is the answer to (c)?

36. An object 10 feet above level ground is shot straight up with an initial velocity of 150 feet per second.

(a) Determine the height $s$ of the object above the ground as a function of time $t$.
(b) Draw a complete graph of the function in (a) indicating what portion of the graph represents the problem situation.
(c) At what time will the object reach its maximum height? What is the maximum height?
(d) At what time $t$ will the object be 300 feet above the ground?

*37. The regular long distance telephone charges $C$ from Columbus, Ohio, to Lancaster, Pennsylvania, are $0.48 for the first minute or fraction of a minute, and $0.28 for each additional minute or fraction of a minute.

(a) Write an equation relating the cost of calling Lancaster from Columbus in terms of time $t$.
(b) Draw a complete graph of the function determined in (a).
(c) What are the domain and range of the function in (a)?
(d) What part of the domain of the function represents the problem situation?
(e) Bill can spend $2.00 for a call from Columbus to his girlfriend in Lancaster. Use the graph in (b) to determine how long Bill can talk to his girlfriend.

*38. A parking garage in downtown Houston, Texas, charges $2 for the first hour and $1 for each additional hour or fraction of an hour, up to a maximum of $12 per 24-hour day. Let $t$ represent the number of hours parked at the garage.

(a) Express the total possible parking charges for one day as a function of time $t$. *Hint:* You will have to consider two cases: One case for maximum charges of $12 and the other for charges less than or equal to the maximum charge of $12.
(b) Draw a complete graph of the function determined in (a).
(c) Use the graph in (b) to determine the time parked if the parking charges are $8.

39. Let the point $P$ be on the line $x + y = 1$. Let the point $Q$ have coordinates $(3, 5)$.

(a) Write the distance $D$ between $P$ and $Q$ as a function of $x$, where $x$ is the $x$-coordinate of the point $P$.
(b) What is the domain of the function $D$?
(c) Draw a complete graph of the function $y = D(x)$.
(d) Use your graph in (c) to determine $x$ if the distance is 10. What are the coordinates of the associated point $P$?

40. A function $f$ is **even** if $f(-x) = f(x)$ for every $x$ in the domain of $f$. Show that the graph of an even function is symmetric with respect to the $y$-axis.

41. Which of the functions are even? Use a graphical argument.
    (a) $f(x) = x^2 + 1$
    (b) $g(x) = 1 - x^3$
    (c) $f(x) = |x|$
    (d) $f(x) = (x - 1)^2$
    (e) $f(x) = \sin x$
    (f) $g(x) = \cos x$

42. A function $f$ is **odd** if $f(-x) = -f(x)$ for every $x$ in the domain of $f$. Show that the graph of an odd function is symmetric with respect to the origin.

43. Which of the functions are odd? Use a graphical argument.
    (a) $f(x) = x^3$
    (b) $f(x) = (x - 1)^3$
    (c) $g(x) = \sqrt{x}$
    (d) $g(x) = 1/x$
    (e) $f(x) = 2 \sin x$
    (f) $g(x) = -3 \cos x$

Draw a complete graph of the function. Determine the domain, range, and equation for any lines of symmetry. Which graphs are symmetric with respect to the origin?

44. $f(x) = |x^2 - 3|$
45. $g(x) = |x^3 - 1|$
46. $h(x) = \sqrt{|x + 5|}$
47. $f(x) = |x^3 - 8x|$
48. $g(x) = |x^2 - 5x + 8|$
49. $h(x) = \left| \dfrac{2}{x - 3} \right|$

## 1.4 Solving Equations

The main purpose of this section is to illustrate how to use a graphing utility to solve equations. This method produces accurate approximate solutions rather than exact solutions. The formulas that give exact solutions to polynomial equations of degrees 2, 3, or 4 involve radicals. However, there are no formulas involving radicals that give the exact solutions to polynomial equations of degree 5 or higher. Niels Henrik Abel (1802–1829), who died when he was only 26 years old, proved that it is not possible to give a formula involving only radicals for the exact solution to a polynomial equation of degree 5 or higher. In fact, there is no known method for finding exact solutions to polynomial equations of degree 5.

The formulas for the exact solutions of polynomial equations of degree 3 or 4 are not very useful unless we approximate the radicals involved. Generally, the best we can do when solving a given equation is to find approximations to its solutions using numerical techniques. M.J. Maron puts it this way in his textbook on numerical analysis:

"Fortunately, one rarely needs *exact* numerical answers. Indeed, in the 'real world' the problems themselves are generally inexact because they are posed in terms of parameters that are only approximate because they are *measured*. What one is likely to require in a realistic situation is not an exact answer, but rather one having a prescribed accuracy."

- **DEFINITION**  The number $a$ is a **solution to an equation** in one variable if we get a true statement when the variable is replaced by $a$ everywhere in the equation.

For example, $-2$ is a solution to the equation $x^3 - x + 6 = 0$ because we get the following true statement when $x$ is replaced by $-2$ in this equation: $(-2)^3 - (-2) + 6 = 0$. Similarly, $\sqrt{3}$ is a solution to $y^2 - 3 = 0$ because $(\sqrt{3})^2 - 3 = 0$. When we have found *all* solutions to an equation we say that we have **solved the equation**. In this chapter we restrict our attention to real number solutions to equations and inequalities. Often we refer to real number solutions more simply as **real solutions**.

One important algebraic procedure to solve an equation is to transform it into an equivalent equation with solutions that are obvious.

- **DEFINITION**  Two equations are said to be **equivalent** if they have exactly the same set of solutions.

For example, the equations $x - 2 = 0$ and $2x - 4 = 0$ are equivalent. Similarly, $x = 3$ and $x + 2 = 5$ are equivalent.

## Equivalent Equations

An equivalent equation is obtained if any one of the following operations is performed.

(1) Add the same real number to, or subtract it from, each side of a given equation.
(2) Add the same polynomial to, or subtract it from, each side of a given equation.
(3) Multiply each side of an equation, or divide each side of an equation, by the same nonzero real number.

## Solutions and x-intercepts

Every equation can be rewritten in an equivalent form with one side equal to zero. For example, an equivalent form of $x^3 + 2x = 1$ is $x^3 + 2x - 1 = 0$. Any equation in one variable $x$ can be rewritten in the form $f(x) = 0$. For the equation $x^3 + 2x = 1$, $f(x)$ is $x^3 + 2x - 1$. Notice that $f$ is a *polynomial function* because $x^3 + 2x - 1$ is a polynomial.

The *graph* of the function $y = f(x)$ can be used to solve the equation $f(x) = 0$. If $x = a$ is a real solution to $f(x) = 0$, then $(a, 0)$ is a point on the graph of $y = f(x)$. Thus, the solution $a$ to $f(x) = 0$ corresponds to the point $(a, 0)$ where the graph of $y = f(x)$ crosses the $x$-axis (Figure 1.4.1). Such points $(a, 0)$ are called the **$x$-intercepts** of $y = f(x)$. We also call $a$ a **zero**, or a **root**, of $f(x)$. The value $a$ is called a zero because it is an $x$-value that makes $f(x)$ equal to zero.

Conversely, if $(a, 0)$ is an $x$-intercept of $y = f(x)$, then $x = a$ is a real solution of $f(x) = 0$. Therefore, the set of all real solutions of $f(x) = 0$ is the same as the set of all first coordinates of the $x$-intercepts of the graph of $y = f(x)$.

**Notation for x-intercept.**  If $(a, 0)$ is an $x$-intercept of $y = f(x)$, we often call $a$ an $x$-intercept. With this agreement, the set of all real solutions of $f(x) = 0$ is the same as the set of all $x$-intercepts of the graph of $y = f(x)$.

1.4 Solving Equations

Figure 1.4.1

Figure 1.4.2

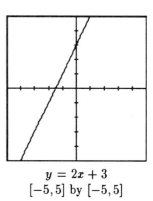

$y = 2x + 3$
$[-5, 5]$ by $[-5, 5]$
Figure 1.4.3

## Linear Equations

- **DEFINITION**  An equation is **linear** in $x$ if it can be written in the form $ax + b = 0$, where $a$ and $b$ are real numbers and $a \neq 0$.

Once we transform a linear equation into the form $ax + b = 0$ it is quite easy to solve the equation. We subtract $b$ from both sides of $ax + b = 0$. Then we divide both sides of the resulting equation by $a$. We can divide both sides by $a$ because, by the definition of linear, $a \neq 0$. The linear equation $ax + b = 0$ has exactly one solution, namely, $-\frac{b}{a}$. Notice that the graph of the equation $y = ax + b$ is a *straight line* that crosses the $x$-axis at $(-\frac{b}{a}, 0)$. (Figure 1.4.2).

- **EXAMPLE 1:** Solve $2x + 3 = 0$.

  SOLUTION: Algebraically, we know that $-\frac{3}{2}$ is the solution. Indeed, the graph of $y = 2x + 3$ shown in Figure 1.4.3 appears to have $x$-intercept $-\frac{3}{2}$. Later in this section we will see how to find an $x$-intercept and, therefore, a solution to an equation with a high degree of accuracy.

The diagram in Figure 1.4.4 demonstrates the correspondence between solutions of equations and $x$-intercepts of graphs.

## Quadratic Equations

- **DEFINITION**  An equation is **quadratic** in $x$ if it can be written in the form $ax^2 + bx + c = 0$, where $a$, $b$, and $c$ are real numbers and $a \neq 0$.

Transforming a quadratic equation into an equivalent form with solutions that are obvious may require more work than transforming linear equations. We use a technique

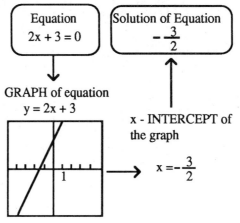

Figure 1.4.4

called **completing the square** to derive the **quadratic formula**, which gives the exact solutions to $ax^2 + bx + c = 0$. The following equations are equivalent.

$$ax^2 + bx + c = 0$$
$$ax^2 + bx = -c \quad \text{(Subtract } c \text{ from each side.)}$$
$$x^2 + \frac{b}{a}x = \frac{-c}{a} \quad \text{(Divide each side by } a.\text{)}$$

We can divide each side by $a$ because, by definition of quadratic, $a \neq 0$. Next we want to add an expression to each side that makes the left side of $x^2 + \frac{b}{a}x = -\frac{c}{a}$ the square of a binomial. Because $(x + t)^2 = x^2 + 2tx + t^2$, we must add the square of half of the coefficient of $x$.

$$x^2 + \frac{b}{a}x = \frac{-c}{a}$$
$$x^2 + \frac{b}{a}x + \left(\frac{b}{2a}\right)^2 = \frac{-c}{a} + \left(\frac{b}{2a}\right)^2 \quad \text{(Add } \left(\frac{b}{2a}\right)^2 \text{ to each side.)}$$
$$\left(x + \frac{b}{2a}\right)^2 = \frac{-c}{a} + \left(\frac{b}{2a}\right)^2 \quad \text{(Factor left side.)}$$
$$\left(x + \frac{b}{2a}\right)^2 = \frac{b^2 - 4ac}{4a^2} \quad \text{(Combine fractions on right side.)}$$

Because the square of $x + \frac{b}{2a}$ equals $\frac{b^2 - 4ac}{4a^2}$, we must have

$$x + \frac{b}{2a} = \frac{\sqrt{b^2 - 4ac}}{2a} \quad \text{or} \quad x + \frac{b}{2a} = -\frac{\sqrt{b^2 - 4ac}}{2a}.$$

Solving for $x$ we have,

$$x = -\frac{b}{2a} + \frac{\sqrt{b^2 - 4ac}}{2a} \quad \text{or} \quad x = -\frac{b}{2a} - \frac{\sqrt{b^2 - 4ac}}{2a}.$$

Generally, the two solutions above are combined into the single expression

$$x = \frac{-b \pm \sqrt{b^2 - 4ac}}{2a}.$$

This last expression is called the **quadratic formula** and gives the solutions to the quadratic equation $ax^2 + bx + c = 0$ in terms of the coefficients $a$, $b$, and $c$.

- **DEFINITION** The expression $b^2 - 4ac$ in the quadratic formula is called the **discriminant** of the quadratic equation $ax^2 + bx + c = 0$.

**Number of Real Solutions.** If $b^2 - 4ac$ is positive, the quadratic equation has exactly two real solutions. If $b^2 - 4ac$ is zero, the quadratic equation has exactly one real solution. If $b^2 - 4ac$ is negative, the quadratic equation has no real solutions.

Figure 1.2.6 from Section 1.2 gives the possible graphs of $y = ax^2 + bx + c$. The $x$-intercepts of $y = ax^2 + bx + c$ give the real solutions to the quadratic equation $ax^2 + bx + c = 0$. Figure 1.4.5 illustrates that the number of $x$-intercepts of $y = ax^2 + bx + c$ with $a > 0$ can be zero, one, or two. We get the same possibilities if the graph of $y = ax^2 + bx + c$ opens downward ($a < 0$). The discriminant $b^2 - 4ac$ of the quadratic equation will be negative, zero, or positive, respectively. (Why?)

- **EXAMPLE 2:** Solve $2x^2 = 1 - 2x$.

SOLUTION: First we rewrite the equation in the form $f(x) = 0$. The desired equivalent equation is $2x^2 + 2x - 1 = 0$, where $f(x) = 2x^2 + 2x - 1$. The quadratic formula gives

$$x = \frac{-2 \pm \sqrt{2^2 - 4(2)(-1)}}{2(2)} = \frac{-2 \pm \sqrt{12}}{4} = \frac{-1 \pm \sqrt{3}}{2}.$$

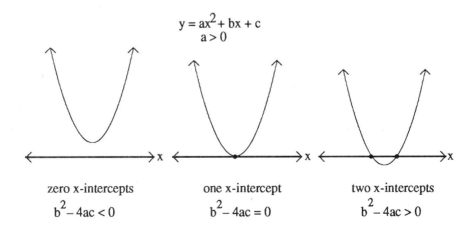

Figure 1.4.5

These exact solutions are generally not as useful as decimal approximations. We can use a calculator to show that, accurate to seven decimal places, $\frac{-1+\sqrt{3}}{2}$ is 0.3660254 and $\frac{-1-\sqrt{3}}{2}$ is $-1.3660254$. The graph of $f(x) = 2x^2 + 2x - 1$ in the $[-5, 5]$ by $[-5, 10]$ viewing rectangle in Figure 1.4.6 suggests that these two values of $x$ are reasonable approximations to the solutions because they are good approximations to the $x$-intercepts.

• **EXAMPLE 3:** Solve $x^2 + 2x - 15 = 0$.

**SOLUTION:** The graph of $y = x^2 + 2x - 15$ in Figure 1.4.7 strongly suggests that the $x$-intercepts are $-5$ and 3. This means that $-5$ and 3 are likely the solutions to the equation. We can confirm this conjecture made graphically either by direct substitution or by factoring the equation.

$$x^2 + 2x - 15 = 0$$
$$(x + 5)(x - 3) = 0$$

Thus, $-5$ and 3 are exact solutions to this equation.

The techniques used to solve quadratic equations can occasionally be applied to polynomial equations of higher degree.

• **EXAMPLE 4:** Solve $x^4 - 5x^2 - 36 = 0$.

**SOLUTION:** The easiest way to solve this equation is to notice that it can be factored; it is quadratic in $x^2$.

$$x^4 - 5x^2 - 36 = 0$$
$$(x^2 + 4)(x^2 - 9) = 0$$

Thus, either $x^2 + 4 = 0$ or $x^2 - 9 = 0$. The equation $x^2 + 4 = 0$ has no real solutions, and the solutions to $x^2 - 9 = 0$ are $-3$ and 3. Therefore, the original equation has two real solutions: $-3$ and 3.

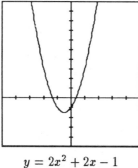

$y = 2x^2 + 2x - 1$
$[-5, 5]$ by $[-5, 10]$

Figure 1.4.6

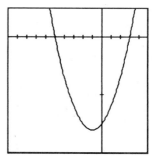

$y = x^2 + 2x - 15$
$[-10, 5]$ by $[-20, 5]$

Figure 1.4.7

# 1.4 Solving Equations

Alternatively, the equation $x^4 - 5x^2 - 36 = 0$ can be viewed as a quadratic equation in the variable $x^2$ by rewriting it in the form

$$(x^2)^2 - 5(x^2) - 36 = 0.$$

Then, the quadratic formula can be applied with $a = 1$, $b = -5$, and $c = -36$, to obtain

$$x^2 = \frac{5 \pm \sqrt{25 + 144}}{2} = \frac{5 \pm 13}{2}.$$

Thus, $x^2 = 9$ or $x^2 = -4$. The equation $x^2 = -4$ has no real solutions, and $-3$ and $3$ are the solutions to $x^2 = 9$. The graph of $f(x) = x^4 - 5x^2 - 36$ in Figure 1.4.8 confirms the real solutions.

## Cubic Equations

- **DEFINITION**  A polynomial equation of degree 3 is called a **cubic equation**.

The quadratic formula always gives exact solutions, which can then be approximated if desired or needed. There are formulas for the exact solutions to polynomial equations of degrees 3 and 4. However, these formulas are quite complicated, and solving equations graphically is so easy that we will *not* use these formulas in this course. To illustrate their complexity we give the solutions to

$$x^3 + a_1 x^2 + a_2 x + a_3 = 0. \qquad (1)$$

First, we set

$$p = a_2 - \frac{a_1^2}{3}, \quad q = \frac{2a_1^3}{27} - \frac{a_1 a_2}{3} + a_3,$$

$$P = \sqrt[3]{-\frac{q}{2} + \sqrt{\frac{p^3}{27} + \frac{q^2}{4}}}, \quad \text{and} \quad Q = \sqrt[3]{-\frac{q}{2} - \sqrt{\frac{p^3}{27} + \frac{q^2}{4}}}.$$

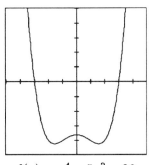

$f(x) = x^4 - 5x^2 - 36$
$[-5, 5]$ by $-50, 50]$
Figure 1.4.8

Then, the three solutions of equation (1) are

$$P+Q-\frac{a_1}{3}, \quad WP+W^2Q-\frac{a_1}{3}, \quad \text{and} \quad W^2P+WQ-\frac{a_1}{3},$$

where $W$ is a complex number with third power equal to 1. These formulas are called the **Cardan formulas**. You will see that a computer-based graphing approach makes the use of these complicated formulas unnecessary.

The Cardan solutions appeared in *Ars Magna* in the year 1545. The graphical techniques in this text represent the first major practical advance in equation-solving techniques in the intervening 400 plus years that are easily available to precalculus students.

## Finding Solutions Graphically Using Zoom-In

Next, we turn our attention to the use of a graphing utility to find solutions to an equation. One advantage of a graphical approach is that we are able to solve very complicated equations. We can solve any equation $f(x) = 0$ graphically provided we can enter $y = f(x)$ into our graphing utility. In particular, polynomial equations of *any* degree are no more difficult to solve than quadratic equations.

To find a solution of $f(x) = 0$ graphically we *trap the corresponding x-intercept* in a viewing rectangle that is small enough to read the value of $x$ as accurately as desired. In fact, we actually create a nested sequence of viewing rectangles, each containing the next, that trap the $x$-intercept (Figure 1.4.9). However, solutions can be read only within the limits of machine precision, that is, only as accurately as the graphing utility permits. Most computers and graphing calculators allow answers to be read accurately to at least 7 or 8 significant figures. This procedure for finding the solutions to $f(x) = 0$ graphically is called **zoom-in**.

- **EXAMPLE 5:** Solve the equation $x^3 + 2x = 1$.

    SOLUTION: First we write the equation in the form $x^3 + 2x - 1 = 0$ and set $f(x) = x^3 + 2x - 1$. The graph of $f$ in the $[-10, 10]$ by $[-10, 10]$ viewing rectangle is given in Figure 1.4.10. You

Figure 1.4.9

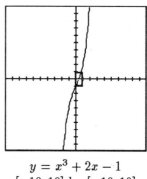

$y = x^3 + 2x - 1$
$[-10, 10]$ by $[-10, 10]$

Figure 1.4.10

can check that the graph in Figure 1.4.10 is a complete graph. Thus, there is exactly one real solution to $x^3 + 2x - 1 = 0$. We can see from this figure that the graph of $f$ crosses the $x$-axis between $x = 0$ and $x = 1$ because the scale marks on the $x$-axis are one unit apart. Remember that scale marks that appear in computer graphs produced by our graphing utility are equally spaced and depend on the viewing rectangle parameters $L$, $R$, $B$, and $T$. Figure 1.4.10 suggests that we can choose the next viewing rectangle to be $[0, 1]$ by $[-1, 1]$. The graph of $f$ in this viewing rectangle is shown in Figure 1.4.11.

The scale marks on the $x$-axis in Figure 1.4.11 are 0.1 units apart. (Why?) We can see that the graph of $f$ crosses the $x$-axis between $x = 0.4$ and $x = 0.5$. Thus, $[0.4, 0.5]$ by $[-0.1, 0.1]$ is a reasonable viewing rectangle to choose, and the graph of $f$ in this viewing rectangle is given in Figure 1.4.12.

Figure 1.4.12 indicates that the graph of $f$ crosses the $x$-axis between $x = 0.45$ and $x = 0.46$. Any number between 0.45 and 0.46 that is used to approximate the actual solution is in error by at most 0.01, the distance between consecutive scale marks on the $x$-axis in Figure 1.4.12. For example, we could say that the solution is 0.453 with error of at most 0.01 because the graph of $f$ in Figure 1.4.12 appears to cross the $x$-axis three-tenths of the way from the scale mark labeled 0.45 to the scale mark labeled 0.46. •

## Error

In Example 5 we found that 0.453 was a solution to $x^3 + 2x - 1 = 0$ with error of at most 0.01. The reason the error is at most 0.01 is that Figure 1.4.12 shows that the *exact solution* lies between consecutive scale marks on the $x$-axis that are 0.01 units apart. In fact, *any number* in the interval $[0.45, 0.46]$, determined by these two scale marks, is a *solution to* $x^3 + 2x - 1 = 0$ *with error of at most* 0.01. For example, the left endpoint 0.45 and the right endpoint 0.46 are also solutions to $x^3 + 2x - 1 = 0$ with error of at most 0.01. Notice that we are *not* saying that the corresponding values of $f(x) = x^3 + 2x - 1$ differ from 0 by at most 0.01. In fact, Figure 1.4.12 shows that for some values of $x$ in $[0.45, 0.46]$ the corresponding values of $f$ differ from 0 by more than 0.01. Notice that the distance between consecutive vertical scale marks is also 0.01.

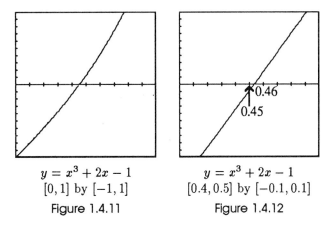

$y = x^3 + 2x - 1$
$[0, 1]$ by $[-1, 1]$

Figure 1.4.11

$y = x^3 + 2x - 1$
$[0.4, 0.5]$ by $[-0.1, 0.1]$

Figure 1.4.12

In a sufficiently small viewing rectangle a portion of the graph of a polynomial function will always appear to be a straight line. You can convince yourself that this statement is true by graphing several polynomial functions in small viewing rectangles that contain a portion of the graph. From Figure 1.4.12, we can see that the graph of $f$ rises as we move from left to right in $0.45 \leq x \leq 0.46$. This means $f(x)$ will differ from zero by at most the larger of the two numbers $|f(0.45)|$ and $|f(0.46)|$ for $x$ in $0.45 \leq x \leq 0.46$. Both $|f(0.45)|$ and $|f(0.46)|$ are less than 0.02 so that 0.02 is a good *estimate* for the error in the value of $f$ when we use a value of $x$ in $0.45 \leq x \leq 0.46$ to approximate the zero of $f$.

- **DEFINITION** Suppose that the graph of $y = f(x)$ in some viewing rectangle crosses the $x$-axis between two consecutive scale marks $a$ and $b$ that are $r$ units apart. If $c$ is any number in the interval determined by the scale marks, then $c$ is said to be a **solution with error of at most $r$** to the equation $f(x) = 0$ (Figure 1.4.13). If the graph of $f$ is always rising or always falling in the viewing rectangle, then the larger of $|f(a)|$ and $|f(b)|$ is a good estimate for the corresponding error bound on the function values $f(x)$. •

Suppose that the graph of $y = f(x)$ in the viewing rectangle $[L, R]$ by $[B, T]$ crosses the $x$-axis but displays no scale marks. Then, any number between $L$ and $R$ would be a solution with error of at most $r = R - L$.

- **EXAMPLE 6:** Solve the equation $x^3 + 2x = 1$ with error of at most 0.0000001. Estimate the error in $f(x) = x^3 + 2x - 1$.

SOLUTION: If we continue the process started in Example 5 we can look at the graph of $f$ in smaller and smaller viewing rectangles. Figure 1.4.14 gives the graph of $f$ in the viewing rectangle $[0.453397, 0.453398]$ by $[-0.000001, 0.000001]$. The actual solution lies between the scale marks labeled 0.4533976 and 0.4533977. Any number between 0.4533976 and 0.4533977 that is used to approximate the actual solution is in error by at most 0.0000001, the distance

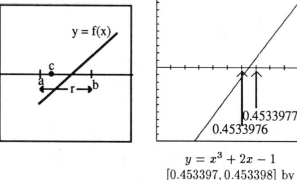

$y = x^3 + 2x - 1$
$[0.453397, 0.453398]$ by
$[-0.000001, 0.000001]$

Figure 1.4.13   Figure 1.4.14

between horizontal scale marks in Figure 1.4.14. Figure 1.4.14 suggests we use 0.45339765 because the graph of $f(x) = x^3 + 2x - 1$ appears to cross the $x$-axis half-way between the scale marks labeled 0.4533976 and 0.4533977. Notice that the graph of $f$ appears to be a straight line in this viewing rectangle. Both $|f(0.4533976)|$ and $|f(0.4533977)|$ are less than 0.0000002. Thus 0.0000002 is a good approximation to the error on $f(x)$. •

It is a good idea for you to think about trying to find the real solution to the equation in Example 6 using the Cardan formulas. This should convince you that finding solutions graphically is both very quick and very accurate.

Generally, we need not approximate solutions as accurately as we did in Example 6. However, this example illustrates that a computer based graphing approach can do more than simply produce rough approximations to solutions; accuracy is limited only by machine precision.

To find all the real solutions to an equation $f(x) = 0$ graphically, we first find a complete graph of $y = f(x)$. Once we know the complete behavior of $f$, we can determine the number and approximate locations of the real solutions to the equation $f(x) = 0$. Then, using the technique illustrated in Examples 5 and 6, we can find each of the real solutions with a high degree of accuracy.

The procedure used in Examples 5 and 6 to trap the solutions is called *zoom-in* because this procedure is quite like using the focus adjustment on a very powerful microscope. Until recently, very specialized numerical techniques were needed to find solutions with this degree of accuracy. Using graphing utilities eliminates, or at least reduces, the need for sophisticated and complicated numerical techniques.

• **EXAMPLE 7:** Find a solution to $x^3 - 5x^2 + 6x - 1 = 0$ between 1 and 2 with error of at most 0.01.

**SOLUTION:** The graph of $f(x) = x^3 - 5x^2 + 6x - 1$ in the standard viewing rectangle (Figure 1.4.15) is complete (why?) and shows that this equation has three real solutions. One of the solutions is between 1 and 2. In order to find a solution with error of at most 0.01 we need a viewing rectangle with distance between scale marks equal to 0.01 or less.

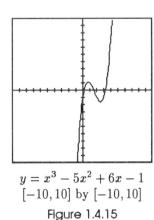

$y = x^3 - 5x^2 + 6x - 1$
$[-10, 10]$ by $[-10, 10]$
Figure 1.4.15

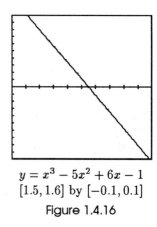

$y = x^3 - 5x^2 + 6x - 1$
$[1.5, 1.6]$ by $[-0.1, 0.1]$
Figure 1.4.16

Figure 1.4.16 gives the graph of $f$ in the $[1.5, 1.6]$ by $[-0.1, 0.1]$ viewing rectangle. Notice that the distance between scale marks in Figure 1.4.16 is 0.01. Thus, 1.555 is a solution to $x^3 - 5x^2 + 6x - 1 = 0$ that is between 1 and 2 and has error of at most 0.01. Can you give other approximations to the solution that are also in error by at most 0.01? •

**Accuracy Agreement.** Unless otherwise stated, from now on when we ask you to *solve an equation*, we mean for you to *approximate the real solutions with error of at most* 0.01. Furthermore, we assume that the numbers provided in Examples and Exercises are exact unless otherwise specified. Some textbooks assume that numbers like 8.7 and 7.9 are accurate only to tenths, and numbers like 6 and −3 are accurate only to units.

Next, we look at an application about making a cardboard box from a rectangular sheet of cardboard. Squares of side length 5 inches are cut from each corner of a rectangular piece of cardboard with width $W$ and length $L$ (Figure 1.4.17). Then, we fold along the dashed lines in Figure 1.4.17 to form a box with no top (Figure 1.4.18). The height of the box is 5, the width is $W - 10$, and the length is $L - 10$.

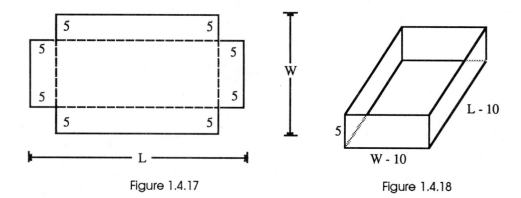

Figure 1.4.17

Figure 1.4.18

# 1.4 Solving Equations

- **EXAMPLE 8:** Suppose that the length of the rectangular piece of cardboard in Figure 1.4.17 is twice its width. Determine the dimensions of the cardboard if the volume of the resulting box is 2040 cubic inches.

  **SOLUTION:** From Figure 1.4.18, the volume of the box is $5(W-10)(L-10)$. Because $L = 2W$ and the volume of the box is 2040 we have $5(W-10)(2W-10) = 2040$. The following equations are equivalent.

  $$5(W-10)(2W-10) = 2040$$
  $$5(W-10)(2)(W-5) = 2040$$
  $$(W-10)(W-5) = 204$$
  $$W^2 - 15W + 50 = 204$$
  $$W^2 - 15W - 154 = 0$$

  We can use the quadratic formula or notice that the above quadratic equation factors as

  $$(W-22)(W+7) = 0.$$

  Both $-7$ and 22 are solutions to this equation. Because width must be positive only $W = 22$ is a solution to the problem. The dimensions of the original piece of cardboard must be 22 inches by 44 inches. Verify that the resulting box has volume 2040 cubic inches. •

## EXERCISES 1-4

1. Solve the equation $x^2 - 3x - 10 = 0$ algebraically. Draw a complete graph of $y = x^2 - 3x - 10$ and compare the $x$-intercepts of the graph with the solutions to the equation.
2. Solve the equation $x^3 - 25x = 0$ algebraically. Draw a complete graph of $y = x^3 - 25x$ and compare the $x$-intercepts of the graph with the solutions to the equation.
3. Solve the equation $|x + 4| = 8$ algebraically. Draw a complete graph of $y = |x + 4| - 8$ and compare the $x$-intercepts of the graph with the solutions to the equation.

Use an *algebraic* method to solve each equation. Check your answer with a graphing utility.

4. $(x-3)(x+2) = 0$
5. $x^2 = 14$
6. $x^2 - 3x + 2 = 0$
7. $x^2 - 2x + 3 = 0$
8. $x^3 - x = 0$
9. $|x - 2| = 6$

Use an algebraic method to solve each equation. Check by substitution into the original equation.

10. $y^2 = 9$
11. $x^2 - 5x + 6 = 30$
12. $3y^3 - 2y^2 + y = 0$
13. Bill thinks $-1$, 2, and 3 are solutions to the equation $x^3 + 2x^2 - 5x - 6 = 0$. Is he right? If not, which of these numbers are solutions? Does this equation have any other real solutions? How can graphing help determine all real solutions?

Use a graphing utility to determine the number of real solutions to each equation.

14. $x^3 - 2x^2 + 3x - 5 = 0$
15. $x^3 - 65x + 10 = 0$

Use factoring to find all real solutions to the equation.

16. $x^3 - 2x^2 - 2x + 4 = 0$
17. $10 - 15x + 6x^2 - 9x^3 = 0$

18. Consider the equation $x^3 - x^2 + x - 3 = 0$. Find a sequence of four viewing rectangles containing each solution. The first viewing rectangle should permit the solution to be read with error of at most 0.1; the second 0.01; the third 0.001; and the last 0.0001.

19. Consider the equation $x^3 - 2x + 3 = 0$. It has one real solution. Find a sequence of four viewing rectangles containing the solution. The first viewing rectangle should permit the solution to be read with error of at most 0.1; the second 0.01; the third 0.001; and the last 0.0001.

20. Find the other two solutions to the equation in Example 7 with error of at most 0.01.

Determine *one positive solution* to each equation $f(x) = 0$ with error of at most 0.01. Then estimate the maximum error in the corresponding $f(x)$ value.

21. $f(x) = x^3 - x - 2$

22. $f(x) = \frac{1}{100}x^3 - x - 2$

23. $f(x) = \frac{1}{10}x^3 - x^2 - 2$

24. $f(x) = x^3 - \frac{1}{10}x$

Solve each equation.

25. $3x^2 - 15x + 8 = 0$

26. $x^3 - 2x^2 + 3x - 1 = 0$

27. $x(x - 25)(x - 35) = 3000$

28. Find three distinct approximations to the one real solution to $x^3 - 10 = 0$ with error of at most 0.01. What is the exact solution?

Solve each equation for $x$ in the given interval with error of at most 0.01.

29. $x^4 - 3x^3 - 6x + 5 = 0$ where $0 \leq x \leq 10$

30. $\dfrac{x^3 - 10x^2 + x + 50}{x - 2} = 0$ where $-10 \leq x \leq 10$

31. $3\sin(x - 5) = 0$ where $0 \leq x \leq 10$

32. $\sqrt[3]{x^2 - 2x + 3} = 0$ where $-10 \leq x \leq 10$

33. Assume it is possible to find the exact solutions to $x^3 + a_1 x^2 + a_2 x + a_3 = 0$. Explain why it is then possible to find the exact solutions to $b_1 x^3 + b_2 x^2 + b_3 x + b_4 = 0$, where $b_1 \neq 0$.

34. The perimeter of some rectangles is 320 inches. Let $x$ be the length of such a rectangle.
    (a) If the length is 20 inches, then what is the area of the rectangle?
    (b) Show that $A = x(160 - x)$ is an algebraic representation of the area of such rectangles.
    (c) Draw a complete graph of the algebraic representation in (b).
    (d) Verify that $(80, 6400)$ is a point on the graph in (c). What meaning do these coordinates have in this problem situation?
    (e) Verify that $(-40, -8000)$ is a point on the graph in (c). What meaning do these coordinates have in this problem situation?
    (f) What values of $x$ make sense in this problem situation? Why? Draw a graph of the problem situation.
    (g) Use zoom-in to determine the area of a rectangle if its length is 83 inches.

35. A storage box of height 25 cm has a volume of 5400 cm³. Let $x$ be the length of the box in centimeters and $y$ the width in centimeters.
    (a) Write $y$ as a function of $x$.
    (b) Draw a complete graph of this equation.
    (c) What part of the graph in (b) represents the problem situation?
    (d) Use the graph in (b) to determine the width of the box if the length is 24 cm.

## 1.4 Solving Equations • 55

36. A laboratory keeps two acid solutions on hand. One is 20% acid and the other is 35% acid. An order is received for 25 liters of a 26% acid solution. How much 20% acid solution and how much 35% acid solution should be used to fill this order?

37. The formula $F = \frac{9}{5}C + 32$ gives the Fahrenheit ($F$) temperature as a function of the Celsius ($C$) temperature. Solve this equation for $C$.

38. The formula for the area of a trapezoid (Figure 1.4.19) is $A = \frac{1}{2}h(b_1 + b_2)$. Solve this equation for $b_1$.

39. $S = P(1+rn)$ is an algebraic representation of the value $S$ after $n$ years of an initial investment of $P$ dollars that earns simple interest at rate $r$ per year. *Note:* For a simple interest rate of 10%, $r$ is 0.10.
    (a) Solve this equation for $n$.
    (b) How many years are required for an investment earning 8% simple interest to *triple* in value?

40. A box with no top is made from a rectangular sheet of cardboard that is three times as long as it is wide by removing 6-inch squares from each corner and folding up the sides as suggested by Figures 1.4.17 and 1.4.18. Determine the dimensions of the cardboard sheet if the volume of the resulting box is 3000 cubic inches.

41. A semicircle is placed on one side of a square so that its diameter coincides with a side of the square (Figure 1.4.20). Find the side length of the square if the total area of the square plus the semicircle is 200 square units.

42. The owner of the Olde Time Ice Cream Shop pays $1000 per month for fixed expenses such as rent, lights, and wages. Ice cream cones are sold at $0.75 each, of which $0.40 goes for ice cream, cone, and napkin. How many cones must be sold to break even?

43. For a certain make of car it is determined that when the car is traveling at a speed of $r$ mph the stopping distance $D$ in feet is given by $D = r + \frac{r^2}{19.85}$.
    (a) Draw a complete graph of $y = D(r)$.
    (b) What are the domain and range of the algebraic representation? What values of $r$ make sense in this problem situation?
    (c) Use the graph in (a) to estimate the speed of the car if the stopping distance is 300 feet.

Figure 1.4.19      Figure 1.4.20

44. An 8 1/2 inch by 11 inch piece of paper contains a picture with uniform border. The distance from the edge of the paper to the picture is $x$ inches on all sides (Figure 1.4.21).
    (a) Express the area $A$ of the picture as a function of $x$.
    (b) Draw a complete graph of the function $y = A(x)$.
    (c) What are the domain and range of the function in (a)? What values of $x$ make sense in this problem situation?
    (d) Use the graph in (b) to estimate the uniform border if the area of the picture is 50 square inches.

45. The distribution of velocities $v$ (in cm/sec) in the laminar flow of arterial blood is given by the equation $v = 1.19 - (1.85 \times 10^4)r^2$ where $r$ is the distance (in cm) of the blood layer from the center of the artery.
    (a) Draw a complete graph of $y = v(r)$.
    (b) What are the domain and range of the algebraic representation? What values of $r$ make sense in this problem situation?
    (c) If the blood layer velocity is 0.975 cm/sec, use the graph in (a) to estimate the distance of the layer from the center of the artery.

46. A toy company makes $x$ toys each day at a total cost of $\frac{1}{4}(x - 20)^2 + 10$ dollars. The company receives $0.67 for each toy sold.
    (a) Write the daily profit of the company as a function of $x$.
    (b) What are the fixed costs?
    (c) Draw a complete graph of the profit function in (a).
    (d) What is the domain of the function in (a)? What values of $x$ make sense in this problem situation?
    (e) Use the graph in (c) to determine what level of daily production the company must maintain to break even.

A single-commodity open market is driven by the supply and demand principle. Economists have determined that supply curves are usually increasing (as the price increases, the sellers increase

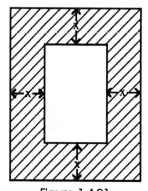

Figure 1.4.21

production) and demand curves are usually decreasing (as the price increases, the consumer buys less). For each demand equation, find the price $p$ if the production level is 120 units, and find the production level if the price is $2.30. Draw a complete graph of each equation and check your answers with a graphing utility.

47. $p = 15 - 0.023x$
48. $p = 100 - 0.0015x^2$
49. Draw a complete graph of the function $f(x) = x^3 - 2x + 3$.
    (a) How many real solutions are there to the equation $f(x) = 0$?
    (b) How many real solutions are there to the equation $f(x) = 3$?
    (c) Estimate the endpoints of the interval of $x$ values where the values of $f$ are decreasing as $x$ is increasing.
50. Solve the equation: $|x| + |x - 3| = 6$.
51. Solve the equation $|x| - |x - 6| = 0$ using the distance interpretation of absolute value. Check with a graphing utility.

## 1.5 Solving Systems of Equations

Sometimes it is convenient, or even necessary, to give an algebraic representation of a problem situation that consists of a system of functions or a system of equations. In this section we focus on solving these systems graphically. We solve systems of equations graphically by using zoom-in to trap each point of intersection in a viewing rectangle small enough to read the coordinates of the point as accurately as desired. In Section 1.4, we found solutions to an equation in one variable with error less than any prescribed number. This was accomplished by trapping the $x$-intercept between consecutive scale marks on the $x$-axis that were at a distance less than or equal to the prescribed error. If a given graphing utility did not draw scale marks, then to satisfy the error condition we had to choose the viewing rectangle $[L, R]$ by $[B, T]$ so that the difference $R - L$ was less than or equal to the prescribed error.

In an analogous way, we can prescribe a maximum error for solutions to systems of equations in two variables. Solutions of systems of equations in two variables are pairs of numbers, one for each of the two variables. Thus, it seems technically possible to prescribe the error in each variable. However, we have to be careful because these errors are generally *not independent*. The first example illustrates this point. When we ask you to give a simultaneous solution to a system of equations with error at most $r$, we mean that *each* coordinate should be estimated with error less than $r$.

The graphing utility we use to draw the graphs in this text overlays a lattice using the horizontal and vertical scale marks. For example, Figure 1.5.1 shows the lattice in the standard viewing rectangle with the distance between scale marks equal to one unit in both the horizontal and vertical direction.

The lattice in Figure 1.5.1 subdivides the viewing rectangle $[-10, 10]$ by $[-10, 10]$ into 400 unit squares. If we can determine that a specific point lies in one of these unit squares, then this lattice permits us to read the $x$- and $y$-coordinates of the point with error at most one. If scale marks but *no* lattice appear in a viewing rectangle, then it is usually possible to use the scale marks to estimate a smaller rectangle that contains a specific point. This smaller rectangle can be used to determine the error in a given approximation. If neither a lattice nor scale marks appear in a given viewing rectangle $[L, R]$ by $[B, T]$, then the $x$-coordinate of any point in the viewing rectangle can be read with error at most $R - L$, and the $y$-coordinate with error at most $T - B$. Most graphing calculators do *not* have the grid feature. You need to use the scale marks to read answers with a prescribed error on a graphing calculator. For this reason, we rarely use the grid feature of our graphing utility.

Suppose we want to find the length and width of a rectangle with perimeter 100 and area 300 (Figure 1.5.2). Let $L$ and $W$ stand for the length and width of such a rectangle, respectively. Then, the following system of equations is an algebraic representation of the problem.

$$2L + 2W = 100 \quad \text{(perimeter)}$$
$$LW = 300 \quad \text{(area)}$$

A solution to this system of equations is a pair of values, one for $L$ and one for $W$, that is a solution to each of the equations.

- **DEFINITION**  A pair of real numbers is a **solution to a system of equations in two variables** if and only if the pair of numbers is a solution to each of the equations. These solutions are called **simultaneous solutions** to the system of equations. When we have found all solutions to the system of equations we say that we have **solved the system of equations**. •

The graph of an equation in two variables is the set of solutions to the equation considered as points in the plane. A simultaneous solution to a system of equations is a common point, or point of intersection, of the graphs of the equations in the system.

We solve systems of equations graphically by finding the points of intersection of the graphs of the equations in the system. Algebraic techniques can be used to solve systems of equations.

- **EXAMPLE 1:** Find the simultaneous solutions to $\begin{cases} 2y + 2x = 100, \\ xy = 300. \end{cases}$

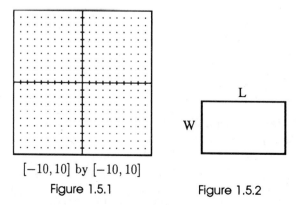

$[-10, 10]$ by $[-10, 10]$
Figure 1.5.1

Figure 1.5.2

## 1.5 Solving Systems of Equations

SOLUTION: To graph these equations with a graphing utility, you will need to solve each equation for $y$ and rewrite the system as follows.

$$y = 50 - x$$
$$y = \frac{300}{x}$$

Figure 1.5.3 gives complete graphs of the two equations in the viewing rectangle $[-60, 60]$ by $[-60, 60]$. We can see from Figure 1.5.3 that the graphs have two points in common. Thus, we should expect to find two simultaneous solutions. In Example 3 we will use a graphing zoom-in procedure to find the coordinates of points of intersection. In this example we illustrate an algebraic procedure called the substitution method.

$$2L + 2W = 100 \quad \text{(perimeter)}$$
$$LW = 300 \quad \text{(area)}$$

In the substitution method we solve one equation for a variable and then substitute the expression for that variable into the other equation. Solving the first equation for $y$ gives $y = 50 - x$. Substituting for $y$ in the second equation gives the following.

$$xy = 300$$
$$x(50 - x) = 300$$
$$x^2 - 50x + 300 = 0$$

Using the quadratic formula we find the following values for $x$.

$$x = \frac{50 \pm \sqrt{(-50)^2 - 4(300)}}{2}$$
$$= \frac{50 \pm 10\sqrt{13}}{2}$$
$$= 25 \pm 5\sqrt{13}$$

Accurate to hundredths, $25 + 5\sqrt{13}$ is 43.03 and $25 - 5\sqrt{13}$ is 6.97.

$[-60, 60]$ by $[-60, 60]$
Figure 1.5.3

The corresponding values of $y$ can be found from the equation $y = 50 - x$. If $x = 25 + 5\sqrt{13}$, then $y = 50 - (25 + 5\sqrt{13}) = 25 - 5\sqrt{13}$. If $x = 25 - 5\sqrt{13}$, then $y = 50 - (25 - 5\sqrt{13}) = 25 + 5\sqrt{13}$. Thus, the two solutions are $(25 + 5\sqrt{13}, 25 - 5\sqrt{13})$ and $(25 - 5\sqrt{13}, 25 + 5\sqrt{13})$. In decimal form accurate to hundredths, the two solutions are $(43.03, 6.97)$ and $(6.97, 43.03)$. Each coordinate of the solution in decimal form is in error of at most 0.01. Remember, unless otherwise specified, we find solutions with error of at most 0.01. •

- **EXAMPLE 2**: Find the exact dimensions of a rectangle with perimeter 100 and area 300.

    **SOLUTION**: From Example 1 we know that $L = 25 + 5\sqrt{13}$ and $W = 25 - 5\sqrt{13}$, or $L = 25 - 5\sqrt{13}$ and $W = 25 + 5\sqrt{13}$. These two solutions produce two rectangles (Figure 1.5.4).

    Actually the two rectangles are congruent (really the same rectangle). Thus there is a unique (one and only one) solution and the dimensions are $25 - 5\sqrt{13}$ by $25 + 5\sqrt{13}$. In decimal form, the dimensions are 6.97 by 43.03 with error of at most 0.01. •

- **EXAMPLE 3**: Find the simultaneous solutions to $\begin{cases} y = -x^3 + 3x^2 + x - 3, \\ y = -2x^2 + 5. \end{cases}$

    **SOLUTION**: According to our agreement about error, we must determine the coordinates of each solution with error at most 0.01. Figure 1.5.5 gives the graph of these two equations in the standard viewing rectangle. This figure suggests there are two points of intersection. However, there is a third point of intersection in the fourth quadrant that is outside of this viewing rectangle. The third point of intersection shows up in the viewing rectangle $[-10, 10]$ by $[-100, 100]$ (Figure 1.5.6). Figure 1.5.6 gives a complete graph of each equation. (Why?) Thus, there are *three* solutions to this system of equations.

    We can estimate from Figure 1.5.5 that the point of intersection with smallest $x$-coordinate lies in the $[-2, -1]$ by $[1, 3]$ viewing rectangle. The graphs of the two equations

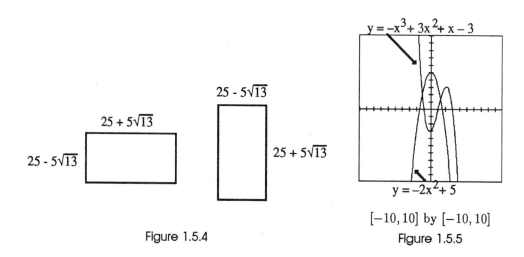

Figure 1.5.4

$[-10, 10]$ by $[-10, 10]$
Figure 1.5.5

### 1.5 Solving Systems of Equations

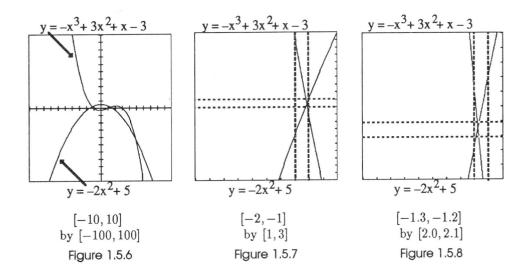

$[-10, 10]$
by $[-100, 100]$
Figure 1.5.6

$[-2, -1]$
by $[1, 3]$
Figure 1.5.7

$[-1.3, -1.2]$
by $[2.0, 2.1]$
Figure 1.5.8

in this viewing rectangle are given in Figure 1.5.7. Figure 1.5.7 suggests we draw the graphs of the two equations in the $[-1.3, -1.2]$ by $[2.0, 2.1]$ viewing rectangle (Figure 1.5.8).

The scale marks in Figure 1.5.8 are 0.01 units apart in both the horizontal and vertical directions. Thus, any point in the $[-1.22, -1.21]$ by $[2.03, 2.04]$ rectangle gives the coordinates of this point of intersection with error at most 0.01. We read the point of intersection to be $(-1.218, 2.035)$. Similarly, we can find that the other solution in Figure 1.5.5 is $(1.350, 1.356)$ with error at most 0.01.

We were able to keep the horizontal and vertical distance between scale marks the same for the first two points of intersection. This will *not* be possible for the remaining point of intersection. Figure 1.5.6 suggests that we look at the graphs in the $[4, 5]$ by $[-50, -40]$ viewing rectangle. Notice that the two graphs drop very sharply near this third point of intersection, because small changes in $x$ produce very large changes in $y$ near this point. Continuing as before, the next two viewing rectangles found will likely be $[4.8, 4.9]$ by $[-43, -42]$ and $[4.86, 4.87]$ by $[-42.4, -42.3]$. You will likely find it difficult to determine the coordinates of this point of intersection with error at most 0.01 in the latter viewing rectangle. We can estimate the coordinates of the point of intersection to be $(4.868, -42.39)$ from the graphs in $[4.864, 4.869]$ by $[-42.42, -42.36]$ (Figure 1.5.9). In this figure the horizontal distance between scale marks is 0.001 and the vertical distance is 0.01. *We need error at most* 0.001 *in the x-coordinate in order to obtain error at most* 0.01 *in the y-coordinate.* If you try to keep both horizontal and vertical distances between scale marks the same, you will end up with graphs like the one in Figure 1.5.10. Notice it is very hard to read the coordinates of the point of intersection from Figure 1.5.10. •

### The Box Problem

Squares of side length $x$ are cut from each corner of a rectangular piece of cardboard that is 20 inches wide by 25 inches long (Figure 1.5.11). Then the cardboard is folded along the dashed line segments in Figure 1.5.11 to form a box with no top. The height of this box is $x$. The base of the box has dimensions $20 - 2x$ by $25 - 2x$. Finally, the volume of the box is $V(x) = x(20 - 2x)(25 - 2x)$. $V$ is an algebraic representation of this problem situation.

- **EXAMPLE 4:** Draw a complete graph of the function $V(x) = x(20-2x)(25-2x)$ and indicate what part of the graph represents the box problem. Determine the domain and range of $V$.

    **SOLUTION:** $V(x)$ is a polynomial of degree 3. Thus, the domain of the function $V$ is the set of all real numbers. The graph of $V$ in the $[-5, 15]$ by $[-1000, 1000]$ viewing rectangle (Figure 1.5.12) is a complete graph. (Why?) The range of $V$ is also the set of all real numbers. However, $V$ represents the volume of the box constructed above only for $x$ between 0 and 10. (Why?) The graph of $V$ in Figure 1.5.13 is a graph of the box problem. •

- **EXAMPLE 5:** Find the side length of the square that must be cut out from a 20-inch by 25-inch piece of cardboard to form a box with volume 500 cubic inches.

    **SOLUTION:** We need to find a value of $x$ for which the volume is 500, that is, $V(x) = 500$. Remember that $V(x) = 500$ means that the point $(x, 500)$ is on the graph of $V$. Notice also that the point $(x, 500)$ lies on the horizontal line $y = 500$. Thus, a solution to this problem is the first coordinate of a simultaneous solution to the following system of equations.

    $$y = V(x) = x(20 - 2x)(25 - 2x)$$
    $$y = 500$$

The graphs of these two equations in the $[-5, 15]$ by $[-1000, 1000]$ viewing rectangle are given in Figure 1.5.14. The horizontal line $y = 500$ intersects the graph of $V$ three times, but only two are possible values for the side length of the square. (Why?) Figure 1.5.15

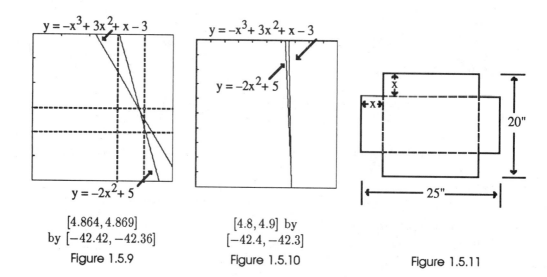

[4.864, 4.869] by [−42.42, −42.36]

Figure 1.5.9

[4.8, 4.9] by [−42.4, −42.3]

Figure 1.5.10

Figure 1.5.11

1.5 Solving Systems of Equations

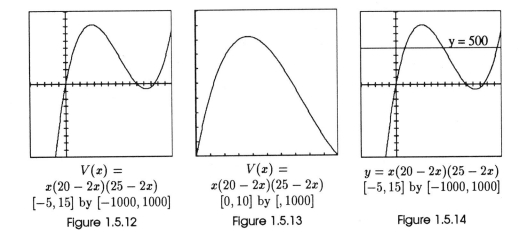

$V(x) =$
$x(20 - 2x)(25 - 2x)$
$[-5, 15]$ by $[-1000, 1000]$
Figure 1.5.12

$V(x) =$
$x(20 - 2x)(25 - 2x)$
$[0, 10]$ by $[, 1000]$
Figure 1.5.13

$y = x(20 - 2x)(25 - 2x)$
$[-5, 15]$ by $[-1000, 1000]$
Figure 1.5.14

is a geometric representation for only those values of $x$ that satisfy the problem situation. Notice that these two graphs intersect twice.

We can use zoom-in to determine the two values of $x$. Figure 1.5.16 gives the graphs in the $[1.2, 1.3]$ by $[490, 510]$ viewing rectangle. The distance between horizontal scale marks is 0.01. The graph of $y = V(x)$ crosses the horizontal line $y = 500$ at some value of $x$ in $1.27 \leq x \leq 1.28$. The graph of $V$ appears to be a straight line in $1.27 \leq x \leq 1.28$. The graph in Figure 1.5.16 suggests that $V(1.27)$ is about 498 and $V(1.28)$ is about 501. We can compute to see that $V(1.27)$ is actually a little larger than 498 and $V(1.28)$ is a little less than 501. Thus, for $x$ between 1.27 and 1.28 the corresponding values of $V$ range between 498 and 501. Therefore, Figure 1.5.16 permits us to read a solution to this system of equations with error at most 0.01 in the first coordinate and with error at most 2 in the second coordinate. It follows that any value of $x$ between 1.27 and 1.28 used for the side length of the square cut out gives a solution to the problem with error of at most 0.01. Figure 1.5.16 suggests we use 1.277 as the value of $x$.

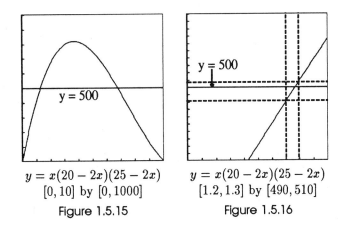

$y = x(20 - 2x)(25 - 2x)$
$[0, 10]$ by $[0, 1000]$
Figure 1.5.15

$y = x(20 - 2x)(25 - 2x)$
$[1.2, 1.3]$ by $[490, 510]$
Figure 1.5.16

Suppose we also want the value of $V$ to be in error by at most 0.01. Then, we must zoom-in more because the error in $V$ from Figure 1.5.16 can be as much as 2 cubic inches. It doesn't help to change $B$ and $T$ and fix $L$ and $R$ in the viewing rectangle of Figure 1.5.16. For example, the graph of the system of equations in the $[1.2, 1.3]$ by $[499.9, 500.1]$ viewing rectangle (Figure 1.5.17) does not permit us to read the error in the second coordinate for $x$ between 1.27 and 1.28 because the graph of $V$ goes outside the viewing rectangle for $x$ between 1.27 and 1.28.

If we want to increase the accuracy of $V$ we must also increase the accuracy of $x$. The graph of the system of equations in the $[1.276, 1.277]$ by $[499.9, 500.1]$ viewing rectangle (Figure 1.5.18) permits us to read a solution with error in the second coordinate at most 0.03 because, for $x$ between 1.2767 and 1.2768, the value of $V$ differs from 500 by at most 0.03. Again, this can be checked by direct computation using the fact that the graph of $V$ appears to be a straight line in $1.2767 \leq x \leq 1.2768$, or estimated from Figure 1.5.18. Thus, $x = 1.27678$ gives a solution to the problem with error in the value of $V$ at most 0.03 and error in the value of $x$ at most 0.0001, the distance between horizontal scale marks in Figure 1.5.18. •

Solving the equation $V(x) = 500$ in Example 5 graphically is very easy. Most of the discussion in Example 5 was concerned with the accuracy of the solution. To appreciate the power of a graphical approach, you should try to solve $V(x) = x(20 - 2x)(25 - 2x) = 500$ algebraically. You will find that the Cardan cubic equation formulas will be needed!

The next example illustrates that exact solutions can *sometimes* be found when we solve graphically.

• **EXAMPLE 6:** Find the simultaneous solutions to $\begin{cases} y = x^3 - x \\ y = 3x \end{cases}$

SOLUTION: The graphs of these two equations in the $[-5, 5]$ by $[-10, 10]$ viewing rectangle shown in Figure 1.5.19 are complete. Figure 1.5.19 strongly suggests that the three points

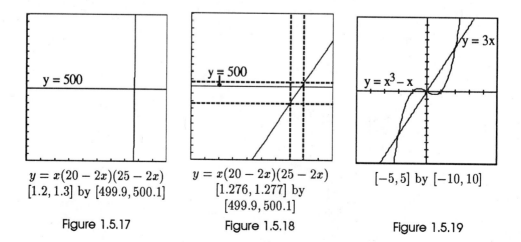

$y = x(20 - 2x)(25 - 2x)$
$[1.2, 1.3]$ by $[499.9, 500.1]$

Figure 1.5.17

$y = x(20 - 2x)(25 - 2x)$
$[1.276, 1.277]$ by $[499.9, 500.1]$

Figure 1.5.18

$[-5, 5]$ by $[-10, 10]$

Figure 1.5.19

## 1.5 Solving Systems of Equations

of intersection are: $(-2, -6)$, $(0, 0)$, and $(2, 6)$. Exactness of an answer *cannot* be established completely graphically. You can verify directly by substitution that these three pairs are solutions to the system of equations. Thus, we have found all solutions to this system of equations.

We can also find these solutions by solving the system algebraically using the substitution method. Substituting the value of $y$ from the first equation into the second equation we obtain $x^3 - x = 3x$. Now we have one equation in one variable that we solve as follows.

$$x^3 - x = 3x$$
$$x^3 - 4x = 0$$
$$x(x^2 - 4) = 0$$
$$x(x - 2)(x + 2) = 0$$

We see that $x$ is $-2$, $0$, or $2$. The corresponding values of $y$ are $-6$, $0$, and $6$, respectively. Thus, the three solutions are $(-2, -6)$, $(0, 0)$, and $(2, 6)$. •

Actually, the equations in Example 6 were selected so that we could use an algebraic method of solution. Most of the time we will *not* be able to solve such systems algebraically. Frequently you will want to use the strategy of approximating a solution graphically and then substitute to verify your solution. However, a graphical approach will always work provided the equations can be graphed with a graphing utility.

• **EXAMPLE 7:** Determine the simultaneous solution to $\begin{cases} y = \cos x \\ y = x^2 \end{cases}$ with $x$ between 0 and 1.

**SOLUTION:** The graphs of the equations in Figure 1.5.20 indicate that there are two points of intersection. When trigonometry is studied later in the book, we will find that Figure 1.5.20 gives a complete graph of each function. (If you are using a graphing calculator, be sure it is in radian mode.) Figure 1.5.21 gives a closer look at the desired point of intersection. We can use zoom-in to determine that the coordinates of the point of intersection are $(0.82, 0.68)$. Thus, the desired solution for $0 < x < 1$ is the pair of numbers $x = 0.82$ and $y = 0.68$. •

$[-10, 10]$ by $[-2, 2]$

Figure 1.5.20

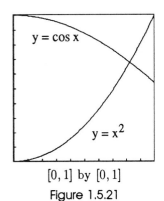

$[0, 1]$ by $[0, 1]$

Figure 1.5.21

## EXERCISES 1-5

Use the substitution method to find the simultaneous solutions to each system of equations.

1. $2x - 3y = -23$
   $x + y - 6 = 0$

2. $3x + y = 20$
   $x - 2y = 10$

Use the elimination method to solve each system of equations.

3. $x - y = 10$
   $x + y = 6$

4. $2x + y = 10$
   $x - 2y = -5$

5. Graph the equations $E_1$, $E_2$, and $E_3$ in the standard viewing rectangle, $[-10, 10]$ by $[-10, 10]$.

$$E_1 : x - 2y = 4$$
$$E_2 : 4y - 2x = 12$$
$$E_3 : 4y - 2x = -8$$

(a) Describe geometrically the simultaneous solution to the system of equations $E_1$ and $E_2$.

(b) Describe geometrically the simultaneous solution to the system of equations $E_1$ and $E_3$.

Use an algebraic method to determine the simultaneous solution to each system. Use a graphing utility and zoom-in to check your answer.

6. $x - y = 4$
   $2x + y = 14$

7. $2x - 3y = 33$
   $5x + 2y = 35$

8. $2x + 2y = 40$
   $xy = 100$

9. $y = x^3 - x$
   $y = x$

10. Consider the system of equations $\begin{cases} y = x^3 - 2x^2 + 3x - 5, \\ y = 5x. \end{cases}$

    Find each simultaneous solution $(x, y)$ to the system and determine a maximum error in $x$ to ensure the maximum error in $y$ is at most 0.01. For each simultaneous solution, specify a viewing rectangle (give the parameters $L$, $R$, $B$, and $T$) for which the solution can be read with an error of at most 0.01.

Solve each system. (Remember *both* $x$ and $y$ must have an error of at most 0.01.)

11. $2x - 5y = 6$
    $y = 6 - x^2$

12. $y = x^2 - 4$
    $y = 6 - x^2$

13. $y = x^3$
    $y = 4 - x^2$

14. $y = 2x^2 - 3x - 10$
    $y = \dfrac{1}{x}$

15. $y = 16x - x^3$
    $y = 10 - x^2$

16. $y = 2x + 10 - x^2$
    $y = 7x^3 + 13x^2 - 9x + 1$

17. $y = \sin x$
    $y = 2 - x^2$

18. $y = 2 - \cos x$
    $y = x^3$

19. $y = \sin x$
    $y = x^3 - x$

20. $y = \dfrac{1}{x}$
    $y = 1 + \cos x$

21. A candy recipe calls for 2.25 times as much brown sugar as white sugar. If 23 ounces of sugar are required, how much of each type is needed?

22. Find the dimensions of a rectangular cornfield with perimeter 220 yards and area 3000 square yards.

23. Hank can row a boat one mile upstream (against the current) in 24 minutes. He can row the same distance downstream in 13 minutes. If both the rowing speed and current speed are constant, find Hank's rowing speed and the speed of the current.

24. Four hundred fifty-two tickets were sold for a high school basketball game. There were two prices of tickets: student at $0.75 and nonstudent at $2.00. How many tickets of each type were sold if the total proceeds from the sale of the tickets were $429?

25. Let $H(x) = ax + b$ be a polynomial function of degree one. If $H(2) = -3$ and $H(-4) = -5$, determine $a$ and $b$.

26. A function $T$ describing the behavior of a certain physical phenomenon is known to be a polynomial function of degree two; that is, $T(x) = ax^2 + bx + c$. It is also known that $T(0) = 5$, $T(1) = -2$, and $T(3) = 6$. Determine the function $T$.

27. A scholarship fund earns income from three investments at simple interest rates of 5%, 6%, and 10%. The annual income available for scholarships from the three investments is $1000. If the amount invested at 5% is five times the amount invested at 6%, and the total amount invested at 5% and 6% is equal to the amount invested at 10%, determine how much is invested at each rate.

28. The total value of 17 coins consisting of nickels, dimes, and quarters is $2.95. There are twice as many quarters as nickels. Determine the number of nickels, dimes, and quarters.

29. The sum of the digits of a certain two-digit number is 16. If the digits are reversed, the original number is increased by 18. What is the original number?

30. A cyclist peddling against a headwind travels 22 miles down a straight road in 2 hours. The return trip with the wind takes only 1.6 hours. If the speed of the cyclist and the speed of the wind are both constant, find both the speed of the cyclist and the speed of the wind.

31. Let $x$ be the side length of the square as illustrated in Figure 1.5.11 that must be cut out from each corner of a 30-inch-by-40-inch piece of cardboard to form a box with no top.
    (a) Express the volume of the box as a function of $x$.
    (b) Write two equations that can be solved simultaneously to determine the size of the square that must be cut out to produce a box with volume 1200 cubic inches.
    (c) Draw complete graphs of each equation in the same viewing rectangle.
    (d) How many simultaneous solutions are there to your system in (b)? Which of these solutions are also solutions to the problem situation? Explain.
    (e) Use zoom-in to determine all values of $x$ that produce a box of volume 1200 cubic inches. State the maximum error in $x$ in order to have the maximum error in the volume of at most 0.01 cubic inches.

Do the following for each pair of indicated supply and demand equations.

(i) Graph both the supply and demand equations on the same coordinate system in the *first* quadrant; $p$ is the price (vertical axis), and $x$ represents the number of units of the commodity produced.

(ii) Determine the equilibrium price, that is, the price where supply is equal to demand.

32. Supply: $10 + 0.1x^2$.

    Demand: $\dfrac{50}{1 + 0.1x}$.

33. Supply: $5 + 0.014x^2$.

    Demand: $\dfrac{133}{1 + 0.025x}$.

34. Sam makes two initial investments of $1000 each. The first earns 10% simple interest and the second earns interest using 5% *simple discount*. After $n$ years, the value of the first investment is $S = 1000(1 + 0.10n)$ and the value of the second investment is $S = 1000/(1 - 0.05n)$.

    (a) Draw the graph of each equation in the viewing rectangle [0, 30] by [0, 6000].

    (b) Is there a time when the future values of both investments are equal?

    (c) What happens to the simple discount investment value at the end of 19 years? 19.5 years? 19.9 years? 19.99 years? exactly 20 years? Could your results explain why the simple discount model is used for only short periods of time in actual practice? *Note:* Normally, interest is paid at the *end* of a term. In the case of simple discount, interest is paid *in advance*, at the beginning of the term.

35. Assume that the graph of $x^2 + y^2 = r^2$ is a circle of radius $r$ centered at the origin.

    (a) Sketch the graphs of $x^2 + y^2 = 16$ and $x + y = 2$ on the same coordinate system.

    (b) Determine the number of simultaneous solutions to the system $\begin{cases} x^2 + y^2 = 16, \\ x + y = 2. \end{cases}$

    (c) Determine the solutions with error at most 0.01 using zoom-in with a graphing utility. *Hint:* Write $x^2 + y^2 = 16$ as $y = \sqrt{16 - x^2}$ or $y = -\sqrt{16 - x^2}$.

    (d) Determine the simultaneous solutions algebraically.

    (e) Identify the possibilities for the number of simultaneous solutions to the system

    $$x^2 + y^2 = r^2$$
    $$ax + by + c = 0.$$

    Sketch graphs illustrating each possibility.

36. Assume that the graph of $x^2 + y^2 = r^2$ is a circle of radius $r$ centered at the origin.

    (a) Sketch the graphs of $x^2 + y^2 = 25$ and $y = x^2 - 16$ on the same coordinate system.

    (b) Determine the number of simultaneous solutions to the system $\begin{cases} x^2 + y^2 = 25, \\ y = x^2 - 16. \end{cases}$

    (c) Determine the solutions with error at most 0.01 using zoom-in with a graphing utility.

    (d) Determine the simultaneous solutions algebraically.

(e) Identify the possibilities for the number of simultaneous solutions to the system
$$x^2 + y^2 = r^2,$$
$$y = x^2 + k.$$
Sketch graphs illustrating each possibility.

*37. Determine the $y$-intercepts of all lines of the form $y = x + b$ that are *tangent* to the circle $x^2 + y^2 = 16$.

*38. Matrices can be used to represent and solve systems of linear equations. Compare the result of each step in the matrix solution with the corresponding algebraic result.

ALGEBRAIC                MATRIX REPRESENTATION

$3x + y = 14$

$2x - 3y = 2$           $\begin{pmatrix} 3 & 1 & 14 \\ 2 & -3 & 2 \end{pmatrix}$

Multiply Row 1 by $\frac{1}{3}$.

$x + \frac{1}{3}y = \frac{14}{3}$

$2x - 3y = 2$           $\begin{pmatrix} 1 & \frac{1}{3} & \frac{14}{3} \\ 2 & -3 & 2 \end{pmatrix}$

Add $-2$ times Row 1 to Row 2.

$x + \frac{1}{3}y = \frac{14}{3}$

$-\frac{11}{3}y = -\frac{22}{3}$      $\begin{pmatrix} 1 & \frac{1}{3} & \frac{14}{3} \\ 0 & -\frac{11}{3} & -\frac{22}{3} \end{pmatrix}$

Multiply Row 2 by $-\frac{3}{11}$.

$x + \frac{1}{3}y = \frac{14}{3}$

$y = 2$                 $\begin{pmatrix} 1 & \frac{1}{3} & \frac{14}{3} \\ 0 & 1 & 2 \end{pmatrix}$

Add $-\frac{1}{3}$ times Row 2 to Row 1.

$x = 4$

$y = 2$                 $\begin{pmatrix} 1 & 0 & 4 \\ 0 & 1 & 2 \end{pmatrix}$

Use the illustrated matrix method to find the simultaneous solutions to each system.

(a) $x - 3y = 1$
    $x + y = 5$

(b) $2x - 5y = 13$
    $3x + 2y = -12$

(c) $2x - 6y = 8$
    $5x + 3y = 15$
    $3x + y - z = 10$

(d) $x - y + 2z = 3$
    $2y - z = -3$
    $x + 2y = 0$

## 1.6 Solving Inequalities Algebraically and Graphically

In this section we solve inequalities involving absolute value and introduce an algebraic technique that is used to solve certain higher order inequalities. This technique is called the sign chart method and is particularly useful with inequalities involving polynomials that are easily factored. However, solving inequalities with a graphing utility is almost always the most reasonable and preferred method to use with higher order inequalities. In general, algebraic techniques, like the sign chart method, are very limited because the expressions involved in the inequality must be relatively simple.

• **DEFINITION** The number $a$ is a **solution to an inequality** in one variable if we get a true statement when the variable is replaced by $a$ everywhere in the inequality. When we have found all solutions to an inequality we say that we have **solved the inequality**. •

For example, $x = 120$ is a solution of the inequality $0.2x + 15 \leq 100$ because $0.2(120) + 15 = 39 \leq 100$. Solving an inequality algebraically is similar to solving an equation algebraically. The goal is to transform a given inequality into an equivalent one with solutions that are obvious.

• **DEFINITION** Two inequalities are said to be **equivalent** if they have exactly the same set of solutions. •

**Equivalent Inequalities.** An equivalent inequality is obtained if any one of the following operations is performed.

(1) Add the same real number to, or subtract it from, each side of the inequality.
(2) Add the same polynomial to, or subtract it from, each side of the inequality.
(3) Multiply each side of the inequality, or divide each side of the inequality, by the same positive number.
(4) Multiply each side of the inequality, or divide each side of the inequality, by the same negative number *and* reverse the inequality sign.

Notice that both $5 > 2$ and $5(3) > 2(3)$ are true statements. However, we must reverse the sign of the inequality if we multiply each side by the same negative number. Check that both $5 > 2$ and $5(-3) < 2(-3)$ are true.

A real number $x$ is a solution to the inequality $|x| < 2$ if and only if the distance from the origin of the point with coordinate $x$ is less than 2. These are precisely the real numbers between $-2$ and 2, not including either 2 or $-2$. Therefore, the solution to the inequality $|x| < 2$ is the set of all real numbers $x$ satisfying $-2 < x < 2$.

## 1.6 Solving Inequalities Algebraically and Graphically

The inequality $|x| > 4$ has for its solution the real numbers with distance from the origin greater than 4. Thus, the solution to $|x| > 4$ consists of all real numbers $x$ satisfying $x < -4$ or $x > 4$.

Inequalities that involve absolute value have the equivalent representations illustrated in Figure 1.6.1 ($a > 0$).

In interval notation, $|x| < a$ is $(-a, a)$ and $|x| \leq a$ is $[-a, a]$. It is not possible to write $|x| > a$, or $|x| \geq a$ as a single interval. $|x| > a$ consists of the real numbers in either the interval $(-\infty, -a)$, or the interval $(a, \infty)$. $|x| \geq a$ consists of the real numbers in either the interval $(-\infty, -a]$ or the interval $[a, \infty)$.

- **DEFINITION** We use the notation $(a, b) \cup (c, d)$ to indicate the collection of all real numbers that belong to $(a, b)$ or $(c, d)$ or both. The symbol $\cup$ is read **union**.

Thus, we can include interval notation among the equivalent representations of $|x|$ where $a > 0$ given in Figure 1.6.1.

| Absolute Value | Interval Notation |
|---|---|
| $|x| < a$ | $(-a, a)$ |
| $|x| \leq a$ | $[-a, a]$ |
| $|x| > a$ | $(-\infty, -a) \cup (a, \infty)$ |
| $|x| \geq a$ | $(-\infty, -a] \cup [a, \infty)$ |

- **EXAMPLE 1:** Solve $|x - 2| < 3$ both algebraically and graphically. Give a number-line graph of the solution.

SOLUTION: The inequality $|x - 2| < 3$ is equivalent to $|x - 2| - 3 < 0$. We use the graph of $f(x) = |x - 2| - 3$ to solve the inequality graphically. Figure 1.6.2 gives a complete graph

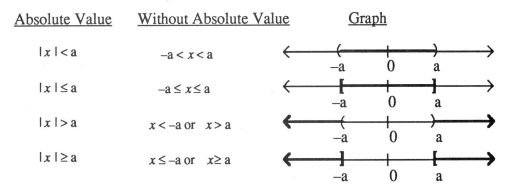

Figure 1.6.1

of $f(x) = |x - 2| - 3$. The graph appears to cross the $x$-axis at $x = -1$ and $x = 5$. This can be confirmed by direct substitution. The graph of $f$ is *below* the $x$-axis for all $x$ between $-1$ and 5. Thus, the solution is $-1 < x < 5$, that is, the interval $(-1, 5)$.

Next we give an algebraic solution. Remember that $|a| < 3$ is equivalent to $-3 < a < 3$. Replace $a$ by $x - 2$ and we obtain the following equivalent inequalities.

$$|x - 2| < 3$$
$$-3 < x - 2 < 3$$
$$-1 < \phantom{x-2} x \phantom{-2} < 5$$

A number-line graph of the solution is given in Figure 1.6.3. •

We return to an application studied in Section 1.3. The height $s$ at any time $t$ of an object shot straight up from a point $s_0$ feet above level ground with initial velocity $v_0$ is given by the model $s = -16t^2 + v_0 t + s_0$. Notice that the coefficient of $t$ is $v_0$ and the constant term is $s_0$. The coefficient of $t^2$ is one-half of the directed acceleration due to gravity.

• **EXAMPLE 2:** Suppose that a baseball is thrown straight up from level ground with initial velocity 80 feet per second (Figure 1.6.4).

(a) Determine an algebraic representation of the problem situation.
(b) Draw a complete graph of the algebraic representation in (a). What are the domain and range of the algebraic representation?
(c) What part of the graph in (b) represents the problem situation? Draw a geometric representation of the problem situation.
(d) When will the baseball be above ground?
(e) When will the baseball be at least 64 feet above ground?

SOLUTION:

(a) The height above ground of the baseball at any time $t$ is given by $s = -16t^2 + 80t$. This function is an algebraic representation of the problem situation.

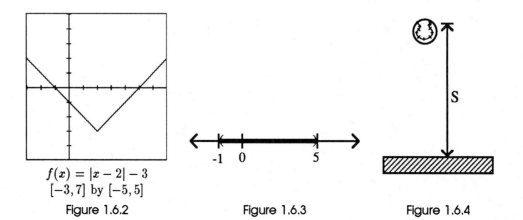

Figure 1.6.2
$f(x) = |x - 2| - 3$
$[-3, 7]$ by $[-5, 5]$

Figure 1.6.3

Figure 1.6.4

(b) A complete graph of the function $s$ is given in Figure 1.6.5. The domain of $s$ is $(-\infty, \infty)$. To find the range of $s$ we can use zoom-in to find the $y$-coordinate of the highest point of the graph of $s$. We must be careful when we zoom in because it is possible to produce viewing rectangles where the graph of $s$ will appear so flat that we are unable to distinguish the high point from the rest of the points of the graph in the viewing rectangle. We need to choose viewing rectangles that are wider than they are tall. The graph of $s$ in Figure 1.6.6 clearly highlights the high point of $s$. We can read from this graph that the coordinates of the high point are approximately $(2.5, 100)$. Using the formula for the coordinates of the vertex given in Section 1.2, you will find that the coordinates of the highest point of the graph of $s$ are exactly $(2.5, 100)$. Thus, the range of $s$ is $(-\infty, 100]$.

(c) Notice that $s = 0$ for $t = 0$ and $t = 5$. The distance $s$ is positive for $0 < t < 5$. Thus, the graph in Figure 1.6.5 represents the problem situation only for $0 \leq t \leq 5$. Figure 1.6.7 gives a geometric representation that is also a graph of the problem situation.

(d) The baseball is above ground when $s > 0$. It follows from (c) that the baseball will be above ground for any time between 0 and 5 seconds.

(e) The baseball will be at least 64 feet above ground provided that $s \geq 64$. To determine the values of $t$ for which $s \geq 64$ we overlay the graph of $y = 64$ in Figure 1.6.7. The result is given in Figure 1.6.8. The solution to the inequality $s \geq 64$ consists of those values of $t$ for which the graph of $s$ lies above or coincides with the graph of $y = 64$. It appears from Figure 1.6.8 that $s \geq 64$ for $1 \leq t \leq 4$. You should verify directly that $s$ is 64 when $t$ is 1 or 4. Thus, the baseball will be at least 64 feet above ground for any time from 1 to 4 seconds after it is thrown. •

- **EXAMPLE 3:** Solve (d) and (e) of Example 1 algebraically.

SOLUTION:

(d) To solve the inequality $-16t^2 + 80t > 0$ algebraically, first factor $-16t^2 + 80t$ to get the equivalent inequality $-16t(t-5) > 0$. The conditions of the problem situation

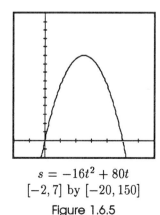

$s = -16t^2 + 80t$
$[-2, 7]$ by $[-20, 150]$
Figure 1.6.5

$s = -16t^2 + 80t$
$[2.4, 2.6]$ by $[99.9, 100.1]$
Figure 1.6.6

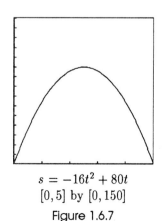

$s = -16t^2 + 80t$
$[0, 5]$ by $[0, 150]$
Figure 1.6.7

$s = -16t^2 + 80t$
$[0, 5]$ by $[0, 150]$
Figure 1.6.8

require that $t \geq 0$. Because we are looking for values of $t$ for which the baseball is above the ground, $-16t(t-5)$ cannot be zero, and the factors $-16t$ and $t-5$ must be both positive or both negative. However, $-16t$ is negative because $t \geq 0$. Thus, $t - 5$ must also be negative and it follows that $0 < t < 5$.

(e) The inequality $-16t^2 + 80t \geq 64$ is equivalent to the following inequalities.

$$-16t^2 + 80t - 64 \geq 0$$
$$t^2 - 5t + 4 \leq 0$$
$$(t-1)(t-4) \leq 0$$

The product $(t-1)(t-4)$ is positive for $t > 4$ because each factor is positive for $t > 4$. Similarly, the product $(t-1)(t-4)$ is positive for $t < 1$ because each factor is negative for $t < 1$. If $t$ is between 1 and 4, then $t - 1$ is positive and $t - 4$ is negative. Thus, the product $(t-1)(t-4)$ is negative for $1 < t < 4$. Finally, the product is zero for $t = 1$ or $t = 4$. Thus, the solution to the inequality is $1 \leq t \leq 4$, or $[1, 4]$. •

## Sign Chart

We will organize the argument used to solve the inequality $(t-1)(t-4) \leq 0$ in Example 3 in the form of a **sign chart**. The linear expression $x - 2$ is zero for $x = 2$, positive for $x > 2$, and negative for $x < 2$. Every linear polynomial $ax + b$, where $a \neq 0$, is zero for exactly one value of $x$ and *opposite* in sign to the right and left of that value. This is the fact we use to fill in the sign chart.

The solution to $(t-1)(t-4) \leq 0$ consists of those values of $t$ for which $(t-1)(t-4) < 0$ together with the values of $t$ for which $(t-1)(t-4) = 0$. The values of $t$ that make each factor equal to zero determine the range for $t$ in the following sign chart.

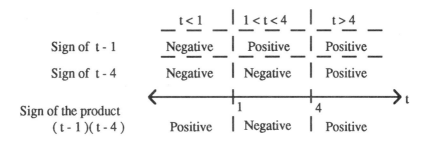

Notice that we use $t = 1$ and $t = 4$, the two values of $t$ for which $(t-1)(t-4) = 0$, to divide the number line into the three intervals $t < 1$, $1 < t < 4$, and $t > 4$. Each row of the sign chart above the number line gives the sign of one of the two factors in each of the intervals $t < 1$, $1 < t < 4$, and $t > 4$. The first line of the sign chart tells us that $t - 1$ is negative for $t < 1$ and positive for $1 < t < 4$ and $t > 4$.

The last line of the sign chart gives the sign of the product of the two factors in these three intervals. The last line of the sign chart tells us that the product $(t-1)(t-4)$ is negative for $1 < t < 4$ and positive for $t < 1$ or $t > 4$. In addition, $(t-1)(t-4)$ is zero for $t = 1$ and $t = 4$. Thus, the solution to $(t-1)(t-4) \leq 0$ is $1 \leq t \leq 4$.

The sign of $t-1$ is fixed in each of the intervals $t < 1$, $1 < t < 4$, and $t > 4$. Thus, the sign of $t-1$ for $t$ in these intervals can be determined by evaluating the expression $t-1$ for exactly one value of $t$ in each of these intervals. Now, 5 is in the interval $t > 4$ and $t-1$ is positive if $t = 5$. Thus, $t-1$ is positive in $t > 4$. Next, 2 is in the interval $1 < t < 4$ and $t-1$ is positive if $t = 2$. Thus, $t-1$ is positive in $1 < t < 4$. Finally, 0 is in the interval $t < 1$ and $t-1$ is negative if $t = 0$. Thus, $t-1$ is negative in $t < 1$.

We can simplify the above argument even further. To solve the inequality $f(x) \geq 0$, $f(x) > 0$, $f(x) \leq 0$, or $f(x) < 0$, where $f$ is a polynomial, we first solve the corresponding equation $f(x) = 0$. Then the solutions are used to partition the number line into one more interval than the number of solutions. The inequality is either always true or always false in each interval. Therefore, testing the inequality with just one number in the interval identifies whether or not the inequality is true for that interval. We can summarize the above argument for $f(t) = (t-1)(t-4) \leq 0$ in another chart that is also called a **sign chart**.

| Interval | $t < 1$ | $1 < t < 4$ | $t > 4$ |
|---|---|---|---|
| Test-value of $t$ | 0 | 2 | 5 |
| Value of $f(t) = (t-1)(t-4)$ | 4 | $-2$ | 4 |
| Sign of $f(t) = (t-1)(t-4)$ | + | − | + |

The last line of the chart tells us that $f(t) < 0$ for $1 < t < 4$ and $f(t) > 0$ for $t < 1$ or $t > 4$. Thus, the solution to $(t-1)(t-4) \leq 0$ is $1 \leq t \leq 4$.

The sign chart is useful whenever we are able to *factor* the polynomial $f$ completely, or determine all of the zeros by other means. The *Intermediate Value Theorem* introduced in Section 1.9.1 is the basis for a sign chart method.

- **EXAMPLE 4:** Use a sign chart to solve $2x^2 + 9x - 5 > 0$, and draw a number-line graph of the solution. Check your answer with a graphing utility.

  SOLUTION: First we factor $2x^2 + 9x - 5 = (2x - 1)(x + 5)$, and then we make a sign chart. Notice that $2x^2 + 9x - 5 = 0$ for $x = \frac{1}{2}$ and $x = -5$. These numbers are used to determine the intervals on the number line.

  | Interval | $x < -5$ | $-5 < x < \frac{1}{2}$ | $x > \frac{1}{2}$ |
  |---|---|---|---|
  | Test value of $x$ | $-6$ | $0$ | $1$ |
  | Value of $2x^2 + 9x - 5$ | $13$ | $-5$ | $6$ |
  | Sign of $2x^2 + 9x - 5$ | $+$ | $-$ | $+$ |

  The values of $x$ that make $(2x-1)(x+5)$ equal to 0 are not solutions to $(2x-1)(x+5) > 0$. The last line of the sign chart shows that the solution to the inequality $(2x - 1)(x + 5) > 0$ is $x < -5$ or $x > \frac{1}{2}$. In interval notation, the solution to the original inequality is $(-\infty, -5) \cup (\frac{1}{2}, \infty)$. A number-line graph of the solution is given in Figure 1.6.9.

  A complete graph of $f(x) = 2x^2 + 9x - 5$ is given in Figure 1.6.10. It appears that the graph of $f$ crosses the $x$-axis at $-5$ and $\frac{1}{2}$. This can be confirmed by direct substitution. Thus, the graph of $f(x) = 2x^2 + 9x - 5$ is above the $x$-axis for $x < -5$ or $x > \frac{1}{2}$. This geometrically confirms the solution $(-\infty, -5) \cup (\frac{1}{2}, \infty)$ found algebraically. •

- **EXAMPLE 5:** Solve $x(x + 3)(x - 1) \geq 0$ both algebraically and graphically.

  SOLUTION: The expression $x(x + 3)(x - 1)$ is equal to 0 for $x = 0, -3,$ or $1$. Be sure you take the factor $x$ into account when making the sign chart.

  | Interval | $x < -3$ | $-3 < x < 0$ | $0 < x < 1$ | $x > 1$ |
  |---|---|---|---|---|
  | Test value of $x$ | $-4$ | $-1$ | $0.5$ | $2$ |
  | Value of $x(x+3)(x-1)$ | $-20$ | $4$ | $-0.875$ | $10$ |
  | Sign of $x(x+3)(x-1)$ | $-$ | $+$ | $-$ | $+$ |

Figure 1.6.9

The solution to $x(x+3)(x-1) > 0$ is $-3 < x < 0$ or $x > 1$. The product $x(x+3)(x-1)$ is 0 for $x = -3, 0$, or 1. Therefore, the solution to $x(x+3)(x-1) \geq 0$ in interval notation is $[-3, 0] \cup [1, \infty)$.

The graph of $f(x) = x(x+3)(x-1)$ in Figure 1.6.11 is a complete graph. This graph is above or on the $x$-axis for $-3 \leq x \leq 0$ or $x \geq 1$, geometrically confirming the solution found algebraically. •

The sign chart method can also be used with inequalities of the form $f(x) \geq 0$, $f(x) > 0$, $f(x) \leq 0$, or $f(x) < 0$ with $f$ the quotient of two polynomials provided that we can determine all the zeros of the two polynomials. Such functions $f$ are called **rational functions**. In this situation, the zeros of the two polynomials of $f$ are used to partition the number line, and then we proceed as we did with polynomials. Notice that the zeros of the polynomial in the denominator of $f$ are *not* in the domain of $f$. The next example illustrates the procedure for rational functions.

- EXAMPLE 6: Use a sign chart to solve

$$\frac{x-4}{2x+5} \leq 0 .$$

SOLUTION: The zeros of the polynomials in the numerator and denominator of the rational function $f(x) = \frac{x-4}{2x+5}$ are 4 and $-2.5$. These numbers are used to determine the intervals on the number line in the following sign chart.

| Interval | $x < -2.5$ | $-2.5 < x < 4$ | $x > 4$ |
|---|---|---|---|
| Test value of $x$ | $-3$ | 0 | 5 |
| Value of $\frac{x-4}{2x+5}$ | 7 | $-\frac{4}{5}$ | $\frac{1}{15}$ |
| Sign of $\frac{x-4}{2x+5}$ | $+$ | $-$ | $+$ |

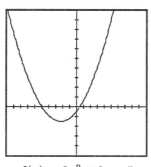

$f(x) = 2x^2 + 9x - 5$
$[-10, 10]$ by $[-50, 100]$
Figure 1.6.10

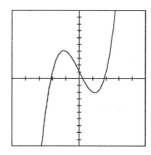

$f(x) = x(x+3)(x-1)$
$[-5, 5]$ by $[-10, 10]$
Figure 1.6.11

The last line of the sign chart tells us that $\frac{x-4}{2x+5} < 0$ for $-2.5 < x < 4$. Notice that $\frac{x-4}{2x+5}$ is zero for $x = 4$ and undefined for $-2.5$. Thus, the solution to $\frac{x-4}{2x+5} \leq 0$ is $-2.5 < x \leq 4$, or $(-2.5, 4]$. •

In Example 5, we solved an inequality that was given in factored form. The next example involves a function that cannot be factored. This example illustrates both the power of a graphical approach and the limitation of algebraic procedures. *It is extremely rare to find a real-world inequality in which the expression can be factored algebraically to apply a sign chart method.* We use zoom-in to determine the endpoints of the intervals involved in the solution. Unless we specify otherwise, these endpoints should be determined with error of at most 0.01.

- **EXAMPLE 7:** Solve the inequality $x^3 < 4x - 1$.

SOLUTION: This inequality can be rewritten in the form $x^3 - 4x + 1 < 0$. Let $f(x) = x^3 - 4x + 1$. The solution to the inequality consists of those values of $x$ for which the graph of $f$ lies below the $x$-axis (Figure 1.6.12). This graph is a complete graph of $f$. We can use zoom-in to determine that the zeros of $f$ are $-2.11$, $0.25$, and $1.86$ with error less than $0.01$. Thus, the solution to this inequality is $(-\infty, -2.11) \cup (0.25, 1.86)$. •

- **Agreement** We used round parentheses at the endpoints of the intervals in Example 7 because the inequality < means strictly less than. We would use square brackets if the inequality symbol was ≤. We use this convention even though the endpoints we report in solutions are approximations. That is, we act as if the endpoints are exact when deciding whether to use round parentheses or square brackets.

- **EXAMPLE 8:** A box is formed by removing squares of side length $x$ from each corner of a rectangular piece of cardboard 20 inches wide by 30 inches long. Determine $x$ so that the volume of the resulting box is at least 800 cubic inches.

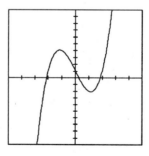

$f(x) = x^3 - 4x + 1$
$[-5, 5]$ by $[-10, 10]$
Figure 1.6.12

## 1.6 Solving Inequalities Algebraically and Graphically

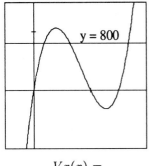

$Vx(x) =$
$x(20 - 2x)(30 - 2x)$
$[-5, 20]$ by $[-1000, 1500]$
Figure 1.6.13

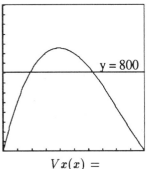

$Vx(x) =$
$x(20 - 2x)(30 - 2x)$
$[0, 10]$ by $[0, 1500]$
Figure 1.6.14

SOLUTION: You may want to refer to Example 4 of Section 1.5 to see how to form the box. The volume of the box is $V(x) = x(20 - 2x)(30 - 2x)$. Figure 1.6.13 gives a complete graph of $V$ and the horizontal line $y = 800$. The graph of $V$ represents the problem situation only for values of $x$ between 0 and 10. (Why?) Thus, the graph of $V$ in Figure 1.6.14 is a geometric representation that is also a graph of the problem situation. The solution to the problem consists of those values of $x$ for which the graph of $V$ is on or above the graph of $y = 800$. We can use zoom-in to determine that the $x$-coordinates of the two points of intersection of $V$ and $y = 800$ between $x = 0$ and $x = 10$ are 1.88 and 6.36 with error at most 0.01. The solution to the problem is the interval $[1.88, 6.36]$, that is, the volume of the resulting box is at least 800 cubic inches, provided that the side length $x$ of the square cut out satisfies $1.88 \leq x \leq 6.36$. •

## EXERCISES 1-6

Use interval notation to describe the intervals depicted graphically.

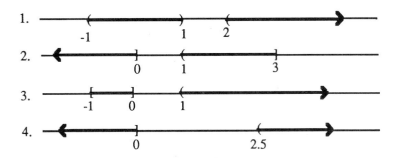

Use a number line to graph the intervals.

5. $[-4, 2), [3, 7)$
6. $x \leq 2$ or $x > 5$
7. $(-\infty, 2), [5, \infty)$
8. $x > -3$ or $x < -6$

**Complete the table.**

|  | Interval Notation | Inequality Notation | English Phrase |
|---|---|---|---|
| Example: | $(-\infty, 3) \cup (5, 7)$ | $x < 3$ or $5 < x < 7$ | The set of all real numbers less than 3 or between 5 and 7 |
| 9. |  | $x > 5$ or $x < -3$ |  |
| 10. | $[-2, 3) \cup (6, \infty)$ |  |  |
| 11. |  |  | The set of all real numbers between $-3$ and 2 or greater than 6 |

**Complete the table.**

|  | Interval Notation | Absolute Value Inequality Notation | English Phrase |
|---|---|---|---|
| Example: | $(-3, 3)$ | $|x| < 3$ | The set of all real numbers less than 3 units from the origin |
| 12. |  | $|x| \geq 5$ |  |
| 13. | $(-1, 7)$ |  |  |
| 14. |  |  | The set of all real numbers greater than 2 units from $-5$ |
| 15. | $(-\infty, 1] \cup [6, \infty)$ |  |  |

16. Solve $|x-2| < 4$ algebraically. Write the solution as an interval. Graph $y = |x-2|-4$. Compare the values of $x$ for which the graph of $y = |x-2|-4$ is *below* the $x$-axis with the algebraic solution to $|x-2| < 4$.

17. Solve $|2x-6| > 8$ algebraically. Write the solution as an interval. Graph $y = |2x-6|-8$. Compare the values of $x$ for which the graph of $y = |2x-6| = 8$ is *above* the $x$-axis with the algebraic solution to $|2x-6| > 8$.

Solve the inequalities algebraically. Write your answers in interval notation. Check your answers with a graphing utility.

18. $|x-3| < 2$
19. $|x-3| \geq 5$
20. $|3-x| < 8$
21. $\dfrac{x-3}{|x+2|} < 0$

Consider the graph of $y = f(x)$ in Figure 1.6.15.

22. (a) Compare $f(2)$ and $-2$.
    (b) Is $x = 2$ a solution to $f(x) < -2$?
    (c) Use the graph to solve $f(x) < -2$.

23. (a) Compare $f(-3)$ and $0$.
    (b) Is $x = -3$ a solution to $f(x) > 0$?
    (c) Use the graph to solve $f(x) > 0$.

In Exercises 24 through 38 use a sign chart to solve the inequalities. Write the solution in interval notation. Use a graphing utility to check your answers.

24. $(x-1)(x+2) < 0$
25. $x(x-3) \leq 0$
26. $x^2 - 5x + 6 \geq 0$
27. $x^2 - 4x - 21 < 0$
28. $2x^2 + 5x \geq 3$
29. $x^2 < x$
30. $x^4 + 3x^2 < 4$
31. $\dfrac{x+3}{x-1} < 0$
32. $\dfrac{x+3}{x-1} \geq 0$
33. $x(x-4)(x+2) \geq 0$
34. $(x+1)(x-2)(x+4) < 0$
35. $(x-2)^3 > 0$
36. $(x-2)^2(x-1) \leq 0$
37. $\dfrac{x+9}{x+4} \leq 0$
38. $\dfrac{3x-10}{4-x} > 0$

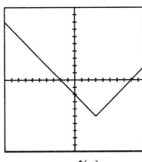

$y = f(x)$
$[-10, 10]$ by $[-10, 10]$
Figure 1.6.15

39. Solve the inequality $x^3 - 2x^2 - 5x + 7 \geq 2x + 1$ by determining the values of $x$ for which the graph of $f(x) = x^3 - 2x^2 - 5x + 7$ lies above the graph of $g(x) = 2x + 1$. Compare this method with that given in the example.

Solve each inequality.

40. $x^2 - 5x + 3 > 0$
41. $2 - 8x + 3x^2 \leq 0$
42. $3x^3 - 8x^2 - 5x + 6 \leq 0$
43. $x^3 - 9x^2 + 6x + 55 < 0$
44. $|x^2 - 1|(x - 2) \geq 3$
45. $2x + 1 > x^3 + 2x^2 - 3x + 5$
46. $3x^2 - x + 6 \leq 2x^3 - x^2 + 3x - 4$
47. $5 \sin x < 2$ for $0 \leq x \leq 10$

48. An object is propelled upward from a tower 200 feet in height with initial velocity 100 feet per second.
    (a) Determine an algebraic representation of the problem situation.
    (b) Draw a complete graph of the algebraic representation.
    (c) What are the domain and range of the algebraic representation?
    (d) What part of the graph in (b) represents the problem situation?
    (e) When will the object be above ground?
    (f) When will the object hit the ground?
    (g) What is the maximum height attained by the object?

49. An object is propelled upward from a tower 155 feet in height with initial velocity 275 feet per second.
    (a) Determine an algebraic representation of the problem situation.
    (b) Draw a complete graph of the algebraic representation.
    (c) What are the domain and range of the algebraic representation?
    (d) What part of the graph in (b) represents the problem situation?
    (e) When will the object be above ground?
    (f) When will the object hit the ground?
    (g) What is the maximum height attained by the object?

A single-commodity open market is driven by the supply and demand principle. Economists have determined that supply curves are usually increasing (as the price increases, the sellers increase production) and the demand curves are decreasing (as the price increases, the consumers buy less). For each pair of indicated supply and demand equations do the following.
   (i) Draw complete graphs of both the supply and demand equations on the same coordinate system. $p$ is the price (vertical axis) and $x$ represents the number of units of the commodity produced.
   (ii) What portion of the graphs in (i) are complete graphs of the problem situation?
   (iii) Determine the *equilibrium price*, that is, the price where supply is equal to demand.
   (iv) How many items must be produced for the supply to be greater than the demand?

50. Supply: $p = 5 + \dfrac{2}{13}x$

    Demand: $p = 10 - \dfrac{1}{20}x^2$

51. Supply: $p = \dfrac{1}{20}x + 1$

    Demand: $p = \dfrac{10}{x}$

52. A swimming pool with dimensions 20 feet by 30 feet is surrounded by a sidewalk of uniform width $x$. Find the possible widths of the sidewalk if the total area of the sidewalk is to be greater than 200 square feet but less than 360 square feet.
   (a) Write an equation that gives the area of the sidewalk as a function of $x$.
   (b) Write an inequality that is an algebraic representation of this problem situation.
   (c) Solve the inequality in (b) algebraically, and then determine the solution to the problem situation.
   (d) Use a graphing utility to check your answer to (c).

53. Consider the simple interest formula $S = P(1 + rn)$ (Exercise 39, Section 1.4). Find the rate of simple interest necessary for an investor to accumulate at least $30,000 in 18 years by investing $5000 today.
   (a) Write an inequality that is an algebraic representation of this problem situation.
   (b) Draw a complete graph of each side of the inequality in (a) in the same viewing rectangle.
   (c) Determine the solution to this problem situation both algebraically and geometrically.

54. Equal squares are removed from the four corners of a 22-inch-by-29-inch rectangular sheet of cardboard. The sides are turned up to make a box with no lid. Find the possible lengths of the sides of the removed squares if the volume of the box is to be less than 2000 cubic inches.
   (a) Write an inequality that is an algebraic representation of this problem situation.
   (b) Draw a complete graph of each side of the inequality in (a) in the same viewing rectangle. Draw a complete graph of the problem situation.
   (c) Determine the solution to this problem situation geometrically.

55. Sam makes two initial investments of $1000 each. The first earns 10% simple interest, and the second earns interest using 5% *simple discount*. After $n$ years, the value of the first investment is $S = 1000/(1 + 0.10n)$ and the value of the second investment is $S = 1000/(1 - 0.05n)$.
   (a) Draw a complete graph of each algebraic representation in the viewing rectangle $[0, 30]$ by $[0, 6000]$.
   (b) Is there a time when the future values of both investments are equal?
   (c) Is there a period of time when the simple interest investment is less than the value of the simple discount investment?

## 1.7 Maximum and Minimum Values

Many problem situations require that we find the largest or smallest value of a given function over some portion of its domain. We have encountered such situations in some of the real-world applications in previous sections. In this section we use zoom-in to find the coordinates of points where maximum and minimum values occur. Applications whose solutions involve finding a largest or smallest value of a model are also investigated.

- **EXAMPLE 1:** Draw a complete graph of $f(x) = x^3 - 4x$.

SOLUTION: The graph of $f(x) = x^3 - 4x$ in the viewing rectangle $[-6, 6]$ by $[-10, 10]$ is shown in Figure 1.7.1 and is a complete graph.

The graph of $f(x) = x^3 - 4x$ has a high point (peak) between $x = -2$ and $x = 0$. The value of the function at this point is called a local maximum value of $f$. Similarly, the graph of $f(x) = x^3 - 4x$ has a low point (valley) between $x = 0$ and $x = 2$. The value of $f$ at this point is called a local minimum value of $f$. More precisely we have the following definitions.

- **DEFINITION**  The value of $f$ at $x = a$ is called a **local maximum value** of $f$ if there is an interval $(c, d)$ with $c < a < d$ so that $f(x) \leq f(a)$ for all $x$ in $(c, d)$. The value of $f$ at $x = b$ is called a **local minimum value** of $f$ if there is an interval $(c, d)$ with $c < b < d$ so that $f(x) \geq f(b)$ for all $x$ in $(c, d)$.

When we use the phrase "for all $x$ in $(c, d)$" in the above definition, it is implicitly understood that we mean those $x$'s in $(c, d)$ that are also in the domain of $f$. For example, using the above definition with this understanding about domain, the function $f(x) = \sqrt{x}$ has a local minimum value at $x = 0$.

- **EXAMPLE 2:** Determine the local maximum and local minimum values of the function $f(x) = x^3 - 4x$.

SOLUTION: Refer to Figure 1.7.1. We use zoom-in to trap the low point between $x = 0$ and $x = 2$ in a very small viewing rectangle. The graph in Figure 1.7.1 suggests that we next look at the graph of $f$ in the viewing rectangle $[0, 2]$ by $[-4, -2]$ (Figure 1.7.2). Because the distance between scale marks in Figure 1.7.2 is 0.1, we can estimate that the low point has $x$-coordinate between 1.1 and 1.2 and $y$-coordinate between $-3.1$ and $-3.0$. The graph of $f$ in the viewing rectangle $[1.1, 1.2]$ by $[-3.1, -3.0]$ is given in Figure 1.7.3. Notice that this viewing rectangle is actually a *square*.

The graph of $f$ in Figure 1.7.3 appears relatively flat. In calculus you will see that this is exactly what should be expected when you magnify around such points. If we are not

$f(x) = x^3 - 4x$
$[-6, 6]$ by $[-10, 10]$
**Figure 1.7.1**

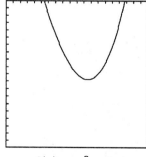

$f(x) = x^3 - 4x$
$[0, 2]$ by $[-4, -2]$
**Figure 1.7.2**

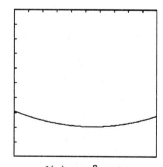

$f(x) = x^3 - 4x$
$[1.1, 1.2]$ by $[-3.1, -3.0]$
**Figure 1.7.3**

careful when we zoom in, we may produce viewing rectangles where the graph of $f$ is so flat that we are unable to distinguish the low point (or high point) from the rest of the points in the viewing rectangle.

What we need to do is choose viewing rectangles that are much wider than they are tall. (Why?) The graph of $f$ in the viewing rectangle $[1.1, 1.2]$ by $[-3.08, -3.07]$ (Figure 1.7.4) more clearly highlights the low point of $f$.

Notice that $L$ and $R$ are the same in Figures 1.7.3 and 1.7.4. The height of the viewing rectangle in Figure 1.7.3 is 0.1, and the height of the viewing rectangle in Figure 1.7.4 is 0.01. In fact, the width of the viewing rectangle in Figure 1.7.4 is ten times its height.

From Figure 1.7.4 we can estimate that the coordinates of the low point are $(1.155, -3.0792)$. The error in the first coordinate is at most 0.01 and the error in the second coordinate is at most 0.001. It was necessary to increase the precision on the second coordinate in order to achieve error at most 0.01 in the first coordinate. Thus, $f$ achieves a local minimum value of $-3.0792$ when $x$ is 1.155.

We can use the graph in Figure 1.7.1 or verify directly that $f(-x) = -f(x)$ to show that the graph of $f(x) = x^3 - 4x$ is symmetric with respect to the origin. (Why?) This means that the coordinates of the point where the local maximum value occurs, with error at most 0.01, are $(-1.155, 3.0792)$. (Why?) •

## Polynomials of Degree 3

The graph of the function $g = -f$ where $f(x) = x^3 - 4x$ is the reflection of the graph of $f$ through the $x$-axis (Figure 1.7.5). We can see from this graph that $g$ has a local minimum value at $x = a$ and a local maximum value at $x = b$. It can be shown that every polynomial function of degree 3 has behavior like one of the following functions: $y = x^3$, $y = -x^3$, $y = x^3 - 4x$, or $y = -x^3 + 4x$. Thus it follows that every polynomial function of degree 3 has either one local maximum value and one local minimum value like the function in Example 1 or no local maximum or minimum value like $y = x^3$.

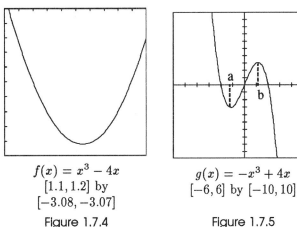

$f(x) = x^3 - 4x$
$[1.1, 1.2]$ by
$[-3.08, -3.07]$

Figure 1.7.4

$g(x) = -x^3 + 4x$
$[-6, 6]$ by $[-10, 10]$

Figure 1.7.5

Figure 1.7.6

## Polynomials of Degree 4 or 5

A polynomial of degree 4 has local maximum and minimum behavior like one of the four graphs in Figure 1.7.6. In Figure 1.7.6, the lowest point of the third graph or the highest point of the fourth graph need not be in the order suggested. A value of a function that is either a local maximum value or a local minimum value of the function is called a **local extreme value** of the function. As suggested by Figure 1.7.6, a polynomial of degree 4 has either *one* or *three* local extreme values.

A polynomial of degree 5 has local maximum and minimum behavior like one of the six graphs in Figure 1.7.7. Again, the relative position of the high or low points in the last four graphs need not be in the order suggested. As suggested by Figure 1.7.7, a polynomial of degree 5 has zero, two, or four local extreme values. You will be expected to use the information in Figures 1.7.6 and 1.7.7 in the exercises.

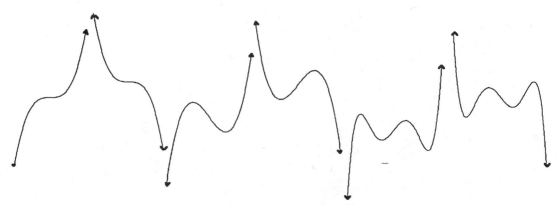

Figure 1.7.7

## 1.7 Maximum and Minimum Values

- **EXAMPLE 3:** An object is shot straight up (launched) from the top of a building 200 feet tall with initial velocity 64 feet per second.
    - (a) When will the object hit the ground?
    - (b) When does the object reach its maximum height above ground, and what is this maximum height?

**SOLUTION:** The height of the object above ground $t$ seconds after launch is $s = -16t^2 + 64t + 200$. The graph of $s$ in the $[-10, 10]$ by $[-400, 400]$ viewing rectangle is shown in Figure 1.7.8. Notice that this is *not* a graph of the trajectory (flight path) of the object. (Why?)

- (a) The object will hit the ground when $s = 0$. From Figure 1.7.8, it appears that the object will hit the ground when $t$ is approximately 6. We can zoom in and find that this value of $t$ is 6.06 with error at most 0.01. Thus, the object will hit the ground about 6.06 seconds after launch.
- (b) From Figure 1.7.8 we can see that the object will reach its maximum height when $t$ is approximately 2. We can zoom in and find that the coordinates of the highest point are $(2, 264)$ with error at most 0.01. We can complete the square to write $y = -16(t - 2)^2 + 264$ and see that the coordinates of the vertex are exactly $(2, 264)$. Thus, the object reaches its maximum height of 264 feet above ground 2 seconds after launch. •

- **EXAMPLE 4:** Squares of side length $x$ are removed from a 20-inch by 25-inch piece of cardboard (Figure 1.7.9). A box with no top is formed by folding along the dashed line segments in Figure 1.7.9. Determine $x$ so that the box has maximum possible volume, and find this maximum volume.

**SOLUTION:** The height of the box is $x$, and the base of the box has dimensions $20 - 2x$ by $25 - 2x$. Thus, the volume of the box is $V(x) = x(20 - 2x)(25 - 2x)$. The graph of the function $V(x) = x(20 - 2x)(25 - 2x)$ in the viewing rectangle $[-5, 20]$ by $[-1000, 1000]$ is

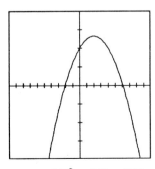

$s = -16t^2 + 64t + 200$
$[-10, 10]$ by $[-400, 400]$

Figure 1.7.8

Figure 1.7.9

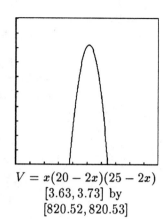

$V = x(20 - 2x)(25 - 2x)$
$[-5, 20]$ by $[-1000, 1000]$

Figure 1.7.10

$V = x(20 - 2x)(25 - 2x)$
$[3.63, 3.73]$ by
$[820.52, 820.53]$

Figure 1.7.11

shown in Figure 1.7.10. The side length $x$ must be a number between 0 and 10 because $x$, $20 - 2x$, and $25 - 2x$ all have to be positive. (Why?)

We can see from the graph in Figure 1.7.10 that $V$ has a local maximum value between $x = 0$ and $x = 10$. The $x$-coordinate of this high point gives the value of $x$ that will produce a box with the maximum possible volume. The $y$-coordinate of this high point gives the value of the maximum possible volume. We use zoom-in to find the coordinates of the high point between $x = 0$ and $x = 10$.

The graph of $V$ in the $[3.63, 3.73]$ by $[820.52, 820.53]$ viewing rectangle (Figure 1.7.11) permits us to read the coordinates of the high point as $(3.68, 820.5282)$ with error at most 0.01 in either coordinate. Actually, it was necessary to find the second coordinate with error at most 0.001 in order to determine the first coordinate with error at most 0.01.

Thus, if the side length of the removed squares is 3.68 inches, we produce a box with maximum volume. The corresponding maximum value of the volume is 820.5282 cubic inches. •

## • EXERCISES 1-7

Draw a complete graph of each polynomial function. Specify a viewing rectangle that displays a *complete* graph of the function.

1. $f(x) = x^2 - x + 3$
2. $y = 12x^2 - 20x + 60$
3. $A = \left(\frac{x}{4}\right)^2 + \left(\frac{500 - x}{4}\right)^2$
4. $A = x(8 - x^2)$
5. $V = x(30 - 2x)(40 - 2x)^2$
6. $g(x) = 2x^3 - 6x^2 + 3x - 5$
7. $V(x) = 200 - 20x^2 + 0.01x^3$
8. $f(x) = 2x^5 - 10x^4 + 3x^3 + 2x^2 - 10x + 5$

## 1.7 Maximum and Minimum Values

Determine the *exact* values of all local maximum and minimum values and the corresponding values of $x$ that give these values.

9. $y = 30 - 3x + 7x^2$
10. $g(x) = |10 - 7x + x^2|$
11. $y = \sqrt{x-1}$
12. $f(x) = 3 - 2|4x + 7|$

Draw a complete graph of the function. Determine all real zeros. Determine all local maximum and minimum values and the corresponding values of $x$ that give these values.

13. $y = 10 - x^2$
14. $f(x) = 2x^2 - 6x + 10$
15. $g(x) = x^3 - 10x^2$
16. $y = 2x^3 - x^2 + x - 4$
17. $f(x) = x^4 - 4x^2 - 3x + 12$
18. $y = |2x^3 - 4x^2 + x - 3|$
19. $f(x) = x^3 - 2x^2 + x - 30$
20. $V(x) = x(34 - 2x)(53 - 2x)$
21. $T(x) = |20x^3 + 2x^2 - 10x + 5|$
22. $V(x) = x(22 - 2x)(8 - 2x)$
23. $f(x) = 12 - x + 3x^2 - 2x^3$
24. $y = x^5 - 3x^2 + 3x - 6$

25. An object is shot straight up from the top of a building 300 feet tall with initial velocity 40 feet per second.
    (a) Determine an algebraic representation of the height above ground of the object as a function of time.
    (b) When will the object hit the ground? Assume it does not land on the building.
    (c) When does the object reach its maximum height above ground, and what is this maximum height?

26. A large apartment rental company has 2500 units available, and 1900 are currently rented at $450 per month. A market survey indicates that each $15 decrease in monthly rent will result in 20 new tenants.
    (a) Determine a function $R$ which represents the total rental income realized by the company where $x$ is the number of $15 decreases in monthly rent (that is, if $x = 3$, then the monthly rent is $405). Assume that all rents are adjusted to the new figure.
    (b) Draw a complete graph of the function $y = R(x)$.
    (c) Determine the rent that will yield the rental company the maximum monthly income.

27. A rectangular area is to be fenced against an existing wall. The three sides of the fence must total 1050 feet in length. Find the dimensions of the maximum area that can be enclosed. What is the maximum area?

28. A 300 inch piece of wire is cut into two pieces. Each piece of wire is used to make a square. Let $x$ be the length of one piece of the wire.
    (a) Determine an algebraic representation that gives the total area of the two squares as a function of $x$.
    (b) Draw a complete graph of the algebraic representation.
    (c) What portion of the graph of the algebraic representation represents the problem situation?

Figure 1.7.12          Figure 1.7.13

(d) Find the lengths of the two pieces of wire that produce two squares of maximum total area.

(e) Find the lengths of the two pieces of wire that produce two squares of minimum total area.

29. The total daily revenue of a lemonade stand at a state fair is given by the equation $R = x \cdot p$ where $x$ is the number of glasses of lemonade sold daily and $p$ is the price of one glass of lemonade. Assume the price of the lemonade is given by the "supply" equation $p = 2 + 0.002x - 0.0001x^2$.
   (a) Determine an algebraic representation that gives the total daily revenue of the lemonade stand as a function of $x$.
   (b) Draw a complete graph of the algebraic representation.
   (c) Draw the graph of the problem situation.
   (d) Determine the number of glasses of lemonade to be sold to produce maximum daily revenue. What is the maximum daily revenue?

30. A rectangular sheet of metal 37 inches wide is to be made into a trough to feed hogs by turning up sides of equal length perpendicular to the sheet (Figure 1.7.12). Find the length of the turned-up sides that give the trough the maximum cross-sectional area. What is the maximum cross-sectional area?

31. A cardboard box with no lid is to be constructed from a 55 inch by 86 inch piece of flat cardboard by removing equal squares from each corner and folding up the sides. Find the dimensions of the box with maximum volume. What is the maximum volume?

32. An apartment building is built in the shape of a box with square cross section and with a triangular prism forming the roof as shown in Figure 1.7.13. The total height of the building is 30 feet and the sum of the length and width of the building is 100 feet.
   (a) Determine an algebraic representation of the volume $V$ of the building as a function of $x$.
   (b) Draw a complete graph of the algebraic representation.
   (c) What values of $x$ represent the problem situation? Draw a complete graph of the problem situation.
   (d) What are the dimensions of the building with maximum volume?
   (e) What is the maximum volume?

## 1.8 Increasing and Decreasing Functions

Polynomial functions of higher degrees are used to approximate values of complicated functions. Calculators and computers often use such approximations to calculate the values of their built-in functions. In this section we determine where functions are increasing and decreasing.

It is known that every polynomial of degree 2 can be obtained from $x^2$ by appropriate use of reflection, vertical stretching or shrinking, horizontal shifting, and vertical shifting. It is *not* true that all polynomials of degree 3 can be obtained from $x^3$ using such techniques. A complete graph of $f(x) = x^3$ is given in Figure 1.8.1. This graph shows that the function $f(x) = x^3$ is an increasing function; that is, the function values $f(x)$ increase as the values of $x$ increase. This means as we walk along the graph in the direction of increasing $x$, that is, from left to right, we climb. More precisely, we have the following definition.

- **DEFINITION**   A function $f$ is **increasing on an interval** $J$ if for all $x_1, x_2$ in $J$ with $x_2 > x_1$ we have $f(x_2) > f(x_1)$ (Figure 1.8.2).

The function $f(x) = x^3$ is increasing on the interval $(-\infty, \infty)$. In fact, $f$ is increasing on any interval.

A complete graph of $f(x) = x^4$ is given in Figure 1.8.3. This function is decreasing on the interval $(-\infty, 0]$ and increasing on the interval $[0, \infty)$.

- **DEFINITION**   A function $f$ is said to be **decreasing on an interval** $J$ if for all $x_1, x_2$ in $J$ with $x_2 > x_1$ we have $f(x_2) < f(x_1)$ (Figure 1.8.4).

If the inequality signs between $f(x_1)$ and $f(x_2)$ in the previous two definitions are not strict inequalities, then we have the following definitions.

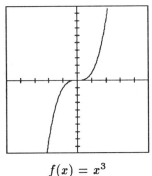

$f(x) = x^3$
$[-5, 5]$ by $[-10, 10]$
Figure 1.8.1

Figure 1.8.2

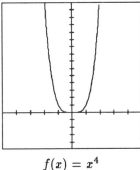

$f(x) = x^4$
$[-5, 5]$ by $[-5, 15]$
Figure 1.8.3

Figure 1.8.4

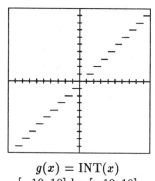
$g(x) = \text{INT}(x)$
$[-10, 10]$ by $[-10, 10]$
Figure 1.8.5

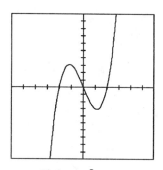
$f(x) = x^3 - 4x$
$[-6, 6]$ by $[-10, 10]$
Figure 1.8.6

- **DEFINITION**  A function $f$ is **nondecreasing on an interval** $J$ if for all $x_1$, $x_2$ in $J$ with $x_2 > x_1$ we have $f(x_2) \geq f(x_1)$. A function $f$ is **nonincreasing on an interval** $J$ if for all $x_1$, $x_2$ in $J$ with $x_2 > x_1$ we have $f(x_2) \leq f(x_1)$. •

- **EXAMPLE 1:** Let $f(x) = x^4$, and let $J$ be any interval. What can be said about the increasing or decreasing behavior of $f$ on $J$?

  SOLUTION: Let $a$ and $b$ be the left and right endpoints of the interval $J$, respectively. Refer to Figure 1.8.3. If $b \leq 0$, then $f$ is decreasing on $J$. If $a \geq 0$, then $f$ is increasing on $J$. If $a < 0$ and $b > 0$, then $f$ is neither increasing nor decreasing on $J$. For example, $f$ is neither increasing nor decreasing on $[-1, 1]$. •

- **EXAMPLE 2:** Let $g(x) = \text{INT}(x)$ and let $J$ be any interval. What can be said about the increasing or decreasing behavior of $g$ on $J$?

  SOLUTION: The graph of the step function $g(x) = \text{INT}(x)$ is given in Figure 1.8.5. We see from this figure that $g$ is *nondecreasing* on every interval $J$. Furthermore, $g$ is *not* an increasing function on any interval, nor is $g$ a decreasing function on any interval. $g$ is nonincreasing on every interval of the form $[n, n+1)$, $n$ an integer. •

In the previous section we found the coordinates of the points at which $f(x) = x^3 - 4x$ had local maximum and minimum values. This information can be used to determine the intervals on which $f$ is increasing and the intervals on which $f$ is decreasing.

- **EXAMPLE 3:** Determine the intervals on which $f(x) = x^3 - 4x$ is increasing and the intervals on which $f$ is decreasing.

  SOLUTION: In the previous section we showed that the graph of $f$ in Figure 1.8.6 was complete. We also showed that the coordinates of the points at which $f$ had local maximum

## 1.8 Increasing and Decreasing Functions

and minimum values were $(-1.155, 3.0792)$ and $(1.155, -3.0792)$, respectively, with error of at most 0.01. Therefore, $f$ is increasing on the intervals $(-\infty, -1.155]$ and $[1.155, \infty)$ and is decreasing on $[-1.155, 1.155]$.

- **EXAMPLE 4:** Draw a complete graph of each inequality.
  (a) $y < x^3$  (b) $y \geq x^4$

SOLUTION: The graph of $y < x^3$ consists of all points below but not on the graph of $y = x^3$ (Figure 1.8.7). The graph of $y \geq x^4$ consists of all points above or on the graph of $y = x^4$ (Figure 1.8.8).

- **Agreement** Unless specified otherwise, when we ask you to determine the intervals on which a function is increasing or decreasing we want the intervals to be as inclusive as possible.

The graph of a function can be used to find the intervals on which the function is increasing or decreasing. Often the endpoints of these intervals occur at points where the function has a local maximum or minimum value. In this section, we will be content to estimate the endpoints of these intervals. In the previous section we determined the endpoints with greater accuracy using zoom-in. In calculus, you will be able to learn more about the increasing and decreasing behavior of functions and how to determine the endpoints of intervals over which a function is increasing or decreasing.

- **EXAMPLE 5:** Estimate the $x$-coordinates of the points at which $f(x) = x^3 - 4x^2 - 4x + 16$ has local extremum values. Then determine the intervals on which $f$ is increasing and on which $f$ is decreasing.

SOLUTION: The graph of $f$ in the $[-5, 5]$ by $[-10, 20]$ viewing rectangle is shown in Figure 1.8.9. We can use this graph to estimate that the $x$-coordinate of the point at which the

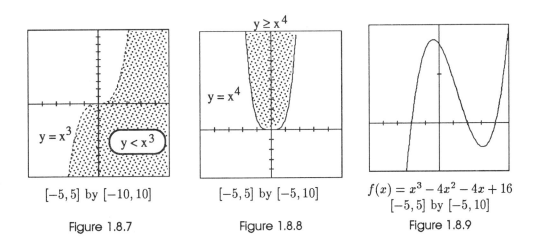

Figure 1.8.7 — $[-5, 5]$ by $[-10, 10]$

Figure 1.8.8 — $[-5, 5]$ by $[-5, 10]$

Figure 1.8.9 — $f(x) = x^3 - 4x^2 - 4x + 16$, $[-5, 5]$ by $[-5, 10]$

graph of $f$ has a local maximum value is $-0.5$. Similarly, the $x$-coordinate of the point at which the graph of $f$ has a local minimum value is 3. Then, $f$ is increasing on $(-\infty, -0.5]$ and $[3, \infty)$ and decreasing on $[-0.5, 3]$.   •

• **EXAMPLE 6:** Draw a complete graph of $y = -2(x+3)^3 + 1$ and determine the intervals on which the function is increasing or decreasing.

**SOLUTION:** The graph of this function can be obtained from the graph of $y = x^3$ by first vertically stretching by a factor of 2 to get $y = 2x^3$, then reflecting through the $x$-axis to get $y = -2x^3$, next shifting horizontally left 3 units to get $y = -2(x+3)^3$, and finally shifting vertically upward one unit to get $y = -2(x+3)^3 + 1$ (Figure 1.8.10). This can be confirmed by using a graphing utility to draw the graph of $y = -2(x+3)^3 + 1$ (Figure 1.8.11).

We can see from Figure 1.8.11 that $y = -2(x+3)^3 + 1$ is a decreasing function on $(-\infty, \infty)$. Alternatively, we can analyze the steps used in Figure 1.8.10 to see that $y = -2(x+3)^3 + 1$ is decreasing on $(-\infty, \infty)$. From Figure 1.8.10 we can see that both $y = x^3$ and $y = 2x^3$ are increasing on $(-\infty, \infty)$. However, the function $y = -2x^3$ that results after reflecting $y = 2x^3$ through the $x$-axis is decreasing on $(-\infty, \infty)$. Applying a horizontal shift 3 units left to $y = -2x^3$ followed by a vertical shift one unit up gives a function with the same increasing and decreasing behavior as $y = -2x^3$. Thus, $y = -2(x+3)^3 + 1$ is also decreasing on $(-\infty, \infty)$.   •

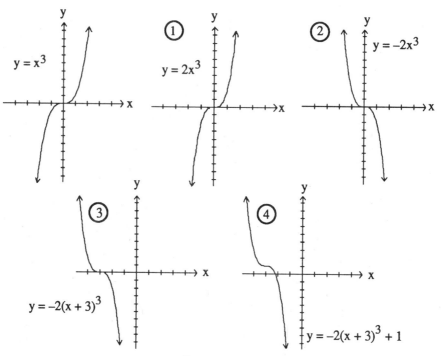

Figure 1.8.10

# 1.8 Increasing and Decreasing Functions

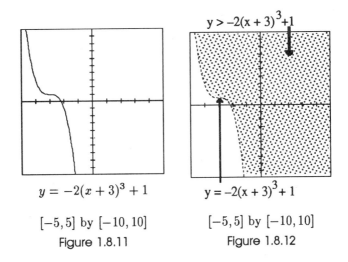

$y = -2(x + 3)^3 + 1$

$[-5, 5]$ by $[-10, 10]$

Figure 1.8.11

$y = -2(x + 3)^3 + 1$

$[-5, 5]$ by $[-10, 10]$

Figure 1.8.12

In Example 6 we used the geometric transformations needed to obtain the graph of $y = -2(x+3)^3 + 1$ from $y = x^3$ to reason that $y = -2(x+3)^3 + 1$ is decreasing on $(-\infty, \infty)$. This reasoning can be extended to obtain the following theorem.

- **THEOREM 1**    The function $y = a(x - h)^3 + k$ is
  (a) increasing on $(-\infty, \infty)$ if $a > 0$.    (b) decreasing on $(-\infty, \infty)$ if $a < 0$.

In Example 5 we showed that the polynomial function $f(x) = x^3 - 4x^2 - 4x + 16$ is neither increasing nor decreasing on $(-\infty, \infty)$. This fact together with Theorem 1 shows that $f(x) = x^3 - 4x^2 - 4x + 16$ *cannot* be obtained from $y = x^3$ using vertical stretching or shrinking, reflecting, horizontal shifting, or vertical shifting. Contrast this with the fact that every quadratic function can be obtained from $y = x^2$ by using these geometric transformations.

We can use the graph of $y = -2(x + 3)^3 + 1$ to determine the graph that results if $=$ is replaced by $>$.

- **EXAMPLE 7:** Draw a complete graph of $y > -2(x + 3)^3 + 1$.

    **SOLUTION:** The graph of $y > -2(x + 3)^3 + 1$ consists of all points above but not on the graph of $y = -2(x + 3)^3 + 1$ (Figure 1.8.12).    •

- ## EXERCISES 1-8

List the transformations that will produce the graph of each function from the graph of $y = x^3$. Specify an order in which the transformations could be applied and then sketch the graph of the function without using a graphing utility. Use a graphing utility to check your answer.

1. $y = x^3 - 4$
2. $y = (x - 4)^3$
3. $y = 3(x + 4)^3 - 5$
4. $y = -2(x - 5)^3 + 3$

5. $y = -(x-2)^3 - 2$
6. $y = 2 + 3(x+4)^3$

Determine all local maximum and minimum values. Determine the intervals on which each function is *increasing* and the intervals on which each function is *decreasing*. *Warning:* There is *hidden* behavior in problems 11 and 12.

7. $g(x) = x(x-3)$
8. $P(x) = \dfrac{1}{2-x}$
9. $t(x) = x(x-2)(x+3)$
10. $g(x) = x^3 - 4x^2 + 4x$
11. $f(x) = x^3 - 2x^2 + x - 30$
12. $f(x) = 20x^3 + 2x^2 - 10$
13. $g(x) = \text{INT}(x-2)$
14. $T(x) = |x - 5|$

Draw a complete graph of each inequality.

15. $y \le x^3 - 4x$
16. $y > \dfrac{1}{x-2}$
17. $y > |2x - 1|$
18. $y \ge 2x^3 - 3x^2 + 2x - 3$
19. Characterize the end behavior of $f(x) = ax^n$ for
    (a) $a$ positive, $n$ odd
    (b) $a$ positive, $n$ even
    (c) $a$ negative, $n$ odd
    (d) $a$ negative, $n$ even

Draw a complete graph. Determine all local maximum and minimum values. Determine all real zeros. Determine the intervals on which the function is increasing and the intervals on which the function is decreasing.

20. $f(x) = 2 - 3x + x^2 - x^3$
21. $f(x) = (x-1)^2 x^3$
22. $g(x) = 3x^4 - 5x^3 + 2x^2 - 3x + 6$
23. $T(x) = 2x^5 - 3x^4 + 2x^3 - 3x^2 + 7x - 4$
24. Draw the graph of $f(x) = 33x^3 - 100x^2 + 101x + 5$ in each viewing rectangle.
    (a) $[-10, 10]$ by $[-1000, 1000]$
    (b) $[-1, 1]$ by $[-100, 100]$
    (c) $[0.9, 1.1]$ by $[20, 50]$
25. The function $f$ in Exercise 24 appears to be increasing on any interval. *It is not*. There is *hidden behavior* that can be determined by zooming in near the point $(1, f(1))$. Find a viewing rectangle that exhibits the *hidden behavior*.
26. It can be shown that $f(x) = x^3 + x + 1$ is an increasing function on any interval. Show that it is impossible to find real numbers $a$, $b$, and $c$ so that $x^3 + x + 1 = a(x-b)^3 + c$.
27. A rectangular area, with one side against an existing wall, is to be enclosed by three sides of fencing totaling 335 feet in length. Let $x$ be the side length of the fence perpendicular to the existing wall.
    (a) Determine an algebraic representation of the enclosed area as a function of $x$.
    (b) Draw a complete graph of the algebraic representation.
    (c) What portion of the graph of the algebraic representation represents the problem situation?
    (d) Determine the intervals on which the algebraic representation is increasing or decreasing.

# 1.8 Increasing and Decreasing Functions

28. A 300-inch piece of wire is cut into two pieces. Each piece of wire is used to make a square. Let $x$ be the length of one piece of the wire.
    (a) Determine an algebraic representation of the total area of the two squares as a function of $x$.
    (b) Draw a complete graph of the algebraic representation.
    (c) What portion of the graph of the algebraic representation represents the problem situation?
    (d) Determine the intervals on which the algebraic representation is increasing or decreasing.

29. An object is shot straight up from the top of a building 186 feet tall with initial velocity 80 feet per second. Let $t$ be the number of seconds elapsed since the object was launched.
    (a) Determine an algebraic representation of the height of the object above ground level as a function of $t$.
    (b) Draw a complete graph of the algebraic representation.
    (c) What portion of the graph of the algebraic representation represents the problem situation?
    (d) Determine the intervals on which the algebraic representation is increasing or decreasing.

30. A box with no lid is made from a rectangular sheet of cardboard 50 inches by 85 inches by removing squares of equal side length from each corner and turning up the sides. Let $x$ be the side length of the removed squares.
    (a) Determine an algebraic representation of the volume of the box as a function of $x$.
    (b) Draw a complete graph of the algebraic representation.
    (c) What portion of the graph of the algebraic representation represents the problem situation?
    (d) Determine the intervals on which the algebraic representation is increasing or decreasing.

31. A rectangular sheet of metal 20 inches long is to be made into a gutter by turning up sides of equal length perpendicular to the sheet. Let $x$ be the height of the turned-up sides.
    (a) Determine an algebraic representation of the cross-sectional area of the gutter as a function of $x$.
    (b) Draw a complete graph of the algebraic representation.
    (c) What portion of the graph of the algebraic representation represents the problem situation?
    (d) Determine the intervals on which the algebraic representation is increasing or decreasing.

32. The total daily revenue of a lemonade stand at a state fair is given by the equation $R = x \cdot p$ where $x$ is the number of glasses of lemonade sold daily and $p$ is the price of one glass of lemonade. Assume the price of the lemonade is given by the supply equation $p = 1.2x - 0.01x^2$.
    (a) Determine an algebraic representation of the total daily revenue of the lemonade stand as a function of $x$.
    (b) Draw a complete graph of the algebraic representation.
    (c) Draw a complete graph of the problem situation.
    (d) Determine the intervals on which the algebraic representation is increasing or decreasing.

## 1.9 Continuity and End Behavior

In this section we introduce the concepts of continuity and end behavior of functions. An important consequence of continuity called the Intermediate Value Property is studied. We have made intuitive use of the Intermediate Value Property in earlier sections to find real zeros as well as other values of functions.

It is not an accident that graphs of polynomial functions obtained so far have no breaks in them. Some functions have graphs with breaks in them. For example, the graph of $f(x) = \frac{1}{x}$ obtained in Section 1.1 has a break at $x = 0$ (Figure 1.9.1). In calculus, you will be given a precise pointwise definition of continuity. In this textbook we will be content with an intuitive idea of the meaning of continuity. We begin by giving a rough idea of what we mean by continuity on an interval.

- **Understanding**  A function is said to be **continuous on the interval $J$** if, for all $a$ and $b$ in $J$, it is possible to trace the graph of the function between $x = a$ and $x = b$ with a pencil without lifting the pencil from the paper.

For example, $f(x) = \frac{1}{x}$ is continuous on the intervals $(0, \infty)$ and $(-\infty, 0)$, but it is *not* continuous on $(-\infty, \infty)$. In fact, $f$ is not continuous on any interval containing zero. However, it is common to say that $f(x) = \frac{1}{x}$ is continuous on its domain $D = (-\infty, 0) \cup (0, \infty)$. Notice that the above understanding about continuity does not hold for $f(x) = \frac{1}{x}$ on $D$, but does on each interval that makes up $D$. In calculus, you may find that there are conventions about continuity that seem to violate our above understanding.

We give the following theorem without proof.

- **THEOREM 1**  Every polynomial function is continuous on $(-\infty, \infty)$.

In light of the above theorem we make the following definition.

$f(x) = \frac{1}{x}$
$[-10, 10]$ by $[-10, 10]$
Figure 1.9.1

# 1.9 Continuity and End Behavior

- **DEFINITION**  A function is said to be continuous if and only if it is continuous on the interval $(-\infty, \infty)$.

Thus, we say that polynomial functions are continuous functions. In the next example we consider a function with breaks that are different than the break in $f(x) = \frac{1}{x}$ at $x = 0$.

- **EXAMPLE 1:** Determine the largest intervals on which the function $f(x) = \text{INT}(x)$ is continuous.

**SOLUTION:** Figure 1.9.2 shows the graph of $f(x) = \text{INT}(x)$ obtained in Section 1.3. Examine the graph on the interval $[3, 4)$. The open circle at the point $(3, 2)$ indicates that $f(3) \neq 2$, and the closed circle at the point $(3, 3)$ indicates that $f(3) = 3$. For all values of $x$ in the interval $[3, 4)$, $f(x) = 3$. Similarly, for all values of $x$ in the interval $[n, n+1)$ where $n$ is an integer, $f(x) = n$. We could illustrate that $f(n) = n$ and $f(n) \neq n - 1$ by using a closed circle at the point $(n, n)$ and an open circle at the point $(n, n-1)$. Several are illustrated in Figure 1.9.2.

The graph has a break at each integer value of $x$. The function $f(x) = \text{INT}(x)$ is continuous on each interval $[n, n+1)$ where $n$ is an integer. On any interval that contains an integer as an *interior* point the function is not continuous. For example, $f$ is not continuous on the interval $[-2.1, -1.5]$ or on the interval $[0, 2.5]$. Thus, the largest intervals on which $f$ is continuous are of the form $[n, n+1)$, where $n$ is an integer.

Notice that the union of the intervals on which $f(x) = \text{INT}(x)$ is continuous is $(-\infty, \infty)$, the domain of $f$, but even in calculus we do *not* consider this function continuous on $(-\infty, \infty)$.

The function $y = \frac{1}{x}$ is not defined at $x = 0$, and its graph has a break in it at $x = 0$. We say that $y = \frac{1}{x}$ has a *point of discontinuity* at $x = 0$. This discontinuity is sometimes called an "infinite" discontinuity because of the behavior of the function near $x = 0$.

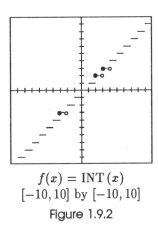

$f(x) = \text{INT}(x)$
$[-10, 10]$ by $[-10, 10]$

Figure 1.9.2

The graph of $y = \text{INT}(x)$ has a break at every integer value of $x$. These points are also called points of discontinuity. This type of discontinuity is sometimes called a "jump" discontinuity because of the type of break at integer values of $x$.

Another type of point of discontinuity that can occur is illustrated by the following piecewise defined function whose graph appears in Figure 1.9.3:

$$f(x) = \begin{cases} \frac{x^2-1}{x-1}, & \text{for } x \neq 1, \\ 3, & \text{for } x = 1. \end{cases}$$

The graph of $f$ appears to have a hole punched out at $(1,2)$ and the point moved up to $(1,3)$. This is also a point of discontinuity and is sometimes called a "removable" discontinuity because the break in the graph can be repaired by replacing the missing point.

We have the following understanding about breaks in graphs of functions in this textbook.

- **Understanding** The function $f$ has a **point of discontinuity** or is **discontinuous** at $x = a$ if one of the following three conditions hold:

    1. The function is not defined at $x = a$. For example, $f(x) = \frac{1}{x}$ at $x = 0$ (Figure 1.9.1)
    2. The graph of the function has a break at $x = a$. For example, $f(x) = \text{INT}(x)$ at any integer value of $x$ (Figure 1.9.2).
    3. The graph of the function has a hole in it at $x = a$. For example,
    $$f(x) = \begin{cases} \frac{x^2-1}{x-1}, & \text{for } x \neq 1, \\ 3, & \text{for } x = 1, \end{cases}$$
    at $x = 1$ (Figure 1.9.3).

With the above understanding, $y = \frac{1}{x}$ has exactly one point of discontinuity and it is at $x = 0$. Similarly, the function whose graph appears in Figure 1.9.3 has only one point of discontinuity, and it is at $x = 2$. However, $y = \text{INT}(x)$ has a point of discontinuity at and only at each integer value of $x$.

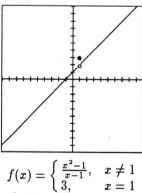

$$f(x) = \begin{cases} \frac{x^2-1}{x-1}, & x \neq 1 \\ 3, & x = 1 \end{cases}$$
$[-10, 10]$ by $[-10, 10]$

Figure 1.9.3

### 1.9 Continuity and End Behavior

Piecewise-defined functions can have breaks similar to the greatest integer function.

- **EXAMPLE 2:** Determine the domain, the range, the points of discontinuity, and the largest intervals on which each function is continuous.

  (a) $f(x) = \begin{cases} 2x+4 & \text{for } x \leq -1 \\ x^2 & \text{for } x > -1 \end{cases}$

  (b) $g(x) = \sqrt{x+1}$

SOLUTION:

(a) Figure 1.9.4 gives a complete graph of $f$. The domain and the range are the set of all real numbers. The graph has a break, or jump, at $x = -1$. Notice $f$ is continuous on the intervals $(-\infty, -1]$ and $(-1, \infty)$, and $f$ is *discontinuous* at $x = -1$.

(b) Figure 1.9.5 gives a complete graph of $g$. The domain of $g$ is $[-1, \infty)$, and the range of $g$ is $[0, \infty)$. Notice that $g$ is continuous on the interval $[-1, \infty)$, and $g$ is discontinuous at each value of $x$ less than $-1$, because $g$ is not defined at those values. We leave the issue of continuity or discontinuity at endpoints of intervals like $x = -1$ to calculus. In this textbook, we will *not* say that $g$ has a point of discontinuity at $x = -1$. •

By Theorem 1, polynomial functions are continuous on the set of all real numbers, that is, on $(-\infty, \infty)$. Thus, by the definition following Theorem 1, polynomial functions are called continuous functions. For example, linear functions $f(x) = ax + b$ and quadratic functions $g(x) = ax^2 + bx + c$ are continuous functions. One important consequence of continuity is the Intermediate Value Property.

- **Intermediate Value Property** If a function $f$ is continuous on $[a, b]$, then $f$ assumes every value between $f(a)$ and $f(b)$. In other words, if $L$ is any number between $f(a)$ and $f(b)$, then there is a value $c$ between $a$ and $b$ such that $f(c) = L$ (Figure 1.9.6).

$f(x) = \begin{cases} 2x+4 & \text{for } x \leq -1 \\ x^2 & \text{for } x > -1 \end{cases}$

Figure 1.9.4

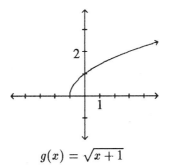

$g(x) = \sqrt{x+1}$

Figure 1.9.5

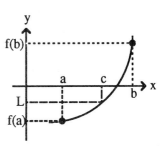

Figure 1.9.6

- **EXAMPLE 3:** Let $f(x) = 3x - 4$. Determine $c$ on $[-10, 10]$ so that $f(c) = 15$.

  **SOLUTION:** Notice that $f(-10) = -34$, $f(10) = 26$, and $f$ is continuous on $[-10, 10]$. Furthermore, $-34 < 15 < 26$. Thus the Intermediate Value Property assures us that there must be a $c$ satisfying $-10 \le c \le 10$ with $f(c) = 15$. We find $c$ by solving the following equation:

  $$f(c) = 15$$
  $$3c - 4 = 15$$
  $$3c = 19$$
  $$c = \frac{19}{3}.$$

  Thus $\frac{19}{3}$ is a number between $-10$ and $10$ with $f(\frac{19}{3}) = 15$ (Figure 1.9.7).

- **EXAMPLE 4:** Let $f(x) = x^2 + 3x - 4$. Determine $c$ on $[-2, 3]$ so that $f(c) = -3$.

  **SOLUTION:** Notice that $f(-2) = -6$, $f(3) = 14$, and $f$ is continuous on $[-2, 3]$. Furthermore, $-6 < -3 < 14$. Thus, the Intermediate Value Property assures us that there is a $c$ between $-2$ and $3$ for which $f(c) = -3$. To determine $c$ we solve the following quadratic equation:

  $$f(c) = -3$$
  $$c^2 + 3c - 4 = -3$$
  $$c^2 + 3c - 1 = 0.$$

  The quadratic formula gives $c = \frac{-3 \pm \sqrt{13}}{2}$. Accurate to tenths, the two possible values for $c$ are $-3.3$ and $0.3$. Only $0.3$ is between $-2$ and $3$. Thus, $\frac{-3+\sqrt{13}}{2}$ is a number between $-2$ and $3$ with $f\left(\frac{-3+\sqrt{13}}{2}\right) = -3$ (Figure 1.9.8).

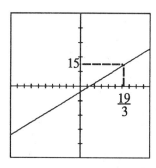

$f(x) = 3x - 4$
$[-10, 10]$ by $[-50, 50]$
Figure 1.9.7

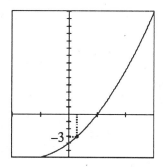

$f(x) = x^2 + 3x - 4$
$[-2, 3]$ by $[-6, 14]$
Figure 1.9.8

## 1.9 Continuity and End Behavior

The Intermediate Value Property is often the basis for approximating zeros of continuous functions. For example, consider the function $f(x) = x^5 + 2x - 1$. Because $f$ is continuous on $(-\infty, \infty)$, it is continuous on every interval. In particular, $f$ is continuous on $[0,1]$, and you can check that $f(0) = -1$ and $f(1) = 2$. Applying the Intermediate Value Property to $f$ on $[0,1]$ we can conclude that $f(c) = 0$ for some $c$ between 0 and 1, because 0 is between $-1$ and 2 (Figure 1.9.9). Next, we compute the value of $f$ at the midpoint 0.5 of $[0,1]$. Because $f(0.5)$ is positive, the Intermediate Value Property applied to the interval $[0, 0.5]$ shows that $f$ has a zero between 0 and 0.5 (Figure 1.9.10). If we compute the value of $f$ at the midpoint 0.25 of $[0, 0.5]$, we will find that this zero of $f$ lies in $[0.25, 0.5]$. By repeated application of the Intermediate Value Property, we can find this zero as accurately as we like. This process, called the **bisection method**, is a numerical zoom-in procedure.

Computer and graphing utility techniques to find zeros and local maximum and minimum values of functions are based on the Intermediate Value Property. We have seen how to find zeros and local maximum and minimum values of functions using zoom-in. Before the easy availability of technology it was necessary to do much of this work by brute force. The power of modern technology allows today's precalculus students to investigate functions that were not accessible to precalculus students in the past.

### End Behavior Using Zoom-Out

- **DEFINITION**    The behavior of a function for values of $x$ large in absolute value is called the **end behavior** of the function.

For purposes of drawing graphs and understanding the general behavior of functions, it is often enough to know that the end behavior of a given function is very much like the end behavior of some simpler function.

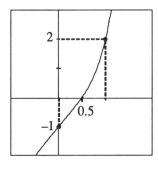

$f(x) = x^5 + 2x - 1$
$[-1, 2]$ by $[-2, 3]$

**Figure 1.9.9**

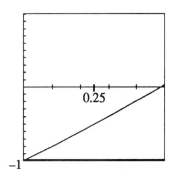

$f(x) = x^5 + 2x - 1$
$[0, 0.5]$ by $[-0.1, 0.1]$

**Figure 1.9.10**

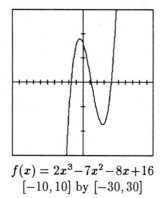

$f(x) = 2x^3 - 7x^2 - 8x + 16$
$[-10, 10]$ by $[-30, 30]$
Figure 1.9.11

- **EXAMPLE 5:** Show that the end behavior of $f(x) = 2x^3 - 7x^2 - 8x + 16$ is the same as the end behavior of $g(x) = 2x^3$.

  **SOLUTION:** The graph of $f$ in the viewing rectangle $[-10, 10]$ by $[-30, 30]$ is a complete graph and suggests the behavior of $f$ for values of $x$ large in absolute value (Figure 1.9.11). It appears the values of $f$ are positive and increasing if the values of $x$ are greater than or equal to 4 and increasing. We say $f(x)$ approaches infinity as $x$ approaches infinity, and we write $f(x) \to \infty$ as $x \to \infty$. Figure 1.9.11 also suggests the values of $f$ are negative and decreasing if the values of $x$ are less than or equal to $-2$ and decreasing. In this case we say $f(x)$ approaches negative infinity as $x$ approaches negative infinity, and we write $f(x) \to -\infty$ as $x \to -\infty$.

  Table 1 demonstrates that the end behavior of $f(x) = 2x^3 - 7x^2 - 8x + 16$ is similar to the end behavior of $y = 2x^3$. Compare the values of $2x^3$ and $f(x)$ for $x$ equal to 20, 50, 100, $-20$, $-50$, and $-100$.

  Notice as the absolute value of $x$ increases, the values in the third column of Table 1 become negligible in comparison to the values in the second column. This means we can approximate the value of $f$ by the value of $2x^3$ when $x$ is far away from zero. Another way

| $x$ | $2x^3$ | $-7x^2 - 8x + 16$ | $f(x)$ | $\frac{f(x)}{2x^3}$ |
|---|---|---|---|---|
| 20 | 16,000 | $-2,944$ | 13,056 | 0.816 |
| 50 | 250,000 | $-17,884$ | 232,116 | 0.928 |
| 100 | 2,000,000 | $-70,884$ | 1,929,216 | 0.965 |
| $-20$ | $-16,000$ | $-2,624$ | $-18,624$ | 1.164 |
| $-50$ | $-250,000$ | $-17,084$ | $-267,084$ | 1.068 |
| $-100$ | $-2,000,000$ | $-69,184$ | $-2,069,184$ | 1.035 |

Table 1

to arrive at the same conclusion is to divide $f(x)$ by $2x^3$.

$$\frac{f(x)}{2x^3} = 1 - \frac{7}{2x} - \frac{4}{x^2} + \frac{8}{x^3}$$

Now, we can see that $\frac{f(x)}{2x^3}$ gets closer and closer to 1 as the absolute value of $x$ becomes larger and larger, because each term of the right-hand side of the above equation after the first term approaches zero as the absolute value of $x$ approaches infinity. We can also say $\frac{f(x)}{2x^3} \to 1$ as $|x| \to \infty$. The last column of Table 1 gives the value of $\frac{f(x)}{2x^3}$ for some values of $x$. Thus, $f(x)$ is approximately equal to $2x^3$ for values of $x$ large in absolute value.

We describe the *end behavior* of the function $f(x) = 2x^3 - 7x^2 - 8x + 16$ of Example 5 as follows: $f(x) \to \infty$ as $x \to \infty$ and $f(x) \to -\infty$ as $x \to -\infty$. We call $2x^3$ an *end behavior model* of $f$. It can be shown that $2x^2 + 3$ is another end behavior model of $f$ because $\frac{f(x)}{2x^2+3} \to 1$ as $|x| \to \infty$. Thus, an end behavior model of a function is not unique. We try to choose an end behavior model that is as simple as possible.

---

• **DEFINITION**  The function $g$ is an **end behavior model** for the function $f$ if and only if

$$\frac{f(x)}{g(x)} \to 1 \quad \text{as} \quad |x| \to \infty.$$

Information about the end behavior of a function can be obtained by drawing its graph in very large viewing rectangles. For example, the graph of $f(x) = 2x^3 - 7x^2 - 8x + 16$ in the viewing rectangle $[-30, 30]$ by $[-3000, 3000]$ is shown in Figure 1.9.12. This graph appears to be much like the graph of $y = 2x^3$ in the same viewing rectangle (Figure 1.9.13).

The graphs in Figures 1.9.12 and 1.9.13 strongly suggest that $y = 2x^3$ is an end behavior model of $f(x) = 2x^3 - 7x^2 - 8x + 16$. This confirms the numerical evidence in Table 1 showing that the value of $\frac{f(x)}{2x^3}$ is approximately 1 for $x$ large in absolute value.

The notation introduced in Example 5 is helpful when describing the end behavior of certain functions.

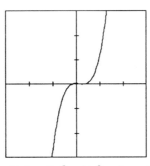

$f(x) = 2x^3 - 7x^2 - 8x + 16$
$[-30, 30]$ by $[-3000, 3000]$
Figure 1.9.12

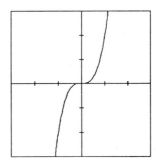

$y = 2x^3$
$[-30, 30]$ by $[-3000, 3000]$
Figure 1.9.13

- **DEFINITION**   Let $f$ be a function.
    1. If the values of $f$ are positive and increase without bound for positive values of $x$ that increase without bound, we say that **$f(x)$ approaches infinity as $x$ approaches infinity**, and we write $f(x) \to \infty$ as $x \to \infty$.
    2. If the values of $f$ are negative and the values of $|f|$ increase without bound for positive values of $x$ that increase without bound, we say that **$f(x)$ approaches negative infinity as $x$ approaches infinity**, and we write $f(x) \to -\infty$ as $x \to \infty$.
    3. If the values of $f$ are positive and increase without bound for negative values of $x$ for which $|x|$ increases without bound, we say that **$f(x)$ approaches infinity as $x$ approaches negative infinity**, and we write $f(x) \to \infty$ as $x \to -\infty$.
    4. If the values of $f$ are negative and the values of $|f|$ increase without bound for negative values of $x$ for which $|x|$ increases without bound, we say that **$f(x)$ approaches negative infinity as $x$ approaches negative infinity**, and we write $f(x) \to -\infty$ as $x \to -\infty$.

- **EXAMPLE 6:** Describe the end behavior of $f(x) = x^4$ and $g(x) = -x^4$.

    SOLUTION: The graph of $f(x) = x^4$ in Figure 1.9.14 indicates that $f(x) \to \infty$ as $x \to \infty$ and that $f(x) \to \infty$ as $x \to -\infty$. The graph of $g(x) = -x^4$ in Figure 1.9.15 indicates that $g(x) \to -\infty$ as $x \to \infty$ and $g(x) \to -\infty$ as $x \to -\infty$.

Notice that the end behavior of $y = x^2$ and $y = x^4$ is the same. However, $y = x^2$ is not an end behavior model for $y = x^4$ because $\frac{x^4}{x^2}$ does *not* approach 1 as $|x|$ approaches infinity.

The process of determining end behavior of a function graphically requires viewing the graph of a function in a nested sequence of viewing rectangles each contained in the next.

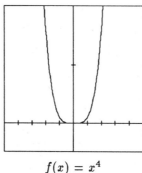

$f(x) = x^4$
$[-5, 5]$ by $[-5, 20]$

Figure 1.9.14

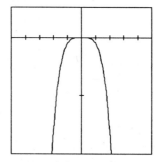

$g(x) = -x^4$
$[-5, 5]$ by $[-20, 5]$

Figure 1.9.15

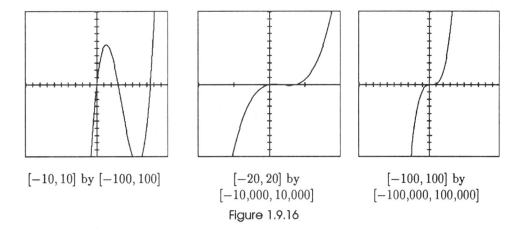

[−10, 10] by [−100, 100]    [−20, 20] by [−10,000, 10,000]    [−100, 100] by [−100,000, 100,000]

Figure 1.9.16

Figure 1.9.16 shows three possible viewing rectangles beyond the standard viewing rectangle for a certain function. Notice that the end behavior appears to be like $y = x^3$. The process of viewing a graph in increasingly larger viewing rectangles is called **zoom-out**.

The viewing rectangle of Figure 1.9.12 can be viewed as the last step in a *zoom-out* process applied to $f(x) = 2x^3 - 7x^2 - 8x + 16$. This graph gives strong evidence that $y = 2x^3$ is an end behavior model for $f$. Generally, we only report the last viewing rectangle used when we zoom out.

- EXAMPLE 7: Graphically show that $g(x) = -0.5x^4$ is an end behavior model of $f(x) = -0.5x^4 + 5x^3 - 10x + 1$. Determine the end behavior of $f$.

SOLUTION: The graphs of $f$ and $g$ in $[-200, 200]$ by $[-100,000, 100,000]$ (Figure 1.9.17) give strong geometric evidence that these functions have the same end behavior. It can be shown that $\frac{f(x)}{g(x)} \to 1$ as $|x| \to \infty$. Notice that the end behavior of $f$ is $f(x) \to -\infty$ as $|x| \to \infty$. •

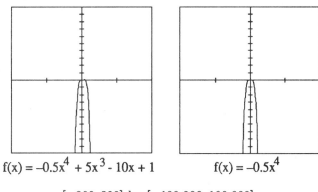

$f(x) = -0.5x^4 + 5x^3 - 10x + 1$    $f(x) = -0.5x^4$

[−200, 200] by [−100,000, 100,000]

Figure 1.9.17

You might be tempted to say that $y = -x^4$, or $y = -2x^4$ appear to be end behavior models for the function $f(x) = -0.5x^4+5x^3-10x+1$ of Example 7 because their graphs seem to overlap in some viewing rectangles and they have the same end behavior. However, in the viewing rectangles $[-40, 40]$ by $[-1,000,000, 100,000,000]$ the graphs of $f$, $y = -x^4$, and $y = -2x^4$ are clearly separated. Notice also that $\frac{f(x)}{-x^4} \to 0.5$ and $\frac{f(x)}{-2x^4} \to 0.25$ as $|x| \to \infty$. Thus, both numerically and graphically, it is possible to determine that $y = -0.5x^4$ is an end behavior model for $f(x) = -0.5x^4 + 5x^3 - 10x + 1$ and that $y = -x^4$ or $y = -2x^4$ are *not* end behavior models for $f$.

It is also true that any polynomial function of degree 4 with the coefficient of $x^4$ negative will have the same end behavior as the function $f$ of Example 7. However, only those polynomial functions of degree 4 with leading coefficient $-0.5$ are *end behavior models* for $f$. (Why?) The term with largest exponent is the simplest end behavior model of a polynomial function.

**SUMMARY**   The polynomial function $f(x) = a_n x^n + a_{n-1} x^{n-1} + \cdots + a_0$ of degree $n \geq 1$ has end behavior model $y = a_n x^n$. This means the values of $a_n x^n$ give good approximations to the values of $f$ for values of $x$ large in absolute value. Furthermore, $f$ has the following end behavior:

(a) If $a_n > 0$ and $n$ is even, then $f(x) \to \infty$ as $|x| \to \infty$.
(b) If $a_n > 0$ and $n$ is odd, then $f(x) \to \infty$ as $x \to \infty$ and $f(x) \to -\infty$ as $x \to -\infty$.
(c) If $a_n < 0$ and $n$ is even, then $f(x) \to -\infty$ as $|x| \to \infty$.
(d) If $a_n < 0$ and $n$ is odd, then $f(x) \to -\infty$ as $x \to \infty$ and $f(x) \to \infty$ as $x \to -\infty$.

- **EXAMPLE 8:** Find the maximum possible area of a rectangle that has its base on the $x$-axis and its upper two vertices on the graph of the equation $y = 4 - x^2$ (Figure 1.9.18).

**SOLUTION:** Let the coordinates of the upper right-hand corner of the rectangle be $(x, y)$. Then, the width of the rectangle is $2x$, and the height of the rectangle is $y$. Because the point $(x, y)$ lies on the graph of $y = 4 - x^2$, the height $y$ of the rectangle can also be

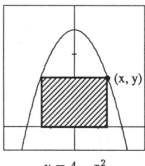

$y = 4 - x^2$
$[-3, 3]$ by $[-1, 3]$

Figure 1.9.18

### 1.9 Continuity and End Behavior

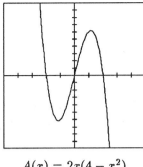

$A(x) = 2x(4 - x^2)$
$[-5, 5]$ by $[-10, 10]$
Figure 1.9.19

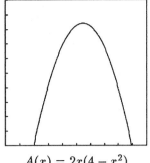

$A(x) = 2x(4 - x^2)$
$[1.1, 1.2]$ by $[6.15, 6.16]$
Figure 1.9.20

expressed as $4 - x^2$. Thus, the area of the rectangle can be written as a function of $x$: $A(x) = 2xy = 2x(4 - x^2)$.

From Figure 1.9.18 we can see that the only values of $x$ that make sense in the problem situation are $0 < x < 2$. (Why?) A complete graph of $A$ is given in Figure 1.9.19. We can see from this graph that $A$ has a local maximum value between $x = 0$ and $x = 2$. Because the algebraic representation of $A$ represents the problem situation only for values of $x$ between 0 and 2, this local maximum value of $A$ must be the solution to the problem. Using zoom-in we can determine that the value of $x$, with error at most 0.01, producing the largest value for the area of the rectangle is 1.155 (Figure 1.9.20). The area of the corresponding rectangle with maximum area is 6.1585.

## EXERCISES 1-9

In Exercises 1–8, determine the domain, range, and the values of $x$ for which the function given by the graph is discontinuous. Assume each graph is complete.

1.

2.

3.
4.
5.
6.
7.
8.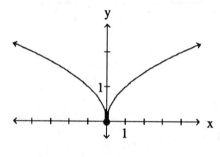

Determine the domain, range, and the points of discontinuity of each function.

9. $f(x) = \text{INT}\,(x+2)$

10. $g(x) = 2x^2 - 3x + 6$

11. $V(x) = 4x^3 - 200x^2 + 30x$

12. $f(x) = \begin{cases} |x-2| & \text{for } x < -1 \\ 2x & \text{for } x \geq -1 \end{cases}$

13. $y = \begin{cases} -\dfrac{1}{x} & \text{for } x < 0 \\ x^2 + 1 & \text{for } x \geq 0 \end{cases}$

14. $y = \dfrac{2}{x-3}$

15. Overlay the graphs of $f(x) = -5x^3 + 10x^2 - 3x + 15$ and $g(x) = -5x^3$ in each of the viewing rectangles. Which viewing rectangle *best* suggests that $g$ is an end behavior model for $f$?

   (a) $[-10, 10]$ by $[-100, 100]$
   (b) $[-10, 10]$ by $[-1000, 1000]$
   (c) $[-10, 10]$ by $[-3000, 3000]$
   (d) $[-10, 10]$ by $[-5000, 5000]$

(e) $[-100, 100]$ by $[-100{,}000, 100{,}000]$     (f) $[-100, 100]$ by $[-100, 100]$

Graph the polynomial functions $f$ and $g$ in a viewing rectangle that graphically illustrates both that the end behavior of the two functions is the same, and that $g$ is an end behavior model for $f$. That is, choose a viewing rectangle in which the graphs appear to be almost identical.

16. $f(x) = 3x^3 - 2x^2 + x - 20$; $g(x) = 3x^3$
17. $f(x) = -2x^3 - x^2 + 3x + 5$; $g(x) = -2x^3$
18. $f(x) = 3x^4 - 2x^3 + x^2 - x + 5$; $g(x) = 3x^4$
19. $f(x) = -5x^4 + 2x^3 - x + 3$; $g(x) = -5x^4$
20. For the function $f$ in Exercise 18 write $\frac{f(x)}{3x^4}$ in the form $\frac{f(x)}{3x^4} = 1 + r_1(x) + r_2(x) + r_3(x) + r_4(x)$.
    (a) Graph $y = r_1(x)$, $y = r_2(x)$, $y = r_3(x)$, and $y = r_4(x)$ in the same viewing rectangle. Zoom out and investigate what happens to $r_i(x)$ as $x \to \infty$ or as $x \to -\infty$.
    (b) Explain how the result of (a) shows that $y = 3x^4$ is an end behavior model for $f$.
21. For the function $f$ in Exercise 19 write $\frac{f(x)}{-5x^4}$ in the form $\frac{f(x)}{-5x^4} = 1 + r_1(x) + r_2(x) + r_3(x)$.
    (a) Graph $y = r_1(x)$, $y = r_2(x)$, and $y = r_3(x)$ in the same viewing rectangle. Zoom out and investigate what happens to $r_i(x)$ as $x \to \infty$ or as $x \to -\infty$.
    (b) Explain how the result of (a) shows that $y = -5x^4$ is an end behavior model for $f$.
22. Consider the linear function $f(x) = 3x - 12$. Verify that 5 is a number between $f(-10)$ and $f(10)$. Determine a value of $c$ so that $f(c) = 5$ and $-10 \leq c \leq 10$.
23. Consider the linear function $f(x) = 4x - 7$. Verify that 0 is a number between $f(1)$ and $f(2)$. Determine a value of $c$ so that $f(c) = 0$ and $1 \leq c \leq 2$.
24. Consider the quadratic function $f(x) = 2x^2 + 4x - 10$. Verify that 50 is a number between $f(0)$ and $f(10)$. Determine a value of $c$ so that $f(c) = 50$ and $0 \leq c \leq 10$.
25. Consider the linear function $f(x) = ax + b$. Let $L$ be a number between $f(-10)$ and $f(10)$. Determine a value of $c$ so that $f(c) = L$ and $-10 \leq c \leq 10$.

Assume the graph of $y = f(x)$, given in Figure 1.9.21 is complete. Complete each statement.

26. As $x \to \infty$, $f(x) \to$ ?

$[-10, 10]$ by $[-100, 100]$

Figure 1.9.21

27. As $x \to -\infty$, $f(x) \to$ ?

Assume the graph of $y = f(x)$, given in Figure 1.9.22 is complete. Complete each statement.

28. As $x \to \infty$, $f(x) \to$ ?
29. As $x \to -\infty$, $f(x) \to$ ?
*30. As $x \to 3$ through values of $x$ satisfying $x > 3$, $f(x) \to$ ?
*31. As $x \to 3$ through values of $x$ satisfying $x < 3$, $f(x) \to$ ?
32. A rectangle has its base on the $x$-axis and its upper two vertices on the graph of the equation $y = 12 - x^2$. Find the maximum possible area of such a rectangle and the dimensions of the rectangle with maximum area.
33. The total daily revenue of Chris's Cookie Shop is given by the equation $R = x \cdot p$, where $x$ is the number of pounds of cookies sold and $p$ is the price of one pound of cookies. Assume the price per pound of the cookies is given by the supply equation $p = 0.2 + 0.01x - 0.00001x^2$.
    (a) Determine for what positive values of $x$ the revenue function is increasing.
    (b) Draw a complete graph of the revenue function $R$.
    (c) What values of $x$ make sense in the problem situation?
    (d) Determine the maximum possible daily revenue of Chris's Cookie Shop and the number of pounds of cookies she needs to sell to achieve maximum revenue.
34. A box is made by cutting equal squares from the four corners of a 16 inch by 28 inch piece of material. Determine the size of the square that must be cut out to produce a box with maximum volume. What is the corresponding maximum volume?
*35. It is shown in numerical analysis that there is a unique polynomial of degree less than or equal to $n$ that interpolates (the graph goes through) $n+1$ given points. These polynomials are called *Lagrange* polynomials. Such interpolating polynomials are important tools in understanding data generated from many kinds of scientific experiments and statistical analyses. Let $L(x) = 2x^3 - 11x^2 + 12x + 5$ be a Lagrange interpolating polynomial for four data points.
    (a) Draw a complete graph of $y = L(x)$.

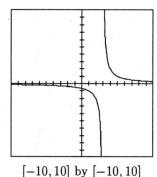

$[-10, 10]$ by $[-10, 10]$

Figure 1.9.22

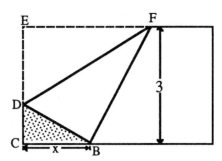

Figure 1.9.23

(b) Suppose an experiment generates the four data points $(-1,-20)$, $(0,5)$, $(1,8)$, and $(2,1)$. Show that $L$ interpolates these four points.

(c) Suppose it is later determined that a fifth data point $(1.5, 6)$ is generated by the experiment. What would have been the predicted outcome if $x = 1.5$ was used in the Lagrange interpolating polynomial $L(x)$?

36. Suppose one corner $E$ of a long, flat, rectangular piece of material 3 feet wide is folded over along a line $DF$ until the corner $E$ touches the opposite edge at $B$ (Figure 1.9.23). Let $A$ be the area of the shaded triangle $BCD$. Let $x$ be the distance between the vertex $C$ and point $B$. It can be shown that the area of triangle $BCD$ is $A = \frac{3}{4}x - \frac{1}{12}x^3$. What is the largest possible area $A$?

37. In Exercise 36, prove that $A = \frac{9x - x^3}{12}$. *Hint*: Observe that triangles $DEF$ and $BDF$ are congruent. Then, if the distance from $D$ to $C$ is $y$, the Pythagorean theorem can be used to determine $y$ as a function of $x$. Finally, substitution for $y$ in the area formula $A = \frac{1}{2}xy$ will give the desired result.

*38. Consider the function $f(x) = 2x^3 - 7x^2 - 8x + 16$ used in Table 1. How large should the absolute value of $x$ be so that

(a) $\left| \frac{f(x)}{2x^3} - 1 \right| < 0.1$?  (b) $\left| \frac{f(x)}{2x^3} - 1 \right| < 0.01$?  (c) $\left| \frac{f(x)}{2x^3} - 1 \right| < 0.001$?

## • KEY TERMS

Listed below are the key terms, vocabulary, and concepts in this chapter. You should be familiar with their definitions and meanings as well as be able to illustrate each of them.

| | |
|---|---|
| Absolute value function | Complete graph |
| Accuracy agreement | Completing the square |
| Algebraic representation | Constant term |
| Axes | Continuity |
| Cartesian coordinate system | Cubic equation |
| Coefficient | Decreasing function |

Degree of a polynomial
Discriminant
Domain
End behavior
Equivalent equations
Equivalent inequalities
Error
Folding symmetry
Function
Geometric representation
Graph of an equation
Graph of a function
Graph of a problem situation
Graph of a relation
Graphing utility
Greater than
Greater than or equal to
Greatest integer function
Horizontal shift
Increasing function
Inequality
Integer
Intermediate Value Property
Interval notation
Irrational number
Leading coefficient
Less than
Less than or equal to
Linear equation
Linear function
Linear polynomial
Local extremum value
Local maximum value
Local minimum value
Model
Nondecreasing function

Nonincreasing function
Parabola
Piecewise-defined function
Point of discontinuity
Polynomial
Polynomial equation
Polynomial function
Quadrant
Quadratic equation
Quadratic formula
Quadratic function
Quadratic polynomial
Range
Rational number
Real number
Real number line
Reflection through the $x$-axis
Relation
Sign chart
Simultaneous solutions
Solution to an equation
Solution to an inequality
Solution to a system of equations
Square root function
Standard viewing rectangle
Substitution method
Understood domain
Vertex of a parabola
Vertical line test
Vertical shift
Vertical stretch
Vertical shrink
Viewing rectangle
$x$-intercept
Zoom-in

# REVIEW EXERCISES

This section contains representative problems from each of the previous six sections. You should use these problems to test your understanding of the material covered in this chapter.

1. Graph the points on the same coordinate plane. Identify the quadrant containing each point.

    (a) $(1, 4)$      (b) $(0, -4)$      (c) $(-\frac{1}{3}, -2)$      (d) $(-3, \frac{1}{2})$

2. The point $(-3, T)$ is on the graph of $y = 5x^2 - 8$. What is $T$?
3. The point $(a, 50)$ is on the graph of $y = 2x^2$. What is $a$?
4. (a) Complete Table 1 for the graph of the relation in Figure R.1.
    (b) What are the domain and range of this relation?

Draw a complete graph of each equation. Do not use a graphing utility. Choose an appropriate scale for each axis.

5. $y - 5 = \frac{1}{2}x$      6. $x - y = 6$

7. $y = 200x - 10$      8. $y = 2x^2 + 1000$

Draw a complete graph of each equation.

9. $D = 3N^2 - 30$      10. $V = L^2(L - 20)$

11. What are the domain and range of the equation given in Exercise 10?
12. Compute each value.

    (a) $|-2.5|$      (b) $|2 - 6|$      (c) $|-(5-2)|$

Draw a complete graph of each equation and specify its domain and range. Give an equation of the line of symmetry and indicate which equations define $y$ as a function of $x$.

13. $y = 5 - |x|$      14. $y = |x - 3|$

Table 1

Figure R.1

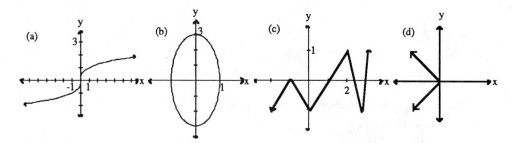

Figure R.2

15. Which of the graphs in Figure R.2 are graphs of functions?
    Use a graph to determine whether each relation is a function.
16. $y = |x - 1|$
17. $y = \sin x + 3$
18. (a) Let $f(x) = x^2 + 2$. Find $f(0)$, $f(1)$, $f(3)$, $f(-5)$, $f(t)$, $f(-t)$, $f\left(-\frac{1}{t}\right)$, $f(a+h)$, and $\frac{f(a+h)-f(a)}{h}$.
    (b) Let $g(x) = \frac{1}{2-x}$. Find $g(2)$, $g(0)$, $g(-2)$, $g(a)$, $g\left(\frac{1}{a}\right)$ and $g(a+h)$.
19. Consider the graph of the function $y = f(x)$ given in Figure R.3. Estimate the indicated values.
    (a) $f(0)$, $f(-1)$, $f(2)$
    (b) $x$ if $f(x) = 0$
    (c) $x$ if $f(x) = 2$
20. Choose a viewing rectangle [L, R] by [B, T] that will include all of the indicated points.
    (a) $(-9, 12)$, $(2, -8)$ and $(3, 20)$
    (b) $(20, -11)$, $(18, 156)$ and $(-11, 2)$
21. Which of the following viewing rectangles gives the best complete graph of $y = x^2 - 30x - 100$?
    (a) $[-10, 5]$ by $[-10, 5]$
    (b) $[-10, 10]$ by $[-10, 10]$

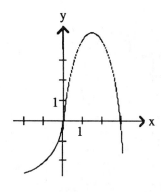

Figure R.3

(c) $[25, 50]$ by $[-500, 1000]$      (d) $[-10, 50]$ by $[-400, 800]$
(e) $[-100, 100]$ by $[-10, 10]$

22. Which of the following viewing rectangles gives the best complete graph of $y = 200 - 10x - x^3$?
    (a) $[-10, 10]$ by $[-10, 10]$      (b) $[-50, 50]$ by $[-10, 10]$
    (c) $[-10, 10]$ by $[-50, 50]$      (d) $[-10, 10]$ by $[-1000, 1000]$
    (e) $[-100, 100]$ by $[-10, 10]$

Draw a complete graph of each function. Explain why the graph is a complete graph.

23. $g(x) = x^2 - 40$      24. $f(x) = 10x^3 - 400x$      25. $t(x) = \left(\dfrac{x}{4}\right)^2 + \left(\dfrac{200 - x}{4}\right)^2$

26. $y = \sqrt{x - 3}$      27. $y = -2 - \sqrt{x + 3}$      28. $y = \dfrac{1}{2 - x}$

29. Determine the domain and range of the function in Exercise 23.
30. Determine the domain and range of the function in Exercise 24.
31. Determine the domain and range of the function in Exercise 26.
32. Determine a viewing rectangle which gives a complete graph of $f(x) = 50 - 25x + 100x^2 - x^3$. Sketch a complete graph of the function. Do not use a graphing utility. Determine the domain and range.

33. $f(x) = \begin{cases} x + 7, & x \leq -3 \\ x^2 + 1, & x > -3 \end{cases}$      34. $f(x) = \begin{cases} 4 - x^2, & x < 1 \\ \sqrt{x + 2}, & x \geq 1 \end{cases}$

35. A family takes a 450-mile trip. Let $R$ be their average speed in miles per hour and $t$ be the total time of the trip in hours.
    (a) Write an equation that shows how the time of the trip depends on the average rate of speed traveled.
    (b) Draw a complete graph of this equation.
    (c) What part of the graph in (b) represents the problem situation, that is, the family trip?
    (d) Use the graph in (b) to determine the total time $t$ if the average speed is 40 mph.

36. The sum of the length and the width of some rectangles is 150 inches. Let $x$ be the length of such a rectangle.
    (a) If the length is 25 inches, then what is the area of the rectangle?
    (b) Show that $A = x(150 - x)$ is an algebraic representation that gives the area of such rectangles in terms of their length.
    (c) Draw a complete graph of the algebraic representation and specify its domain.
    (d) Verify that $(10, 1400)$ is a point on the graph in (c). What meaning do these coordinates have in this problem situation?
    (e) Verify that $(-10, -1600)$ is a point on the graph in (c). What meaning do these coordinates have in this problem situation?
    (f) What values of $x$ make sense in this problem situation? Draw a graph of the problem situation.

37. A school club buys a scientific calculator for $18.25 to use as a raffle prize for a fund raising project. The club charges $0.50 per raffle ticket.
   (a) Determine an algebraic representation that gives the club's profit (or loss) $P$ as a function of the number $n$ of tickets sold.
   (b) Draw a complete graph of the algebraic representation.
   (c) What are the domain and range of the algebraic representation?
   (d) What part of the graph in (b) represents the problem situation? *Hint:* You cannot sell half a ticket!
   (e) Use the graph in (b) to determine the minimum number of tickets that must be sold for the club to realize a profit.

38. Sherri invests $20,000, part at 6% simple interest and the remainder at 7.5% simple interest.
   (a) Write an equation that shows how the total interest Sherri receives in one year depends on the amount she invests at 6%.
   (b) Draw a complete graph of the algebraic representation in (a).
   (c) What portion of the graph in (b) represents the problem situation?
   (d) Use the graph in (b) to estimate the amount invested at each rate if Sherri receives $1300 interest in one year.
   (e) Is there a maximum amount of interest Sherri can receive in one year?

39. A bike manufacturer determines the annual cost $C$ of making $x$ bikes to be $85 per bike plus $75,000 in fixed overhead costs.
   (a) Write the total annual cost as a function of the number of bikes made.
   (b) Draw a complete graph of the algebraic representation in (a).
   (c) What are the domain and range of the function?
   (d) What portion of the graph in (b) represents possible total annual costs of producing the bikes?
   (e) Use the graph in (b) to determine the number of bikes made if the total cost is $143,000.
   (f) If $(A, B)$ is a point on the graph representing the problem situation, then what does $A$ represent in the problem situation?

40. The bike manufacturer in Exercise 39 determines that he can sell each bike for $98.00.
   (a) Write an equation that gives the total annual revenue $R$ as a function of the number of bikes produced and sold.
   (b) Draw a complete graph of the function in (a).
   (c) What are the domain and range of the function?
   (d) What portion of the graph in (b) represents possible total annual revenue from selling the bikes?
   (e) Use the graph in (b) to determine the number of bikes made if the total revenue is $107,800.
   (f) If $(A, B)$ is a point on the graph representing the problem situation, then what does $B$ represent in the problem situation?

# Review Exercises

41. A three-sided rectangular fence with a perimeter of 72 feet is constructed against a barn. Let $x$ be the length of the fence parallel to the barn.
    (a) Write an equation for the area enclosed by the fence as a function of $x$.
    (b) What are the domain and range of the function in (a)?
    (c) Draw a complete graph of the function in (a).
    (d) What values of $x$ make sense in this problem situation? Draw a graph of the problem situation.
    (e) Use the graph in (c) to determine the length of the fence parallel to the barn if the area enclosed is 200 square feet.
    (f) Is there a largest area that can be enclosed by the fence?

42. One hundred twenty feet of fencing is cut into two unequal lengths, and each piece is used to make a square enclosure. Let $x$ be the perimeter of the smaller enclosure.
    (a) Express the total area $A$ of both enclosures as a function of $x$.
    (b) Draw a complete graph of the function in (a).
    (c) What is the domain of the function $A$? What numbers in the domain represent possible lengths of the smaller piece of fence?
    (d) If $A = 6500$ square feet, use the graph in (b) to approximate the smaller length of fencing.

43. A 12-inch by 15-inch poster pasted on a cardboard sheet is surrounded by a region (border) of uniform width $x$.
    (a) Determine the area $A$ of the border as a function of $x$.
    (b) Draw a complete graph of the function $y = A(x)$.
    (c) What is the domain of the function $A$?
    (d) What values in the domain are possible widths? Draw a graph of the problem situation.
    (e) Use a graph to determine the width of the border if its area is 50 square inches.

44. The annual profit $P$ of a baby food manufacturer is determined by the formula $P = R - C$, where $R$ is the total revenue generated from selling $x$ jars of baby food and $C$ is the total cost of making and selling $x$ jars of baby food. Each jar sells for $0.60 and costs $0.45 to make. The fixed costs of making and selling the baby food are $83,000 annually.
    (a) Express the company's annual profit as a function of $x$.
    (b) Draw a complete graph of the function determined in (a).
    (c) How many jars of baby food must be sold for the company to "break even?"
    (d) How many jars of baby food must be sold for the company to make a profit of $10,000?

Use an *algebraic* method to solve each equation. Check your answers by substitution into the original equation.

45. $\frac{1}{3}x + 2 = 3x + \frac{1}{5}$
46. $3(2 - x) - 2(3x + 7) = x + 2$
47. $x^2 - 4x - 21 = 0$
48. $|2 - 3x| = 7$
49. $2x^4 - 4x^2 - 6 = 0$

Solve each equation.

50. $3x^2 - 4x + 2 = 0$
51. $2x^3 - 4x + 7 = 0$
52. $4x^3 + 60x^2 - 103x - 65 = 0$

Use a graphing utility to determine the number of real solutions to each equation.

53. $2x^3 - 4x^2 + 3x - 9 = 0$
54. $x^3 - 31x + 2 = 0$

Use a number line to graph the intervals.

55. $(-2, 4]$
56. $1 \le x < 7$

Complete the table.

|  | Interval Notation | Inequality Notation | English Phrase |
|---|---|---|---|
| 57. | $[-4, 2)$ |  |  |
| 58. |  | $x < 4$ or $15 \le x < 20$ |  |
| 59. |  |  | The set of all real numbers between $-4$ and $1$ or greater than $7$ |

Consider the graphs of $y = f(x)$ and $y = g(x)$ in Figure R.4.

60. Is $x = -1$ a solution to
    (a) $f < g$?
    (b) $f = g$?
    (c) $f > g$?
    (d) Use the graph to solve $f(x) > g(x)$.

Use an *algebraic* method to solve each inequality. Write your answers in interval notation. Check your answers with a graphing utility.

61. $3x + 5 \le x - 4$
62. $\dfrac{x-1}{2} - \dfrac{2x+1}{5} > 1$

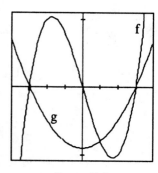

Figure R.4

63. $\dfrac{1}{4} \le \dfrac{4x-1}{12} < \dfrac{11}{6}$
64. $|3x-2| \ge 7$

Use a sign chart to solve each inequality. Write the solution in interval notation. Check your answers with a graphing utility.

65. $x^2 + 20 < x$
66. $10 + x - 2x^2 < 0$
67. $|x-1|(2x-4) \le 0$
68. $x(x+2)(2x-3) > 0$
69. $\dfrac{4x-8}{x+5} < 0$
70. $\dfrac{5-3x}{x-1} \ge 0$

Solve each inequality.

71. $3x^2 - 5x + 1 \ge 0$
72. $x^3 - 4x^2 - 2x + 3 < 0$
73. $2x^2 + x - 6 \le x^3 - x^2 + 4x$
74. The formula for the volume of a cone is $V = \frac{1}{3}\pi r^2 h$. Solve this equation for $r$.
75. A storage box of height 25 cm has a volume of 10,500 cm³. Let $x$ be the length of the box in centimeters and $y$ the width in centimeters.
    (a) Write $y$ as a function of $x$.
    (b) Determine the length if the width is 12 cm.
76. Consider the equation $3x^3 + 8x^2 + 24x - 9 = 0$. It has one real solution. Find a sequence of four viewing rectangles containing the solution. The first viewing rectangle should permit the solution to be read with error of at most 0.1, the second 0.01, the third 0.001, and the last 0.0001.
77. The length of a certain rectangle is five inches greater than its width. Find the possible widths of the rectangle if its perimeter is less than 300 inches.
78. A publishing company prints $x$ books each week at a total cost of $\frac{1}{2}(x-100)^2 + 400$ dollars. The company receives $15.25 for each book sold. Assume all books printed are sold.
    (a) Write an equation expressing the weekly profit of the company as a function of $x$.
    (b) What are the fixed costs?
    (c) Draw a complete graph of the profit function in (a).
    (d) What is the domain of the function in (a)? What portion of the graph in (c) is the graph of the problem situation?
    (e) Use the graph in (c) to determine what level of weekly production the company must maintain to break even.
79. An object is propelled upward from a tower 200 feet above ground with initial velocity 80 feet per second.
    (a) Determine an algebraic representation of the problem situation.
    (b) Draw a complete graph of the algebraic representation.
    (c) What are the domain and range of the algebraic representation?
    (d) What part of the graph in (b) represents the problem situation?
    (e) When will the object be above ground?
    (f) When will the object hit the ground?

(g) What height above ground will the object reach?

(h) At what time will the object reach maximum height above ground?

80. The annual profit $P$ of a candy manufacturer is determined by the formula $P = R-C$, where $R$ is the total revenue generated from selling $x$ pounds of candy and $C$ is the total cost of making and selling $x$ pounds of candy. Each pound of candy sells for $5.15 and costs $1.26 to make. The fixed costs of making and selling the candy are $34,000 annually. How many pounds of candy must be produced and sold for the company to make a profit of at least $42,000 for the year?

   (a) Express the company's annual profit as a function of $x$.

   (b) Write an inequality that gives an algebraic representation of this problem situation.

   (c) Draw a complete graph of each side of the inequality in (b). Draw a graph of the problem situation.

   (d) Solve this problem situation both algebraically and geometrically.

81. Use the distance formula representation of absolute values to solve $|x - 6| - |x| > 0$.

82. A park with dimensions 125 feet by 230 feet is surrounded by a sidewalk of uniform width $x$. Find the possible widths of the sidewalk if the total area of the sidewalk is to be greater than 2900 square feet but less than 3900 square feet.

   (a) Write an equation that gives the area of the sidewalk as a function of $x$.

   (b) Write an inequality that is an algebraic representation of this problem situation.

   (c) Solve the inequality in (b) algebraically, and then determine the solution to the problem situation.

   (d) Use a graphing utility to check your answer to (c).

Determine the intervals on which each function is *increasing* and the intervals on which each function is *decreasing*. Determine all real zeros.

83. $g(x) = (x - 2)(x - 5)$  
84. $f(x) = |x^2 - 5x + 3|$

85. $f(x) = x^3 - x^2 + 2x - 5$  
86. $f(x) = 3 - 2x^3 + 2x^4$

Draw a complete graph of each inequality.

87. $y > x^2 - 3x + 2$  
88. $y < 3x^3 - 2x^2 + x - 1$

89. $y \geq 1 - (x - 2)^2$

Use the sign chart method to *algebraically* solve each inequality. Check your answer with a graphing utility.

90. $x^2(x + 2)(x - 5) > 0$  
91. $x^3 - 9x \leq 0$

Solve each inequality.

92. $x^3 - 2x + 2 < 0$  
93. $2x^3 - x^2 + 3x - 4 \geq 0$

Draw a complete graph of each polynomial function. Specify the viewing rectangle. Determine all real zeros.

94. $y = 5x^3 - 2x - 17$  
95. $T(x) = 5 + 6x + 2x^3 - 2x^4$

Determine the intervals on which each function is *increasing* and the intervals on which each function is *decreasing*.

96. $y = 10 - 3x - x^2$  
97. $f(x) = x^3 - 5x^2 + 2x - 4$

# Review Exercises

Determine all local maximum and minimum values, the corresponding values of $x$ that give these values, and draw a complete graph.

98. $g(x) = |x^2 - 6x + 7|$

99. $y = 2x^3 + 4x^2 - 16$

100. $V(x) = x(22 - 2x)(30 - 2x)$

101. $f(x) = 2x^4 - 8x^2 + x - 3$

102. A dress store is marking all items down 20% for its "spring fling" sale.
    (a) Draw a complete graph that shows the relationship between the original price and the sale price.
    (b) If the sale price of an item is $74.00, use a graph to determine its original price.

103. A rectangle is 2 feet longer than it is wide. If each side is increased by 2 feet, the area of the new rectangle is 288 square feet. Find the dimensions of the original rectangle.

104. A cardboard box with no lid is constructed from a 40 inch by 55 inch piece of cardboard by removing equal squares of side length $x$ from each corner and folding up the sides. Determine $x$ so that the volume of the resulting box is at least 6750 cubic inches.

105. An object is shot straight up (launched) from the top of a building 200 feet tall with initial velocity 70 feet per second. Let $t$ be the number of seconds since the object was launched.
    (a) Determine an algebraic representation that gives the height of the object above ground level as a function of $t$.
    (b) Draw a complete graph of the algebraic representation.
    (c) What portion of the graph of the algebraic representation represents the problem situation?
    (d) Determine the time when the object is 225 feet above the ground.
    (e) Determine the time when the object is more than 225 feet above the ground.

106. A large apartment rental company has 3000 units available, and 2643 are currently rented at $500 per month. A market survey indicates that each $20 decrease in monthly rent will result in ten new tenants.
    (a) Determine a function $R$ which represents the total rental income realized by the company where $x$ is the number of $20 decreases in monthly rent (i.e., if $x = 3$, then the monthly rent is $440). Assume that all rents are adjusted to the new figure.
    (b) Draw a complete graph of the function $y = R(x)$.
    (c) Determine the rent that will yield the rental company the maximum monthly income.

107. The total daily revenue of a cotton candy stand at a state fair is given by the equation $R = x \cdot p$ where $x$ is the number of bags of cotton candy sold daily and $p$ is the price of one bag of cotton candy. Assume the price of the cotton candy is given by the "supply" equation $p = 1.5x - 0.03x^2$.
    (a) Determine an algebraic representation that gives the total daily revenue of the cotton candy stand as a function of $x$.
    (b) Draw a complete graph of the algebraic representation.
    (c) Draw a graph of the problem situation.
    (d) Determine the number of bags of cotton candy to be sold to produce maximum daily revenue. What is the maximum daily revenue?

$[-10, 10]$ by $[-10, 10]$     $[-30, 30]$ by $[-10, 10]$
Figure R.5                     Figure R.6

In Exercises 108–109, determine the domain, range, and the values of $x$ for which the function given by the graph is discontinuous. Assume the graph is complete.

108.

109.
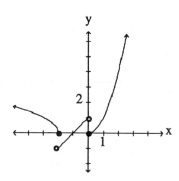

Determine the domain, range, and the points of discontinuity of each function.

110. $V(x) = 5x^3 - 25x^2 + 30x + 9$

111. $g(x) = \begin{cases} x^2 - 1, & x < 2 \\ \dfrac{3x}{2}, & x \geq 2 \end{cases}$

112. $f(x) = \text{INT}(x + 2)$

   Assume the graph of $y = f(x)$ in Figure R.5 is complete.

113. $y = \dfrac{1}{x + 5}$

   Assume the graph of $y = f(x)$ in Figure R.6 is complete.

114. As $x \to \infty$, $f(x) \to ?$

115. As $x \to -\infty$, $f(x) \to ?$

116. As $x \to \infty$, $f(x) \to ?$

117. Graph the polynomial functions $f(x) = 5x^5 + 2x^2 - 4x + 7$ and $g(x) = 5x^5$ in a viewing rectangle that graphically illustrates that the end behavior of the two functions are the same. That is, a viewing rectangle in which the two graphs are nearly identical.

# 2  Trigonometric Functions

## • Introduction

In this chapter we begin the study of trigonometry. More than 2000 years ago the Greeks started the development of this branch of mathematics to make precise measurements of the angles and sides of triangles. This theory is still extremely important today and is extensively used in numerous applications including space science.

Two units of measure for angles are used in this textbook. One is called *degrees* and the other *radians*. Radian measure is introduced in Section 2.5. We will see that an angle with measure 360° also has measure $2\pi$ radians. In this textbook, we almost always give degree measurements in decimal form.

We start with a brief discussion of angles. Then, right triangle trigonometry is studied in the second section of this chapter and applications that involve right triangles are investigated. We define the inverse relation of a relation. Inverses are extremely important in the study of mathematics. The six basic trigonometric functions of acute angles are defined. Then, these definitions are extended to arbitrary angles and complete graphs of the basic trigonometric functions are obtained. Calculators are used to compute values of trigonometric functions, and graphing utilities are used to draw complete graphs of functions of the form $y = af(bx + c) + d$ where $f$ is one of the six basic trigonometric functions.

There are important relationships among the trigonometric functions. These relationships are called identities, and their study begins in this chapter. We solve equations, find local extrema, zeros, asymptotes, and determine the increasing and decreasing behavior of trigonometric functions. Applications with models that involve trigonometric functions are studied. In particular, we begin the study of sinusoidal functions, which are important in applications.

## 2.1 Angle

In this section we define angle and specify the measure of an angle in degrees. We say what is meant by acute angle, right angle, central angle, and an angle in standard position. The terms complementary angles, supplementary angles, quadrantal angle, and coterminal angles are defined.

In geometry, an angle is defined as the union of two rays with a common endpoint. In trigonometry, an angle is defined in terms of a rotation.

- **DEFINITION** Let $\ell_1$ be a ray with initial point $O$. Let $\ell_1$ be rotated counterclockwise about 0 to a new position given by the ray $\ell_2$ in Figure 2.1.1. This determines the **angle** with **initial side** $\ell_1$, **terminal side** $\ell_2$, and the rotation specified by Figure 2.1.1. The amount of rotation is the **measure** of the angle. The point $O$ is called the **vertex** of the angle. •

In trigonometry the measure of an angle can be positive or negative. If the rotation of $\ell_1$ to $\ell_2$ is counterclockwise, then the measure of the angle is positive (Figure 2.1.2). If the rotation of $\ell_1$ to $\ell_2$ is clockwise, then the measure of the angle is negative. The measure of an angle determined by one complete counterclockwise rotation is defined to be 360 degrees. We use the notation 360° to stand for 360 degrees. Notice that the initial side and terminal side of a 360° angle coincide (Figure 2.1.3). The measure of an angle given by one-half of a complete counterclockwise rotation is 180° (Figure 2.1.3). (Why?) An angle that is formed by $\frac{1}{360}$ of a complete rotation counterclockwise has measure one **degree**. A **minute** is $\frac{1}{60}$ of a degree, and a **second** is $\frac{1}{60}$ of a minute. Because of the accuracy of calculators, angles are now commonly measured in degrees in decimal form rather than the historical use of degrees, minutes and seconds.

- **DEFINITION** An angle is said to be **acute** if its measure is between 0° and 90°. An angle with measure 90° is called a **right angle**. •

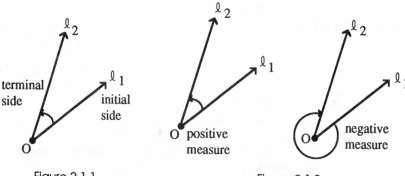

Figure 2.1.1  Figure 2.1.2

## 2.1 Angle

Figure 2.1.3

- **DEFINITION**  An angle in the coordinate plane is said to be in **standard position** if the vertex is at the origin and the initial side is along the positive $x$-axis (Figure 2.1.4).

  - **EXAMPLE 1:** Determine the measure and draw the angle specified by each rotation in standard position.

    (a) $\frac{3}{4}$ rotation, counterclockwise      (b) $\frac{1}{2}$ rotation, clockwise

    (c) 1.5 rotation, counterclockwise      (d) $\frac{2}{3}$ rotation, clockwise

    **SOLUTION:** The measure of the angles in (a)–(d) are $270°$, $-180°$, $540°$, and $-240°$, respectively. The angles are drawn in Figure 2.1.5.

    Notice that the terminal side of the angles in (a)–(c) of Example 1 lie along a coordinate axis. An angle in standard position with terminal side along a coordinate axis is called a **quadrantal angle**. An angle in standard position is said to be in a given quadrant if its terminal side lies in that quadrant.

  - **EXAMPLE 2:** Determine the quadrant of the following angles in standard position with specified measure.

    (a) $2268°$      (b) $-1008°$      (c) $630°$

Figure 2.1.4

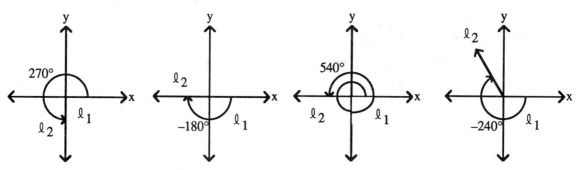

Figure 2.1.5

SOLUTION:

(a) One complete rotation is 360°. Now, $\frac{2268}{360} = 6.3$. Thus, 2268° is 6.3 complete counterclockwise rotations. The positive $y$-axis is 0.25 of a counterclockwise rotation, and the negative $x$-axis is 0.5 of a counterclockwise rotation. It follows that 0.3 of a rotation must be in the second quadrant. Therefore, 2268° is in the second quadrant.

(b) Now, $\frac{-1008}{360} = -2.8$. Thus, $-1008°$ is 2.8 complete clockwise rotations. In a clockwise direction, 0.8 of a rotation is in the first quadrant. (Why?) Thus, $-1008°$ is in the first quadrant.

(c) $\frac{630}{360} = 1.75$. Thus, the terminal side of 630° is along the negative $y$-axis. Therefore, 630° is a quadrantal angle and does *not* lie in any quadrant. •

Specifying the initial and terminal side of an angle does *not* determine a unique angle. For example, angles in standard position with measure 60° or 420° have the same terminal side (Figure 2.1.6). (Why?)

Figure 2.1.6

Figure 2.1.7

## 2.1 Angle

- **DEFINITION**    Two angles in standard position with the same terminal sides are said to be **coterminal**.

Coterminal angles differ by an integer multiple of 360°. For example, 180° and −180° are coterminal because $180° - (-180°) = 360°$ (Figure 2.1.7). Similarly, 1125° and 45° are coterminal because $1125° - 45° = 1080° = 3(360°)$ (Figure 2.1.7). Notice that $-180° - 180° = -360°$ and $45° - 1125° = -1080° = (-3)(360°)$. Thus, it does not matter in which order the subtraction is performed to test for coterminal angles. We can summarize in the following way.

- **THEOREM 1**    Let the measure of an angle be $\theta$. Then, the measure of any angle coterminal with $\theta$ is of the form $\theta + n \cdot 360°$ where $n$ is an integer.

An immediate consequence of Theorem 1 is given by the following theorem.

- **THEOREM 2**    If $\theta$ is the measure of an angle, then there is an integer $n$ and a measure $\theta'$ such that $\theta = \theta' + n \cdot 360°$, where $0 \leq \theta' < 360°$.

- **EXAMPLE 3:** Give the measures of three angles coterminal with an angle of measure −60°.

SOLUTION: Any angle coterminal with an angle of measure −60° must have measure $-60° + n \cdot 360°$. The angles of measure 300°, 1020°, and −780° where we have chosen $n$ to be 1, 3, and −2, respectively, are coterminal with an angle of measure −60°. Of course, other integer values for $n$ could have been selected.

### Central Angles

If the vertex of an angle is the center of a circle, then the angle is called a **central angle**. If the measure of a central angle is known, then the length of the portion of the circumference of a circle intercepted by the angle can be determined.

- **EXAMPLE 4:** Determine the length of the arc of a circle of radius 2 intercepted (subtended) by a central angle of measure
    (a) 45°.          (b) 112°.

SOLUTION:
(a) The circumference of a circle with radius 2 is $4\pi$ because the circumference of a circle of radius $r$ is $2\pi r$. A central angle of 45° subtends $\frac{45}{360} = \frac{1}{8}$ of the circumference (Figure 2.1.8). Thus, the length of the arc is $\frac{1}{8}(4\pi) = \frac{\pi}{2}$ units.
(b) The length of the arc in this case is $\frac{112}{360}(4\pi) = \frac{56\pi}{45} = 3.091$ units.

- **THEOREM 3**    Let $S$ be the length of the arc of a circle of radius $r$ subtended by a central angle of measure $\theta°$. Then, $S = \frac{\theta}{360}(2\pi r)$.

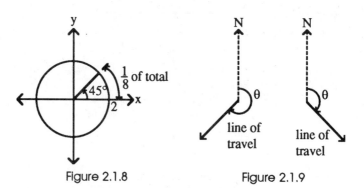

Figure 2.1.8   Figure 2.1.9

Notice that the length of the arcs in Example 4 depend on the radius of the circle. If we divide the length of the arc by the length of the radius we get a new measure of the angle that is called *radian measure*. We study radian measure in Section 2.5.

## Navigation

In navigation, the **course**, or **path**, of an object is sometimes given as the measure of the *clockwise* angle $\theta$ that the **line of travel** makes with *due north* (Figure 2.1.9). By convention, the angle measure $\theta$ is positive even though the angle rotation is clockwise.

We also say that the "line of travel" has **bearing** $\theta$.

- **EXAMPLE 5**: The captain of a boat steers a course of 35° from home port. Use a drawing to illustrate the path of the boat.

**SOLUTION**: We draw a ray from home port in the due north direction. The boat travels in the direction given by the ray obtained by rotating the north ray 35° clockwise (Figure 2.1.10). •

If an object at $B$ is sighted from the point $A$, the **bearing** from $A$ to $B$ is the same as the *line of travel* from $A$ to $B$. (Figure 2.1.11).

Figure 2.1.10   Figure 2.1.11

## 2.1 Angle

### • EXERCISES 2-1

Determine the measure and draw the angle in standard position.

1. $\frac{1}{2}$ counterclockwise rotation
2. $\frac{1}{3}$ clockwise rotation
3. $\frac{4}{3}$ counterclockwise rotation
4. $\frac{5}{3}$ clockwise rotation
5. 2.5 counterclockwise rotations
6. 3.5 clockwise rotations

Determine the quadrant of the terminal side of the angle in standard position.

7. $-160°$
8. $280°$
9. $452°$
10. $-827°$
11. $1150°$
12. $-455°$

Determine the measure of an angle $\theta$ coterminal with the given angle satisfying the specified condition.

13. $48°$, $360° \leq \theta \leq 720°$
14. $110°$, $0° \leq \theta \leq -360°$
15. $-15°$, $180° \leq \theta \leq 540°$
16. $-250°$, $360° \leq \theta \leq 720°$

Determine four *different* angles, two with positive measures and two with negative measures, coterminal with the angle.

17. $55°$
18. $-22°$
19. $410°$
20. $-150°$

Assume the given point is on the terminal side of an angle $\theta$ in standard position where $0° \leq \theta \leq 360°$. Determine $\theta$.

21. $(-1, 0)$
22. $(0, 5)$
23. $(3, 3)$
24. $(-2, 2)$
25. $(5, -5)$
26. $(10, 0)$

Assume the given point is on the terminal side of an angle $\theta$ in standard position. Determine the distance $r$ from the point to the origin.

27. $(2, 2)$
28. $(5, -2)$
29. $(-2, 3)$
30. $(4, 5)$
31. $(-2, -8)$
32. $(-4, 6)$

Determine a formula for the measure of *all* angles coterminal with the angle of specified measure.

33. $37°$
34. $-125°$
35. $-72°$
36. $458°$

Determine the length of the arc of a circle with the specified radius subtended by a central angle of the given measure.

37. $r = 2$, $\theta = 30°$
38. $r = 3.75$, $\theta = 122°$
39. $r = 5.76$, $\theta = 155°$
40. $r = 20.55$, $\theta = 72°$

Consider a family of circles with center at the origin each with the specified radius $r$ (Figure 2.1.12). Let $\theta$ be a central angle with initial side on the positive $x$-axis. Compute the ratio $s/r$ of the arc length $s$ to the radius $r$ for each circle.

41. $\theta = 50°$, $r_1 = 1$, $r_2 = 2$, $r_3 = 2.5$ and $r_4 = 5$

Figure 2.1.12

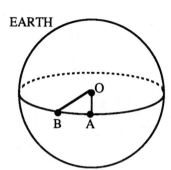

Figure 2.1.13

42. $\theta = 38°$, $r_1 = 1$, $r_2 = 1.75$, $r_3 = 5$ and $r_4 = 7.2$

## Nautical Miles

Let $A$ and $B$ be two points on the earth's surface. Suppose further that $A$ and $B$ lie on a circle with center $O$ located at the center of the earth. We assume the earth is a sphere. If $\angle AOB$ has measure one minute ($\frac{1}{60}$ of a degree), then the distance between $A$ and $B$ on the circumference of the circle is *one* **nautical mile** (Figure 2.1.13). Assume the diameter of the earth is 7,952 statute miles. (A **statute mile** is a standard mile of 5280 feet.)

43. How many feet are in one nautical mile?
44. Points $A$ and $B$ are 257 nautical miles apart. How far apart are $A$ and $B$ in *statute* miles?
45. Draw a diagram to illustrate a navigational course of 35°, 128°, and 310°, respectively.
46. At a certain time an airplane is between two signal towers that lie on an east–west line. The bearing from the plane to each tower is 340° and 37°, respectively. Use a drawing to show the exact location of the plane.
47. Let $A$ be the origin of a coordinate plane. Draw a diagram showing the location of the point $B$ if the bearing from $A$ to $B$ is 22°, 185°, and 292°, respectively.

Two rational numbers, $\frac{a}{b}$ and $\frac{c}{d}$ ($b \neq 0$, $d \neq 0$), are equal when $ad = bc$. That is, if $b \neq 0$ and $d \neq 0$, $\frac{a}{b} = \frac{c}{d}$ if and only if $ad = bc$. Technically the equal sign between $\frac{a}{b}$ and $\frac{c}{d}$ is different from the equal sign between $ad$ and $bc$ because they are defined relative to different sets of objects (in this case, different sets of numbers).

48. Devise a definition of equality of angles that involves coterminality (e.g., we want 35° = −325° = 395° = …).
49. Explain how reducing a rational number to lowest terms (e.g., $\frac{6}{8} = \frac{3}{4}$) is similar to finding an angle $\theta'$ equal (in the sense of Exercise 48) to $\theta$ with $0° \leq \theta' < 360°$.

## 2.2 Right Triangle Trigonometry

In this section we define the six trigonometric functions of an acute angle. These functions can be evaluated by using the lengths of the sides of a *right triangle* containing the acute angle or by using a calculator. A triangle is called a **right triangle** if one of its angles has measure 90°. The sum of the measures of the three angles of *any* triangle is 180°. This means that *each* of the other two angles of a right triangle is acute and the sum of their measures is 90° (Figure 2.2.1). We use the notation $\angle A$, $\angle B$, and $\angle C$ to represent the name of the angle of the triangle at the vertices $A$, $B$, and $C$, respectively. We will study applications that can be modeled with right triangles in Section 2.3. In this section we find the measure of the remaining parts of a right triangle, given the measures of two sides or the measures of one side and one of the angles.

Let $\theta$ be any acute angle. We can construct a right triangle containing $\theta$ by choosing a point $B$ on the terminal side of $\theta$ and drawing a line segment from $B$ that is *perpendicular* to the initial side of $\theta$ (Figure 2.2.2). The triangle $ABC$ is a right triangle with $\angle A = \theta$. Let the side opposite $A$ have length $a$, the side opposite $B$ have length $b$, and the side opposite $C$ have length $c$. We call $BC$ the **side opposite** to $\theta$, $AC$ the **side adjacent** to $\theta$, and $AB$ the **hypotenuse** of the right triangle in Figure 2.2.2. Notice that $BC$ is the side adjacent to $\angle B$, $AC$ is the side opposite $\angle B$, and the measure of $\angle B$ is 90° minus the measure of $\theta$.

- **Agreement**  We use the symbol $\theta$ to represent both the name of the angle and the measure of the angle. Thus, the measure of $\angle B$ is $90° - \theta$. We often label the side opposite an angle with the lower-case letter corresponding to the angle name. For example, the side $BC$ of the triangle in Figure 2.2.2 will often be called the *side a*.

- **DEFINITION**  The six trigonometric functions of the acute angle $\theta$ in Figure 2.2.2 are defined as follows:

$$\mathbf{sine}\ \theta = \frac{\text{length of opposite side}}{\text{length of hypotenuse}} = \frac{a}{c},$$

$\angle C = 90°$
$\angle A + \angle B = 90°$
Figure 2.2.1

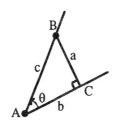

Figure 2.2.2

$$\text{cosecant } \theta = \frac{\text{length of hypotenuse}}{\text{length of opposite side}} = \frac{c}{a},$$

$$\text{cosine } \theta = \frac{\text{length of adjacent side}}{\text{length of hypotenuse}} = \frac{b}{c},$$

$$\text{secant } \theta = \frac{\text{length of hypotenuse}}{\text{length of adjacent side}} = \frac{c}{b},$$

$$\text{tangent } \theta = \frac{\text{length of opposite side}}{\text{length of adjacent side}} = \frac{a}{b},$$

$$\text{cotangent } \theta = \frac{\text{length of adjacent side}}{\text{length of opposite side}} = \frac{b}{a}.$$

We often use the abbreviation *trig* for *trigonometry* or *trigonometric*. In addition, we use the following abbreviations for the trig functions:

$$\sin \theta = \text{sine } \theta \qquad \csc \theta = \text{cosecant } \theta$$
$$\cos \theta = \text{cosine } \theta \qquad \sec \theta = \text{secant } \theta$$
$$\tan \theta = \text{tangent } \theta \qquad \cot \theta = \text{cotangent } \theta$$

Next, we give some basic properties of the trig functions of acute angles.

• **THEOREM 1**   If $\theta$ is an acute angle, then

(a) $\sin \theta = \dfrac{1}{\csc \theta}$, $\csc \theta = \dfrac{1}{\sin \theta}$   (b) $\cos \theta = \dfrac{1}{\sec \theta}$, $\sec \theta = \dfrac{1}{\cos \theta}$

(c) $\tan \theta = \dfrac{1}{\cot \theta}$, $\cot \theta = \dfrac{1}{\tan \theta}$   (d) $\tan \theta = \dfrac{\sin \theta}{\cos \theta}$, $\cot \theta = \dfrac{\cos \theta}{\sin \theta}$

**PROOF**   From the above definition we have the following:

$$\frac{1}{\csc \theta} = \frac{1}{\frac{c}{a}} = \frac{a}{c} = \sin \theta.$$

Taking reciprocals of each side of the above equation gives the following:

$$\frac{1}{\csc \theta} = \sin \theta,$$

$$\frac{1}{\frac{1}{\csc \theta}} = \frac{1}{\sin \theta},$$

$$\csc \theta = \frac{1}{\sin \theta}.$$

Thus, (a) is true. The remaining proofs are left as exercises.

In Section 2.4 we will see that Theorem 1 is true even if $\theta$ is not acute, provided that the denominators of the fractions are *not* zero.

## 2.2 Right Triangle Trigonometry

• **THEOREM 2** If $\theta$ is an acute angle, then

(a) $0 < \sin\theta < 1$,  (b) $0 < \cos\theta < 1$,

(c) $\csc\theta > 1$,  (d) $\sec\theta > 1$.

**PROOF** Refer to Figure 2.2.2. We know that $c = \sqrt{a^2 + b^2}$. It follows that $0 < a < c$ and $0 < b < c$. (Why?) Thus, $0 < \frac{a}{c} < 1$ and $0 < \frac{b}{c} < 1$. That is, $0 < \sin\theta < 1$ and $0 < \cos\theta < 1$. Next (c) follows from (a) and $\csc\theta = \frac{1}{\sin\theta}$. Similarly, (d) follows from (b) and $\sec\theta = \frac{1}{\cos\theta}$.

• **DEFINITION** Two angles are said to be **complementary** if the sum of their measures is $90°$.

The two acute angles of a right triangle are complementary. Notice in Figure 2.2.2 that the side adjacent to $\angle B$ is the side opposite $\angle A$, and the side opposite $\angle B$ is the side adjacent to $\angle A$. This observation together with the above definition of the trig functions give the following *identities* about complementary angles. We will give a definition of identity in Section 2.4.

**Identities for Complementary Angles** The following equations are true for any acute angle $\theta$:

$$\sin(90° - \theta) = \cos\theta, \qquad \csc(90° - \theta) = \sec\theta,$$
$$\cos(90° - \theta) = \sin\theta, \qquad \sec(90° - \theta) = \csc\theta,$$
$$\tan(90° - \theta) = \cot\theta, \qquad \cot(90° - \theta) = \tan\theta.$$

The sine of the complement of $\theta$ is equal to the cosine of $\theta$. Similar statements can be made about the tangent and secant of the complement of $\theta$. (Why?) If we agree that sine, tangent, and secant are the co-functions of cosine, cotangent, and cosecant, respectively, then the above identities can be summarized in the following way:

**Complementary Angles** The trig function of the complement of $\theta$ is equal to the co-function of $\theta$.

• **EXAMPLE 1:** Let $\theta$ be an acute angle with $\sin\theta = \frac{5}{6}$. Find the values of all trig functions at $\theta$.

**SOLUTION:** Figure 2.2.3 gives a right triangle containing an acute angle $\theta$ with opposite side equal to 5 and hypotenuse equal to 6. The Pythagorean theorem gives $b^2 + 5^2 = 6^2$ so that $b = \sqrt{11}$. The definitions of the trig functions give the following for their exact

Figure 2.2.3

values at $\theta$:

$$\sin \theta = \frac{5}{6} \qquad \csc \theta = \frac{6}{5}$$

$$\cos \theta = \frac{\sqrt{11}}{6} \qquad \sec \theta = \frac{6}{\sqrt{11}}$$

$$\tan \theta = \frac{5}{\sqrt{11}} \qquad \cot \theta = \frac{\sqrt{11}}{5}$$

•

The complementary angle identities can be used to compute the values of the trig functions at $90° - \theta$ if the values at $\theta$ are known. For example, using Figure 2.2.3 and Example 1 we have

$$\sin(90° - \theta) = \cos \theta = \frac{\sqrt{11}}{6} \quad \text{and} \quad \cos(90° - \theta) = \sin \theta = \frac{5}{6}.$$

Notice that using the definition of the trig functions gives the same values.

## Using a Calculator

In Example 1 we described an angle $\theta$ by specifying two of the three sides of a right triangle that contained the angle $\theta$. If the measure of an angle is known, then a scientific calculator can be used to compute the values of the trig functions at that angle. Calculators generally only have keys reserved for $\sin \theta$, $\cos \theta$, and $\tan \theta$. To compute $\cot \theta$, $\sec \theta$, and $\csc \theta$, we use the identities $\cot \theta = \frac{1}{\tan \theta}$, $\sec \theta = \frac{1}{\cos \theta}$, $\csc \theta = \frac{1}{\sin \theta}$ given in Theorem 1.

• **EXAMPLE 2:** Find the value of the six trig functions at $42°$.

**SOLUTION:** A scientific calculator will compute directly the sine, cosine, or tangent of any angle $\theta$ by using the appropriate trig key. Some calculators require that $\theta$ be entered before a trig key is pressed. Other calculators require that a trig key be pressed before an angle is entered. Check the owner's manual of your calculator to see how the trig keys on your calculator work. Be sure that your calculator is set in *degree* mode rather than radian mode. Then, use your calculator and the reciprocal relationships of Theorem 1 to compute the trig values at $42°$. The following approximations were obtained using a *Casio* graphing

## 2.2 Right Triangle Trigonometry

calculator:

$$\sin 42° = 0.6691306064, \qquad \csc 42° = \frac{1}{\sin 42°} = 1.49447655,$$

$$\cos 42° = 0.7431448255, \qquad \sec 42° = \frac{1}{\cos 42°} = 1.34563273,$$

$$\tan 42° = 0.9004040443, \qquad \cot 42° = \frac{1}{\tan 42°} = 1.110612515.$$ •

In Example 1 we found the value of all trig functions at $\theta$ where $\sin \theta = \frac{5}{6}$. In Example 3 we see how to approximate the value of $\theta$ with a calculator. In Section 3.3 we will provide detail about why this calculator procedure works.

- **EXAMPLE 3**: Use a calculator to approximate the acute angle $\theta$ if $\sin \theta = \frac{5}{6}$. Then, compute the values of the other trig functions at $\theta$. Compare with the values obtained in Example 1.

  **SOLUTION**: Scientific calculators will compute the *acute* angle $\theta$ given its sine value in decimal form using two keys. You may need to enter the sine value, $5 \div 6$, before or after pressing the two keys. Usually, the two keys are [INV] [SIN], [2nd F] [SIN], or [SHIFT] [SIN]. Again, you need to check your calculator owner's manual to determine which keys to use and whether to enter the sine value before or after pressing the two keys. Be sure that your calculator is set in *degree* mode rather than radian mode. We will discuss radian measure in Section 2.5. With a calculator set in degree mode you will find that the solution $\theta$ to $\sin \theta = \frac{5}{6}$ is approximately $56.44269024°$. Next, we store $56.44269024$ in the memory of the calculator and compute the values of the other trig functions at $\theta$:

$$\sin 56.44269° = 0.8333333, \qquad \csc 56.44269024° = 1.2,$$
$$\cos 56.44269024° = 0.5527707984, \qquad \sec 56.44269024° = 1.809068067,$$
$$\tan 56.44269024° = 1.507556723, \qquad \cot 56.44269024° = 0.663324958.$$ •

If we approximate the exact values for the trig functions at $\theta$ given in Example 1 with a calculator, we will find that they agree with the above values. For example, $\tan \theta = \frac{5}{\sqrt{11}} = 1.507556723$ and $\cos \theta = \frac{\sqrt{11}}{6} = 0.5527707984$.

- **EXAMPLE 4**: Use a calculator to determine the value of $\theta$ if $\sec \theta = 2.3$. Then, compute the values of the other trig functions at $\theta$.

  **SOLUTION**: The secant function is *not* built in on most calculators. In order to find $\theta$ we first compute the value of $\cos \theta$. Now, $\cos \theta = \frac{1}{\sec \theta}$ so that $\cos \theta = 0.4347826087$. We can use the calculator keys [INV] [COS], [2nd F] [COS], or [SHIFT] [COS] to find that $\theta = 64.22853826°$. After storing this number the other trig values at $\theta$ can be computed with a calculator:

$$\sin 64.22853826° = 0.9005354425, \qquad \csc 64.22853826° = 1.110450464,$$
$$\cos 64.22853826° = 0.4347826087, \qquad \sec 64.22853826° = 2.3$$
$$\tan 64.22853826° = 2.071231518, \qquad \cot 64.22853826° = 0.4828045495.$$ •

If we did not want the measure of $\theta$ in Example 4, then the other trig values at $\theta$ could be computed using the procedure of Example 1. However, we often need to determine the measure of angles. Thus, we need to know how to use a calculator to determine the measure of $\theta$ if a trig value of $\theta$ is known.

Before the easy availability of scientific calculators, tables were used to compute the values of the trig functions. In addition, knowing the values of the trig functions for the special angles of 30°, 45°, and 60° was very important. We give the next example for historical reasons. It is still a good idea to commit to memory the relationships of the two triangles given in the following example.

- **EXAMPLE 5:** Use triangles to determine the values of the trig functions at 30°, 45°, and 60°.

**SOLUTION:** Triangle $ABC$ in Figure 2.2.4 is equilateral with the length of each side equal to 2. The measure of each angle is 60°. The line from $B$ perpendicular to $AC$ bisects $AC$ and the angle at $B$. Thus, the length of $AD$ or $DC$ is 1, and the measure of angle $ABD$ or $DBC$ is 30°. The length of $BD$ is $\sqrt{3}$ because $ABD$ is a right triangle. Triangle $ABD$ can be used to find the values of the trig functions at 30° and 60° (Table 1). Some of the values are left for the exercises.

| | |
|---|---|
| $\sin 30° = \dfrac{1}{2}$ | $\csc 30° = 2$ |
| $\cos 30° = \dfrac{\sqrt{3}}{2}$ | $\sec 30° = ?$ |
| $\tan 30° = ?$ | $\cot 30° = \sqrt{3}$ |
| $\sin 60° = \dfrac{\sqrt{3}}{2}$ | $\csc 60° = ?$ |
| $\cos 60° = ?$ | $\sec 60° = 2$ |
| $\tan 60° = \sqrt{3}$ | $\cot 60° = ?$ |

Table 1

The triangle in Figure 2.2.5 is an isoceles right triangle. Each of the non-right angles is 45°, and the length of the hypotenuse is $\sqrt{2}$. Again, some of the values are left for the exercises. (Table 2)

| | |
|---|---|
| $\sin 45° = \dfrac{1}{\sqrt{2}}$ | $\csc 45° = ?$ |
| $\cos 45° = ?$ | $\sec 45° = \sqrt{2}$ |
| $\tan 45° = ?$ | $\cot 45° = 1$ |

Table 2

## 2.2 Right Triangle Trigonometry

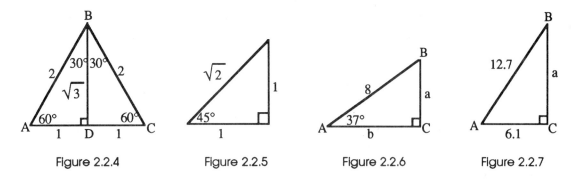

Figure 2.2.4    Figure 2.2.5    Figure 2.2.6    Figure 2.2.7

### Solving Right Triangles

If the measure of two sides of a triangle, or the measure of one side and one angle are known, then the measure of the other parts (sides or angles) of the right triangle can be determined. Determining the measures of the missing parts of a right triangle is often referred to as *solving a right triangle*. Example 6 illustrates this.

- **EXAMPLE 6:** One angle of a right triangle has measure 37° and the hypotenuse has length 8. Find the measure of the remaining parts of the right triangle.

SOLUTION: The right triangle in Figure 2.2.6 has the angle at $A$ with measure 37° and the hypotenuse with length 8. We need to determine $a$, $b$, and $\angle B$. Notice $\angle C$ is the 90° angle so that the measure of $\angle B$ is 53°. (Why?) We can use the following equations to solve for $a$ and $b$:

$$\sin 37° = \frac{a}{8} \qquad \cos 37° = \frac{b}{8}$$

Thus, $a = 8\sin 37°$ and $b = 8\cos 37°$. Accurate to hundredths, $a$ is 4.81 and $b$ is 6.39. Suppose we want to check our estimates for $a$ and $b$. We know that $\sqrt{a^2 + b^2}$ should be 8. (Why?) The calculator we use gives 7.998012253 when we compute $\sqrt{(4.81)^2 + (6.39)^2}$. This number is close enough to 8 for us to be confident that 4.81 and 6.39 are correct. •

There is another way to check the computations in Example 6. If the full calculator approximations are used for $a = 8\sin 37°$ and $b = 8\cos 37°$, then most calculators will give 8 as the value of $\sqrt{(8\sin 37°)^2 + (8\cos 37°)^2}$. This happens because the exact value of $\sqrt{(8\sin 37°)^2 + (8\cos 37°)^2}$ is 8 and most calculators carry extra digits (not shown) so that machine round-off produces 8 when this expression is evaluated.

In the next example the measure of two sides of a right triangle are given, and the measure of the other parts are determined. It does not matter which two sides are given.

- **EXAMPLE 7:** The hypotenuse and one other leg of a right triangle have measure 12.7 and 6.1, respectively. Find the measure of the remaining parts of the right triangle.

SOLUTION: We must determine $a$ and the measures of $\angle A$ and $\angle B$ of Figure 2.2.7. Now, $\cos A = \frac{6.1}{12.7}$. We can use the calculator keys INV COS, 2nd F COS, or SHIFT COS to determine that the measure of $A$ is $61.29°$, accurate to hundredths. We can use the Pythagorean theorem to determine that $a = \sqrt{12.7^2 - 6.1^2}$ is $11.14$, accurate to hundredths. Notice that $\sin A = \frac{a}{12.7}$, so that the equation $a = 12.7 \sin A$ could also be used to find the value of $a$. •

- **Note Concerning Accuracy** In the above example we assumed that the given lengths 12.7 and 6.1 of the sides were exact to produce answers with error at most 0.01. Some textbooks assume that numbers like 12.7 and 6.1 are accurate only to tenths. In this textbook, assume given information is exact unless otherwise specified, and find answers with error at most 0.01.

## EXERCISES 2-2

In this exercise set, all angles are acute angles or right angles.

Find the value of the six trig functions at each angle.

1. $23°$
2. $59°$
3. $38°$
4. $83°$
5. Complete Table 1.
6. Complete Table 2.
7. Approximate the values in the solution to Example 1 and compare with the values of the six trig functions at $\theta = 56.44269024°$ in the solution to Example 3.

Consider the triangle $ABC$ in Figure 2.2.8.

8. Find all six trig function values at $\angle A$.
9. Find all six trig function values at $\angle B$.

Consider the triangle $ABC$ in Figure 2.2.9.

10. Find all six trig function values at $\angle A$.
11. Find all six trig function values at $\angle B$.

Use a right triangle to determine the values of all trig functions at $\theta$.

12. $\sin \theta = \dfrac{3}{5}$
13. $\tan \theta = \dfrac{1}{3}$

Use a calculator to determine $\theta$. Make sure your calculator is in degree mode.

14. $\sin \theta = \dfrac{1}{2}$
15. $\sin \theta = 0.8245$

Figure 2.2.8

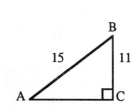
Figure 2.2.9

## 2.3 Applications

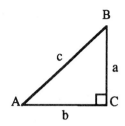

Figure 2.2.10

16. $\cos\theta = \dfrac{4}{5}$
17. $\cos\theta = 0.125$
18. $\tan\theta = 1$
19. $\tan\theta = 3$
20. $\tan\theta = 0.423$
21. $\tan\theta = 2.80$
22. Prove (b) of Theorem 1.
23. Prove (c) of Theorem 1.
24. Prove (d) of Theorem 1.
25. Use a right triangle to explain how $\tan\theta$ could be very large (greater than 1000) or very small (less than 0.001).
26. Determine $\theta$ if $\csc\theta = 2$.
27. Determine $\theta$ if $\sec\theta = 3.81$.
28. Determine $\theta$ if $\cot\theta = \dfrac{3}{5}$.
29. Determine $\theta$ if $\cot\theta = 1.875$.

Solve the right triangle in Figure 2.2.10.

30. $\angle A = 20°$, $a = 12.3$
31. $a = 3$, $b = 4$
32. $\angle A = 41°$, $c = 10$
33. $\angle A = 55°$, $b = 15.58$
34. $b = 5$, $c = 7$
35. $a = 20.2$, $c = 50.75$
36. $a = 2$, $b = 9.25$
37. $a = 5$, $\angle B = 59°$
38. $c = 12.89$, $\angle B = 12.55°$
39. $\angle A = 10.2°$, $c = 14.5$
40. $\angle A = 21.82°$, $b = 13.65$
41. Let $\theta = 10°, 20°, 25°, 48°, 55°$, and $69°$. Compute $\sin^2\theta + \cos^2\theta$ for each value of $\theta$.

## 2.3 Applications

In this section we study applications that can be modeled with right triangles. Solving the applications requires finding the measure of sides or angles of the right triangles.

Suppose that an observer sights the top of an object, say a pole (Figure 2.3.1). The angle that the line of sight makes with the horizontal is called the **angle of elevation**. If the object is below the observer, then the angle that the line of sight makes with the horizontal is called the **angle of depression** (Figure 2.3.2).

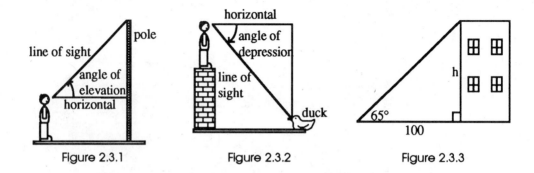

Figure 2.3.1    Figure 2.3.2    Figure 2.3.3

- **Accuracy Agreement**  Unless we specify otherwise all answers will be given accurate to hundredths. Furthermore, we will carry full calculator approximations until all computations are completed. Then, we round all answers to hundredths.

- **EXAMPLE 1:** The angle of elevation of the top of a building from a point 100 feet away from the building on level ground is 65° (Figure 2.3.3). Determine the height $h$ of the building.

  SOLUTION: Notice that $\tan 65° = \frac{h}{100}$ so that $h = 100 \tan 65°$. Accurate to hundredths, $h$ is 214.45 feet.   •

- **EXAMPLE 2:** The angle of depression of a buoy from a point on a lighthouse 130 feet above the surface of the water is 6° (Figure 2.3.4). Find the distance $x$ from the base of the lighthouse to the buoy.

  SOLUTION: The angle $\theta$ in Figure 2.3.4 must be 6°, because the angle of depression is 6°. (Why?) Now, $\tan \theta = \frac{130}{x}$ so that $\tan 6° = \frac{130}{x}$. Solving for $x$ we have $x = \frac{130}{\tan 6°}$. Thus, $x$ is 1236.87 feet.   •

- **EXAMPLE 3:** A guy wire 75 feet long runs from an antenna to a point on level ground 10 feet from the base of the antenna (Figure 2.3.5). Determine the angle $\alpha$ the guy wire makes

Figure 2.3.4    Figure 2.3.5

## 2.3 Applications

with the horizontal, the angle $\beta$ the guy wire makes with the antenna, and the distance $h$ above ground of the point $B$ at which the guy wire is attached to the antenna.

SOLUTION: We can use the calculator keys [INV] [COS], [2nd F] [COS], or [SHIFT] [COS] to determine that $\alpha = 82.34°$ from the equation $\cos \alpha = \frac{10}{75}$. Thus, $\beta$ is $7.66°$. (Why?) We can determine $h$ from the equation $75^2 = 10^2 + h^2$, or from the equation $\tan \alpha = \frac{h}{10}$. We find that $h$ is 74.33 feet if we use the equation $75^2 = 10^2 + h^2$. However, if we use $\tan \alpha = \frac{h}{10}$ with $\alpha = 82.34°$ we would find that $h$ is 74.35. This discrepancy in the second decimal place occurs because $\alpha$ was rounded to hundredths *before* the value of $h$ was computed. If we use the full calculator approximation of 82.33774434 for $\alpha$ we would find that $h$ is 74.33 when rounded. It is a good practice to finish *all* computations before rounding. Thus, the guy wire makes an 82.34° angle with the horizontal, a 7.66° angle with the antenna, and the guy wire is attached to a point 74.33 feet above ground. •

• **EXAMPLE 4**: From the top of a 100-foot building a man observes a car moving toward him. If the angle of depression of the car changes from 22° to 46° during the period of observation, how far does the car travel?

SOLUTION: Let $A$ and $B$ be the positions of the car that correspond to the 22° and 46° angles of depression. Let the man be at point $D$ at the top of the building and let $x$ be the distance traveled by the car (Figure 2.3.6). The angles $\alpha$ and $\beta$ are equal to 22° and 46°, respectively. (Why?)  Let $d$ be the length of $BC$. From triangle $ADC$ we have $\tan \alpha = \frac{100}{x+d}$, or $\tan 22° = \frac{100}{x+d}$. From triangle $BDC$ we have $\tan \beta = \frac{100}{d}$, or $\tan 46° = \frac{100}{d}$. We can rewrite these two equations as follows:

$$x + d = \frac{100}{\tan 22°}, \qquad d = \frac{100}{\tan 46°}.$$

Next, we solve these equations for $x$ by substituting the value of $d$ from the second equation into the first equation:

$$x = \frac{100}{\tan 22°} - d$$
$$= \frac{100}{\tan 22°} - \frac{100}{\tan 46°}$$
$$= 150.94.$$

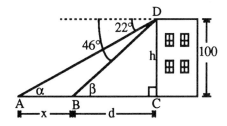

Figure 2.3.6

Thus, the car travels 150.94 feet during the period of observation.

## Distance Formula

We determine a formula for the distance between any two points in the coordinate plane using the Pythagorean relationship.

To find the distance between the points $P(x_1, y_1)$ and $Q(x_2, y_2)$, we form a right triangle as illustrated in Figure 2.3.7. The coordinates of the point $R$ are $(x_2, y_1)$. The distance from $P$ to $R$ is $|x_2 - x_1|$ and the distance from $R$ to $Q$ is $|y_2 - y_1|$. We need to use absolute value because, in general, $Q$ need not be higher than $P$ or to the right of $P$. Now, using the Pythagorean theorem, we can find the distance $D(P, Q)$ between $P$ and $Q$.

$$D(P,Q) = \sqrt{|x_2 - x_1|^2 + |y_2 - y_1|^2}$$
$$= \sqrt{(x_2 - x_1)^2 + (y_2 - y_1)^2}$$

The last step follows because $|x_2 - x_1|^2 = (x_2 - x_1)^2$ and $|y_2 - y_1|^2 = (y_2 - y_1)^2$. In fact, $|a|^2 = a^2$ for any real number $a$. (Why?)

- **Distance Formula.** The distance $D$ between the points $(x_1, y_1)$ and $(x_2, y_2)$ is given by

$$D = \sqrt{(x_2 - x_1)^2 + (y_2 - y_1)^2}.$$

- **EXAMPLE 5:** Find the distance between the points $(-1, -3)$ and $(2, 5)$.

SOLUTION: We use the distance formula:
$$D = \sqrt{(2 - (-1))^2 + (5 - (-3))^2}$$
$$= \sqrt{3^2 + 8^2}$$
$$= \sqrt{73}$$

Thus, the distance between $(-1, -3)$ and $(2, 5)$ is $\sqrt{73}$, or approximately 8.54 units.

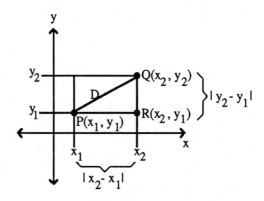

Figure 2.3.7

## 2.3 Applications

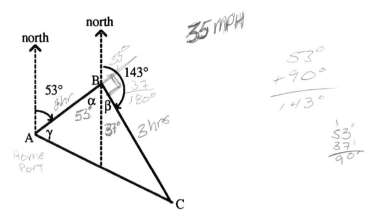

Figure 2.3.8

- **EXAMPLE 6:** A naval boat travels at a speed of 35 mph from its home port on a course of 53° for two hours and then changes to a course of 143° for three hours. Determine the distance from the boat to its home port (Figure 2.3.8). What is the bearing from the boat's home port to its location at the end of five hours?

**SOLUTION:** In Figure 2.3.8 we let $A$ represent the boat's home port, $B$ the position of the boat when it changes course, and $C$ the position of the boat at the end of the trip. Then the angles marked $\alpha$ and $\beta$ in Figure 2.3.8 have measure 53° and 37°, respectively. (Why?) Thus, $\alpha + \beta = 90°$. The triangle $ABC$ is a right triangle with hypotenuse $AC$. $d(A,B) = 70$ miles because the boat travels for two hours on a course of 53° at 35 mph. Similarly, $d(B,C) = 105$ miles. The distance from the boat to $A$ is $d(A,C)$. Now,

$$d(A,C) = \sqrt{70^2 + 105^2} = 126.19.$$

Thus, the boat is 126.19 miles from $A$.

The bearing from $A$ to $C$ is $53° + \gamma$ where $\sin\gamma = \frac{d(B,C)}{d(A,C)} = \frac{105}{126.19} = 0.83$. Using the inverse calculator keys we find $\gamma = 56.31°$. Thus, the bearing is $53° + 56.31° = 109.31°$. •

## EXERCISES 2-3

Find the distance between the points in the plane with the given coordinates.

1. $(0,0)$, $(3,4)$
2. $(-1,2)$, $(2,3)$
3. $(-3,-2)$, $(6,-2)$
4. $(a,b)$, $(2,3)$

Let $f(x) = 2 - x^2$ and $g(x) = \sqrt{6-x}$. Find the distance between the points.

5. $(2, f(2))$, $(-3, g(-3))$
6. $(-3, f(-3))$, $(4, g(4))$

7. The angle of elevation of the top of a building from a point 100 feet away from the building on level ground is 45°. Determine the height of the building.

8. The angle of elevation of the top of a building from a point 80 feet away from the building on level ground is 70°. Determine the height of the building.

9. The angle of depression of a buoy from a point on a lighthouse 120 feet above the surface of the water is 10°. Find the distance from the lighthouse to the buoy.

10. The angle of depression of a buoy from a point on a lighthouse 100 feet above the surface of the water is 3°. Find the distance from the lighthouse to the buoy.

11. The angle of elevation of the top of a building from a point 250 feet away from the building on level ground is 23°. Determine the height of the building.

12. A guy wire 30 meters long runs from an antenna to a point on level ground 5 meters from the base of the antenna. Determine the angle the guy wire makes with the horizontal, the angle the guy wire makes with the antenna, and how far above ground the guy wire is attached to the antenna.

13. A building casts a shadow 130 feet long when the angle of elevation of the sun (measured from the horizon) is 38°. How tall is the building?

14. An observer on the ground is one mile from a building 1200 feet tall. What is the angle of elevation from the observer to the top of the building?

15. The angle of elevation from an observer to the top of a building located 200 feet from the observer is 30° (Figure 2.3.9). The angle of elevation from the observer to the top of a microwave tower (on top of the building) is 40°. What is the height of the tower?

16. From the top of a 100-foot building a man observes a car moving toward him. If the angle of depression of the car changes from 15° to 33° during the period of observation, how far does the car travel?

17. A boat travels at a speed of 30 mph from its home port on a course of 95° for two hours and then changes to a course of 185° for two hours. Determine the distance from the boat to its home port and the bearing from the home port to the boat.

18. A boat travels at a speed of 40 mph from its home base on a course of 65° for two hours and then changes to course of 155° for four hours. Determine the distance from the boat to its home port and the bearing from the home port to the boat.

19. A point on the north rim of the Grand Canyon is 7256 feet above sea level. A point on the south rim directly across from the other point is 6159 feet above sea level. The canyon is 3180

Figure 2.3.9

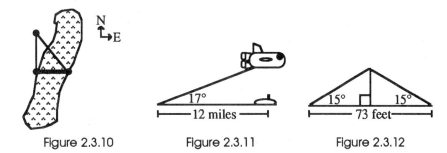

Figure 2.3.10   Figure 2.3.11   Figure 2.3.12

feet wide (horizontal distance) between the two points. What is the angle of depression from the north rim point to the south rim point?

20. A ranger spots a fire from a 73-foot tower at Yellowstone National Park. She measures the angle of depression to be 1°20′. How far is the fire from the tower? (*Note:* 20′ = 20 minutes, and there are 60 minutes in one degree.)

21. A foot bridge is to be constructed along an east–west line across a river gorge (Figure 2.3.10). The bearing of a line of sight from a point 325 feet due north of the west end of the bridge to the east end of the bridge is 117°. How long is the bridge?

22. The angle of elevation of a space shuttle launched from Cape Canaveral is measured to be 17° relative to the point of launch when it is directly over a ship 12 miles down range (Figure 2.3.11). What is the altitude of the shuttle when it is directly over the ship?

23. A truss for a barn roof is constructed as shown in Figure 2.3.12. What is the height of the vertical center span?

24. A hot air balloon over Park City, Utah is 760 feet above the ground. The angle of depression from the balloon to a small lake is 5.25°. How far is the lake from a point on the ground directly under the balloon?

25. The bearings of lines of sight from two points on the shore to a boat are 110° and 100°, respectively (Figure 2.3.13). Assume the two points are 550 feet apart. How far is the boat from the nearest point on shore?

26. Boats $A$ and $B$ leave from ports on opposite sides of a large lake. The ports are on a due east–west line. Boat $A$ steers a course of 105°, and boat $B$ steers a course of 195°. Boat $A$ averages 23 mph and collides with boat $B$ (it was a foggy night). What was boat $B$'s average speed?

27. Figure 2.3.14 shows an acute central angle $\theta$ of a circle with radius 1, and six line segments with lengths $a$ through $f$. Identify each of these lengths with the six trig function values at $\theta$.

Figure 2.3.13

Figure 2.3.14

28. Explain how a right triangle can be used to determine the exact value of $\cos\theta$ and $\tan\theta$ if $\sin\theta = a$.

Explain how a right triangle can be used to determine the exact value of $\sin\theta$ and $\tan\theta$ if $\cos\theta = a$.

30. Determine $\cos\theta$ if $\cos(90° - \theta) = a$.
31. Determine $\sin\theta$ if $\sin(90° - \theta) = a$.

## 2.4 Trigonometric Functions of Any Angle

In this section the trig functions for any angle are defined using the coordinates of a point on the terminal side of the angle. If the measure of an angle is known, a calculator is used to find the values of the trig functions of the angle. We define the reference triangle and reference angle for any nonquadrantal angle and use a calculator to find the measure of the reference angle. As an application of reference angles, we determine the measure of the angle a line makes with the $x$-axis. Fundamental identities about trig functions are established.

Let $\theta$ be a nonquadrantal angle in standard position, and $P(x, y)$ a point other than the origin on the terminal side of $\theta$. Figure 2.4.1 gives one possibility for $\theta$. There are infinitely many possibilities for a rotation that produces the terminal side shown in Figure 2.4.1. Moreover, the point $P(x, y)$ could have been in any of the four quadrants. Let $r$ be the distance from the origin to $P$. Then, $r = \sqrt{x^2 + y^2} \neq 0$ because $(x, y) \neq (0, 0)$.

- **DEFINITION** Let $\theta$ be an angle in standard position, $P(x, y)$ a point other than the origin on the terminal side of $P$, and $r = \sqrt{x^2 + y^2}$. The six trig function values of $\theta$ are defined as follows:

$$\sin\theta = \frac{y}{r}, \qquad \csc\theta = \frac{r}{y} \text{ (if } y \neq 0\text{)},$$

$$\cos\theta = \frac{x}{r}, \qquad \sec\theta = \frac{r}{x} \text{ (if } x \neq 0\text{)},$$

## 2.4 Trigonometric Functions of Any Angle

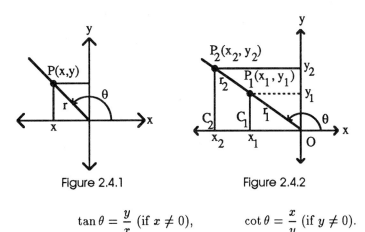

Figure 2.4.1    Figure 2.4.2

$$\tan\theta = \frac{y}{x} \text{ (if } x \neq 0\text{)}, \qquad \cot\theta = \frac{x}{y} \text{ (if } y \neq 0\text{)}.$$

First, we show that the above definition of the six trig functions is meaningful.

- **THEOREM 1**  Let $\theta$ be an angle in standard position. The six trig function values of $\theta$ are independent of the particular point $P$ chosen on the terminal side of $\theta$ to make the definitions.

   **PROOF**  We give the proof for nonquadrantal angles. It can be shown that the theorem is also true for quadrantal angles. Let $P_1(x_1, y_1)$ and $P_2(x_2, y_2)$ be any two points different from $(0,0)$ on the terminal side of $\theta$ (Figure 2.4.2). The triangles $P_1OC_1$ and $P_2OC_2$ are *similar*. Let $r_1 = d(P_1, O)$ and $r_2 = d(P_2, O)$. It follows that
   $$\frac{y_1}{r_1} = \frac{y_2}{r_2},$$
   where $r_1 = \sqrt{x_1^2 + y_1^2}$ and $r_2 = \sqrt{x_2^2 + y_2^2}$. Thus, we obtain the same value for $\sin\theta$ no matter which point on the terminal side of $\theta$ is used to compute $\sin\theta$. Similarly,
   $$\frac{x_1}{r_1} = \frac{x_2}{r_2}, \quad \frac{y_1}{x_1} = \frac{y_2}{x_2}, \quad \frac{r_1}{y_1} = \frac{r_2}{y_2}, \quad \frac{r_1}{x_1} = \frac{r_2}{x_2}, \quad \text{and} \quad \frac{x_1}{y_1} = \frac{x_2}{y_2}.$$
   Therefore, the other trig function values of $\theta$ do not depend on the particular point $P$ chosen on the terminal side of $\theta$ to make the definitions.

If $\theta$ is acute, it can be shown that the new definitions of the six trig functions agree with the definitions given in Section 2.2.

It follows from the definition of the trig functions that the trig function values of any angle *coterminal* with $\theta$ are the same as the trig function values of $\theta$. For example, the trig function values of $400° = 40° + 360°$ and $-320° = 40° - 360°$ are the same as the trig function values of $40°$. The trig functions have the following important property.

- **Circular Property of Trig Functions**  Let $\theta$ be any angle and $k$ any integer. The six trig function values at $\theta$ are equal to the values at $\theta + k(360°)$.

If the coordinates of a point on the terminal side of an angle are given, then the trig function values of the angle can be computed.

- **EXAMPLE 1:** Determine the values of the trig functions at an angle $\theta$ in standard position with the following point on its terminal side:

    (a) $P_1(2, 3)$                            (b) $P_2(-3, 4)$

SOLUTION:

(a) Let $\theta$ be any angle with the point $P_1(2,3)$ on its terminal side (Figure 2.4.3). The distance $r$ from the origin to $P_1$ is $\sqrt{2^2 + 3^2} = \sqrt{13}$. We use the definitions to determine the values of the six trig functions at $\theta$:

$$\sin\theta = \frac{3}{\sqrt{13}}, \qquad \csc\theta = \frac{\sqrt{13}}{3},$$

$$\cos\theta = \frac{2}{\sqrt{13}}, \qquad \sec\theta = \frac{\sqrt{13}}{2},$$

$$\tan\theta = \frac{3}{2}, \qquad \cot\theta = \frac{2}{3}.$$

A calculator could be used to determine decimal approximations to the above values. Notice that all trig function values are positive if the terminal side lies in the first quadrant. (Why?)

(b) Let $\theta$ be any angle with the point $(-3, 4)$ on its terminal side (Figure 2.4.4). The distance $r$ from the origin to $P_2$ is $\sqrt{(-3)^2 + 4^2} = 5$. Again, we use the definitions to determine the values of the six trig functions at $\theta$:

$$\sin\theta = \frac{4}{5}, \qquad \csc\theta = \frac{5}{4},$$

$$\cos\theta = \frac{-3}{5} = -\frac{3}{5}, \qquad \sec\theta = \frac{5}{-3} = -\frac{5}{3},$$

$$\tan\theta = \frac{4}{-3} = -\frac{4}{3}, \qquad \cot\theta = \frac{-3}{4} = -\frac{3}{4}.$$

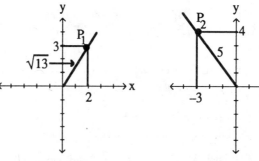

Figure 2.4.3                        Figure 2.4.4

## 2.4 Trigonometric Functions of Any Angle

Notice that $\cos\theta$, $\sec\theta$, $\tan\theta$, and $\cot\theta$ are negative, and only $\sin\theta$ and $\csc\theta$ are positive. This is true for any angle in standard position with terminal side in the second quadrant. (Why?) •

Let $\theta$ be any angle in standard position and $P(x,y)$ a point other than the origin on the terminal side of $\theta$. The distance $r = \sqrt{x^2 + y^2}$ of $P$ from the origin is positive. The sign of $x$ and $y$ depend on the quadrant in which $P$ lies. The sign of a trig function value can be positive or negative and depends on the signs of the $x$- and $y$-coordinates of $P$. Table 1 gives the sign of the value of the trig functions at $\theta$ for those angles where the terminal side of $\theta$ does *not* lie on the $x$- or $y$-axis.

### Signs of the Trig Functions

• **EXAMPLE 2:** Find the quadrant of the angle $\theta$ if $\tan\theta < 0$ and $\sin\theta < 0$.

**SOLUTION:** If $\tan\theta < 0$, then $\theta$ is in Quadrant II or IV. If $\sin\theta < 0$, then $\theta$ is in Quadrant III or IV. Thus, $\theta$ must be in Quadrant IV. •

## Quadrantal Angles

• **EXAMPLE 3:** Determine the six trig function values of $\theta$ if the terminal side of $\theta$ lies on
  (a) the positive $x$-axis.          (b) the positive $y$-axis.
  (c) the negative $x$-axis.          (d) the negative $y$-axis.

**SOLUTION:** We leave (b), (c), and (d) for the exercises.

(a) The point $(1,0)$ is on the terminal side of $\theta$. Thus, $r = \sqrt{1^2 + 0^2} = 1$, and

$$\sin\theta = \frac{y}{r} = \frac{0}{1} = 0, \qquad \csc\theta \text{ (no value)},$$

$$\cos\theta = \frac{x}{r} = \frac{1}{1} = 1, \qquad \sec\theta = 1,$$

$$\tan\theta = \frac{y}{x} = \frac{0}{1} = 0, \qquad \cot\theta \text{ (no value)}, \qquad •$$

| Quadrant of $\theta$ | $\sin\theta$ | $\cos\theta$ | $\tan\theta$ | $\csc\theta$ | $\sec\theta$ | $\cot\theta$ |
|---|---|---|---|---|---|---|
| I | + | + | + | + | + | + |
| II | + | − | − | + | − | − |
| III | − | − | + | − | − | + |
| IV | − | + | − | − | + | − |

Table 1

| $\theta$ coterminal with | $\sin\theta$ | $\cos\theta$ | $\tan\theta$ | $\csc\theta$ | $\sec\theta$ | $\cot\theta$ |
|---|---|---|---|---|---|---|
| 0° | 0 | 1 | 0 | no value | 1 | no value |
| 90° | 1 | 0 | no value | 1 | no value | 0 |
| 180° | 0 | −1 | 0 | no value | −1 | no value |
| 270° | −1 | 0 | no value | −1 | no value | 0 |

Table 2

Example 3 can be summarized as shown in Table 2. (Why?)

From now on we will assume that angles are in standard position unless specified otherwise. Let $P(x,y)$ be a point on the terminal side of a nonquadrantal angle $\theta$ and draw a perpendicular from $P$ to the $x$-axis to form a right triangle. Figure 2.4.5 shows four possibilities for the right triangle corresponding to the quadrant containing the terminal side of $\theta$. The right triangle is called the *reference triangle* determined by $\theta$, and the *acute angle* $\theta'$ made by the terminal side of $\theta$ and the positive or negative $x$-axis is called the *reference angle* determined by $\theta$. Any angle coterminal with $\theta$ produces the same reference angle.

- **EXAMPLE 4:** Sketch the reference triangle and reference angle $\theta'$ determined by each angle $\theta$ of Example 1. Compare the trig function values of $\theta$ and $\theta'$. Determine the measures of $\theta$ and $\theta'$.

SOLUTION:

(a) The reference triangle and reference angle $\theta'$ of an angle $\theta$ determined by $P_1(2,3)$ are shown in Figure 2.4.6. Note $\theta$ and $\theta'$ are coterminal. The side opposite to $\theta'$ has length 3, the side adjacent to $\theta'$ has length 2, and the hypotenuse has length

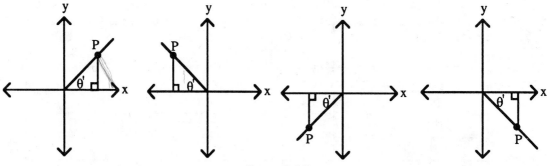

Figure 2.4.5

$\sqrt{13}$. The right triangle definitions of the trig functions give the following:

$$\sin \theta' = \frac{3}{\sqrt{13}}, \qquad \csc \theta' = \frac{\sqrt{13}}{3},$$

$$\cos \theta' = \frac{2}{\sqrt{13}}, \qquad \sec \theta' = \frac{\sqrt{13}}{2},$$

$$\tan \theta' = \frac{3}{2}, \qquad \cot \theta' = \frac{2}{3}.$$

In Example 1(a) we found that

$$\sin \theta = \frac{3}{\sqrt{13}}, \qquad \csc \theta = \frac{\sqrt{13}}{3},$$

$$\cos \theta = \frac{2}{\sqrt{13}}, \qquad \sec \theta = \frac{\sqrt{13}}{2},$$

$$\tan \theta = \frac{3}{2}, \qquad \cot \theta = \frac{2}{3}.$$

Therefore, the trig function values of $\theta$ and $\theta'$ are identical in this case. We can use the calculator keys [INV] [TAN], [2nd F] [TAN], or [SHIFT] [TAN] and the equation $\tan \theta' = \frac{3}{2}$ to show that $\theta'$ is 56.31°. Thus, $\theta$ has measure $56.31° + k(360°)$ for some integer $k$. Of course, we could have used other trig functions of $\theta'$ to determine $\theta'$.

(b) The reference triangle and reference angle $\theta'$ of an angle determined by $P_2(-3, 4)$ are shown in Figure 2.4.7. The side opposite to $\theta'$ has length 4, the side adjacent to $\theta'$ has length 3, and the hypotenuse has length 5. The right triangle definitions of the trig functions give the following:

$$\sin \theta' = \frac{4}{5}, \qquad \csc \theta' = \frac{5}{4},$$

$$\cos \theta' = \frac{3}{5}, \qquad \sec \theta' = \frac{5}{3},$$

$$\tan \theta' = \frac{4}{3}, \qquad \cot \theta' = \frac{3}{4}.$$

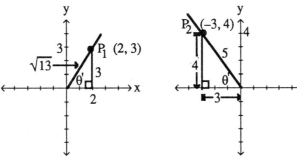

Figure 2.4.6    Figure 2.4.7

In Example 1(b) we found that

$$\sin \theta = \frac{4}{5}, \qquad \csc \theta = \frac{5}{4},$$
$$\cos \theta = -\frac{3}{5}, \qquad \sec \theta = -\frac{5}{3},$$
$$\tan \theta = -\frac{4}{3}, \qquad \cot \theta = -\frac{3}{4}.$$

Notice that $\sin \theta = \sin \theta'$, $\csc \theta = \csc \theta'$, but $\cos \theta = -\cos \theta'$, $\sec \theta = -\sec \theta'$, $\tan \theta = -\tan \theta'$, and $\cot \theta = -\cot \theta'$. We can use a calculator and the equation $\sin \theta' = \frac{4}{5}$ to show that $\theta' = 53.13°$. Then, $\theta$ can be any angle coterminal with $180° - \theta' = 126.87°$. Thus, $\theta$ has measure $126.87° + k(360°)$ for some integer $k$. •

Example 4 can be extended as follows.

• **THEOREM 2**  Except for sign, the trig function values of $\theta$ and its reference angle $\theta'$ are the same. They are the same if $\theta$ is acute. Table 3 gives a complete summary.

• **EXAMPLE 5:** The terminal side of an angle $\theta$ is in the fourth quadrant and lies on the line $y = -2.5x$. Determine all trig function values of $\theta$, the measure of the reference angle $\theta'$, and the measure of $\theta$ if $0° \le \theta \le 360°$.

SOLUTION: The point $(2, -5)$ lies on the line $y = -2.5x$ and is in the fourth quadrant. Thus, $(2, -5)$ is on the terminal side of $\theta$ (Figure 2.4.8). The distance from the origin to $(2, -5)$ is $\sqrt{29}$. Using the reference angle and adjusting for the algebraic sign we obtain

**Quadrant containing $\theta$**

|  | I | II | III | IV |
|---|---|---|---|---|
| $\sin \theta =$ | $\sin \theta'$ | $\sin \theta'$ | $-\sin \theta'$ | $-\sin \theta'$ |
| $\cos \theta =$ | $\cos \theta'$ | $-\cos \theta'$ | $-\cos \theta'$ | $\cos \theta'$ |
| $\tan \theta =$ | $\tan \theta'$ | $-\tan \theta'$ | $\tan \theta'$ | $-\tan \theta'$ |
| $\cot \theta =$ | $\cot \theta'$ | $-\cot \theta'$ | $\cot \theta'$ | $-\cot \theta'$ |
| $\sec \theta =$ | $\sec \theta'$ | $-\sec \theta'$ | $-\sec \theta'$ | $\sec \theta'$ |
| $\csc \theta =$ | $\csc \theta'$ | $\csc \theta'$ | $-\csc \theta'$ | $-\csc \theta'$ |

Table 3

the following:

$$\sin\theta = -\frac{5}{\sqrt{29}}, \qquad \csc\theta = -\frac{\sqrt{29}}{5},$$

$$\cos\theta = \frac{2}{\sqrt{29}}, \qquad \sec\theta = \frac{\sqrt{29}}{2},$$

$$\tan\theta = -\frac{5}{2}, \qquad \cot\theta = -\frac{2}{5}.$$

We can see from Figure 2.4.8 that $\tan\theta' = \frac{5}{2}$. We can use a calculator to determine that the measure of the acute angle $\theta'$ is 68.2°. If $0° \leq \theta < 360°$, then $\theta = 360° - 68.2° = 291.8°$. •

The terminal side of the angle $\theta$ of Example 5 lies in the fourth quadrant. Only $\cos\theta$ and $\sec\theta$ are positive in this case. This is consistent with Table 1. You should check that the signs of the trig function values of the angles in Example 1 also agree with Table 1.

Let $\ell_1$ and $\ell_2$ be two intersecting lines, and $\alpha$ and $\beta$ the two angles determined by $\ell_1$ and $\ell_2$ as illustrated in Figure 2.4.9. Notice that $\alpha + \beta = 180°$. Thus, either $\alpha$ or $\beta$ is acute, or $\alpha = \beta = 90°$.

- **DEFINITION** The smaller, acute angle of Figure 2.4.9, or either angle if $\alpha = \beta = 90°$, is called the **angle between $\ell_1$ and $\ell_2$**. If $\alpha = \beta = 90°$, then $\ell_1$ and $\ell_2$ are perpendicular, and are also said to be **orthogonal** or **normal**. •

By convention the angle between two parallel lines is 0°.

- **EXAMPLE 6**: Determine the angle between the $x$-axis and the line $y = -2x + 3$.

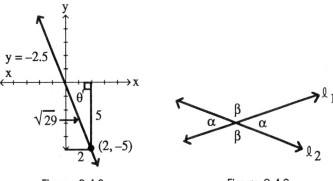

Figure 2.4.8        Figure 2.4.9

SOLUTION: The angle between the $x$-axis and the line $y = -2x + 3$ is the same as the angle between the $x$-axis and the line $y = -2x$. (Why?) Let $\theta$ be the angle $y = -2x$ makes with the positive $x$-axis and $\theta'$ its reference angle (Figure 2.4.10). The point $(-1, 2)$ is on the terminal side of $\theta$. Thus, $\tan \theta = -2$ and $\tan \theta' = 2$. The angle between the $x$-axis and the line $y = -2x$ is the acute angle $\theta'$, the reference angle of $\theta$. We can use a calculator to determine that $\theta'$ is 66.43°. Thus, the angle between the $x$-axis and the line $y = -2x + 3$ is 63.43°. •

If the measure of an angle is known, then a calculator can be used to compute directly the trig function values of the angle. Because only three of the trig functions appear on a calculator, we need to use the following extension of Theorem 1 of Section 2.2. First we give the following definition.

- **DEFINITION**    An equation is called an **identity** if and only if the equation is true for all values of the variables common to the domains of the expressions on each side of the equation. •

- **THEOREM 3**    (Reciprocal Identities) The following equations are identities.

(a) $\sin \theta = \dfrac{1}{\csc \theta}$, $\csc \theta = \dfrac{1}{\sin \theta}$    (b) $\cos \theta = \dfrac{1}{\sec \theta}$, $\sec \theta = \dfrac{1}{\cos \theta}$

(c) $\tan \theta = \dfrac{1}{\cot \theta} = \dfrac{\sin \theta}{\cos \theta}$, $\cot \theta = \dfrac{1}{\tan \theta} = \dfrac{\cos \theta}{\sin \theta}$

PROOF    The proof of this theorem is a direct application of the definitions of the trig functions. We prove the second part of (b) and leave the rest of the proofs for the exercises. Let $P(x, y)$ be a point other than the origin on the terminal side of $\theta$ and $r = \sqrt{x^2 + y^2}$. Then,

$$\sec \theta = \frac{r}{x} = \frac{1}{\frac{x}{r}} = \frac{1}{\cos \theta}.$$

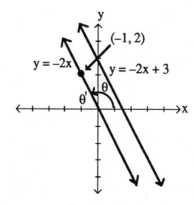

Figure 2.4.10

We use $\csc\theta = \frac{1}{\sin\theta}$, $\sec\theta = \frac{1}{\cos\theta}$, and $\cot\theta = \frac{1}{\tan\theta}$ to compute the values of $\csc\theta$, $\sec\theta$, and $\cot\theta$, respectively, with a calculator.

- **EXAMPLE 7**: Determine all trig function values of each angle.
  (a) 470°
  (b) −670°

  SOLUTION:

  (a) We use the [SIN], [COS], and [TAN] keys together with Theorem 2 to compute the six trig function values of 470°:

  $\sin 470° = 0.9396926208,$ $\qquad \csc 470° = 1.064177772,$

  $\cos 470° = -0.3420201433,$ $\qquad \sec 470° = -2.9238044,$

  $\tan 470° = -2.747477419,$ $\qquad \cot 470° = -0.3639702343.$

  The terminal side of 470° lies in the second quadrant (Figure 2.4.11). Notice that the signs of the above values are consistent with Table 1.

  (b) The terminal side of −670° lies in the first quadrant (Figure 2.4.12). By Table 1, the trig function values of −670° should all be positive. We compute the values with a calculator:

  $\sin -670° = 0.7660444431,$ $\qquad \csc -670° = 1.305407289,$

  $\cos -670° = 0.6427876097,$ $\qquad \sec -670° = 1.555723827,$

  $\tan -670° = 1.191753593,$ $\qquad \cot -670° = 0.8390996309.$

## Trigonometric Identities

There are many important relationships among the trig functions. The fundamental reciprocal identities were given in Theorem 3. Theorem 4 establishes three additional very basic and fundamental identities that should be committed to memory.

- **THEOREM 4**  The following equations are true for any angle $\theta$ in the domain of the corresponding trig functions:

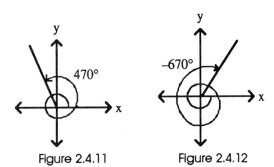

Figure 2.4.11  Figure 2.4.12

(a) $\sin^2\theta + \cos^2\theta = 1$,   (b) $1 + \tan^2\theta = \sec^2\theta$,
(c) $1 + \cot^2\theta = \csc^2\theta$.

**PROOF** Let $P(x,y)$ be a point other than the origin on the terminal side of $\theta$. Then

$$\sin\theta = \frac{y}{r}, \qquad \csc\theta = \frac{r}{y},$$
$$\cos\theta = \frac{x}{r}, \qquad \sec\theta = \frac{r}{x},$$
$$\tan\theta = \frac{y}{x}, \qquad \cot\theta = \frac{x}{y},$$

where $r^2 = x^2 + y^2$. If we divide each side of the equation $x^2 + y^2 = r^2$ by $r^2$ we obtain

$$x^2 + y^2 = r^2$$
$$\frac{x^2}{r^2} + \frac{y^2}{r^2} = \frac{r^2}{r^2}$$
$$\left(\frac{x}{r}\right)^2 + \left(\frac{y}{r}\right)^2 = 1$$
$$\cos^2\theta + \sin^2\theta = 1.$$

Dividing each side of $x^2 + y^2 = r^2$ by $x^2$ ($x \ne 0$) establishes the equation in (b). Similarly, dividing each side of the same equation by $y^2$ ($y \ne 0$) gives the equation in (c).

The equation in Theorem 4(a) is an identity because the common domain of $\sin^2\theta + \cos^2\theta$ and of 1 is the set of all angles, and $\sin^2\theta + \cos^2\theta = 1$ for every angle $\theta$. The equation in Theorem 4(b) is an identity because the common domain of $1 + \tan^2\theta$ and of $\sec^2\theta$ consists of those angles $\theta$ for which $\cos\theta \ne 0$, and $1 + \tan^2\theta = \sec^2$ for those values of $\theta$.

The identities of Theorems 3 and 4 can be used to establish other identities.

• **EXAMPLE 8:** Show algebraically that $\dfrac{1 + \sec\theta}{\sin\theta + \tan\theta} = \csc\theta$ is an identity.

**SOLUTION:** You may be surprised that the complicated expression on the left-hand side of the equation is just another way of writing $\csc\theta$. The two graphs in Figure 2.4.13 support the validity of this identity by suggesting that the graphs of

$$y = \frac{1 + \sec\theta}{\sin\theta + \tan\theta}$$

and $y = \csc\theta$ are identical.

A reasonable start towards verifying the identity algebraically is to rewrite the most complicated side of the equation by converting all functions to sines and cosines using

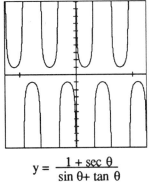
$y = \dfrac{1 + \sec\theta}{\sin\theta + \tan\theta}$

$[-4\pi, 4\pi]$ by $[-10, 10]$

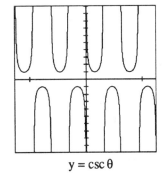
$y = \csc\theta$

$[-4\pi, 4\pi]$ by $[-10, 10]$

Figure 2.4.13

identities and then simplify:

$$\dfrac{1+\sec\theta}{\sin\theta+\tan\theta} = \dfrac{1+\frac{1}{\cos\theta}}{\sin\theta+\frac{\sin\theta}{\cos\theta}} = \dfrac{\frac{1+\cos\theta}{\cos\theta}}{\frac{\sin\theta\cos\theta+\sin\theta}{\cos\theta}}$$

$$= \dfrac{1+\cos\theta}{\sin\theta\cos\theta+\sin\theta} = \dfrac{1+\cos\theta}{(\sin\theta)(1+\cos\theta)}$$

$$= \dfrac{1}{\sin\theta} = \csc\theta.$$

Thus, $\dfrac{1+\sec\theta}{\sin\theta+\tan\theta} = \csc\theta$.

There is a subtle issue in the above example that should be mentioned. As functions, the left-hand side $f(\theta) = \dfrac{1+\sec\theta}{\sin\theta+\tan\theta}$ and the right-hand side $g(\theta) = \csc\theta$ are not the same because their *domains* are different. Notice that $\dfrac{\pi}{2}$ is in the domain of $g$ but *not* in the domain of $f$. The graph of $f$ actually has removable discontinuities at odd multiples of $\dfrac{\pi}{2}$ that are not visible in Figure 2.4.13.

Section 3.2 will be devoted to verifying identities.

# EXERCISES 2-4

Use a calculator unless indicated otherwise. Angles are measured in degrees.

Determine the values of the six trig functions at each angle.

1. 83°
2. −108°
3. −400°
4. 305°
5. Complete (b) and (c) of Example 3.
6. Complete (d) of Example 3.

Explain how the signs of the trigonometric functions are determined in Table 1 for each quadrant.

7. Quadrant I
8. Quadrant II

9. Quadrant III
10. Quadrant IV

Let $\theta$ be an angle in standard position with the point $P$ on the terminal side of $\theta$. Sketch the reference triangle and the reference angle $\theta'$ determined by each angle $\theta$. Find the values of the six trig functions at $\theta$ and $\theta'$. Determine the measure of $\theta'$ and $\theta$.

11. $P = (-3, 0)$
12. $P = (0, -5)$
13. $P = (4, 3)$
14. $P = (-1, 2)$
15. $P = (-4, -6)$
16. $P = (5, -2)$
17. $P = (22, -22)$
18. $P = (-8, 1)$

Sketch the reference triangle and the reference angle $\theta'$ determined by each angle $\theta$.

19. $156°$
20. $-305°$
21. $614°$
22. $213°$

23–26. Determine the trig function values of $\theta$ and $\theta'$ for each angle $\theta$ in Exercises 19–22.

Let $\theta$ be an angle in standard position with terminal side given by the equation. Find the values of the six trig functions at $\theta$.

27. $y = x$ in the third quadrant.
28. $y = 2x$ in the first quadrant.
29. $y = \dfrac{3}{4}x$ in the first quadrant.
30. $y = -x$ in the second quadrant.
31. $y = \dfrac{2}{3}x$ in the third quadrant.
32. $y = -\dfrac{3}{5}x$ in the fourth quadrant.

Determine the measure of the reference angle $\theta'$ for the given angle $\theta$.

33. $106°$
34. $-584°$
35. $-157°$
36. $287°$
37. $214°$
38. $-200°$

Determine the angle between the $x$-axis and each line.

39. $y = 2x$
40. $2x - 3y = 4$
41. $y = -3x + 1$
42. $3x - 5y = -2$

$\theta$ is an angle in standard position. Find the quadrant containing the terminal side of $\theta$.

43. $\sin\theta < 0$ and $\tan\theta > 0$
44. $\cos\theta > 0$ and $\tan\theta < 0$
45. $\tan\theta > 0$ and $\sec\theta < 0$
46. $\sin\theta > 0$ and $\cos\theta < 0$

Verify the identity algebraically and confirm graphically.

47. $\cot\theta \tan\theta = 1$
48. $\sin\theta \sec\theta = \tan\theta$
49. $\sec^2\theta - \sin^2\theta = \cos^2\theta + \tan^2\theta$
50. $\dfrac{1}{\sin^2\theta} + \dfrac{1}{\cos^2\theta} = \dfrac{1}{\sin^2\theta \cos^2\theta}$

51. A wire stretches from the top of a vertical pole to a point on level ground 16 feet from the base of the pole. If the wire makes an angle of $62°$ with the ground, determine the height of the pole and the length of the wire.

52. A lighthouse at $L$ stands three miles from the closest point $P$ along a straight shore. Determine the distance from $P$ to a point $Q$ along the shore if the angle $PQL$ measures $35°$.

53. Using a sextant, a surveyor determines that the angle of elevation of a mountain peak is $35°$. Moving 1000 feet further away from the mountain the surveyor determines the angle of elevation to be $30°$. Determine the height of the mountain.

## 2.4 Trigonometric Functions of Any Angle

|     | $\theta$ | $\sin\theta$ | $\cos\theta$ | $\tan\theta$ | $\cot\theta$ | $\sec\theta$ | $\csc\theta$ |
|-----|----------|--------------|--------------|--------------|--------------|--------------|--------------|
| (a) | 0        |              |              |              |              |              |              |
| (b) | 30°      |              |              |              |              |              |              |
| (c) | 45°      |              |              |              |              |              |              |
| (d) | 60°      |              |              |              |              |              |              |
| (e) | 90°      |              |              |              |              |              |              |
| (f) | 180°     |              |              |              |              |              |              |
| (g) | 270°     |              |              |              |              |              |              |
| (h) | 360°     |              |              |              |              |              |              |

Table 4

54. Prove (a) and (c) of Theorem 4.
55. Use properties of a 30°–60° right triangle and a 45° right triangle (see Figures 2.2.4 and 2.2.5) to determine the *exact* values of each trig function in Table 4.
56. Use properties of a 30°–60° right triangle and a 45° right triangle (see Figures 2.2.4 and 2.2.5) to determine the *exact* values of each trig function in Table 5.

|     | $\theta$ | $\sin\theta$ | $\cos\theta$ | $\tan\theta$ | $\cot\theta$ | $\sec\theta$ | $\csc\theta$ |
|-----|----------|--------------|--------------|--------------|--------------|--------------|--------------|
| (a) | 240°     |              |              |              |              |              |              |
| (b) | 315°     |              |              |              |              |              |              |
| (c) | −30°     |              |              |              |              |              |              |
| (d) | 480°     |              |              |              |              |              |              |
| (e) | −540°    |              |              |              |              |              |              |
| (f) | −135°    |              |              |              |              |              |              |
| (g) | 13230°   |              |              |              |              |              |              |
| (h) | 36000°   |              |              |              |              |              |              |

Table 5

## 2.5 Radian Measure and Graphs of sin x and cos x

In this section we introduce radian measure and convert degree measure to radian measure and vice versa. Using radians, we define the trig functions for real numbers. The graphs of $\sin x$ and $\cos x$ will be drawn and used to solve equations.

### Radian Measure

Recall from Section 2.1 that an angle $\theta$ is called a central angle if its vertex is the center of a circle. Any angle can be made a central angle by placing its vertex at the center of some circle with initial and terminal sides along radii of the circle.

- **DEFINITION**     A central angle with positive measure in a circle of radius $r$ that subtends an arc of length $r$ is said to have measure **1 radian** (Figure 2.5.1).

Because the length of the circumference of a circle of radius $r$ is $2\pi r$, the circumference consists of $2\pi$ segments of length $r$. Thus, the radian measure of the angle of one complete rotation, or revolution counterclockwise, is $2\pi$. An angle has *negative* measure if the rotation that produces the angle is clockwise. We use this convention for angles measured in degrees or radians. The radian measure of the angle of one-half revolution clockwise is $-\pi$. These are special cases of the following more general result.

- **THEOREM 1**     Let the central angle $\theta$ in a circle of radius $r$ subtend an arc of length $s$ (Figure 2.5.2). The radian measure of $\theta$ is $s/r$ if $\theta$ has positive measure, or $-s/r$ if $\theta$ has negative measure.

We can obtain the following relationships from Theorem 1.

- **Degree-Radian Equivalent Measures**

$$360° = 2\pi \text{ radians}, \qquad 180° = \pi \text{ radians},$$

$$1° = \frac{\pi}{180} \text{ radians}, \qquad 1 \text{ radian} = \left(\frac{180}{\pi}\right)°.$$

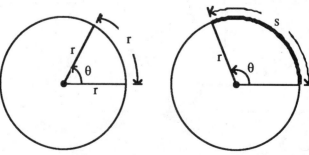

Figure 2.5.1           Figure 2.5.2

## 2.5 Radian Measure and Graphs of sin $x$ and cos $x$

We can use a calculator to see that $1°$ is approximately $0.01745329252$ radians and $1$ radian is approximately $57.59577951°$. To convert radian measure to degree measure we multiply the radian measure by $\frac{180}{\pi}$, and to convert degree measure to radian measure we multiply the degree measure by $\frac{\pi}{180}$.

- **EXAMPLE 1:** Find the radian measure of an angle whose measure is
  - (a) $120°$.
  - (b) $500°$.
  - (c) $-200°$.

  SOLUTION: We know that $1° = \frac{\pi}{180}$ radians. Thus, each degree measure must be multiplied by $\frac{\pi}{180}$.
  
  (a) $\frac{\pi}{180}(120) = \frac{2\pi}{3}$. Thus, $120° = \frac{2\pi}{3}$ radians. We can use a calculator to show that $\frac{2\pi}{3}$ radians is approximately $2.094395102$ radians.
  
  (b) $500° = \frac{\pi}{180}(500)$ radians, which is approximately $8.72664626$ radians.
  
  (c) $-200° = \frac{\pi}{180}(-200)$ radians, which is approximately $-3.490658504$ radians. •

- **EXAMPLE 2:** Find the degree measure of an angle whose measure is
  - (a) $\frac{7\pi}{6}$ radians.
  - (b) $-8$ radians.

  SOLUTION: We know that $1$ radian $= \left(\frac{180}{\pi}\right)°$. Thus, each radian measure must be multiplied by $\frac{180}{\pi}$.
  
  (a) $\frac{7\pi}{6}$ radians $= \frac{180}{\pi}\left(\frac{7\pi}{6}\right) = 210°$.
  
  (b) $-8$ radians $= \frac{180}{\pi}(-8)$, which is approximately $-458.3662361°$. •

**Arc Length** If we use the radian measure $\theta$ of an angle as the name of the angle, then we can rewrite Theorem 1 in the form $|\theta| = \frac{s}{r}$, or $s = r|\theta|$. This equation is not valid if $\theta$ is in degrees. The variable $s$ is the length of the arc subtended by the angle of measure $\theta$ radians.

- **Radian Measure Convention** From now on we will *not* usually write the word *radian* when giving the measure of an angle in radians. For example, if the measure of an angle is 2 radians we usually say that the measure of the angle is 2. We continue to use the established abbreviation for degree measure. That is, $2°$ stands for an angle whose measure is 2 degrees.

- **EXAMPLE 3:** In a circle of radius 3 inches, find the length of the arc subtended by an angle of measure
  - (a) 2.
  - (b) $-60°$.

  SOLUTION:
  
  (a) Now, $r = 3$ and $\theta = 2$ so that $s = r|\theta| = 6$ (Figure 2.5.3). Thus, the length of the arc subtended by an angle of measure 2 radians in a circle of radius 3 inches is 6 inches.

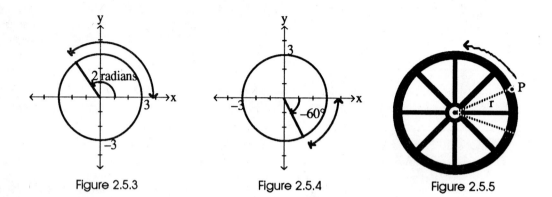

Figure 2.5.3      Figure 2.5.4      Figure 2.5.5

(b) First, we convert $-60°$ to $-\frac{\pi}{3}$ radians. Then, $r = 3$ and $\theta = -\frac{\pi}{3}$ so that $s = r|\theta| = \pi$ (Figure 2.5.4). Thus, the length of the arc subtended by an angle of measure $-60°$ in a circle of radius 3 inches is $\pi$ inches. •

## Angular and Linear Speed

A wheel of radius $r$ is rotating at a constant rate (Figure 2.5.5). Let $P$ be a point on the circumference of the wheel.

• DEFINITION   The **angular speed** of the wheel is the angle formed or swept out in one unit of time by the line segment from the center of the wheel to the point $P$ on the circumference of the wheel. The **linear speed** of the point $P$ is the distance $P$ travels in one unit of time. •

• EXAMPLE 4: A wheel of radius 18 inches is rotating at 850 revolutions per minute (rpm). Determine the

(a) angular speed of the wheel.
(b) linear speed of a point on the circumference of the wheel.

SOLUTION:

(a) The number of revolutions per minute is 850, and each revolution produces an angle of $2\pi$ radians. Thus, the angle formed by the wheel in one minute is $850(2\pi) = 1700\pi$. The angular speed of the wheel is $1700\pi$ radians per minute.
(b) The distance traveled by a point on the wheel in one minute is given by $s = r\theta$ where $r$ is 1.5 feet and $\theta$ is $1700\pi$ radians. Thus, the linear speed of a point on the wheel is 8011.06 feet per minute. •

## Circular Functions

A circle with center at the origin and radius 1 is called a *unit circle* (Figure 2.5.6). We construct a vertical number line tangent to the unit circle at the point $(1, 0)$, that is, a

## 2.5 Radian Measure and Graphs of sin $x$ and cos $x$

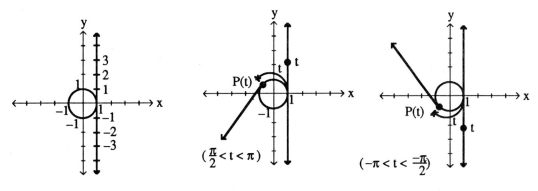

Figure 2.5.6     Figure 2.5.7

line parallel to the $y$-axis through the point $(1, 0)$. Every real number $t$ on the number line corresponds to a point $P(t)$ on the unit circle in the following way. If $t > 0$, we wrap the positive portion of the vertical line in Figure 2.5.6 counterclockwise around the unit circle until the point with coordinate $t$ on the number line meets the unit circle (Figure 2.5.7). We name this point $P(t)$. If $t < 0$, the negative portion of the vertical line is wrapped clockwise around the unit circle.

For example, the points $\frac{\pi}{2}$, $\pi$, $\frac{3\pi}{2}$, and $-\frac{\pi}{2}$ on the vertical number line correspond to the points $(0, 1)$, $(-1, 0)$, $(0, -1)$, and $(0, -1)$, respectively, on the unit circle (Figure 2.5.8).

We know from Theorem 1 that $s = |\theta|$ because the radius of the unit circle is 1. It follows that the radian measure of the central angle $\theta$ determined by the arc of the circle from $(1, 0)$ to $P(t)$ is $t$ (Figure 2.5.9). (Why?) There are infinitely many points on the vertical number line that correspond to the same point on the unit circle. For example, the points $\frac{\pi}{2}$, $\frac{\pi}{2} + 2\pi$, $\frac{\pi}{2} + 4\pi$, and so forth, on the vertical number line correspond to the point $(0, 1)$ on the unit circle. However, there is a one-to-one correspondence between the points of $[0, 2\pi)$ of the vertical number line and the points of the unit circle. This follows because the circumference of a unit circle is $2\pi$.

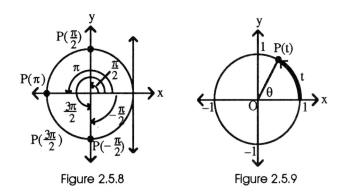

Figure 2.5.8     Figure 2.5.9

Let $P(t) = (x, y)$ be the point on the unit circle corresponding to the real number $t$ (Figure 2.5.10). Let $\theta$ be any angle in standard position with $P(t)$ on its terminal side. In Section 2.4, we gave the following definitions for the trig function values at $\theta$:

$$\sin\theta = \frac{y}{r}, \qquad \csc\theta = \frac{r}{y}\ (y \neq 0),$$

$$\cos\theta = \frac{x}{r}, \qquad \sec\theta = \frac{r}{x}\ (x \neq 0),$$

$$\tan\theta = \frac{y}{x}\ (x \neq 0), \qquad \cot\theta = \frac{x}{y}\ (y \neq 0).$$

Now, $r = \sqrt{x^2 + y^2} = 1$ because $(x, y)$ is on a unit circle. Thus, $y = \sin\theta$ and $x = \cos\theta$. The coordinates of $P(t)$ are the sine and cosine of an angle in standard position with $P(t)$ on its terminal side.

- **DEFINITION** Let $t$ be any real number, and let $P(t) = (x, y)$ be the point on the unit circle corresponding to $t$. We define the six trig function values at $t$ as follows:

$$\sin t = y, \qquad \csc t = \frac{1}{y}\ (y \neq 0),$$

$$\cos t = x, \qquad \sec t = \frac{1}{x}\ (x \neq 0),$$

$$\tan t = \frac{y}{x}\ (x \neq 0), \qquad \cot t = \frac{x}{y}\ (y \neq 0).$$

Because of the previous discussion, we know that these definitions of the trig functions are consistent with the definitions in Section 2.4. Furthermore, it can be shown that the reciprocal identities of Theorem 2 of Section 2.4 hold for the trig functions defined as above. The trig functions are often called **circular functions** because of the above definition that involves the unit circle. Because of the wrapping property illustrated in Figure 2.5.7, the circular property of the trig functions given in Section 2.4 can be extended to the trig functions with angles measured in radians.

Figure 2.5.10

- **Circular Property of Trig Functions** Let $k$ be any integer and $x$ any real number. The six trig function values at $x$ are equal to the values at $x + 2k\pi$.

- **EXAMPLE 5:** Determine the sign of the six trig function values at 4.5.

SOLUTION: An angle of measure 4.5 radians is in Quadrant III. (Why?) Thus, only $\tan 4.5$ and $\cot 4.5$ are positive. The other values are negative. ●

A calculator can be used to compute the trig function values at a real number $t$. In this case, $t$ is also the radian measure of the angle determined by the arc from $(1,0)$ to $P(t)$ on the unit circle. To compute the trig function value at a real number $t$ with a calculator, we must put the calculator in *radian mode*.

- **EXAMPLE 6:** Use a calculator to find the values of the trig functions at

    (a) 4.    (b) $-3.5$.

SOLUTION: Put your calculator in radian mode and then use the trig keys to make the following computations. Remember we now give answers accurate to hundredths unless otherwise specified.

(a) $\sin 4 = -0.76$ $\qquad$ $\csc 4 = -1.32$

$\cos 4 = -0.65$ $\qquad$ $\sec 4 = -1.53$

$\tan 4 = 1.16$ $\qquad$ $\cot 4 = 0.86$

(b) $\sin(-3.5) = 0.35$ $\qquad$ $\csc(-3.5) = 2.85$

$\cos(-3.5) = -0.94$ $\qquad$ $\sec(-3.5) = -1.07$

$\tan(-3.5) = -0.37$ $\qquad$ $\cot(-3.5) = -2.67$ ●

## Graphs of Sin x and Cos x

- **EXAMPLE 7:** Determine the domain, the range, and the zeros of $f(x) = \sin x$ and draw a complete graph. Show that $f(x + 2k\pi) = f(x)$ for any integer $k$.

SOLUTION: Let $x$ be any real number and $P(x)$ the corresponding point on the unit circle (Figure 2.5.11). The coordinates of $P(x)$ are $(\cos x, \sin x)$, and the central angle determined by the arc of the circle from $(1,0)$ to $P(x)$ has radian measure $x$. We have drawn Figure 2.5.11 with $0 < x < \frac{\pi}{2}$. If $x = 0$, then $\sin x = 0$. As $x$ varies from 0 to $\frac{\pi}{2}$, $\sin x$ varies from 0 to 1. Similarly, the following values of $x$ produce the corresponding values of $\sin x$

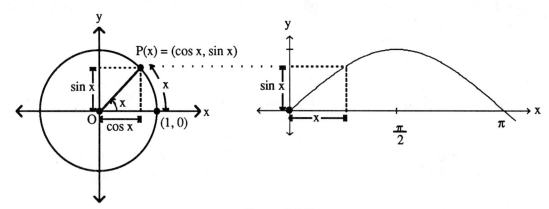

Figure 2.5.11

in the specified order:

$$0 \leq x \leq \frac{\pi}{2}, \qquad 0 \leq \sin x \leq 1,$$

$$\frac{\pi}{2} \leq x \leq \pi, \qquad 1 \geq \sin x \geq 0,$$

$$\pi \leq x \leq \frac{3\pi}{2}, \qquad 0 \geq \sin x \geq -1,$$

$$\frac{3\pi}{2} \leq x \leq 2\pi, \qquad -1 \leq \sin x \leq 0.$$

Notice that $\sin x = 0$ in $[0, 2\pi]$ when $x = 0, \pi$, or $2\pi$.

Because of the circular property of the sine function, as $x$ varies from $2\pi$ to $4\pi$, $\sin x$ varies through the same values in the same order as above:

$$2\pi \leq x \leq \frac{5\pi}{2}, \qquad 0 \leq \sin x \leq 1,$$

$$\frac{5\pi}{2} \leq x \leq 3\pi, \qquad 1 \geq \sin x \geq 0,$$

$$3\pi \leq x \leq \frac{7\pi}{2}, \qquad 0 \geq \sin x \geq -1,$$

$$\frac{7\pi}{2} \leq x \leq 2\pi, \qquad -1 \leq \sin x \leq 0.$$

In fact, as $x$ varies from $2n\pi$ to $(2n+2)\pi$, where $n$ is a positive integer, $\sin x$ varies through the same values in the same order as when $x$ varies from $0$ to $2\pi$. If $x$ is negative, then we mark off an arc of length $|x|$ on the unit circle starting at $(1, 0)$ and move clockwise (Figure 2.5.12).

## 2.5 Radian Measure and Graphs of sin x and cos x

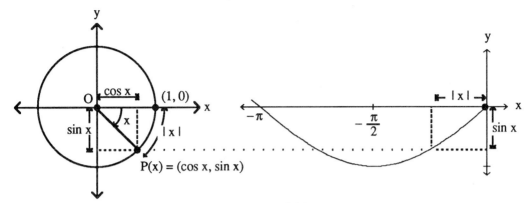

Figure 2.5.12

Notice that the following values of $x$ produce the corresponding values of $\sin x$ in the specified order:

$$0 \geq x \geq -\frac{\pi}{2}, \qquad 0 \geq \sin x \geq -1,$$

$$-\frac{\pi}{2} \geq x \geq -\pi, \qquad -1 \leq \sin x \leq 0,$$

$$-\pi \geq x \geq -\frac{3\pi}{2}, \qquad 0 \leq \sin x \leq 1,$$

$$-\frac{3\pi}{2} \geq x \geq -2\pi, \qquad 1 \geq \sin x \geq 0.$$

The zeros of $f$ in $[-2\pi, 0]$ occur when $x = -2\pi, -\pi$, or $0$. Similarly, as $x$ varies from $(-2n-2)\pi$ to $-2n\pi$ where $n$ is a positive integer, $\sin x$ varies through the same values in the same order as when $x$ varies from $-2\pi$ to $0$. Now, we can see that the range of $f(x) = \sin x$ is $[-1, 1]$ and the domain of $f$ is the set of all real numbers. Moreover, $f(x + 2k\pi) = f(x)$ for any integer $k$. The computer drawn graph of $f(x) = \sin x$ given in Figure 2.5.13 is a complete graph. The zeros of $f$ occur at integer multiples of $\pi$, that is, when $x = k\pi, k$ any integer. •

---

• **DEFINITION**   A function $f$ is said to be **periodic** if there is a positive real number $h$ such that $f(x + h) = f(x)$ for every $x$ in the domain of $f$. The smallest such positive number $h$ is called the **period** of $f$.   •

It follows from Example 7 that $2\pi$ is a candidate for the period of the function $f(x) = \sin x$. It is clear from the graph in Figure 2.5.13 that no smaller number can be the period. Thus, $2\pi$ is the period of $\sin x$. The graph of $f(x) = \sin x$ is completely determined by its graph in the interval $[0, 2\pi]$, because the length of this interval is $2\pi$. Therefore, the graph of $f$ in Figure 2.5.14 is also a complete graph if we indicate that $f$ is periodic with period $2\pi$.

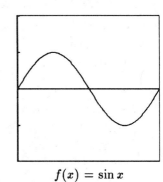

$f(x) = \sin x$
$[-10, 10]$ by $[-2, 2]$
Figure 2.5.13

$f(x) = \sin x$
$[0, 2\pi]$ by $[-2, 2]$
Figure 2.5.14

$f(x) = \sin x$
$[0°, 360°]$ by $[-2, 2]$
Figure 2.5.15

- **EXAMPLE 8:** Find the coordinates of the points at which $f(x) = \sin x$ has a local maximum value or a local minimum value.

  **SOLUTION:** The range of $f$ is $[-1, 1]$ so that $f\left(\frac{\pi}{2}\right) = \sin \frac{\pi}{2} = 1$ is a local maximum value of $f$. Similarly, $f\left(-\frac{\pi}{2}\right) = \sin\left(-\frac{\pi}{2}\right) = -1$ is a local minimum value of $f$. Now, for any integer $k$, the circular property of the trig functions gives $f\left(\frac{\pi}{2} + 2k\pi\right) = f\left(\frac{\pi}{2}\right) = 1$ and $f\left(-\frac{\pi}{2} + 2k\pi\right) = f\left(-\frac{\pi}{2}\right) = -1$. Thus, $f$ has local maximum values at $\frac{\pi}{2} + 2k\pi$ and local minimum values at $-\frac{\pi}{2} + 2k\pi$. •

The domain of $f(x) = \sin x$ in Example 7 is the set of all real numbers. We can interpret radian measures of angles as real numbers. Thus, we can view $x$ in $\sin x$ as an angle. In this case the domain of $f$ consists of all angles with measure in radians. Moreover, if $\theta$ is the equivalent degree measure of $x$ radians, the domain of $f(\theta) = \sin \theta$ consists of all angles with measures in degrees. The graph of $f(\theta) = \sin \theta$ in the $[0°, 360°]$ by $[-2, 2]$ viewing rectangle represents one complete period of $f$ with angle measure in degrees (Figure 2.5.15). Compare this graph with the one in Figure 2.5.14. From now on, we usually display computer drawn trig graphs with real numbers (radians) as viewing rectangle parameters.

Notice that the graph of $f(x) = \sin x$ in Figure 2.5.13 is *symmetric with respect to the origin*. Thus $\sin x$ is an *odd* function and $\sin(-x) = -\sin x$. Alternatively, we can use the definition of the sine function to show that $\sin(-x) = -\sin x$. Let $P(x)$ and $P(-x)$ be the points on the unit circle corresponding to $x$ and $-x$ (Figure 2.5.16). We have drawn Figure 2.5.16 assuming $x > 0$ and $P(x)$ is in the first quadrant. The following argument is valid without these restrictions. If the coordinates of $P(x)$ are $(a, b)$, then the coordinates of $P(-x)$ are $(a, -b)$. We also know that $b = \sin x$ and $-b = \sin(-x)$ by the definition of the trig functions. Replacing $b$ by $\sin x$ in $-b = \sin(-x)$ gives $-\sin x = \sin(-x)$.

- **EXAMPLE 9:** Determine all solutions of $\sin x = 0.4$ in the interval $0 \leq x \leq 2\pi$.

  **SOLUTION:** Figure 2.5.17 gives the graphs of $f(x) = \sin x$ and $g(x) = 0.4$ in the $[0, 2\pi]$ by $[-2, 2]$ viewing rectangle. There are two values of $x$ satisfying $\sin x = 0.4$; one is in the first

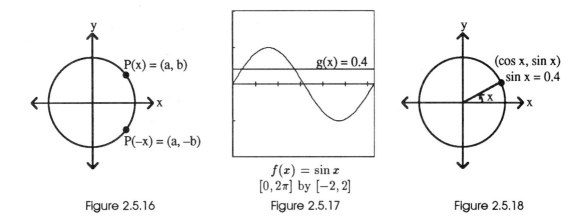

Figure 2.5.16    Figure 2.5.17    Figure 2.5.18

quadrant, and the other is in the second quadrant. (Why?) One way to determine the two values of $x$ is to use zoom-in. Alternatively, we can use a calculator to find the two values of $x$. Suppose $x$ is in the first quadrant (Figure 2.5.18). We can put a calculator in radian mode (Why?) and use the INV SIN, 2nd F SIN, or SHIFT SIN keys to compute directly that $x = 0.41$. Now suppose that $x$ is in the second quadrant (Figure 2.5.19). In this case the reference angle $x'$ is the value found previously using a calculator. Thus, $x' = 0.41$ and $x = \pi - 0.41 = 2.73$ is the second solution of $\sin x = 0.4$ in $0 \leq x \leq 2\pi$. If you obtained $x = 23.58$, check the mode of your calculator. •

We will obtain a complete graph of $\cos x$ by first establishing that $\cos x = \sin(x + \frac{\pi}{2})$ for all $x$. Then, the graph of $y = \cos x$ is simply the graph of $g(x) = \sin x$ shifted horizontally left $\frac{\pi}{2}$ units.

- **THEOREM 2**   $\cos t = \sin(t + \frac{\pi}{2})$ for all $t$.

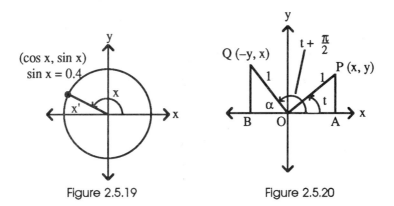

Figure 2.5.19    Figure 2.5.20

**PROOF** We give the proof with $t$ in the first quadrant. The technique used in this case is valid regardless of the quadrant of the angle $t$. Figure 2.5.20 shows the angles $t$ and $t + \frac{\pi}{2}$. We construct the triangles so that the length of $OP$ and $OQ$ are 1, and $OQ$ is perpendicular to $OP$. The measure of the angle $\alpha$ of triangle $OQB$ is $\pi - (t + \frac{\pi}{2}) = \frac{\pi}{2} - t$. Thus, the right triangles $OQB$ and $OPA$ are congruent. (Why?) Let the coordinates of $P$ be $(x, y)$. Then, the coordinates of $Q$ are $(-y, x)$. Now, $x = \cos t$ because the length of $OP$ is 1. Similarly, $x = \sin(t + \frac{\pi}{2})$ because the length of $OQ$ is 1. Therefore, $\cos t = x = \sin(t + \frac{\pi}{2})$.

- **EXAMPLE 10:** Determine the domain, the range, the period, and the zeros of $f(x) = \cos x$ and draw a complete graph. Show that $f(x + 2k\pi) = f(x)$ for any integer $k$.

**SOLUTION:** From Theorem 2 we know that $f(x) = \cos x = \sin(x + \frac{\pi}{2})$. Thus, the graph of $f$ is the graph of $y = \sin x$ shifted horizontally left $\frac{\pi}{2}$ units. The computer drawn graph of $f$ (solid graph) in Figure 2.5.21 confirms this observation and is a complete graph. The period of $f$ must be the same as $\sin x$. Thus, the period of $\cos x$ is $2\pi$. The domain of $f(x) = \cos x$ is the set of all real numbers; the range is $[-1, 1]$. For any integer $k$, $f(x + 2k\pi) = f(x)$ by the circular property of the cosine function. This fact also follows from the equation $f(x) = \sin(x + \frac{\pi}{2})$ and Example 7 as follows:

$$f(x + 2k\pi) = \sin(x + 2k\pi + \frac{\pi}{2})$$
$$= \sin((x + \frac{\pi}{2}) + 2k\pi)$$
$$= \sin(x + \frac{\pi}{2})$$
$$= f(x).$$

The zeros of $f(x) = \cos x$ occur when $x = k\pi - \frac{\pi}{2}$. (Why?) We can rewrite $k\pi - \frac{\pi}{2}$ in the form $(2k - 1)\frac{\pi}{2}$. Thus, the zeros of $\cos x$ occur at odd multiples of $\frac{\pi}{2}$. •

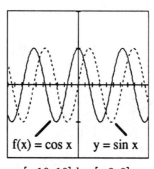

$[-10, 10]$ by $[-2, 2]$
Figure 2.5.21

## 2.5 Radian Measure and Graphs of sin $x$ and cos $x$

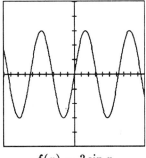

$f(x) = 3 \sin x$
$[-10, 10]$ by $[-5, 5]$
Figure 2.5.22

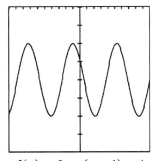

$f(x) = 2\cos(x+1) - 4$
$[-10, 10]$ by $[-8, 0]$
Figure 2.5.23

Notice that the graph of $f(x) = \cos x$ in Figure 2.5.21 is *symmetric with respect to the y-axis* so that it is an even function. In the exercises, you will be asked to show that $f(x) = \cos x$ is an even function using the wrapping property of the trig functions.

- **EXAMPLE 11:** Determine the domain, the range, and the period of each function, and draw a complete graph.

    (a) $f(x) = 3\sin x$      (b) $g(x) = 2\cos(x+1) - 4$

SOLUTION:

(a) The graph of $f(x) = 3\sin x$ can be obtained by applying a vertical stretch factor of 3 to $y = \sin x$. The computer drawn graph of $f$ in Figure 2.5.22 confirms this observation and is a complete graph. The domain of $f$ is the set of all real numbers, the range is $[-3, 3]$, and the period of $f$ is $2\pi$. (Why?)

(b) The graph of $g(x) = 2\cos(x+1) - 4$ can be obtained by applying a vertical stretch factor of 2 to $y = \cos x$, followed by a horizontal shift left one unit and then a vertical shift down 4 units. The graph of $f$ in Figure 2.5.23 is complete. The domain of $f$ is $(-\infty, \infty)$, the range is $[-6, -2]$, and the period is $2\pi$. (Why?) •

The order in which the transformations were applied in Example 11(b) is important. You will see in the exercises that changing the order will not necessarily lead to the same graph.

## • EXERCISES 2-5

Use a calculator unless stated otherwise. Angles are measured in degrees or radians.

Find the equivalent radian measure.

1. 1°
2. 55°
3. 30°
4. 72°
5. −120°
6. −380°

Find the equivalent degree measure of each radian measure.

7. 1
8. 4
9. $\dfrac{3\pi}{4}$
10. $\dfrac{2\pi}{3}$
11. $-2.3$
12. $\dfrac{15\pi}{6}$

Determine the values of the six trig functions at each angle.

13. $200°$
14. 3.8 radians
15. $-2$ radians
16. $-115°$

Sketch a complete graph without using a graphing utility. State the domain, range, and period. Check your answer with a graphing utility.

17. $y = \sin 3x$
18. $g(x) = 2 + \sin x$
19. $y = 2 + \sin(x - 45°)$
20. $y = -1 - \cos(x - \pi)$
21. $y = -1 + 3\cos(x - 2)$
22. $y = 2 - 2\cos(x + 3)$
23. $y = -1 + 3\sin(x - 2)$
24. $y = 2 - 2\sin(x + 180°)$

List, in order, the geometric transformations that can be used to obtain each graph from the graph of $y = \sin x$ or $y = \tan x$.

25. $y = -2\sin(x - 180°)$
26. $y = 2 + \tan\left(x - \dfrac{\pi}{4}\right)$

Use the circular properties of the trig functions to verify the identity.

27. $\cos(-x) = \cos x$ (cosine is an *even* function)
28. $\tan(-x) = -\tan x$ (tangent is an *odd* function)

Use the circular properties of the trig functions to verify each statement. Do not use a calculator. *Hint:* Draw a unit circle and approximate arcs of length 2 and 3 on the circumference from the point $(1, 0)$.

29. $\sin 2 > \sin 3$
30. $\cos 2 > \cos 3$
31. $\sin 10 < 0$
32. $\cos(\sin 3) > 0$

Solve for $x$ if $0 \le x \le 2\pi$.

33. $\sin x = 0.5$
34. $\sin x = -0.6$
35. $\sin x = 0.8$
36. $\sin x = \dfrac{\sqrt{3}}{2}$
37. $\cos x = 0.6$
38. $\cos x = -0.4$

Solve the inequality if $0° \le \theta \le 360°$.

39. $\sin \theta < 0.6$
40. $3\cos \theta < 4$

41. Which of the following statements are true for every $\alpha$ between $0°$ and $90°$ or every $x$ satisfying $0 < x < \dfrac{\pi}{2}$? Justify your conclusion with graphs.

    (a) $\sin(90° - \alpha) = \cos \alpha$
    (b) $\cos\left(\dfrac{\pi}{2} - x\right) = \sin x$
    (c) $\sin(\pi + x) = -\cos x$
    (d) $\tan(90° - \alpha) = \dfrac{1}{\tan \alpha}$

42. Show that no number less than $2\pi$ can be the period of $y = \sin x$.

Let $\theta$ be a central angle of a circle of radius $r$ and $s$ the length of the subtended arc. Find the arc length $s$.

43. $\theta = 22°$, $r = 15$ inches
44. $\theta = 3$ radians, $r = 5.6$ feet

## 2.6 Radian Measure and Graphs of sin $x$ and cos $x$

Figure 2.5.24

A simple winch that is used to lift heavy objects is positioned 10 feet above ground level. Assume that the radius of the winch is $r$ feet as shown in Figure 2.5.24. For the given radius $r$ and winch rotation $\theta$, determine the distance that the object is lifted above ground.

45. $r = 4$ inches, $\theta = 720°$
46. $r = 1$ foot, $\theta = 720°$

Consider a wheel of radius 5 feet rotating at 1200 rpm (revolutions per minute).

47. Determine the angular speed of the wheel.
48. Determine the linear speed of a point on the circumference of the wheel.

Consider a wheel of radius 2.8 feet rotating at 600 rpm.

49. Determine the angular speed of the wheel.
50. Determine the linear speed of a point on the circumference of the wheel.
51. Jeffrey can obtain a speed of 42 mph on his exercise bike in high gear. The bike wheels are 30 inches in diameter, the pedal sprocket is 16 inches in diameter, and the wheel sprocket is 5 inches (in high gear). Find the angular speed of the wheel and of both sprockets. *Note:* The linear speed of a point on the circumference of the wheel is also 42 mph.
52. A regular pentagon is inscribed in a circle of radius 10 inches. What is the perimeter of the pentagon?
53. A regular pentagon is inscribed in a circle. The length of each side of the pentagon is 22 inches. Determine the radius of the circle.
54. Show that applying a vertical stretch factor of 2 to the graph of $y = \sin x$ followed by applying a vertical shift up 3 units is not the same as applying a vertical shift up 3 units to the graph of $y = \sin x$ followed by applying a vertical stretch factor of 2.
55. Determine the two equations of the resulting graphs from Exercise 54. Confirm that they are different graphs.

## 2.6 Graphs of tan x, cot x, sec x, and csc x

In this section we determine the graphs of the other four basic trig functions and we investigate the role of the parameter $a$ in the graph of $y = \sin ax$ and other trig functions. Equations involving trig functions are solved.

In the previous section we found that the period of $\sin x$ or $\cos x$ was $2\pi$. Now, we see that the period of $\tan x$ or $\cot x$ is $\pi$.

• **EXAMPLE 1:** Determine the domain, the range, the zeros, and the asymptotes of $f(x) = \tan x$, and draw a complete graph.

**SOLUTION:** Recall that $\tan x = \frac{\sin x}{\cos x}$. This means that the domain of $f(x) = \tan x$ consists of all real numbers for which $\cos x \neq 0$. (Why?) In the previous section we found that $\cos x = 0$ at all odd multiples of $\frac{\pi}{2}$. Thus, the domain of $f$ consists of all real numbers *not* equal to $(2n+1)\frac{\pi}{2}$, $n$ an integer. The zeros of $f$ occur at the values of $x$ for which $\sin x = 0$. (Why?) In Section 2.5 we showed that $\sin x = 0$ when $x = n\pi$, $n$ an integer.

We know that both $\sin x$ and $\cos x$ are periodic with period $2\pi$. Thus, $\tan x = \frac{\sin x}{\cos x}$ must be periodic. It is natural to conjecture that the period of $\tan x$ is also $2\pi$. However, the graph of $f$ in Figure 2.6.1 suggests that the period of $f$ is $\pi$. This will be verified in the next example. Thus, the graph of $f$ can be obtained by piecing together, end-to-end, copies of the graph of $f$ in $\left[-\frac{\pi}{2}, \frac{\pi}{2}\right]$. Therefore, the graph of $f$ in Figure 2.6.2 is complete and shows three periods of $f$. The graph of $f$ has vertical asymptotes at the odd multiples of $\frac{\pi}{2}$, the zeros of the denominator ($\cos x$) of $f$. This means that $f(x) \to \infty$ or $f(x) \to -\infty$ as $x \to (2n+1)\frac{\pi}{2}$ from the left or right, respectively. •

We can see from Figure 2.6.2 that the graph of $f(x) = \tan x$ is symmetric with respect to the origin. Thus, $f(x) = \tan x$ is an odd function. We can also see that the function $f(x) = \tan x$ is increasing in each interval $\left(-\frac{\pi}{2} + k\pi, \frac{\pi}{2} + k\pi\right)$, $k$ an integer. In the next example we confirm that the period of $f$ is $\pi$.

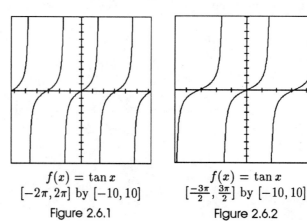

$f(x) = \tan x$
$[-2\pi, 2\pi]$ by $[-10, 10]$

Figure 2.6.1

$f(x) = \tan x$
$\left[\frac{-3\pi}{2}, \frac{3\pi}{2}\right]$ by $[-10, 10]$

Figure 2.6.2

## 2.6 Graphs of $\tan x$, $\cot x$, $\sec x$, and $\csc x$

• **EXAMPLE 2:** Show that $\tan(x + \pi) = \tan x$ for all $x \neq (2k-1)\frac{\pi}{2}$, $k$ an integer.

SOLUTION: Let $P_1(a, b)$ be a point on the terminal side of the angle $x$ (Figure 2.6.3). This figure is drawn with $x$ in the first quadrant. The actual details that follow are valid independent of the quadrant in which $x$ lies. The point $P_2(-a, -b)$ is on the terminal side of $x + \pi$. (Why?) Using the coordinates of $P_1$, we have $\tan x = \frac{b}{a}$. Using the coordinates of $P_2$, we have $\tan(x+\pi) = \frac{-b}{-a} = \frac{b}{a}$. Thus, $\tan(x+\pi) = \tan x$. •

It can be shown that no number smaller than $\pi$ can be a period for $\tan x$. Thus, the period of $\tan x$ is $\pi$.

• **EXAMPLE 3:** Determine the domain, the range, the asymptotes, the zeros, and the period of $f(x) = \cot x$, and draw a complete graph.

SOLUTION: Investigating the graph of $\cot x$ shows that it is very similar to that of $\tan x$. We can use the identity

$$\cot x = \frac{1}{\tan x}$$

to show that period of $f(x) = \cot x$ must be $\pi$:

$$\cot(x+\pi) = \frac{1}{\tan(x+\pi)}$$
$$= \frac{1}{\tan x}$$
$$= \cot x.$$

Furthermore, the range of

$$f(x) = \cot x = \frac{1}{\tan x}$$

is the set of all real numbers. (Why?) We can see that

$$f(x) = \cot x = \frac{\cos x}{\sin x}$$

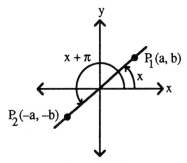

Figure 2.6.3

has a vertical asymptote at any value of $x$ for which $\sin x = 0$. In the previous section we showed that $\sin x = 0$ at any value of $x$ of the form $k\pi$, where $k$ is an integer. Therefore, the domain of $f$ consists of all real numbers different from $k\pi$, $k$ any integer, and the vertical asymptotes are $x = k\pi$. The computer drawn graph of $f(x) = \cot x$ in Figure 2.6.4 is complete and shows four complete periods of the function. •

Now, $\cot x = 0$ if and only if $\cos x = 0$. (Why?) In the previous section we showed that $\cos x = 0$ at any value of $x$ of the form $(2k-1)\frac{\pi}{2}$, $k$ an integer. Thus, $\cot x = 0$ at the odd multiples of $\frac{\pi}{2}$.

• **EXAMPLE 4:** Determine the domain, the range, the asymptotes, the zeros, and the period of $f(x) = \csc x$ and $g(x) = \sec x$, and draw complete graphs.

SOLUTION: We can use the identity

$$\csc x = \frac{1}{\sin x}$$

and Example 3 to see that $f(x) = \csc x$ has a vertical asymptote at $k\pi$, $k$ any integer. Thus, the domain of $f$ consists of all real numbers different from $k\pi$, $k$ any integer. Similarly, we can use the identity

$$\sec x = \frac{1}{\cos x}$$

and Example 1 to see that $g(x) = \sec x$ has vertical asymptotes $x = (2k+1)\frac{\pi}{2}$, $k$ any integer. Thus, the domain of $g$ consists of all real numbers different from $(2k+1)\frac{\pi}{2}$, $k$ any integer. As in Example 2, we can use the above two identities to show that the periods of $f(x) = \csc x = \frac{1}{\sin x}$ and $g(x) = \sec x = \frac{1}{\cos x}$ are $2\pi$. The graphs of $f$ and $g$ in Figure 2.6.5 are complete and show two periods of each function. We can see from these graphs that the ranges of $f$ and $g$ are $(-\infty, -1] \cup [1, \infty)$. Alternatively, we can use the fact that the ranges

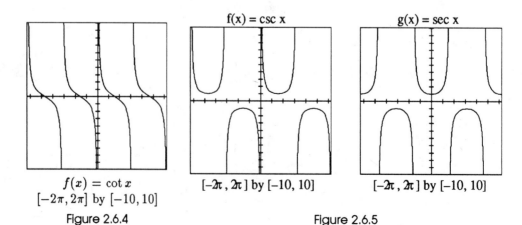

$f(x) = \cot x$
$[-2\pi, 2\pi]$ by $[-10, 10]$

Figure 2.6.4

$[-2\pi, 2\pi]$ by $[-10, 10]$    $[-2\pi, 2\pi]$ by $[-10, 10]$

Figure 2.6.5

of $y = \sin x$ and $y = \cos x$ are $[-1, 1]$, and we can use the identities

$$\csc x = \frac{1}{\sin x}, \qquad \sec x = \frac{1}{\cos x}$$

to see that the ranges of $f$ and $g$ are $(-\infty, -1] \cup [1, \infty)$. Notice that neither $f$ nor $g$ has any real zeros.

## Solving Equations

Let $f$ be any one of the six basic trig functions. The equation $f(x) = a$, where $a$ is a real number, need not have a solution. For example, if $f$ is the sine or cosine function, we can use the graphs in Figure 2.6.6 to see that the equation $f(x) = a$ has solutions if $|a| \le 1$ and *no* solutions if $|a| > 1$. Similarly, Figure 2.6.5 shows that the equations $\sec x = 0.5$ and $\csc x = -0.5$ have no solutions. However, if $f(x) = a$ has a solution, then the circular properties of the trig functions show that the equation has *infinitely many* solutions.

- **Solving Trig Equations** A good strategy to find all solutions is first to solve the equation in an interval of length equal to the period of the trig function. Then add $2k\pi$ ($k\pi$ for $\tan x$ and $\cot x$), where $k$ is any integer, to *each* solution found in the interval of length equal to the period to determine all the solutions.

We can solve the equation $f(x) = a$ with a graphing utility using zoom-in or with a calculator. However, we must be careful when we use a calculator to solve such equations Figures 2.6.2 and 2.6.4 show that the equations $\tan x = a$ and $\cot x = a$ have exactly one solution in any interval of length $\pi$, the period of the tangent or cotangent. Suppose that $f(x)$ is $\sin x$, $\cos x$, $\sec x$, or $\csc x$. The graphs in Figures 2.6.5 and 2.6.6 show that the equations $f(x) = 1$ and $f(x) = -1$ have exactly one solution in any interval of length $2\pi$. However, the equation $f(x) = a$ with $|a| \ne 1$ has either no solutions or exactly two solutions in any interval of length $2\pi$. Calculators give one solution only to $f(x) = a$ when $f(x)$ is $\sin x$, $\cos x$, or $\tan x$ using the 2nd F, SHIFT, or INV key together with the trig key. Furthermore, calculators produce an $x$ in the interval $[-\frac{\pi}{2}, \frac{\pi}{2}]$ for $\sin x = a$ or $\tan x = a$ (Figure 2.6.7). In this case, $|x|$ is the reference angle for any solution to $\sin x = a$ or $\tan x = a$. Calculators

Figure 2.6.6

Figure 2.6.7                    Figure 2.6.8

produce an $x$ in the interval $[0, \pi]$ for $\cos x = a$ (Figure 2.6.8). In this case, $x$ is the reference angle if $0 \leq x \leq \frac{\pi}{2}$ and $\pi - x$ is the reference angle if $\frac{\pi}{2} < x \leq \pi$ for any solution to $\cos x = a$.

- **EXAMPLE 5:** Determine all solutions to the following equations in the interval $[0, 2\pi)$.

    (a) $\tan x = 2.5$       (b) $\cos x = -0.7$       (c) $\csc x = -1.6$

SOLUTION:

(a) We can use Figure 2.6.9 to see that the equation $\tan x = 2.5$ has two solutions in the interval $[0, 2\pi)$. One solution is in the first quadrant, and the other is in the third quadrant. We can *estimate* from Figure 2.6.9 that the two solutions to $\tan x = 2.5$ in $[0, 2\pi)$ are 1.2 and 4.3. A calculator gives only the first quadrant solution 1.19 (Figure 2.6.10). Now, 1.19 is the reference angle for the other solution in the third quadrant (Figure 2.6.11). Thus, the third quadrant solution is $\pi + 1.19$, or 4.33.

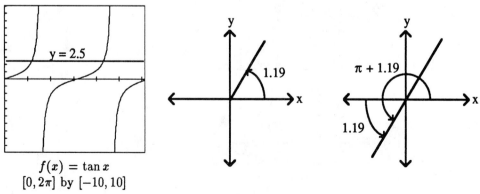

$f(x) = \tan x$
$[0, 2\pi]$ by $[-10, 10]$
Figure 2.6.9                    Figure 2.6.10                    Figure 2.6.11

## 2.6 Graphs of tan $x$, cot $x$, sec $x$, and csc $x$

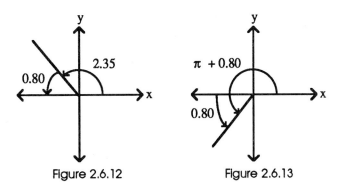

Figure 2.6.12          Figure 2.6.13

(b) We can use Figure 2.6.6 to see that the equation $\cos x = -0.7$ has two solutions in the interval $[0, 2\pi)$. One is in the second quadrant, and the other is in the third quadrant. A calculator gives only the second quadrant solution 2.35 (Figure 2.6.12). The angle $\pi - 2.35 = 0.80$ is the reference angle for the other solution in the third quadrant (Figure 2.6.13). Thus, the third quadrant solution is $\pi + 0.80$, or 3.94.

(c) Figure 2.6.5 shows that the equation $\csc x = -1.6$ has two solutions in the interval $[0, 2\pi)$. One is in the third quadrant, and the other is in the fourth quadrant. To use a calculator to find the solutions we must first observe that $\sin x = \frac{1}{\csc x} = -0.625$. A calculator gives $-0.68$ as the value of $x$. Now, $|-0.68| = 0.68$ is the reference angle for the two solutions we seek. The third quadrant solution is $\pi + 0.68 = 3.82$, and the fourth quadrant solution is $2\pi - 0.68 = 5.61$ (Figure 2.6.14). •

• **EXAMPLE 6:** Determine all solutions to each of the following equations in degrees.

(a) $\sin x = 0.7$  (b) $\sec x = 2.5$  (c) $\cot x = -5$

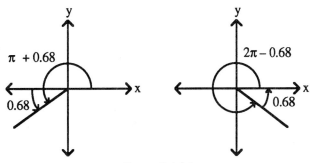

Figure 2.6.14

**SOLUTION:**

(a) sin $x$ is periodic with period $2\pi = 360°$. There are two solutions to this equation in the interval $[0°, 360°)$ (Figure 2.6.15). The two solutions are in the first and second quadrant. A calculator gives $44.43°$ as the solution to sin $x = 0.7$. Now, $44.43°$ is the reference angle for the second quadrant solution. Thus, the second quadrant solution is $180° - 44.43° = 135.57°$. Any other solution is of the form $44.42° + k360°$ or $135.57° + k360°$, $k$ any integer.

(b) First, we convert sec $x = 2.5$ to the equivalent equation cos $x = 0.4$. (How?) Now, cos $x$ is periodic with period $360°$, and the equation cos $x = 0.4$ has two solutions in the interval $[0, 360°)$ (Figure 2.6.16). One solution is in the first quadrant, and the other is in the fourth quadrant. With a calculator we find that $66.42°$ is a solution. This is the first quadrant solution. The solution in the fourth quadrant has $66.42°$ as a reference angle. Thus, the fourth quadrant solution is $360° - 66.42° = 293.58°$. Any other solution is of the form $66.42° + k360°$ or $293.58° + k360°$, $k$ any integer.

(c) First, we must convert cot $x = -5$ to the equivalent equation tan $x = -0.2$. (How?) Now, tan $x$ is periodic with period $\pi = 180°$. Notice that tan $x = a$ has exactly one solution in any interval of length $\pi$ (Figure 2.6.2). Because calculators give values in the interval $[-\frac{\pi}{2}, \frac{\pi}{2}]$ as solutions to tan $x = a$ and the length of the interval $[-\frac{\pi}{2}, \frac{\pi}{2}]$ is $\pi$, we can use the calculator values directly to solve any equation of the form tan $x = a$. A calculator gives $-11.31°$ as the solution to tan $x = -0.2$. Any other solution is of the form $-11.31° + k180°$, $k$ any integer. ●

## Horizontal Stretch or Shrink

Consider trig functions of the form sin $ax$, cos $ax$, and so forth. If $a > 0$, we will show that the effect of the parameter $a$ is to stretch or shrink horizontally the graph of the corresponding basic trig function. If $a < 0$, the effect of the parameter $a$ is to stretch or shrink horizontally the graph of the corresponding basic trig function and then reflect the resulting graph through the $y$-axis.

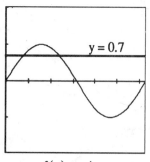

$f(x) = \sin x$
$[0°, 360°]$ by $[-2, 2]$
Figure 2.6.15

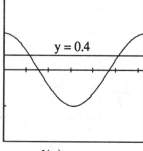

$f(x) = \cos x$
$[0°, 360°]$ by $[-2, 2]$
Figure 2.6.16

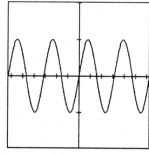

$f(x) = \sin 2x$
$[-2\pi, 2\pi]$ by $[-2, 2]$
Figure 2.6.17

- **EXAMPLE 7:** Draw a complete graph of $f(x) = \sin 2x$ and $g(x) = \sin \frac{1}{3}x$. Show that each function is periodic and determine its period.

SOLUTION: The periodicity of the sine and cosine functions makes it reasonable to expect that $f$ and $g$ are periodic. We will show that the computer drawn graph of $f$ in Figure 2.6.17 is a complete graph. It appears that $f$ is periodic with period $\pi$, one-half the period of $y = \sin x$. As $2x$ varies from 0 to $2\pi$, the values of $\sin 2x$ should produce a complete period of $f(x) = \sin 2x$. If $0 \leq 2x \leq 2\pi$, then $0 \leq x \leq \pi$. Notice that $f(x + \pi) = \sin 2(x + \pi) = \sin(2x + 2\pi) = \sin 2x = f(x)$. No number smaller than $\pi$ can be the period of $f(x) = \sin 2x$ because twice that number would be a period of $\sin x$. (Why?) Thus, the period of $f$ is $\pi$, and the graph in Figure 2.6.17 is a complete graph showing four periods.

Similarly, we can show that the computer drawn graph of $g$ in Figure 2.6.18 is a complete graph. It appears that $y$ is periodic with period $6\pi$, 3 times the period of $y = \sin x$. As $\frac{1}{3}x$ varies from 0 to $2\pi$, the values of $\sin \frac{1}{3}x$ should produce a complete period of $g(x) = \sin \frac{1}{3}x$. If $0 \leq \frac{1}{3}x \leq 2\pi$, then $0 \leq x \leq 6\pi$. Notice that $g(x + 6\pi) = \sin \frac{1}{3}(x + 6\pi) = \sin(\frac{1}{3}x + 2\pi) = \sin \frac{1}{3}x = g(x)$. No number smaller than $6\pi$ can be a period of $g(x) = \sin \frac{1}{3}x$ because $\frac{1}{3}$ of that number would be a period of $\sin x$. (Why?) Thus, the period of $g$ is $6\pi$, and the graph of $g$ in Figure 2.6.18 is complete showing two periods. •

- **EXAMPLE 8:** Draw a complete graph of $f(x) = \sin(-2x)$ and $g(x) = \cos(-2x)$. Show that each function is periodic and determine its period.

SOLUTION: The graph of $f(x) = \sin(-2x)$ can be obtained by reflecting the graph of $y = \sin 2x$ through the $y$-axis. The computer drawn graph of $f$ in Figure 2.6.19 confirms this observation. Similarly, the graph of $g(x) = \cos(-2x)$ can be obtained by reflecting the graph of $y = \cos 2x$ through the $y$-axis. However, the graph of $y = \cos 2x$ is symmetric with respect to the $y$-axis because the cosine function is an even function. Thus, the graph of $g$ and $y = \cos 2x$ should be the same. The graph of $g$ is given in Figure 2.6.20. You should overlay the graph of $y = \cos 2x$ to convince yourself that the two graphs are identical. •

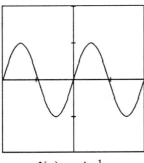

$f(x) = \sin \frac{1}{3}x$
$[-6\pi, 6\pi]$ by $[-2, 2]$

Figure 2.6.18

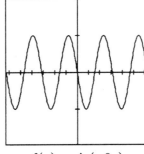

$f(x) = \sin(-2x)$
$[-2\pi, 2\pi]$ by $[-2, 2]$

Figure 2.6.19

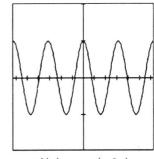

$f(x) = \cos(-2x)$
$[-2\pi, 2\pi]$ by $[-2, 2]$

Figure 2.6.20

The effect of the coefficient 2 of $x$ in $f(x) = \sin 2x$ is to shrink horizontally or compress the graph of $y = \sin x$ by a factor of $\frac{1}{2}$. That is, the portion of the graph of $y = \sin x$ for $0 \leq x \leq 2\pi$ becomes the portion of the graph of $f(x) = \sin 2x$ for $0 \leq x \leq \pi$. This is a *horizontal shrinking* of $y = \sin x$ by a factor of $\frac{1}{2}$. Similarly, the effect of the coefficient $\frac{1}{3}$ of $x$ in $g(x) = \sin \frac{1}{3}x$ is to stretch horizontally or expand the graph of $y = \sin x$ by a factor of 3. That is, the portion of the graph of $y = \sin x$ for $0 \leq x \leq 2\pi$ becomes the portion of the graph of $g(x) = \sin \frac{1}{3}x$ for $0 \leq x \leq 6\pi$. This is a *horizontal stretching* of $y = \sin x$ by a factor of 3.

Example 7 is a special case of the following theorem which we state without proof.

• **THEOREM 1**  Let $a$ be any real number. The function $f(x) = \sin ax$ is periodic with period $\frac{2\pi}{|a|}$.

• **EXAMPLE 9:** Draw a complete graph of $f(x) = \tan 3x$ and $g(x) = \tan \frac{1}{4}x$. Show that each function is periodic and determine its period.

**SOLUTION:** The period of $\tan x$ is $\pi$. As $3x$ varies from 0 to $\pi$, the values of $\tan 3x$ should produce a complete period of $f(x) = \tan 3x$. If $0 \leq 3x \leq \pi$, then $0 \leq x \leq \frac{\pi}{3}$. Thus, the period of $f(x) = \tan 3x$ is $\frac{\pi}{3}$, one-third the period of $y = \tan x$. The graph of $f(x) = \tan 3x$ in Figure 2.6.21 is complete and shows six periods.

As $\frac{1}{4}x$ varies from 0 to $\pi$, the values of $\tan \frac{1}{4}x$ should produce a complete period of $g(x) = \tan \frac{1}{4}x$. If $0 \leq \frac{1}{4}x \leq \pi$, then $0 \leq x \leq 4\pi$. Thus, the period of $g(x) = \tan \frac{1}{4}x$ is $4\pi$, four times the period of $y = \tan x$. The graph of $g(x) = \tan \frac{1}{4}x$ in Figure 2.6.22 is complete and shows two periods. •

Example 9 and Theorem 1 are special cases of the following more general theorem which we give without proof.

• **THEOREM 2**  If $b \neq 0$, the graph of $y = f(bx)$ can be obtained by horizontally stretching or shrinking the graph of $y = f(x)$ by the factor $\frac{1}{|b|}$, followed by a reflection in the

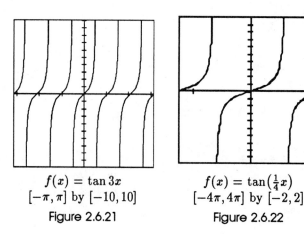

$f(x) = \tan 3x$
$[-\pi, \pi]$ by $[-10, 10]$
Figure 2.6.21

$f(x) = \tan(\frac{1}{4}x)$
$[-4\pi, 4\pi]$ by $[-2, 2]$
Figure 2.6.22

2.6 Graphs of tan $x$, cot $x$, sec $x$, and csc $x$

$y$-axis if $b < 0$. If, in addition, $f$ is periodic with period $p$, then $y = f(bx)$ is periodic with period $\frac{p}{|b|}$.

## EXERCISES 2-6

Use a calculator unless stated otherwise. Angles are measured in degrees or radians.

1. Use a graphing utility to draw the graph of $y = \sin x$ in the viewing rectangles $[0, 2\pi]$ by $[-2, 2]$, $[-2\pi, 0]$ by $[-2, 2]$, $[-8\pi, -6\pi]$ by $[-2, 2]$, and $[10\pi, 12\pi]$ by $[-2, 2]$. (Use $\pi = 3.14159$.)
2. Use a graphing utility to draw the graph of $y = \tan x$ in the viewing rectangles $[-\frac{\pi}{2}, \frac{\pi}{2}]$ by $[-10, 10]$, $[\frac{7\pi}{2}, \frac{9\pi}{2}]$ by $[-10, 10]$, $[-\frac{11\pi}{2}, -\frac{9\pi}{2}]$ by $[-10, 10]$, and $[\frac{71\pi}{2}, \frac{73\pi}{2}]$ by $[-10, 10]$. (Use $\pi = 3.14159$.)

Draw a complete graph without using a graphing utility. Determine the domain, range, period, and asymptotes (if any). Check using a graphing utility.

3. $y = 3 \tan x$
4. $y = -\tan x$
5. $y = \frac{1}{2} \sec x$
6. $y = \sec(-x)$
7. $y = 3 \csc x$
8. $y = 2 \tan x$
9. $y = 3 \sin x$
10. $y = -4 \sin 2x$
11. $y = -3 \tan \frac{1}{2} x$
12. $y = 2 \cot \frac{1}{2} x$
13. $y = 2 \csc x$
14. $y = -2 \sec \frac{1}{2} x$
15. $y = \sin 2x$
16. $y = \sin(-4x)$
17. $y = \sec 2x$
18. $y = \sec\left(-\frac{1}{2}x\right)$
19. $y = \sin \frac{1}{2} x$
20. $y = \tan 2x$
21. $y = 2 \sin 3x$
22. $y = \frac{1}{2} \sin \frac{1}{2} x$
23. $y = 3 \cos\left(-\frac{1}{2}x\right)$
24. $y = 2 \tan \frac{1}{2} x$

Draw a complete graph. Determine the domain, range, period, and asymptotes (if any) of each function.

25. $y = 2 \sin x$
26. $y = -3 + \sin \frac{\pi}{2} x$
27. $y = 2 \tan \pi x$
28. $y = -\tan \frac{\pi}{2} x$
29. $y = 3 \sec 2x$
30. $y = 4 \csc \frac{1}{3} x$

List, in proper order, the geometric transformations that can be used to obtain each graph from the graph of $y = \sin x$, $y = \cos x$, or $y = \tan x$.

31. $y = 2 \sin 3x$
32. $g(x) = \tan \frac{1}{2} x$
33. $y = 3 \cos \pi x$
34. $y = -3 \sin\left(-\frac{\pi}{2}x\right)$

Solve the equation for $-2\pi \leq x \leq 2\pi$.

35. $\sin x = 0.75$
36. $\cos x = 0.3$
37. $\tan x = 3.25$
38. $\cot x = -5.6$

Solve the inequality for $-2\pi \leq x \leq 2\pi$.

39. $2 \sin x > \frac{3}{2} \cos 2x$
40. $-\sin x < 3 \cos \frac{1}{2} x$
41. $2 \tan x < 5$
42. $3 \sec \frac{1}{2} x > \cos x$

Prove the identity.

43. $\cot x = \cot(x + \pi)$
44. $\sec x = \sec(x + 2\pi)$
45. $\csc x = \csc(x + 2\pi)$
46. $\cot x = \cot(x - \pi)$

We know all real solutions to $\sin x = 0.75$ are given by $x + 2k\pi$, $k = 0, \pm 1, \pm 2, \ldots$ where $x$ is 0.85 or 2.29. Use zoom-in to confirm.

47. For $k = 0$ (i.e., for $x = 0.85$ and $x = 2.29$)  48. For $k = 2$
49. For $k = -3$  50. For $k = 5$

Find all real solutions.

51. $\sin x = 0.25$  52. $\tan x = 4$
53. $\cos x = 0.42$  54. $\sec x = 3$
55. $\sin x < 0.15$  56. $\tan x > 2$
57. $\sin 3x = 0.55$  58. $\cos 2x = 0.85$
59. $\csc \frac{1}{2}x \leq 4$  60. $3 \sin \frac{1}{2}x > 2 \cos x$

## 2.7 Trigonometric Graphs

In this section we determine the graph of any function of the form $y = af(bx + c) + d$ where $f$ is one of the six basic trig functions. We also find the local extrema and determine the increasing and decreasing behavior of such functions. An application that involves finding a local extremum of a trigonometric graph will be investigated.

We know that the graph of $y = af(x)$ can be obtained by vertically stretching or shrinking the graph of $y = f(x)$ by the factor $|a|$, followed by a reflection in the $x$-axis if $a < 0$. The graph of $y = f(x + c)$ can be obtained by shifting the graph of $y = f(x)$ horizontally $|c|$ units left if $c > 0$ or right if $c < 0$. Similarly, the graph of $y = f(x) + d$ can be obtained by shifting the graph of $y = f(x)$ vertically $|d|$ units up if $d > 0$ or down if $d < 0$. In the previous section we showed that if $f$ is one of the six basic trig functions, then the function $y = f(bx)$ is periodic with period equal to the period of $f$ divided by $|b|$. In the last section we also found that the graph of $y = f(bx)$ could be obtained by horizontally stretching or shrinking the graph of $y = f(x)$ by the factor $\frac{1}{|b|}$, followed by a reflection in the $y$-axis if $b < 0$. Now, we combine all of this information to determine complete graphs of functions of the form $y = af(bx + c) + d$ where $f$ is one of the six basic trig functions. If $f$ is $\sin x$ or $\cos x$, then $f$ has a largest and a smallest value in the sense of the following definition.

- **DEFINITION** The function value $f(a)$ is called an **absolute maximum** of $f$ if $f(x) \leq f(a)$ for all $x$ in the domain of $f$. The function value $f(a)$ is called an **absolute minimum** of $f$ if $f(x) \geq f(a)$ for all $x$ in the domain of $f$. •

Notice $f(x) = x^2$ has zero as an absolute minimum, and $g(x) = x^3 - 4x$ has neither an absolute minimum nor an absolute maximum.

## 2.7 Trigonometric Graphs

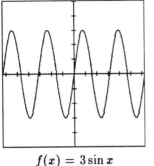

$f(x) = 3\sin x$
$[-2\pi, 2\pi]$ by $[-5, 5]$
Figure 2.7.1

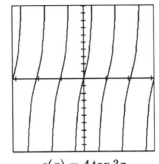

$g(x) = 4\tan 3x$
$[-\pi, \pi]$ by $[-5, 5]$
Figure 2.7.2

- **EXAMPLE 1:** Determine the period, the local maximum and minimum values, the absolute maximum and minimum values, and the increasing and decreasing behavior of each function and draw a complete graph.

  (a) $f(x) = 3\sin 2x$          (b) $g(x) = 4\tan 3x$

  **SOLUTION:**

  (a) A complete graph of $f$ can be obtained by vertically stretching the graph of $y = \sin 2x$ by a factor of 3. From the previous section, we know that the period of $\sin 2x$ is $\frac{2\pi}{2} = \pi$. Thus, the graph of $f$ in Figure 2.7.1 is complete and shows four periods. Because the period of $f$ is $\pi$ its complete behavior can be determined from any interval of length $\pi$, say $[0, \pi]$. In this interval, $f$ has a local maximum of 3 at $x = \frac{\pi}{4}$ and a local minimum of $-3$ at $x = \frac{3\pi}{4}$. (Why?) Because $f$ is periodic, there are infinitely many points at which $f$ has a local maximum value or a local minimum value. In fact, $f$ has a local maximum value of 3 at $\frac{\pi}{4} + k\pi$ and a local minimum value of $-3$ at $\frac{3\pi}{4} + k\pi$ for each integer $k$ because $f$ has period $\pi$. We can also see that 3 is an absolute maximum and $-3$ an absolute minimum of this function. This function alternates increasing and decreasing between consecutive extremum points. That is, if $x_1$, $x_2$, and $x_3$ are consecutive values of $x$ at which $f$ has extremum values, then $f$ increases in $[x_1, x_2]$ and decreases in $[x_2, x_3]$, or vice versa. In fact, $f$ decreases in $[\frac{\pi}{4}+k\pi, \frac{3\pi}{4}+k\pi]$ and increases in $[\frac{3\pi}{4}+k\pi, \frac{\pi}{4}+(k+1)\pi]$ for each integer $k$. (Why?)

  (b) The function $g(x) = 4\tan 3x$ is periodic with period $\frac{\pi}{3}$. (Why?) A complete graph of $g$ can be obtained by vertically stretching the graph of $y = \tan 3x$ by a factor of 4. Thus, the graph of $g$ in Figure 2.7.2 is complete. We can see that $g$ has no local extremum values. Furthermore, $g$ is increasing in each interval $(-\frac{\pi}{6} + k\pi, \frac{\pi}{6} + k\pi)$, $k$ an integer. •

The function $f(x) = 3\sin 2x$ of Example 1(a) has 3 for an absolute maximum value and $-3$ for an absolute minimum value, because $y = \sin 2x$ has 1 for an absolute maximum value and $-1$ for an absolute minimum value.

• **DEFINITION**  Let $f$ be a periodic function with an absolute maximum value of $M$ and an absolute minimum value of $m$. Then, $\frac{1}{2}(M-m)$ is called the **amplitude** of $f$. •

The function $f(x) = 3\sin 2x$ of Example 1(a) has an amplitude of 3. The function $g(x) = 4\tan 3x$ has *no* amplitude because $g$ has neither an absolute maximum value nor an absolute minimum value. The function $y = af(bx+c)+d$ has *no* amplitude if $f$ is the tangent, cotangent, secant, or cosecant.

• **EXAMPLE 2:** Determine the period, the amplitude, the local maximum and minimum values and where they occur, and the increasing and decreasing behavior of each function, and draw a complete graph.

(a) $f(x) = 2\cos\left(x - \frac{\pi}{2}\right)$ 　　　　　(b) $g(x) = 3\sec(x+1)$

**SOLUTION:**

(a) A complete graph of $f$ can be obtained by vertically stretching the graph of $y = \cos x$ by a factor of 2 and then shifting the resulting graph horizontally right $\frac{\pi}{2}$ units. Thus, $f$ is periodic with the same period as $y = \cos x$, and the graph of $f$ in Figure 2.7.3 is complete and shows two periods. The amplitude of $f$ is 2, and $f$ is periodic with period $2\pi$. The complete behavior of $f$ can be determined from any interval of length $2\pi$, say $[0, 2\pi]$. Now, $f$ has a local maximum value of 2 at $\frac{\pi}{2}$ and a local minimum value of $-2$ at $x = \frac{3\pi}{2}$. Thus, $f$ has a local maximum value of 2 at $x = \frac{\pi}{2} + 2k\pi$, a local minimum value of $-2$ at $x = \frac{3\pi}{2} + 2k\pi$, is decreasing in $[\frac{\pi}{2}+2k\pi, \frac{3\pi}{2}+2k\pi]$, and is increasing in $[\frac{3\pi}{2}+2k\pi, \frac{\pi}{2}+2(k+1)\pi]$ for each integer $k$.

(b) A complete graph of $g$ can be obtained by vertically stretching the graph of $y = \sec x$ by a factor of 3 and then shifting the resulting graph horizontally left 1 unit. Thus, $g$ is periodic with period $2\pi$, and the graph of $g$ in Figure 2.7.4 is complete and shows two periods. The function $g$ has *no* amplitude because $g$ has neither an absolute maximum value nor an absolute minimum value. The local extremum values of $y = \sec x$ in $[0, 2\pi]$ are 1 and $-1$, and they occur when $x = 0$ and $x = \pi$,

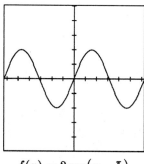

$f(x) = 2\cos\left(x - \frac{\pi}{2}\right)$
$[-2\pi, 2\pi]$ by $[-5, 5]$

Figure 2.7.3

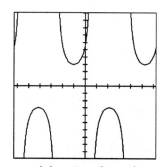

$g(x) = 3\sec(x+1)$
$[-2\pi, 2\pi]$ by $[-10, 10]$

Figure 2.7.4

## 2.7 Trigonometric Graphs

respectively. Thus, using the vertical stretch and horizontal shift mentioned above, the local extremum values of $g(x) = 3\sec(x+1)$ are 3 and $-3$, and these values occur when $x = 0 - 1 = -1$ and $x = \pi - 1$ (about 2.14), respectively. You can confirm these values using zoom-in. Thus, $g$ has a local minimum value of 3 at $x = -1 + 2k\pi$, a local maximum value of $-3$ at $x = \pi - 1 + 2k\pi$, is increasing in $[-1 + 2k\pi, \frac{\pi}{2} - 1 + 2k\pi)$ and $(\frac{\pi}{2} - 1 + 2k\pi, \pi - 1 + 2k\pi]$, and is decreasing in $[\pi - 1 + 2k\pi, \frac{3\pi}{2} - 1 + 2k\pi)$ and $(\frac{3\pi}{2} - 1 + 2k\pi, 2\pi - 1 + 2k\pi]$ for each integer $k$.  •

We sometimes say that the graph of $f(x) = 2\cos(x - \frac{\pi}{2})$ is obtained from the graph of $y = 2\cos x$ by a *phase shift* of $\frac{\pi}{2}$ units, or a *phase shift right* $\frac{\pi}{2}$ units. Similarly, the graph of $g(x) = 3\sec(x+1)$ is said to be obtained from the graph of $y = 3\sec x$ by a *phase shift* of $-1$ units, or a *phase shift left* 1 unit. The complete definitions will be given after the next example.

• **EXAMPLE 3**: Determine the domain, range, period, and amplitude of $f(x) = -4\cot(2x + \frac{\pi}{3})$, and then draw a complete graph.

**SOLUTION**: First we rewrite the expression for $f$ as follows:

$$f(x) = -4\cot 2(x + \frac{\pi}{6}).$$

Now we can see that the graph of $f$ can be obtained by applying, in order, the following transformations to the graph of $y = \cot x$:

1. A vertical stretch by a factor of 4 to obtain $y = 4\cot x$.
2. A reflection in the $x$-axis to obtain $y = -4\cot x$.
3. A horizontal shrink by a factor of $\frac{1}{2}$ to obtain $y = -4\cot 2x$.
4. A horizontal shift left $\frac{\pi}{6}$ units to obtain $f(x) = -4\cot 2(x + \frac{\pi}{6})$.

Further, $f$ has no amplitude because $f$ has neither an absolute maximum nor an absolute minimum. The range of $f$ is the set of all real numbers. The period of $f$ is $\frac{\pi}{2}$, because the period of $\cot x$ is $\pi$. The vertical asymptotes of $f$ are at $x = -\frac{\pi}{6} + k\frac{\pi}{2}$ and $x = \frac{\pi}{2} - \frac{\pi}{6} + k\frac{\pi}{2} = \frac{\pi}{3} + k\frac{\pi}{2}$, $k$ an integer. (Why?) Thus, the domain of $f$ consists of all real numbers different from $-\frac{\pi}{6} + k\frac{\pi}{2}$ or $\frac{\pi}{3} + k\frac{\pi}{2}$, where $k$ is any integer. Thus, the graph of $f$ in Figure 2.7.5 is complete.
The horizontal shift is $\frac{\pi}{6}$ and not $\frac{\pi}{3}$ because $f(x) = -4\cot(2x + \frac{\pi}{3}) = -4\cot 2(x + \frac{\pi}{6})$.  •

---

• **DEFINITION**  Let $f$ be a periodic function. The number $-\frac{c}{b}$ is called the **phase shift** of $y = f(bx + c)$ and is the horizontal shift required in obtaining the graph of $y = af(bx + c) + d$ from the graph of $f$. The phase shift of the function of Example 3 is $-\frac{\pi}{6}$.  •

• **EXAMPLE 4**: Find the domain, range, period, amplitude, and phase shift of $f(x) = 2\sin(3x + 4) - 1$, and draw a complete graph.

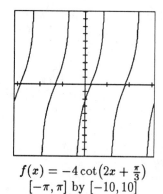

$f(x) = -4\cot\left(2x + \frac{\pi}{3}\right)$
$[-\pi, \pi]$ by $[-10, 10]$
Figure 2.7.5

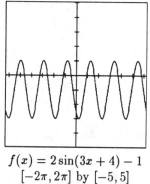

$f(x) = 2\sin(3x + 4) - 1$
$[-2\pi, 2\pi]$ by $[-5, 5]$
Figure 2.7.6

Figure 2.7.7

SOLUTION: We know that this graph should look very much like the graph of $y = \sin x$. We can express $f$ in the following form.

$$f(x) = 2\sin\left[3\left(x + \frac{4}{3}\right)\right] - 1$$

It follows from Theorem 1 of Section 2.6 that $f$ is periodic with period $\frac{2\pi}{3}$. We can also say that the effect of the coefficient 3 of $x$ is to shrink the graph of $y = \sin x$ horizontally by a factor of $\frac{1}{3}$. That is, the portion of the graph of $y = \sin x$ for $0 \le x \le 2\pi$ becomes the portion of the graph of $y = \sin 3x$ for $0 \le x \le \frac{2\pi}{3}$. This means that there are three periods of the graph of $f$ in the interval $[0, 2\pi]$. Next, the graph of $y = \sin 3x$ is shifted horizontally to the left $\frac{4}{3}$ units to obtain the graph of $y = \sin 3\left(x + \frac{4}{3}\right)$. Thus, the phase shift is $-\frac{4}{3}$. Then, the resulting graph is vertically stretched by a factor of 2 to obtain the graph of $y = 2\sin 3\left(x + \frac{4}{3}\right)$. Therefore, the amplitude is 2. Finally, the last graph is shifted vertically down 1 unit to obtain the graph of $f(x) = 2\sin(3x + 4) - 1$. The computer drawn graph of $f$ in Figure 2.7.6 confirms these observations.

The domain of $f$ is all real numbers. Figure 2.7.6 suggests the range is $[-3, 1]$. In general, it will not be necessary to go through the geometric transformations used above to produce the graph of $f$. We can see from the rule for $f$ that the sine wave is centered vertically about the line $y = -1$ and rises 2 units above this line at its highest points and falls 2 units below at its lowest points. (Why?) Then, we draw one complete sine wave with amplitude 2 starting at the point $\left(-\frac{4}{3}, -1\right) = (1.33, -1)$ and ending at the point $\left(-\frac{4}{3} + \frac{2\pi}{3}, -1\right) = (0.76, -1)$ (Figure 2.7.7). Notice that the points $\left(-\frac{4}{3}, -1\right)$ and $\left(-\frac{4}{3} + \frac{2\pi}{3}, -1\right)$ on the graph of $f$ correspond to the points $(0, 0)$ and $(2\pi, 0)$, respectively, on the graph of $y = \sin x$. ●

We can generalize Example 4 as follows.

SUMMARY   Let $f(x) = a\sin(bx + c) + d$. The graph of $f$ is a sine wave centered vertically about the line $y = d$ with one period starting at the point $\left(-\frac{c}{b}, d\right)$ and ending at the point $\left(-\frac{c}{b} + \frac{2\pi}{|b|}, d\right)$. The number $|a|$ is called the *amplitude* of the sine wave. The

## 2.7 Trigonometric Graphs

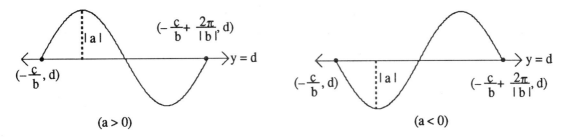

Figure 2.7.8

graph rises $|a|$ units above $y = d$ at its highest points and falls $|a|$ units below $y = d$ at its lowest points. $f$ is periodic with period $\frac{2\pi}{|b|}$. The number $-\frac{c}{b}$ is the *phase shift* and gives the horizontal shift necessary to produce the graph of $f$ from the graph of $y = \sin x$. If $a > 0$, the graph starts out increasing from $\left(-\frac{c}{b}, d\right)$; if $a < 0$, the graph starts out decreasing from $\left(-\frac{c}{b}, d\right)$ (Figure 2.7.8).

The graph of $g(x) = a\cos(bx + c) + d$ is related to the graph of $y = \cos x$ in exactly the same way that the graph of $h(x) = a\sin(bx + c) + d$ is related to the graph of $y = \sin x$. This is also true for the trig functions $\sec x$ and $\csc x$, except that there is no amplitude for these two functions.

The graph of $f(x) = a\tan(bx + c) + d$ is related to the graph of $y = \tan x$ in the same way that the graph of $g(x) = a\sin(bx + c) + d$ is related to the graph of $y = \sin x$ with two exceptions. The period of $f(x) = a\tan(bx + c) + d$ is $\frac{\pi}{|b|}$, where $\pi$ is the period of $y = \tan x$, and $f$ has no amplitude.

### Sinusoids

- **DEFINITION** Any function of the form $f(x) = a\sin(bx + c) + d$, where $a, b, c,$ and $d$ are real numbers, is called a **sinusoid**.

For example, the function $f$ of Example 4 is a sinusoid. In Theorem 2 of Section 2.5 we showed that $\cos x = \sin\left(x + \frac{\pi}{2}\right)$. From this it follows that any function of the form $f(x) = a\cos(bx + c) + d$ is also a sinusoid. There are other more complicated expressions that turn out to be sinusoidal. Sinusoids will be studied in more detail in Chapter 3. It turns out that the function of the next example is a sinusoid.

- **EXAMPLE 5:** A cross section of a tunnel is a semicircle of radius 20 feet. The interior surfaces of the tunnel form a rectangular opening as illustrated in Figure 2.7.9. Determine the dimensions of a tunnel opening with maximum cross-sectional area, and determine the value of the maximum area.

SOLUTION: Let $P(x, y)$ be the upper right-hand corner of the rectangle and $\theta$ the angle $OP$ makes with the positive $x$-axis (Figure 2.7.9). The area $A$ of the rectangle is $2xy$. Notice that

Figure 2.7.9

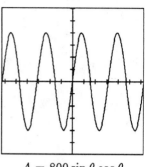

$A = 800 \sin \theta \cos \theta$
$[-2\pi, 2\pi]$ by $[-600, 200]$
Figure 2.7.10

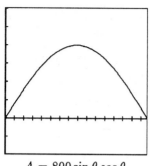

$A = 800 \sin \theta \cos \theta$
$[0, \frac{\pi}{2}]$ by $[-600, 600]$
Figure 2.7.11

$\cos \theta = \frac{x}{20}$ and $\sin \theta = \frac{y}{20}$. Thus, $A = 2xy = 800 \sin \theta \cos \theta$. Figure 2.7.10 gives a complete graph of $A = 800 \sin \theta \cos \theta$. The values of $\theta$ that make sense in this problem situation are $0 < \theta < \frac{\pi}{2}$ (Figure 2.7.11). Notice that $A$ is 0 when $x = \frac{\pi}{2}$. We can use zoom-in to estimate that the coordinates of the highest point of $A$ in $[0, \frac{\pi}{2}]$ are $(0.785, 400)$ with error at most 0.01. Thus, $x = 20\cos(0.785) = 14.15$ and $y = 20\sin(0.785) = 14.15$. It turns out that the exact coordinates of the highest point of $A$ between $\theta = 0$ and $\theta = \frac{\pi}{2}$ are $(\frac{\pi}{4}, 400)$. Thus, $x = 20\cos\frac{\pi}{4}$ and $y = 20\sin\frac{\pi}{4}$ for the rectangle with maximum cross-sectional area. The exact value of $\cos\frac{\pi}{4}$ or $\sin\frac{\pi}{4}$ is $\frac{1}{\sqrt{2}}$, or $\frac{\sqrt{2}}{2}$. Therefore, $x = y = 10\sqrt{2}$. The dimensions of the rectangle with maximum area are $20\sqrt{2}$ feet by $10\sqrt{2}$ feet, and the maximum area is 400 square feet. •

In Chapter 3 we will show that $\sin\theta\cos\theta = \frac{1}{2}\sin 2\theta$. This means that the algebraic representation of Example 5 is a sinusoid.

### EXERCISES 2-7

Sketch a complete graph without using a graphing utility. Check your answer with a graphing utility.

1. $y = 2 + \sin\left(x - \frac{\pi}{4}\right)$
2. $y = -1 - \cos(x - \pi)$
3. $y = -1 + 3\sin(x - 2)$
4. $y = 2 - 2\sin(x + 3)$
5. $f(x) = 2 - 3\sin(2x)$
6. $g(x) = -2 + \tan\left(\frac{1}{2}x\right)$
7. $y = 2 + 3\cos\left(\frac{x-1}{2}\right)$
8. $y = -2 - 4\sin(2x + 6)$
9. $T(x) = -3 + 2\sin(2x - \pi)$
10. $y = 2 - 3\cos(4x - 2\pi)$

Determine the domain, range, period, and asymptotes (if any) of each function in the specified Exercise.

11. Exercise 3
12. Exercise 5
13. Exercise 6
14. Exercise 8

## 2.7 Trigonometric Graphs

Determine where the function is increasing and decreasing and all local extrema for $x$ in $[0, 2\pi]$.

15. $y = 3 \sin 2x$
16. $y = 1 + 2\cos(x - \pi)$
17. $y = -\cot(x - \pi)$
18. $y = 2\sec\left(x + \frac{\pi}{2}\right)$
19. $y = 2 - 3\sin(4x - \pi)$
20. $y = -3 + 2\cos\left(\frac{1}{3}x - \frac{\pi}{6}\right)$

Find all real solutions.

21. $2 = 4 \sin 3x$
22. $3 = 2 \sin 3x$
23. $\cos \frac{1}{2}x = 0.24$
24. $4 \sin 2x < x$

Find all real solutions for $x$ in $[0, 5]$.

25. $5\sin\left(\frac{1}{4}x - \pi\right) = 3\cos\left(x + \frac{\pi}{2}\right)$
26. $\tan(x - \pi) = 3 \sin 2x$
27. $3 \sin x > 2 \cos(x - 1)$
28. $3 \sin^2 x > 2.65$ [Note: $\sin^2 x = (\sin x)^2$]

Write an equation of a sinusoid with given amplitude $A$, phase shift $B$, and period $P$, and draw a complete graph.

29. $A = 3$, $B = \frac{\pi}{2}$, $P = \pi$
30. $A = \frac{1}{3}$, $B = 2$, $P = 4$
31. $A = 2$, $B = -\frac{\pi}{4}$, $P = 4\pi$
32. $A = 5$, $B = -1$, $P = 1$

List the geometric transformations which if applied in order to $y = \sin x$ or $y = \tan x$ will produce the indicated graph.

33. $y = 3 - 4\sin(2x - \pi)$
34. $2 + 3\sin\left(\frac{1}{2}x + \frac{\pi}{2}\right)$
35. $y = 2 - \tan(x - \pi)$
36. $1 + 2\tan(2x - \pi)$

## Frequency

The *frequency* of a periodic function is the reciprocal of its period. One *cycle* of a periodic function is the graph in one period. It follows that the frequency represents the number of cycles of a periodic function in an interval of length *one unit*. Determine the period and frequency.

37. $y = 2 \sin 3x$
38. $y = -4 \cos \frac{1}{2}x$
39. $y = 3 \sin \frac{\pi}{2}x$
40. $y = 2 \cos 2\pi x$

## Harmonic Motion

It can be shown that $y = A \cos Bt$ is a sinusoidal model for the bouncing motion of an object hung from the end of a spring, or *simple harmonic motion*. Assume a coordinate system has been introduced so the vertical axis of the spring coincides with the $y$-axis and the rest position of the object is at the origin. Further assume that at $t = 0$ the object has been stretched to a position $(0, A)$ when $A < 0$ and released (when $t > 0$). Let $y$ be the vertical position of the object as a function of time $t$. We assume there is no friction to stop the motion once it is started.

An object on a spring oscillates between $y = -15$ and $y = 15$, and the frequency of oscillation is $\frac{1}{4}$ of a cycle per second.

41. Determine the time interval of one complete cycle (i.e., the period) and then determine $A$ and $B$ of the algebraic representation.

42. Draw a complete graph of the algebraic representation.

43. What part of the graph represents the problem situation?
44. At what times is the object 5 units from the rest position?

An object on a spring oscillates between $y = -28$ and $y = 28$, and the *frequency* of oscillation is 3 cycles per second.

45. Determine the time interval of one complete cycle (i.e., the period) and then determine $A$ and $B$ of the algebraic representation.
46. Draw a complete graph of the algebraic representation.
47. What part of the graph represents the problem situation?
48. At what times is the object 5 units from the rest position?

## Damped Motion

Actual oscillatory motion of an object on a spring is affected by friction so eventually the object returns to an at rest position. Such motion is called damped.

49. Draw a complete graph of $y = (20 - x)\cos(x - 3)$.
50. Explain how a portion of the graph in Exercise 49 could be a model for the motion of an object hung from a stretched spring. (This motion is called *dampened* oscillatory motion.)
51. Determine the length of a belt running around two wheels of radii 22 inches and 15 inches, respectively. Assume the wheels are 50 inches apart. (*Hint:* See Figure 2.7.12. Compute $d$, then compute $\angle \alpha$.)
52. Determine the length of a belt running around two wheels of radii 18 inches and 11 inches, respectively. Assume the wheels are 36 inches apart. (*Hint:* See Figure 2.7.12. Compute $d$, then compute $\angle \alpha$.)

A playground merry-go-round 16 feet in diameter rotates at 20 rpm (revolutions per minute).

53. Determine the angular velocity in radians per second.
54. Determine the speed (in mph) of a boy located at a point 2 feet from the center.
55. Determine the speed (in mph) of a girl located at the edge of the merry-go-round.

Figure 2.7.12

56. Solve Example 5 using a non-trig algebraic representation for the cross-sectional area. (*Hint:* $x^2 + y^2 = 400$.)

# • KEY TERMS

Listed below are the key terms, vocabulary, and concepts in this chapter. You should be familiar with their definitions and meanings as well as be able to illustrate each of them.

| | |
|---|---|
| Acute angle | Periodic functions |
| Amplitude | Perpendicular (orthogonal) |
| Angle | Phase shift |
| Angular speed | Positive and negative measure |
| Arc length | Quadrantal angle |
| Bearing | Radians |
| Central angle | Reciprocal identities |
| Circular property of trig functions | Right angle |
| Complementary angles | Right triangle |
| Coterminal angles | Second |
| Degrees | Sinusoids |
| Horizontal shifting | Solving right triangles |
| Horizontal shrinking | Standard position |
| Hypotenuse | Supplementary angle relationships |
| Identities | Supplementary angles |
| Initial side | Terminal side |
| Linear speed | Trigonometric identities |
| Minute | Trigonometric functions |
| Opposite side | Vertex of an angle |
| Period | Zeros of periodic functions |

# • REVIEW EXERCISES

This section contains representative problems from each of the previous seven sections. You should use these problems to test your understanding of the material covered in this chapter.

Find the value of the six trig functions at

1. $32°$
2. 2 radians

| | |
|---|---|
| $\sin 30° = ?$ | $\csc \frac{\pi}{2} = ?$ |
| $\cos 270° = ?$ | $\sec 270° = ?$ |
| $\tan 135° = ?$ | $\cot \frac{\pi}{4} = ?$ |
| $\sin \frac{5\pi}{6} = ?$ | $\csc 210° = ?$ |
| $\cos \frac{2\pi}{3} = ?$ | $\sec 330° = ?$ |
| $\cot 135° = ?$ | $\cot \frac{5\pi}{4} = ?$ |

Table 1

Figure R.1

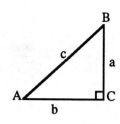

Figure R.2

3. Complete Table 1. Find *exact* values.

Use properties of a 30°–60° right triangle and a 45° right triangle to determine the *exact* values of each trig function in Table 2.

10. Consider the triangle $ABC$ in Figure R.1. Find all six trig functions of $\angle A$.

11. Use a right triangle to determine the values of all trig functions at $\theta$ with $\cos\theta = \frac{5}{7}$.

12. Use the [INV] [COS], [2nd F] [COS] or [SHIFT] [COS] keys on your calculator to determine $0° \le \theta \le 90°$ with $\cos\theta = \frac{5}{13}$. Make sure your calculator is in degree mode.

13. Show that $\cot\theta = \dfrac{\cos\theta}{\sin\theta}$.

14. Determine $\theta$ if $\cot\theta = \dfrac{12}{5}$.

Solve the right triangle in Figure R.2.

15. $\angle A = 35°$, $c = 15$

16. $b = 8$, $c = 10$

17. Find the equivalent radian measure of $60°$.

18. Find the equivalent degree measure of $\frac{3\pi}{4}$ radians.

19. Determine the measure and draw the angle in standard position.

    (a) $\frac{3}{4}$ counterclockwise rotation
    (b) $2\frac{1}{2}$ counterclockwise rotations

20. Let $\theta$ be an angle in standard position with the point $P$ on the terminal side of $\theta$. Find the values of the six trig functions at $\theta$ if $P = (-3, -2)$.

21. Determine $\theta$ in Exercise 14 if $0° \le \theta \le 360°$.

22. Determine *all* values of $\theta$ that satisfy the conditions of Exercise 20.

23. Let $\theta$ be an angle in standard position with terminal side given by the equation $y = \frac{4}{3}x$ in the first quadrant. Find the values of the six trig functions at $\theta$.

Verify the identity.

24. $1 + \cot^2\theta = \dfrac{1}{\sin^2\theta}$

25. $\dfrac{\sin^2\theta}{1-\cos\theta} = 1 + \cos\theta$

26. Use the circular properties of the trig functions to verify $\cos 1 > \cos 4$. Do not use a calculator.

# Review Exercises

| | $\theta$ | $\sin\theta$ | $\cos\theta$ | $\tan\theta$ | $\cot\theta$ | $\sec\theta$ | $\csc\theta$ |
|---|---|---|---|---|---|---|---|
| 4. | 0 | | | | | | |
| 5. | $-\frac{\pi}{6}$ | | | | | | |
| 6. | $\frac{3\pi}{4}$ | | | | | | |
| 7. | 60° | | | | | | |
| 8. | −135° | | | | | | |
| 9. | 300° | | | | | | |

Table 2

27. Determine the angle between the $x$-axis and the line $2x - 7y = 5$.

$\theta$ is an angle in standard position. Find the quadrant containing $\theta$ if $0° \leq \theta \leq 360°$.

28. $\sin\theta < 0$ and $\tan\theta < 0$
29. $\cos\theta < 0$ and $\csc\theta > 0$
30. Find an acute angle $\alpha$ satisfying the condition $\cos(90° - \alpha) = \cos\alpha$.
31. Determine the measure of an angle $\theta$ coterminal with the given angle satisfying the specified condition.
    (a) 34°; $360° \leq \theta \leq 720°$
    (b) $\frac{-3\pi}{4}$; $0 \leq \theta \leq 2\pi$
32. A central angle $\theta$ subtends an arc of length $s$ on a circle of diameter 10 units. Determine $s$ for the following central angles.
    (a) $\theta = \frac{\pi}{4}$
    (b) $\theta = 100°$

Find *all* real solutions to each equation.

33. $\cos x = \frac{\sqrt{3}}{2}$
34. $\tan x = 0$

Sketch a complete graph without using a graphing utility. Check your answer with a graphing utility.

35. $y = \sin(x + 45°)$
36. $y = 3 + 2\cos x$
37. $y = -\sin(x + \frac{\pi}{2})$
38. $y = -2 - 3\sin(x - \pi)$
39. Determine a complete graph, domain, range, and period of the function $y = 3\sin 2x$.
40. Sketch a complete graph of $y = 1 + 2\cos 3x$ without using a graphing utility. Check your answer with a graphing utility. Determine the domain, range, and period.
41. Determine the domain, range, period, and asymptotes (if any) of the function $y = 1 + 2\sec 3x$.

List, in proper order, the geometric transformations that can be used to obtain each graph from the graph of $y = \sin x$ or $y = \tan x$.

42. $y = 4 + \sin(2x - 1)$
43. $g(x) = -5 + \tan(\frac{1}{2}x)$

Solve the equation given that $0 \leq x \leq 2\pi$.

44. $\sin x = 0.45$

45. $1 - x = \cos x$

46. Solve the inequality $3 \cos x < x + 2$.

True or false? Give an argument based on graphs.

47. $\sin(x + \pi) = \sin x$

48. $\cos\left(x + \frac{\pi}{2}\right) = -\cos x$

49. If $\cos \theta < 0$ and $\tan \theta < 0$, then $\theta$ is in which quadrant?

50. The angle of elevation of the top of a building from a point 100 meters away from the building on level ground is 78°. Determine the height of the building.

51. A tree casts a shadow 51 feet long when the angle of elevation of the sun (measured from the horizon) is 25°. How tall is the tree?

52. From the top of a 150 foot building a man observes a car moving toward him. If the angle of depression of the car changes from 18° to 42° during the observation, how far does the car travel?

53. A lighthouse $L$ stands 4 miles from the closest point $P$ along a straight shore. Find the distance from $P$ to a point $Q$ along the shore if the angle $PLQ$ measures 22°.

54. An airplane at a certain time is between two signal towers that are on line with bearing 0° (along a north–south line). The bearing between the plane and each tower is 23° and 128° respectively. Use a drawing to show the exact location of the plane.

55. The bearings of lines of sight from a boat to two points on the shore are 115° and 123°, respectively. Assume the two points are 855 feet apart. How far is the boat from the nearest point on shore if the shore is a straight line?

Sketch the reference triangle and the reference angle $\theta'$ determined by each angle. Determine the reference angle and the values of the six trig functions at $\theta'$ and $\theta$.

56. 128°

57. −204°

# 3 Analytic Trigonometry

## • Introduction

In this chapter we continue the study of trig functions, find their local extreme values and zeros, and determine their end behavior. We see how to write certain trig functions as sinusoidal functions, that is, as functions of the form $a\sin(bx+c)+d$. Applications involving sinusoidal functions will be investigated.

Trigonometric identities involving double-angles and half-angles will be established. The inverse trig functions are defined and their complete graphs determined. Equations and inequalities involving trig functions are solved both algebraically and graphically. The law of sines and the law of cosines are established and used in applications. Two formulas for the area of a triangle that use the measure of its angles and sides will be obtained. A formula for the angle between two lines is determined.

## 3.1 More on Trigonometric Graphs

The investigation of graphs of trig functions continues in this section. We see how to rewrite certain sums of trig functions as sinusoidal functions, and we determine the end behavior and local extrema of functions that involve trig functions. We investigate an application that has a model involving a trig function.

In Section 2.7 we indicated that a sinusoid is any function that can be written in the form $f(x) = a\sin(bx+c)+d$, where $a$, $b$, $c$, and $d$ are real numbers. The graph of a sinusoid is shaped like the graph of the sine function. (Why?) We showed in Theorem 2 of Section 2.5 that $\cos x = \sin(x + \frac{\pi}{2})$. Thus, $f(x) = \cos x$ is a sinusoid. We investigate other sinusoidal functions in Examples 1 and 2.

In the Exercises you will show that the following theorem is a special case of Theorem 2 that appears later in this section.

• **THEOREM 1**   For all real numbers $a$ and $b$, there exist real numbers $A$ and $\alpha$ such that
$$a \sin x + b \cos x = A \sin(x + \alpha).$$

More precisely, we have the following:

(1) If $a = 0$, then $\alpha = \frac{\pi}{2}$ and $A = b$.
(2) If $a \neq 0$, then $-\frac{\pi}{2} \leq \alpha \leq \frac{\pi}{2}$ with $\tan \alpha = \frac{b}{a}$. If $a > 0$, then $A = \sqrt{a^2 + b^2}$, and if $a < 0$, then $A = -\sqrt{a^2 + b^2}$.

• **EXAMPLE 1:** Determine the domain, range and period of $f(x) = 2 \sin x + 5 \cos x$, and draw a complete graph.

SOLUTION: The graph in Figure 3.1.1 appears to be the graph of a sinusoid. We can estimate from this figure that the amplitude is 5.4, the period $2\pi$, and the phase shift 1.2 radians left. This information can be confirmed geometrically by overlaying the graph of $y = 5.4 \sin(x + 1.2)$ in Figure 3.1.1. Theorem 1 confirms algebraically that $f$ is a sinusoid and that the graph in Figure 3.1.1 is complete.  •

• **EXAMPLE 2:** Determine $A$ and $\alpha$ so that $A \sin(x + \alpha) = 2 \sin x + 5 \cos x$.

SOLUTION: From Theorem 1 we have $A = \sqrt{2^2 + 5^2} = \sqrt{29}$ and $\tan \alpha = \frac{5}{2}$. Thus, $\alpha = 1.19$. Accurate to tenths, $A = 5.4$, and $\alpha = 1.2$.  •

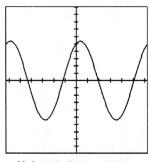

$f(x) = 2 \sin x + 5 \cos x$
$[-2\pi, 2\pi]$ by $[-10, 10]$
Figure 3.1.1

## 3.1 More on Trigonometric Graphs

Notice that the values of $A$ and $\alpha$ found in Example 2 are consistent with the approximations found in Example 1. We can now be sure that the function of Example 1 is a sinusoid. That is,

$$f(x) = 2\sin x + 5\cos x = \sqrt{29}\sin(x + \alpha)$$

where $-\frac{\pi}{2} \leq \alpha \leq \frac{\pi}{2}$ and $\tan \alpha = 2.5$.

- **EXAMPLE 3:** Determine $A$ and $\alpha$ so that $A\sin(x + \alpha) = -3\sin x + 2\cos x$ and confirm geometrically.

  **SOLUTION:** From Theorem 1 we have $A = -\sqrt{(-3)^2 + 2^2} = -\sqrt{13}$ and $\tan \alpha = -\frac{2}{3}$. Thus, $\alpha = -0.59$. The graph of $y = -3\sin x + 2\cos x$ in Figure 3.1.2 appears to be a sinusoid with period $2\pi$, amplitude 3.6, and phase shift 0.6 radians right. We can overlay the graph of $y = -\sqrt{13}\sin(x - 0.59)$ in Figure 3.1.2 to confirm geometrically that $-3\sin x + 2\cos x = -\sqrt{13}\sin(x - 0.59)$. •

- **EXAMPLE 4:** Show that $f(x) = 3\sin(2x - 1) + 4\cos(2x + 3)$ is a sinusoid.

  **SOLUTION:** The graph in Figure 3.1.3 suggests that $f$ is a sinusoid with period $\pi$, amplitude 6.6, and phase shift 0.7 radians right. This information can be confirmed geometrically by overlaying the graph of $y = 6.6\sin 2(x - 0.7)$ in Figure 3.1.3. An algebraic confirmation is provided by Theorem 2. •

Example 4 is a special case of the following theorem which will be proved in the next section.

- **THEOREM 2**     For all real numbers $a, b, d, h$, and $k$, there exist real numbers $A$ and $\alpha$ such that
$$a\sin(bx + h) + d\cos(bx + k) = A\sin(bx + \alpha).$$

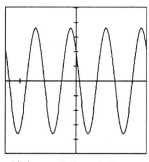

$f(x) = -3\sin x + 2\cos x$

$[-4\pi, 4\pi]$ by $[-5, 5]$

Figure 3.1.2

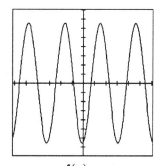

$f(x) = 3\sin(2x-1) + 4\cos(2x+3)$

$[-2\pi, 2\pi]$ by $[-8, 8]$

Figure 3.1.3

More precisely, we have the following:

(1) If $a\cos h - d\sin k = 0$ then $\alpha = \frac{\pi}{2}$ and $A = a\sin h + d\cos k$.
(2) If $a\cos h - d\sin k \neq 0$, then $-\frac{\pi}{2} \leq \alpha \leq \frac{\pi}{2}$ with $\tan\alpha = \frac{a\sin h + d\cos k}{a\cos h - d\sin k}$ and $A = \pm\sqrt{(a\sin h + d\cos k)^2 + (a\cos h - d\sin k)^2}$. $A$ and $a\cos h - d\sin k$ have the same sign.

We can use Theorem 2 to determine $A$ and $\alpha$ so that $3\sin(2x-1) + 4\cos(2x+3) = A\sin(2x+\alpha)$. This confirms that the function of Example 4 is a sinusoid. In the Exercises, you will be asked to use the formulas of Theorem 2 to determine the values of $A$ and $\alpha$.

The functions in the next two examples are not sinusoids.

• **EXAMPLE 5:** Determine the domain, the range, the period, the local maximum and minimum values in one period, and draw a complete graph of $f(x) = \sin 2x + \cos 3x$.

**SOLUTION:** We show that the graph in Figure 3.1.4 is a complete graph. The domain of $f$ is the set of all real numbers. (Why?) The function $f$ is periodic because both $\sin 2x$ and $\cos 3x$ are periodic. Certainly $2\pi$ is a candidate for the period of $f$. Notice that $y = \sin 2x$ is periodic with period $\pi$ and $y = \cos 3x$ is periodic with period $\frac{2\pi}{3}$. Thus, the period of $f$ must be a positive integer multiple of both $\pi$ and $\frac{2\pi}{3}$. Their least common multiple is $2\pi$, and so $f$ has period $2\pi$.

The graph of $f$ in the $[0, 2\pi]$ by $[-3, 3]$ viewing rectangle displays one complete period of $f$ (Figure 3.1.5). We need to find the smallest and largest value of $f$ in this figure to determine the range of $f$. Notice that $f$ has six extreme values in Figure 3.1.5. We do not include the endpoints in our consideration about extreme values of $f$ because, when the entire graph is considered, $f$ does *not* have extreme values at the endpoints.

The absolute maximum of $f$ in $[0, 2\pi]$ is 1.91 and occurs when $x$ is 4.11. The other two local maximum values are 1.22 and 0.22 and occur when $x$ is 0.22 and 1.91, respectively. Similarly, the absolute minimum of $f$ in $[0, 2\pi]$ is $-1.91$ and occurs when $x$ is 5.32. The other two local minimum values are $-0.22$ and $-1.22$ and occur when $x$ is 1.23 and 2.92,

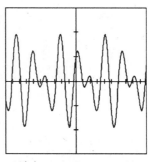

$f(x) = \sin 2x + \cos 3x$
$[-10, 10]$ by $[-3, 3]$

Figure 3.1.4

respectively. Thus, the range of $f$ is $[-1.91, 1.91]$ with error at most 0.01 in the endpoints of the interval. •

Figure 3.1.5 shows that $f(x) = \sin 2x + \cos 3x$ has six zeros in the interval $[0, 2\pi]$. Zoom-in could be used to find these zeros. They can also be determined algebraically, but several trig identities are needed to get started. In Section 3.2 we establish the necessary identities and find the zeros algebraically.

• EXAMPLE 6: Determine the domain, the range and the period of $f(x) = \sin \frac{x}{2} + \sin \frac{x}{3}$, and draw a complete graph.

SOLUTION: The domain of $f$ is the set of all real numbers. Notice that $\sin \frac{x}{2}$ is periodic with period $4\pi$ and $\sin \frac{x}{3}$ is periodic with period $6\pi$. (Why?) The period of $f$ is the least common multiple of $4\pi$ and $6\pi$. (Why?) Thus, the period of $f$ is $12\pi$. The graph of $f$ in Figure 3.1.6 is a complete graph displaying two complete periods and confirms these observations.

The absolute maximum of $f$ in $[0, 12\pi]$ is 1.91 and occurs when $x$ is 3.62. The absolute minimum of $f$ in $[0, 2\pi]$ is $-1.91$ and occurs when $x$ is 34.08. Thus, the range of $f$ is $[-1.91, 1.91]$. •

The function in the next example has interesting end behavior. Its graph turns out to be symmetric with respect to the $y$-axis. Recall that a graph of an equation is symmetric with respect to the $y$-axis if the equation is unchanged when $x$ is replaced by $-x$. We have also seen that a graph of a function $f$ is symmetric with respect to the $y$-axis if $f(-x) = f(x)$, that is, if $f$ is an even function.

• EXAMPLE 7: Determine the end behavior and draw a complete graph of $y = x \sin x$. Show that the graph is symmetric with respect to the $y$-axis.

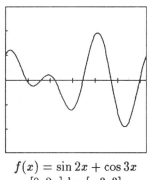

$f(x) = \sin 2x + \cos 3x$
$[0, 2\pi]$ by $[-3, 3]$
Figure 3.1.5

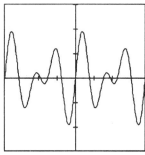

$f(x) = \sin \frac{x}{2} + \sin \frac{x}{3}$
$[-12\pi, 12\pi]$ by $[-3, 3]$
Figure 3.1.6

SOLUTION: Notice that the equation for this function is unchanged when $x$ is replaced by $-x$. We use the fact that $\sin x$ is an odd function, that is, $\sin(-x) = -\sin x$:

$$y = x \sin x$$
$$= (-x)\sin(-x)$$
$$= (-x)(-\sin x)$$
$$= x \sin x$$

Thus, $f$ is an even function, and its graph is symmetric with respect to the $y$-axis. You may need to experiment with different viewing rectangles to convince yourself that the graph in Figure 3.1.7 is a complete graph. The amplitude of the sine-like wave gets larger and larger as $|x| \to \infty$. In fact, the graph bounces back and forth between the lines $y = x$ and $y = -x$ as illustrated in Figure 3.1.8.

Thus, the end behavior of $y = x \sin x$ is a sine-like wave with amplitude that approaches $\infty$ as $|x| \to \infty$. •

The graph of the function $f$ in the next example is symmetric with respect to the origin because it is an odd function. That is, $f(-x) = -f(x)$. Moreover, this function and the function of Example 9 have very interesting behavior near $x = 0$.

• **EXAMPLE 8**: Determine the end behavior and the behavior near $x = 0$ of $f(x) = \sin \frac{1}{x}$, and draw a complete graph.

SOLUTION: The graph of $f$ in Figure 3.1.9 suggests that the end behavior of $f$ is the horizontal line $y = 0$. We can use zoom-out to confirm this conjecture. That is, $f$ has the $x$-axis as a horizontal asymptote. This figure also shows that the behavior of $f$ near $x = 0$ is peculiar. The graph of $f$ for $-0.1 \leq x \leq 0.1$ in Figure 3.1.10 gives a closer look at the behavior of $f$ near $x = 0$.

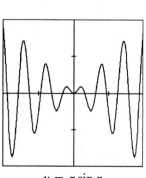

$y = x \sin x$
$[-20, 20]$ by $[-20, 20]$

Figure 3.1.7

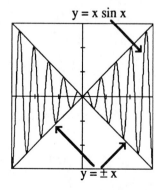

$y = x \sin x$
$y = \pm x$
$[-30, 30]$ by $[-30, 30]$

Figure 3.1.8

$f(x) = \sin \frac{1}{x}$
$[-2\pi, 2\pi]$ by $[-2, 2]$

Figure 3.1.9

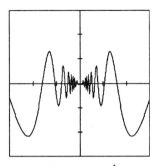

$f(x) = \sin \frac{1}{x}$
$[-0.1, 0.1]$ by $[-2, 2]$
Figure 3.1.10

$f(x) = x \sin \frac{1}{x}$
$[-10, 10]$ by $[-2, 2]$
Figure 3.1.11

$f(x) = x \sin \frac{1}{x}$
$[-0.3, 0.3]$ by $[-0.3, 0.3]$
Figure 3.1.12

The function $f(x) = \sin \frac{1}{x}$ takes on all values between $-1$ and $1$ in smaller and smaller intervals near $x = 0$ and $f(0)$ does not exist. For example, $f(x) = \sin \frac{1}{x}$ takes on all values between $-1$ and $1$ in the interval $\left[\frac{1}{12\pi}, \frac{1}{10\pi}\right]$. We can see this as follows. If $\frac{1}{12\pi} \leq x \leq \frac{1}{10\pi}$, then $12\pi \geq \frac{1}{x} \geq 10\pi$. Thus, $\sin \frac{1}{x}$ takes on the same values in the interval $\frac{1}{12\pi} \leq x \leq \frac{1}{10\pi}$ that $\sin x$ does in the interval $10\pi \leq x \leq 12\pi$. Notice that $\frac{1}{12\pi}$ is approximately $0.0265$ and $\frac{1}{10\pi}$ is approximately $0.0318$. As $x$ approaches $0$ from either side, the graph of $f$ consists of linking together end-to-end complete sine-like waves that have smaller and smaller periods. The graphs of $f$ in Figures 3.1.9 and 3.1.10 constitute a complete graph of $f$. •

The function in the next example also has a horizontal asymptote. Its graph is symmetric with respect to the $y$-axis because it is an even function.

• **EXAMPLE 9:** Determine the end behavior, the behavior near $x = 0$, and draw a complete graph of $f(x) = x \sin \frac{1}{x}$.

SOLUTION: The graph of $f$ in Figure 3.1.11 shows that $f(x) \to 1$ as $|x| \to \infty$. That is, $f$ has the line $y = 1$ as a horizontal asymptote. As in the previous example, the behavior near $x = 0$ is peculiar. The graph of $f$ in Figure 3.1.12 gives a closer look near $x = 0$.

As we saw in Example 8, $\sin \frac{1}{x}$ takes on all values between $-1$ and $1$ in smaller and smaller intervals near $x = 0$. The effect of multiplying $\sin \frac{1}{x}$ by $x$ is to cause the graph of $f$ to bounce back and forth between $y = x$ and $y = -x$ near $x = 0$ (Figure 3.1.13). Therefore, the graphs of $f$ in Figures 3.1.11 and 3.1.12 constitute a complete graph of $f$. •

• **EXAMPLE 10:** Determine the domain and the range of $f(x) = 2^{\sin x}$, and draw a complete graph.

SOLUTION: The graph of $f$ in Figure 3.1.14 suggests that the domain of $f$ is the set of all real numbers. The range of $f$ appears to be about $\left[\frac{1}{2}, 2\right]$. The function $f$ is periodic with period $2\pi$ because $\sin x$ is periodic with period $2\pi$. (Why?) Because $\sin x$ takes on all values between $-1$ and $1$ and $2^x$ is an increasing function, $f$ takes on all values between $2^{-1}$

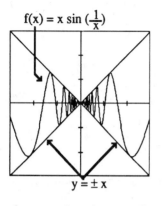

$f(x) = x \sin\left(\frac{1}{x}\right)$

$y = \pm x$

$[-0.3, 0.3]$ by $[-0.3, 0.3]$

Figure 3.1.13

$f(x) = 2^{\sin x}$

$[-10, 10]$ by $[-4, 4]$

Figure 3.1.14

Figure 3.1.15

and $2^1$. Thus, the domain of $f$ is the set of all real numbers, and the range of $f$ is $\left[\frac{1}{2}, 2\right]$. Therefore, the graph of $f$ in Figure 3.1.14 is a complete graph.

A hole is drilled through the center of a sphere of radius 10 (Figure 3.1.15). The portion of the sphere drilled out consists of a right circular cylinder and two spherical caps. Let $r$ and $h$ be the radius and height, respectively, of the cylinder drilled out. We want to determine the volume of the hole in the form of a right circular cylinder that is created by the drilling. Draw a line (radius of sphere) from the center of the sphere to the top of the cylinder as illustrated in Figure 3.1.15. Let $\theta$ be the angle that this radius makes with the horizontal.

- **EXAMPLE 11:** A hole is drilled through the center of a sphere as illustrated in Figure 3.1.15.

    (a) Express the volume $V$ of the right circular cylinder as a function of the angle $\theta$ defined above.
    (b) Draw a complete graph of $V$. What portion of the graph represents the problem situation?
    (c) Determine the dimensions of a cylinder of the largest possible such hole that can be created in the sphere. What is the maximum volume?

SOLUTION:

(a) The volume of the cylinder is $V = \pi r^2 h$. Refer to Figure 3.1.15. Notice that $\sin \theta = (h/2)/10$ and $\cos \theta = \frac{r}{10}$. Thus, $h = 20 \sin \theta$, $r = 10 \cos \theta$ and $V = 2000\pi \sin \theta \cos^2 \theta$.

(b) The graph of $V$ in the $[-10, 10]$ by $[-4000, 4000]$ viewing rectangle is shown in Figure 3.1.16. It can be shown that $V$ is periodic with period $2\pi$. Thus, this graph is a complete graph. We can see from Figure 3.1.15 that $0 < \theta < \frac{\pi}{2}$. Therefore, only the portion of Figure 3.1.16 between 0 and $\frac{\pi}{2}$ represents the problem situation (Figure 3.1.17).

### 3.1 More on Trigonometric Graphs

$V = 2000\pi \sin\theta \cos^2\theta$
$[-10, 10]$ by
$[-4000, 4000]$
Figure 3.1.16

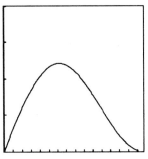

$V = 2000\pi \sin\theta \cos^2\theta$
$[0, \frac{\pi}{2}]$ by
$[0, 4000]$
Figure 3.1.17

(c) We can use zoom-in to determine that coordinates of the highest point of the graph of $V$ in $0 \le \theta \le \frac{\pi}{2}$ are $(0.615, 2418.399)$ with error of at most 0.01. Thus, the maximum volume is 2418.399 cubic units and occurs when $\theta$ is 0.615 radians. We can use $h = 20\sin\theta$ and $r = 10\cos\theta$ to determine that $h$ is 11.54 and $r$ is 8.17. •

## EXERCISES 3-1

Draw a complete graph and show that it is a sinusoid $y = A\sin(bx + \alpha)$ by estimating $A$, $b$, and $\alpha$ and overlaying the graph of $y = A\sin(bx + \alpha)$ to check your estimated values.

1. $y = 3\sin x + 2\cos x$
2. $y = -5\sin x + 3\cos x$
3. $y = -\sin(x + \pi) + \cos x$
4. $y = 4\sin 2x - 3\cos 2x$

Use the formulas in Theorem 2 to verify the values of amplitude and phase shift observed in Example 4.

5. Determine the amplitude factor.
6. Determine the phase shift factor.

Determine the domain, range and period of the function, and draw a complete graph. Determine all local and absolute maximum and minimum values in the interval $[0, 2\pi]$ of each function.

7. $f(x) = \sin x + \cos 2x$
8. $g(x) = 2\sin x + 3\cos 2x$
9. $f(x) = \sin 3x + \cos x$
10. $t(x) = 3\sin 2x - \cos x$

Draw a complete graph and determine the end behavior of the function.

11. $y = \dfrac{\sin x}{x}$
12. $y = x^2 \sin x$
13. $y = (-x)\sin\dfrac{1}{x}$
14. $y = x^2 \sin\dfrac{1}{x}$
15. $y = \ln|\sin x|$
16. $y = e^{\sin x}$

Draw a complete graph of the function. Examine the graph for symmetry. Indicate any horizontal or vertical asymptotes.

17. $y = |\sin x|$

18. $y = \dfrac{x}{\sin x}$

19. $y = \left| x \sin \dfrac{1}{x} \right|$

20. $y = \dfrac{\sin x}{x^2 - 4}$

A right circular cylinder is inscribed in a sphere of radius 20.

21. Express the volume $V$ of the cylinder as a function of the angle $\theta$.
22. Draw a complete graph of $V$. Which portion of the graph represents the problem situation?
23. Determine the dimensions of a cylinder of maximum possible volume that can be inscribed in the sphere. What is the maximum volume?
24. Find *all* maximum and minimum values of $g(x) = 3\cos x + 3\sin x$ and the associated values of $x$ where they occur.
25. Determine an algebraic (non-trigonometric) representation of the problem situation of Example 11. Draw its complete graph and the graph of the problem situation. Find the maximum volume of the right circular cylinder drilled out and compare with the solution given in Example 11.

The Sinusodial Equation $T = A + B\sin(\frac{\pi}{6}\{[360 - (360 - t)]/30\} + C)$ is a model for the mean daily temperature ($T$) as a function of time $t$ measured in days. (Adapted from "Seasonal Temperature Patterns of Selected Cities in and Around Ohio." *Ohio Journal of Science* (1) pp. 8–10, 1986.) $A$ is the mean yearly temperature, $B$ is $\frac{1}{2}$ of the total mean temperature variation, and $C$ is a phase shift factor for the beginning of a temperature cycle. In Columbus, Ohio, the mean yearly temperature is 58° F and the total mean temperature variation is 56°. Assume the temperature on April 1 is the mean yearly temperature, so $C = 0$, and $t$ is the number of days past April 1.

26. Determine a trigonometric representation of the problem situation.
27. Draw a complete graph of the trigonometric representation.
28. What portion of the graph represents the problem situation?
29. About what month and day is it the hottest and coldest during the year in Columbus?

Investigate graphically what happens to the values of $f(x)$ as $x \to 0$.

30. $f(x) = \dfrac{\sin x}{x}$

31. $f(x) = \dfrac{1 - \cos x}{x^2}$

Figure 3.1.18

32. $f(x) = \dfrac{1 - \cos x}{\sin x}$       33. $f(x) = x \cot x$

34. $f(x) = \dfrac{1 - \cos x}{x^2 + x}$

A rocket is launched straight up from ground level at 200 feet per second. A tracking device is located at a point $P$ that is 2055 feet from the launch point (Figure 3.1.18).

35. Write $\angle P = \theta$ as a function of time $t$.
36. Determine the angle $\theta$ 15 seconds after launch.

## 3.2 Trigonometric Identities

In this section we establish important trig identities about sums and differences of angles. These formulas will be used to obtain the *double-angle* and *half-angle* formulas. We use these formulas to help solve an equation algebraically and to show that the slopes of perpendicular lines are negative reciprocals.

In Section 2.4 we established the following basic identities.

- **Basic Identities**

$$\tan \theta = \frac{\sin \theta}{\cos \theta} \qquad \sec \theta = \frac{1}{\cos \theta} \qquad \sin^2 \theta + \cos^2 \theta = 1$$
$$\cot \theta = \frac{\cos \theta}{\sin \theta} \qquad \csc \theta = \frac{1}{\sin \theta} \qquad 1 + \tan^2 \theta = \sec^2 \theta$$
$$\qquad\qquad\qquad\qquad\qquad\qquad\qquad\qquad 1 + \cot^2 \theta = \csc^2 \theta$$

### The Addition Formulas

Let $\alpha$ and $\beta$ be two angles in standard position. Let $A$ and $B$ be the points on the unit circle on the terminal side of $\alpha$ and $\beta$, respectively (Figure 3.2.1). By definition of the trig functions, the coordinates of $A$ are $(\cos \alpha, \sin \alpha)$ and the coordinates of $B$ are $(\cos \beta, \sin \beta)$.

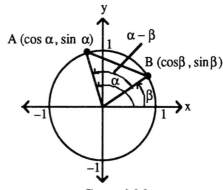

Figure 3.2.1

If we place the angle $\alpha - \beta$ in standard position, the point $P$ with coordinates $(\cos(\alpha - \beta), \sin(\alpha - \beta))$ is on the terminal side of $\alpha - \beta$ and also on the unit circle (Figure 3.2.2).

• **THEOREM 1**   For all angles $\alpha$ and $\beta$, $\cos(\alpha - \beta) = \cos\alpha \cos\beta + \sin\alpha \sin\beta$.

**PROOF**   Refer to Figures 3.2.1 and 3.2.2. The lengths of the line segments $AB$ and $CP$ are equal. We obtain a formula for $\cos(\alpha - \beta)$ by setting the square of the lengths of the line segments $AB$ and $CP$ equal to each other and simplifying the resulting equation. First we compute the square of $d(A, B)$:

$$d^2(A, B) = (\cos\alpha - \cos\beta)^2 + (\sin\alpha - \sin\beta)^2$$
$$= \cos^2\alpha - 2\cos\alpha\cos\beta + \cos^2\beta + \sin^2\alpha - 2\sin\alpha\sin\beta + \sin^2\beta$$
$$= (\sin^2\alpha + \cos^2\alpha) + (\sin^2\beta + \cos^2\beta) - 2\cos\alpha\cos\beta - 2\sin\alpha\sin\beta$$
$$= 2 - 2\cos\alpha\cos\beta - 2\sin\alpha\sin\beta.$$

The last step in the above computation is valid because $\sin^2\alpha + \cos^2\alpha = 1$ and $\sin^2\beta + \cos^2\beta = 1$. Next, we compute the square of $d(C, P)$:

$$d^2(C, P) = (1 - \cos(\alpha - \beta))^2 + (0 - \sin(\alpha - \beta))^2$$
$$= 1 - 2\cos(\alpha - \beta) + \cos^2(\alpha - \beta) + \sin^2(\alpha - \beta)$$
$$= 2 - 2\cos(\alpha - \beta).$$

Finally, we simplify the equation $d^2(C, P) = d^2(A, B)$.

$$d^2(C, P) = d^2(A, B)$$
$$2 - 2\cos(\alpha - \beta) = 2 - 2\cos\alpha\cos\beta - 2\sin\alpha\sin\beta$$
$$-2\cos(\alpha - \beta) = -2(\cos\alpha\cos\beta + \sin\alpha\sin\beta)$$
$$\cos(\alpha - \beta) = \cos\alpha\cos\beta + \sin\alpha\sin\beta$$

In Section 2.2 we showed that $\sin(90° - \theta) = \cos\theta$ and $\cos(90° - \theta) = \sin\theta$ for any *acute* angle $\theta$. In Example 1 we see that these formulas are also true for angles that are

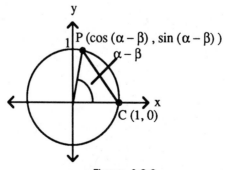

Figure 3.2.2

## 3.2 Trigonometric Identities

*not* acute. We will also establish a relationship for **supplementary angles**; that is, two angles with sum $\pi$, or $180°$.

- **EXAMPLE 1:** Let $\theta$ be any angle. Show that

    (a) $\cos(\frac{\pi}{2} - \theta) = \sin\theta$.  
    (b) $\sin(\frac{\pi}{2} - \theta) = \cos\theta$.  
    (c) $\cos(\pi - \theta) = -\cos\theta$.  
    (d) $\sin(\pi - \theta) = \sin\theta$.

**SOLUTION:**

(a) Notice that the graph of $y = \cos(\frac{\pi}{2} - \theta)$ in Figure 3.2.3 appears to be identical to the graph of $y = \sin\theta$. If we replace $\alpha$ by $\frac{\pi}{2}$ and $\beta$ by $\theta$ in $\cos(\alpha - \beta) = \cos\alpha\cos\beta + \sin\alpha\sin\beta$, then we obtain the following formula:

$$\cos\left(\frac{\pi}{2} - \theta\right) = \cos\frac{\pi}{2}\cos\theta + \sin\frac{\pi}{2}\sin\theta$$
$$= (0)\cos\theta + (1)\sin\theta$$
$$= \sin\theta.$$

(b) In the formula $\cos\left(\frac{\pi}{2} - \theta\right) = \sin\theta$ from (a), we replace $\theta$ by $\frac{\pi}{2} - \theta$ to obtain the following:

$$\cos\left(\frac{\pi}{2} - \theta\right) = \sin\theta$$
$$\cos\left(\frac{\pi}{2} - \left(\frac{\pi}{2} - \theta\right)\right) = \sin\left(\frac{\pi}{2} - \theta\right)$$
$$\cos\theta = \sin\left(\frac{\pi}{2} - \theta\right).$$

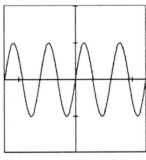

$y = \cos(\frac{\pi}{2} - \theta)$
$[-4\pi, 4\pi]$ by $[-2, 2]$
Figure 3.2.3

(c) Replace $\alpha$ by $\pi$ and $\beta$ by $\theta$ in the formula for $\cos(\alpha - \beta)$:

$$\cos(\alpha - \beta) = \cos\alpha\cos\beta + \sin\alpha\sin\beta$$
$$\cos(\pi - \theta) = \cos\pi\cos\theta + \sin\pi\sin\theta$$
$$= (-1)\cos\theta + (0)\sin\theta$$
$$= -\cos\theta.$$

(d) In the formula $\cos\left(\frac{\pi}{2} - \theta\right) = \sin\theta$ from (a), we replace $\theta$ by $\pi - \theta$ to obtain the following:

$$\cos\left(\frac{\pi}{2} - \theta\right) = \sin\theta$$
$$\cos\left(\frac{\pi}{2} - (\pi - \theta)\right) = \sin(\pi - \theta)$$
$$\cos-\left(\frac{\pi}{2} - \theta\right) = \sin(\pi - \theta)$$
$$\cos\left(\frac{\pi}{2} - \theta\right) = \sin(\pi - \theta) \qquad (\cos(-x) = \cos x)$$
$$\sin\theta = \sin(\pi - \theta).$$

We can use Theorem 1 and Example 1 to obtain formulas for $\cos(\alpha + \beta)$, $\sin(\alpha + \beta)$, and $\sin(\alpha - \beta)$.

• **THEOREM 2**   Let $\alpha$ and $\beta$ be any two angles. Then,

(a) $\cos(\alpha + \beta) = \cos\alpha\cos\beta - \sin\alpha\sin\beta$.
(b) $\sin(\alpha + \beta) = \sin\alpha\cos\beta + \cos\alpha\sin\beta$.
(c) $\sin(\alpha - \beta) = \sin\alpha\cos\beta - \cos\alpha\sin\beta$.

**PROOF**

(a) Replacing $\beta$ by $-\beta$ in Theorem 1, we obtain

$$\cos(\alpha - \beta) \equiv \cos\alpha\cos\beta + \sin\alpha\sin\beta$$
$$\cos(\alpha - (-\beta)) = \cos\alpha\cos(-\beta) + \sin\alpha\sin(-\beta)$$
$$\cos(\alpha + \beta) = \cos\alpha\cos\beta - \sin\alpha\sin\beta.$$

The last step follows because $\cos(-\beta) = \cos\beta$ and $\sin(-\beta) = -\sin\beta$.

(b) From Example 1(a) we have $\sin(\alpha + \beta) = \cos\left(\frac{\pi}{2} - (\alpha + \beta)\right)$. Next, we apply Theorem 1:

$$\sin(\alpha + \beta) = \cos\left(\frac{\pi}{2} - (\alpha + \beta)\right)$$
$$= \cos\left(\left(\frac{\pi}{2} - \alpha\right) - \beta\right)$$
$$= \cos\left(\frac{\pi}{2} - \alpha\right)\cos\beta + \sin\left(\frac{\pi}{2} - \alpha\right)\sin\beta$$
$$= \sin\alpha\cos\beta + \cos\alpha\sin\beta.$$

The last step follows from Example 1(a) and 1(b).
(c) Replace $\beta$ by $-\beta$ in the formula from (b).

$$\sin(\alpha + \beta) = \sin \alpha \cos \beta + \cos \alpha \sin \beta$$
$$\sin(\alpha + (-\beta)) = \sin \alpha \cos(-\beta) + \cos \alpha \sin(-\beta)$$
$$\sin(\alpha - \beta) = \sin \alpha \cos \beta - \cos \alpha \sin \beta.$$

The addition formulas for sine and cosine can be used to establish addition formulas for the tangent function.

• **THEOREM 3** Let $\alpha$ and $\beta$ be any two angles. Then,

(a) $\tan(\alpha + \beta) = \dfrac{\tan \alpha + \tan \beta}{1 - \tan \alpha \tan \beta}$, provided $\tan \alpha \tan \beta \neq 1$.

(b) $\tan(\alpha - \beta) = \dfrac{\tan \alpha - \tan \beta}{1 + \tan \alpha \tan \beta}$, provided $\tan \alpha \tan \beta \neq -1$.

**PROOF** Part (b) follows from (a) by replacing $\beta$ by $-\beta$ and using $\tan(-\beta) = -\tan \beta$. We establish (a) using $\tan \theta = \dfrac{\sin \theta}{\cos \theta}$ and the addition formulas for sine and cosine:

$$\tan(\alpha + \beta) = \frac{\sin(\alpha + \beta)}{\cos(\alpha + \beta)}$$
$$= \frac{\sin \alpha \cos \beta + \cos \alpha \sin \beta}{\cos \alpha \cos \beta - \sin \alpha \sin \beta}.$$

Next, we divide the numerator and denominator of the right-hand side of the above equation by $\cos \alpha \cos \beta$:

$$\tan(\alpha + \beta) = \frac{\sin \alpha \cos \beta + \cos \alpha \sin \beta}{\cos \alpha \cos \beta - \sin \alpha \sin \beta}$$
$$= \frac{\dfrac{\sin \alpha \cos \beta}{\cos \alpha \cos \beta} + \dfrac{\cos \alpha \sin \beta}{\cos \alpha \cos \beta}}{\dfrac{\cos \alpha \cos \beta}{\cos \alpha \cos \beta} - \dfrac{\sin \alpha \sin \beta}{\cos \alpha \cos \beta}}$$
$$= \frac{\tan \alpha + \tan \beta}{1 - \tan \alpha \tan \beta}$$

We summarize as follows:

• **The Addition Formulas**

$$\sin(\alpha + \beta) = \sin \alpha \cos \beta + \cos \alpha \sin \beta, \qquad \cos(\alpha + \beta) = \cos \alpha \cos \beta - \sin \alpha \sin \beta,$$
$$\sin(\alpha - \beta) = \sin \alpha \cos \beta - \cos \alpha \sin \beta, \qquad \cos(\alpha - \beta) = \cos \alpha \cos \beta + \sin \alpha \sin \beta,$$
$$\tan(\alpha + \beta) = \frac{\tan \alpha + \tan \beta}{1 - \tan \alpha \tan \beta} \quad (\tan \alpha \tan \beta \neq 1),$$
$$\tan(\alpha - \beta) = \frac{\tan \alpha - \tan \beta}{1 + \tan \alpha \tan \beta} \quad (\tan \alpha \tan \beta \neq -1).$$

Prior to the widespread availability of calculators, the addition formulas were often used to compute exact values of trig functions. We illustrate this past use of the addition formulas in Example 2 for historical reasons.

- **EXAMPLE 2**: Compute the *exact* trig function values of 15°.

    SOLUTION: Notice that $15° = 60° - 45°$. In Section 2.2, we determined the exact values of the trig functions at 60° and 45° (Figure 3.2.4).

    Now, we apply the addition formulas for sine and cosine and use the exact values given in Figure 3.2.4:

$$\sin 15° = \sin(60° - 45°)$$
$$= \sin 60° \cos 45° - \cos 60° \sin 45°$$
$$= \left(\frac{\sqrt{3}}{2}\right)\left(\frac{1}{\sqrt{2}}\right) - \left(\frac{1}{2}\right)\left(\frac{1}{\sqrt{2}}\right)$$
$$= \frac{\sqrt{3}-1}{2\sqrt{2}};$$
$$\cos 15° = \cos(60° - 45°)$$
$$= \cos 60° \cos 45° + \sin 60° \sin 45°$$
$$= \left(\frac{1}{2}\right)\left(\frac{1}{\sqrt{2}}\right) + \left(\frac{\sqrt{3}}{2}\right)\left(\frac{1}{\sqrt{2}}\right)$$
$$= \frac{\sqrt{3}+1}{2\sqrt{2}};$$
$$\tan 15° = \frac{\sin 15°}{\cos 15°} = \frac{\frac{\sqrt{3}-1}{2\sqrt{2}}}{\frac{\sqrt{3}+1}{2\sqrt{2}}} = \frac{\sqrt{3}-1}{\sqrt{3}+1}.$$

The cosecant, secant, and cotangent of 15° are the reciprocals of the sine, cosine, and tangent of 15°, respectively. •

Figure 3.2.4

## 3.2 Trigonometric Identities

With calculators we simply approximate the values of the trig functions of 15° directly. Even with the exact values in Example 2, we would likely approximate these numbers with a calculator if they were to be used in other calculations. The addition formulas are still important today, as the next example illustrates.

• **EXAMPLE 3:** Let $\ell_1$ be a line that is neither horizontal nor vertical. Show that the slope of a line $\ell_2$ that is perpendicular to $\ell_1$ is the negative reciprocal of the slope of $\ell_1$.

**SOLUTION:** Without any loss of generality we may assume that $\ell_1$ and $\ell_2$ intersect at the origin as illustrated in Figure 3.2.5.

If $\theta$ is the angle that $\ell_1$ makes with the positive $x$-axis, then $90 + \theta$ is the angle $\ell_2$ makes with the positive $x$-axis. The tangent of an angle is the ratio of the length of the opposite side to the length of the adjacent side. Thus, $\tan \theta = \frac{b}{a}$. However, $\frac{b}{a}$ is also the slope of the line $\ell_1$. We need to show that $\tan(90° + \theta)$ is the negative reciprocal of $\tan \theta$. (Why?) You might be tempted to use the formula for $\tan(\alpha + \beta)$, but the tangent function is not defined at 90°. We use $\tan(90° + \theta) = \frac{\sin(90° + \theta)}{\cos(90° + \theta)}$:

$$\sin(90° + \theta) = \sin 90° \cos \theta + \cos 90° \sin \theta$$
$$= (1) \cos \theta + (0) \sin \theta$$
$$= \cos \theta;$$
$$\cos(90° + \theta) = \cos 90° \cos \theta - \sin 90° \sin \theta$$
$$= (0) \cos \theta - (1) \sin \theta$$
$$= -\sin \theta.$$

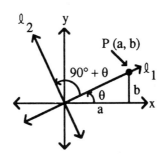

Figure 3.2.5

Next, we compute $\tan(90° + \theta)$:

$$\tan(90° + \theta) = \frac{\sin(90° + \theta)}{\cos(90° + \theta)}$$

$$= \frac{\cos\theta}{-\sin\theta}$$

$$= -\frac{1}{\frac{\sin\theta}{\cos\theta}}$$

$$= -\frac{1}{\tan\theta}.$$

Therefore, the slope of $\ell_2$ is the negative reciprocal of the slope of $\ell_1$. ●

## Double-Angle Formulas

The double-angle formulas are important special cases of the addition formulas that permit us to write $\sin 2\theta$, $\cos 2\theta$, and $\tan 2\theta$ in terms of $\sin\theta$, $\cos\theta$, and $\tan\theta$.

● **THEOREM 4**   Let $\theta$ be any angle. Then, the following **double-angle** formulas are true:

(a) $\sin 2\theta = 2\sin\theta\cos\theta$     (b) $\cos 2\theta = \cos^2\theta - \sin^2\theta$
$\hspace{5cm} = 1 - 2\sin^2\theta$
$\hspace{5cm} = 2\cos^2\theta - 1$

(c) $\tan 2\theta = \dfrac{2\tan\theta}{1 - \tan^2\theta}$

**PROOF**

(a) Notice that the graphs of $y = \sin 2\theta$ and $y = 2\sin\theta\cos\theta$ in Figure 3.2.6 appear to be identical. This can be confirmed by graphing both functions in the same viewing rectangle. Let $\alpha = \beta = \theta$ in the formula for $\sin(\alpha + \beta)$

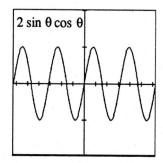

$[-2\pi, 2\pi]$ by $[-2, 2]$
Figure 3.2.6

## 3.2 Trigonometric Identities

and we obtain the following:
$$\sin(\alpha + \beta) = \sin\alpha\cos\beta + \cos\alpha\sin\beta$$
$$\sin(\theta + \theta) = \sin\theta\cos\theta + \cos\theta\sin\theta$$
$$\sin 2\theta = 2\sin\theta\cos\theta.$$

(b) Let $\alpha = \beta = \theta$ in the formula for $\cos(\alpha + \beta)$:
$$\cos(\alpha + \beta) = \cos\alpha\cos\beta - \sin\alpha\sin\beta$$
$$\cos(\theta + \theta) = \cos\theta\cos\theta - \sin\theta\sin\theta$$
$$\cos 2\theta = \cos^2\theta - \sin^2\theta.$$

We can use the identity $\sin^2\theta + \cos^2\theta = 1$ to replace $\cos^2\theta$ by $1 - \sin^2\theta$ in the above equation:
$$\cos 2\theta = \cos^2\theta - \sin^2\theta$$
$$= (1 - \sin^2\theta) - \sin^2\theta$$
$$= 1 - 2\sin^2\theta.$$

The last part of (b) as well as (c) are left for the Exercises.

### Half-Angle Formulas

The double-angle formulas can be used to obtain formulas that permit us to write $\sin\frac{1}{2}\theta$, $\cos\frac{1}{2}\theta$, $\tan\frac{1}{2}\theta$ in terms of $\sin\theta$, $\cos\theta$, and $\tan\theta$.

• **THEOREM 5**   Let $\theta$ be any angle; then the following **half-angle formulas** are true:

(a) $\sin\frac{1}{2}\theta = \pm\sqrt{\dfrac{1-\cos\theta}{2}}$   (b) $\cos\frac{1}{2}\theta = \pm\sqrt{\dfrac{1+\cos\theta}{2}}$

(c) $\tan\frac{1}{2}\theta = \dfrac{\sin\theta}{1+\cos\theta}$

The signs in (a) and (b) are chosen according to the quadrant of $\frac{1}{2}\theta$.

**PROOF**   We use the following two identities:
$$\sin^2\theta + \cos^2\theta = 1,$$
$$\cos^2\theta - \sin^2\theta = \cos 2\theta.$$

(a) Subtracting, we obtain
$$2\sin^2\theta = 1 - \cos 2\theta$$
$$\sin^2\theta = \frac{1-\cos 2\theta}{2}$$
$$\sin\theta = \pm\sqrt{\frac{1-\cos 2\theta}{2}}.$$

Now, replacing $\theta$ by $\frac{1}{2}\theta$ gives the desired formula.

(b) Adding, we obtain the following:
$$2\cos^2\theta = 1 + \cos 2\theta$$
$$\cos^2\theta = \frac{1 + \cos 2\theta}{2}$$
$$\cos\theta = \pm\sqrt{\frac{1 + \cos 2\theta}{2}}.$$

Again, replacing $\theta$ by $\frac{1}{2}\theta$ gives the desired formula.

(c) First, we multiply the numerator and denominator of the right-hand side of $\tan\frac{1}{2}\theta = \sin\frac{1}{2}\theta / \cos\frac{1}{2}\theta$ by $2\cos\frac{1}{2}\theta$:

$$\tan\frac{1}{2}\theta = \frac{\sin\frac{1}{2}\theta}{\cos\frac{1}{2}\theta} = \frac{(\sin\frac{1}{2}\theta)(2\cos\frac{1}{2}\theta)}{(\cos\frac{1}{2}\theta)(2\cos\frac{1}{2}\theta)}$$
$$= \frac{2\sin\frac{1}{2}\theta\cos\frac{1}{2}\theta}{2\cos^2\frac{1}{2}\theta}.$$

The numerator can be simplified as follows:
$$2\sin\frac{1}{2}\theta\cos\frac{1}{2}\theta = \sin 2(\tfrac{1}{2}\theta) = \sin\theta.$$

The denominator can be simplified by replacing $\theta$ by $\frac{1}{2}\theta$ in Theorem 4(b):
$$\cos 2\theta = 2\cos^2\theta - 1$$
$$\cos\theta = 2\cos^2\frac{1}{2}\theta - 1$$
$$1 + \cos\theta = 2\cos^2\frac{1}{2}\theta.$$

Thus,
$$\tan\frac{1}{2}\theta = \frac{\sin\theta}{1 + \cos\theta}.$$

The proof of Theorem 2 in Section 3.1 is an application of the formulas for sine and cosine.

- **EXAMPLE 4**: Prove Theorem 2 of Section 3.1.

SOLUTION: We need to show that for all real numbers $a$, $b$, $d$, $h$, and $k$, there exist real numbers $A$ and $\alpha$ such that
$$a\sin(bx + h) + d\cos(bx + k) = A\sin(bx + \alpha). \tag{1}$$

We use the addition formulas for sine and cosine to expand and simplify each side of equation (1). We can rewrite the right-hand side of equation (1) as follows:
$$A\sin(bx + \alpha) = A(\sin bx \cos\alpha + \cos bx \sin\alpha)$$
$$= (A\cos\alpha)\sin bx + (A\sin\alpha)\cos bx. \tag{2}$$

Similarly, we can rewrite the left-hand side of equation (1) as follows:

$$a\sin(bx+h) + d\cos(bx+k) = a(\sin bx \cos h + \cos bx \sin h) + d(\cos bx \cos k - \sin bx \sin k)$$
$$= (a\cos h - d\sin k)\sin bx + (a\sin h + d\cos k)\cos bx. \qquad (3)$$

Comparing equations (2) and (3), we obtain the following equations:

$$A\cos\alpha = a\cos h - d\sin k, \qquad (4)$$
$$A\sin\alpha = a\sin h + d\cos k. \qquad (5)$$

If the right-hand side of equation (4) is zero, we choose $\alpha = \frac{\pi}{2}$. Then, equation (4) is satisfied because $\cos\frac{\pi}{2} = 0$. Equation (5) will be satisfied if we choose $A = a\sin h + d\cos k$ because $\sin\frac{\pi}{2} = 1$. If the right-hand side of equation (4) is not zero, then we divide each side of equation (5) by the corresponding side of equation (4) to obtain

$$\tan\alpha = \frac{A\sin\alpha}{A\cos\alpha} = \frac{a\sin h + d\cos k}{a\cos h - d\sin k}. \qquad (6)$$

Now, we choose $\alpha$ so that equation (6) is satisfied and $-\frac{\pi}{2} < \alpha < \frac{\pi}{2}$. Next, we square each side of equations (4) and (5) and add to obtain the following result:

$$A^2\cos^2\alpha + A^2\sin^2\alpha = (a\cos h - d\sin k)^2 + (a\sin h + d\cos k)^2$$
$$A^2(\cos^2\alpha + \sin^2\alpha) = (a\cos h - d\sin k)^2 + (a\sin h + d\cos k)^2$$
$$A^2 = (a\cos h - d\sin k)^2 + (a\sin h + d\cos k)^2$$
$$A = \pm\sqrt{(a\cos h - d\sin k)^2 + (a\sin h + d\cos k)^2}. \qquad (7)$$

Now, $\cos\alpha > 0$ because $-\frac{\pi}{2} < \alpha < \frac{\pi}{2}$. Thus, it follows from equation (4) that $A$ and $a\cos h - d\sin k$ have the same sign. Therefore, if $a\cos h - d\sin k > 0$, we use the $+$ sign in equation (7), and if $a\cos h - d\sin k < 0$, we use the $-$ sign in equation (7). •

In Section 3.1 we observed that the function $f(x) = \sin 2x + \cos 3x$ was periodic of period $2\pi$. Thus, the graph of $f$ in Figure 3.2.7 is complete. We use the identities of this section to help us find the zeros of $f(x) = \sin 2x + \cos 3x$ algebraically.

$f(x) = \sin 2x + \cos 3x$
$[0, 2\pi]$ by $[-3, 3]$

Figure 3.2.7

- **EXAMPLE 5:** Find all the solutions of $\sin 2x + \cos 3x = 0$ that are in the interval $[0, 2\pi]$.

SOLUTION: First we write $\cos 3x$ in terms of $\sin x$ and $\cos x$:
$$\cos 3x = \cos(2x + x)$$
$$= \cos 2x \cos x - \sin 2x \sin x$$
$$= (1 - 2\sin^2 x)\cos x - 2\sin x \cos x \sin x$$
$$= \cos x - 2\sin^2 x \cos x - 2\sin^2 x \cos x$$
$$= \cos x - 4\sin^2 x \cos x.$$

Now, we simplify the original equation as follows:
$$\sin 2x + \cos 3x = 0$$
$$2\sin x \cos x + (\cos x - 4\sin^2 x \cos x) = 0$$
$$(\cos x)(2\sin x + 1 - 4\sin^2 x) = 0.$$

Therefore, either $\cos x = 0$ or $2\sin x + 1 - 4\sin^2 x = 0$. The solutions of $\cos x = 0$ in $[0, 2\pi]$ are $\frac{\pi}{2}$ and $\frac{3\pi}{2}$. The other equation can be rewritten
$$4\sin^2 x - 2\sin x - 1 = 0.$$

Notice that this equation is quadratic in $\sin x$. The quadratic formula gives the following:
$$\sin x = \frac{2 \pm \sqrt{4 + 16}}{8}$$
$$= \frac{2 \pm 2\sqrt{5}}{8}$$
$$= \frac{1 \pm \sqrt{5}}{4}.$$

The solutions to $\sin x = (1 + \sqrt{5})/4$ in $[0, 2\pi]$ accurate to hundredths are 0.94 and 2.20. The solutions to $\sin x = (1 - \sqrt{5})/4$ in $[0, 2\pi]$ are 3.46 and 5.97. Therefore, the *six* solutions to $\sin 2x + \cos 3x = 0$ in $[0, 2\pi]$ accurate to hundredths are 0.94, 1.57 $\left(\frac{\pi}{2}\right)$, 2.20, 3.46, 4.71 $\left(\frac{3\pi}{2}\right)$, and 5.97. This information is confirmed by the graph in Figure 3.2.7. •

## EXERCISES 3-2

Simplify (express in terms of a single trig function).

1. $\sin(\theta + 90°)$
2. $\tan(180° + x)$
3. $\cos\left(\frac{\pi}{2} + x\right)$
4. $\cot(\theta + 2\pi)$
5. $\dfrac{1 + \tan x}{1 + \cot x}$
6. $\dfrac{\sec x + \tan x}{\sec x + \tan x - \cos x}$

Use algebraic means to verify the following identities.

7. $\tan x + \cot x = \sec x \cdot \csc x$
8. $\tan x + \cot x = 2\csc 2x$

9. $\tan x \cdot \sin x + \cos x = \sec x$

10. $\cos^4 x - \sin^4 x = \cos 2x$

11. $\dfrac{1 + \cos 2x}{\sin 2x} = \dfrac{\sin 2x}{1 - \cos 2x}$

12. $\cot x - \tan x = 2 \cot 2x$

13. $\dfrac{\tan x + 1}{\cot x + 1} = \tan x$

14. $\csc x - \cot x = \tan \dfrac{x}{2}$

Use graphs to investigate whether the equation is an identity. If not, determine a *counterexample*.

15. $\dfrac{\sin t - \cos t}{\cos t} + 1 = \tan t$

16. $\csc(x + \pi) = -\sec x$

17. $\cos(3\pi + x) = -\cos x$

18. $\dfrac{1}{\tan t} + \dfrac{\sin t}{\cos t - 1} = -\csc t$

Verify the following identities by graphing each side of the equation and comparing the graphs. Then verify the identity algebraically.

19. $1 + \cot^2 \theta = \csc^2 \theta$

20. $1 - 2\sin^2 x = 2\cos^2 x - 1$

21. $\cos^2 \theta + 1 = 2\cos^2 \theta + \sin^2 \theta$

22. $\sin \theta + \cos \theta \cot \theta = \csc \theta$

Determine each *exact* value. Check your answers with a calculator.

23. $\cos \dfrac{\pi}{12}$, $\sin \dfrac{5\pi}{12}$

24. $\tan \dfrac{9\pi}{8}$, $\cos 22.5°$

If $\sin x = \dfrac{2}{3}$ and $\dfrac{\pi}{2} < x < \pi$, then determine the trig value.

25. $\cos x$, $\sin x$

26. $\tan 2x$, $\cos \dfrac{x}{2}$

If $\cos x = -\dfrac{1}{2}$ and $\pi < x < \dfrac{3\pi}{2}$, then determine the trig value.

27. $\sec x$

28. $\sin x$

29. $\tan 2x$

30. $\cos \dfrac{x}{2}$

Find the domain and range of the function, and draw a complete graph.

31. $f(x) = \sec x^2$

32. $g(x) = x^2 \sin\left(x - \dfrac{\pi}{2}\right)$

33. $g(x) = 3 \sin^2 x$

34. $f(x) = \dfrac{\tan x}{x^2}$

Write each as an expression involving only $\sin \theta$ and $\cos \theta$.

35. $\sin 2\theta + \cos \theta$

36. $\sin 2\theta + \cos 3\theta$

37. $\sin 3\theta + \cos 2\theta$

38. $\sin 4\theta + \cos 3\theta$

Solve the equation for $x$ in the interval $[0, 2\pi]$. Find exact answers when possible. Check your answer using a graphing utility.

39. $\sin x = \dfrac{1}{2}$

40. $\cos x = 0$

41. $\sin x = \dfrac{\sqrt{3}}{2}$

42. $\tan x = 1$

43. $\sin^2 x - 1 = 0$

44. $2\sin^2 x + \sin x - 1 = 0$

45. $\sin^2 x - 2\sin x = 0$

46. $\cos 2x + \sin 3x = 0$

Determine the quadrant of the angle $\theta$ satisfying the specified conditions.

47. $\cos t > 0$ and $\sin 2t < 0$

48. $\cos t < 0$ and $\sin 2t > 0$

Draw the graph of the equation. If it is a sinusoid, give the period, amplitude and phase shift.

49.  $y = 3\sin x + 5\sin(x+2)$ 	50.  $y = 3\sin(2x-1) + 5\sin(2x+3)$

51.  Use the addition formula for sine to prove Theorem 1 of Section 3.1 directly.

## 3.3 Inverse Relations

In this section we define the inverse relation of a given relation. Recall that a relation is any set of ordered pairs of real numbers. We show that the graph of an inverse relation can be obtained by reflecting the graph of the relation through the line $y = x$. If a relation is given by an algebraic expression, then we are able to determine the inverse relation algebraically. We show that the inverse of a one-to-one function is a function.

Roughly speaking, *inverse* means *undoing*. Suppose that we are given a certain process and know the effect of that process on an object. The inverse process, if there is one, is what has to be done to get back where we started. For example, *cubing* is a process that is *undone* by taking the principal *cube root*.

$$\text{Cubing} \quad \text{Cube Root}$$
$$2 \xrightarrow{\text{(do)}} 8 \xrightarrow{\text{(undo)}} 2$$

Can you think of others?

We can think of a relation $R$ as a *correspondence* between the domain of $R$ and the range of $R$. For example $\{(-1, 2), (0, 4), (3, 8)\}$ is a relation with domain $\{-1, 0, 3\}$, range $\{2, 4, 8\}$, and correspondence $-1 \to 2, 0 \to 4$, and $3 \to 8$. What should undoing mean for a relation $R$?

$$\text{Domain of } R \quad \text{Range of } R$$
$$a \xrightarrow{R} b \xrightarrow{?} a$$

Whatever the inverse is, it should send $b$ back to $a$. For our example, the correspondence for the inverse should be $2 \to -1, 4 \to 0$, and $8 \to 3$. This is the motivation for the following definition.

- **DEFINITION**  The **inverse relation** of a relation $R$ consists of those ordered pairs of real numbers obtained by interchanging the first and second entries of the ordered pairs in $R$. That is, $(a, b)$ belongs to the inverse of $R$ if and only if $(b, a)$ belongs to $R$. The inverse relation is denoted by $R^{-1}$. The $-1$ in $R^{-1}$ is not to be interpreted as an exponent. That is, $R^{-1} \neq \frac{1}{R}$ when $R$ is a relation. •

- **EXAMPLE 1:** Show that the inverse of a relation that has a vertical line for its graph is a relation that has a horizontal line for its graph and vice versa.

    SOLUTION: First we consider relations that have vertical lines for their graphs. We start with a particular case and then generalize our argument to all relations that have vertical

## 3.3 Inverse Relations

lines for graphs. The relation $x = 3$ has a graph that is a vertical line which consists of all ordered pairs of the form $(3, b)$, where $b$ is any real number. Thus, the inverse relation consists of the set of all ordered pairs of the form $(b, 3)$, where $b$ is any real number. The graph of the inverse relation is the horizontal line $y = 3$ (Figure 3.3.1). Similarly, we can show that the graph of the inverse of the relation $x = a$ that has a vertical line for its graph is the relation $y = a$ that has a horizontal line for its graph.

Now we consider relations that have horizontal lines for their graphs. Again, we begin with a particular case and then generalize. The relation $y = -3$ has a graph that is a horizontal line which consists of all ordered pairs of the form $(a, -3)$, where $a$ is any real number. The inverse relation consists of all ordered pairs of the form $(-3, a)$, where $a$ is any real number. The graph of the inverse relation is the vertical line $x = -3$ (Figure 3.3.2). In general, the graph of the inverse relation of the relation $y = a$ that has a horizontal line for its graph is the relation $x = a$ that has a vertical line for its graph. •

Example 1 can be extended to lines that are not horizontal or vertical by using the concept of slope.

• **EXAMPLE 2**: Draw a complete graph of $y = 2x + 3$ and its inverse relation. Write an equation for the inverse relation.

**SOLUTION**: A complete graph of $y = 2x + 3$ is a straight line with slope 2. There is a one-to-one correspondence between the points on a complete graph of $y = 2x + 3$ and the points on a complete graph of the inverse of $y = 2x + 3$. Let $(a, b)$ and $(c, d)$ be any two distinct points on the graph of the inverse of $y = 2x + 3$. Then $(b, a)$ and $(d, c)$ are the corresponding distinct points on the graph of $y = 2x + 3$. We must have $\frac{c-a}{d-b} = 2$, because the slope of the line $y = 2x + 3$ is 2 and $(b, a)$ and $(d, c)$ are on this line. Now $a \neq c$ and $b \neq d$ because the line $y = 2x + 3$ is neither horizontal nor vertical. If we take the reciprocal of each side of $\frac{c-a}{d-b} = 2$ we find that $\frac{d-b}{c-a} = \frac{1}{2}$. This means that the points $(a, b)$ and $(c, d)$ are on a line with slope $\frac{1}{2}$. (Why?) Thus, a complete graph of the inverse of $y = 2x + 3$ is also a line.

The points $(3, 0)$ and $(7, 2)$ are on the graph of the inverse because $(0, 3)$ and $(2, 7)$ are on the graph of $y = 2x + 3$. We can use the points $(0, 3)$ and $(2, 7)$ to draw a complete

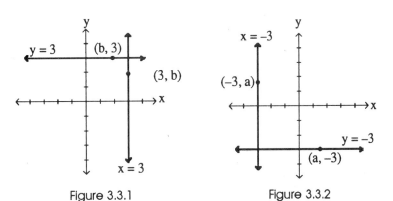

Figure 3.3.1    Figure 3.3.2

graph of $y = 2x + 3$ and the points $(3,0)$ and $(7,2)$ to draw a complete graph of the inverse (Figure 3.3.3).

Next, we find an equation for the inverse of $y = 2x + 3$. First, we use the two points $(3,0)$ and $(7,2)$ of the inverse to compute the slope:

$$m = \frac{2-0}{7-3} = \frac{2}{4} = \frac{1}{2}.$$

Now, we use the point-slope formula to write an equation:

$$y - 0 = \frac{1}{2}(x - 3)$$
$$2y = x - 3$$
$$x = 2y + 3.$$

Notice that the equation $x = 2y + 3$ for the inverse can be obtained by interchanging $x$ and $y$ in the equation $y = 2x + 3$. If we solve for $y$ in terms of $x$ we find that $y = \frac{1}{2}x - \frac{3}{2}$ is another form of the equation for the inverse of $y = 2x + 3$. In this form we can see that the inverse is a straight line with slope $\frac{1}{2}$ and $y$-intercept $-\frac{3}{2}$ as suggested by Figure 3.3.3. •

We give a generalization of Example 2 without proof.

• **THEOREM 1**  If $R$ is the relation determined by a given equation or inequality in two variables, then $R^{-1}$ is the relation determined by interchanging the two variables in the equation or inequality for the relation $R$.

The next theorem gives a geometric relationship between a point $(a, b)$ on the graph of a relation and the corresponding point $(b, a)$ on the graph of the inverse of the relation.

• **THEOREM 2**  The points $(a, b)$ and $(b, a)$ are symmetric with respect to the line $y = x$.

**PROOF**  Without any loss of generality we can assume that $a > 0$, $b > 0$, and $a > b$. Let $C$ be the point common to the line $y = x$ and the line through the

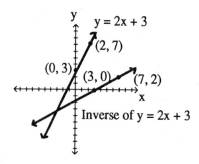

Figure 3.3.3

## 3.3 Inverse Relations

points $A(a, b)$ and $B(b, a)$ (Figure 3.3.4). The slope of the line segment $AB$ is $\frac{b-a}{a-b} = -1$. The slope of the line $y = x$ is 1. Therefore, $AB$ is perpendicular to $OC$. Let the coordinates of $C$ be $(c, c)$. Then, we determine the length of $AC$ and $BC$ as follows:

$$d(A, C) = \sqrt{(a-c)^2 + (b-c)^2},$$
$$d(B, C) = \sqrt{(b-c)^2 + (a-c)^2}.$$

Thus, the length of $AC$ is equal to the length of $BC$. Therefore, the points $A(a, b)$ and $B(b, a)$ are symmetric with respect to the line $y = x$.

We can draw the following conclusion about the graph of the inverse of a relation from Theorem 2.

- **THEOREM 3**  The graph of the inverse of a relation can be obtained by reflecting the graph of the relation through the line $y = x$.

- **EXAMPLE 3:** Draw a complete graph of the function $f(x) = x^2$ and its inverse relation $f^{-1}$. Write an equation for $f^{-1}$. Find the domain and range of $f$ and $f^{-1}$. Is $f^{-1}$ a function?

**SOLUTION:** A complete graph of $f(x) = x^2$ is given in Figure 3.3.5. The domain is $(-\infty, \infty)$, and the range is $[0, \infty)$. A complete graph of the inverse can be obtained in the following way. First draw a graph of $f(x) = x^2$ on a square piece of paper. Be sure to label the axes. Then, hold the line $y = x$ fixed and rotate the paper $180°$ about the line. The top of the paper will become the bottom of the paper and vice versa. Now, hold the paper up to a light and you will see the graph displayed in Figure 3.3.6. Notice that the axes have switched position.

We can check a few points to confirm that the graph in Figure 3.3.6 is the graph of the inverse of $f(x) = x^2$. Notice that $(-2, 4)$, $(0, 0)$, and $(2, 4)$ are three points on the graph of $f(x) = x^2$. The corresponding points $(4, -2)$, $(0, 0)$, and $(4, 2)$ are points on the graph of $f^{-1}$.

Figure 3.3.4

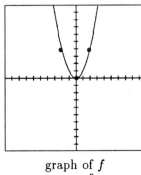

graph of $f$
$y = x^2$
$[-10, 10]$ by $[-10, 10]$
Figure 3.3.5

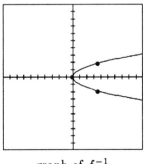

graph of $f^{-1}$
$x = y^2$
$[-10, 10]$ by $[-10, 10]$
Figure 3.3.6

An equation can be obtained for $f^{-1}$ by setting $y = f(x)$ and interchanging $x$ and $y$ in the equation $y = x^2$. Thus, $x = y^2$ is an equation for the inverse of $y = x^2$. The graph of $x = y^2$ is a parabola with the $x$-axis as its axis of symmetry. Parabolas will be studied in more detail in Chapter 6. The domain of the inverse is $[0, \infty)$ and the range is $(-\infty, \infty)$. Notice that the domain and range of $y = x^2$ and its inverse $x = y^2$ are interchanged.

Finally, the vertical line test shows that $f^{-1}$ is *not* a function. Thus, some functions have inverses that are *not* functions. •

The information found in Example 3 about the domain and range of a relation and its inverse is true in general.

---

• **THEOREM 4**  Let $R$ be any relation. Then domain of $R^{-1}$ = range of $R$ and range of $R^{-1}$ = domain of $R$.

---

Notice that the points $(4, -2)$ and $(4, 2)$ on the graph of the inverse of $y = x^2$ (Figure 3.3.6) correspond to the points $(-2, 4)$ and $(2, 4)$ of the graph of $y = x^2$ (Figure 3.3.5) under reflection with respect to the line $y = x$. The points $(-2, 4)$ and $(2, 4)$ lie on the horizontal line $y = 4$. The reason the inverse of $y = x^2$ is not a function is that there are horizontal lines that intersect the graph of $y = x^2$ in two points. This suggests the following theorem which we state without proof.

---

• **THEOREM 5**  $f^{-1}$ is a function if and only if every horizontal line intersects the graph of $f$ in at most one point.

---

• **EXAMPLE 4:** Let $g$ be the function with domain $[0, \infty)$ defined by the rule $g(x) = x^2$. Show that $g^{-1}$ is a function and find an algebraic rule for $g^{-1}$.

SOLUTION: A complete graph of $g$ is given in Figure 3.3.7. Notice that the range of $g$ is also $[0, \infty)$. It is important to observe that $g$, *by definition*, has no value when $x$ is negative even though the rule for $g$ makes sense for $x$ negative. Every horizontal line intersects the

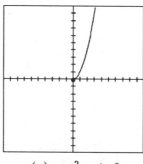

$g(x) = x^2$, $x \geq 0$
$[-10, 10]$ by $[-10, 10]$

Figure 3.3.7

## 3.3 Inverse Relations

graph of $g$ in either one or zero points. Thus, $g^{-1}$ is a function. The graph of $g^{-1}$ is given in Figure 3.3.8.

The domain and range of $g^{-1}$ are $[0, \infty)$. If we let $y = g(x)$, then we can interchange $x$ and $y$ in the equation $y = x^2$ and obtain the equation $x = y^2$ for $g^{-1}$. Because $x \geq 0$ and $y \geq 0$ we can write the equation for $g^{-1}$ as $y = \sqrt{x}$. Thus, $g^{-1}(x) = \sqrt{x}$.

There is another way to describe those functions with inverses that are also functions.

- **DEFINITION**  A function $f$ is said to be **one-to-one** if no two distinct ordered pairs of $f$ have the same second entry.

In Example 3 we found that $f(x) = x^2$ is not one-to-one because $(-2, 4)$ and $(2, 4)$ are two distinct ordered pairs of $f$ with the same second entry. Saying that the function $f$ is one-to-one is equivalent to saying that every horizontal line intersects the graph of $f$ in at most one point.

**Horizontal Line Test**  A function $f$ is one-to-one if and only if every horizontal line intersects the graph of $f$ in at most one point.

This fact together with Theorem 5 shows that the following is a theorem.

- **THEOREM 6**  Let $f$ be a function. $f^{-1}$ is a function if and only if $f$ is one-to-one.

In some textbooks the notation $f^{-1}$ is used only for functions $f$ whose inverses are also functions.

- **EXAMPLE 5:** Determine whether the following functions are one-to-one.

    (a) $f(x) = -3x + 4$    (b) $g(x) = x^3 - 4x$

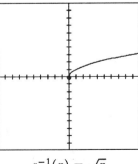

$g^{-1}(x) = \sqrt{x}$
$[-10, 10]$ by $[-10, 10]$
Figure 3.3.8

SOLUTION: A complete graph of $f(x) = -3x + 4$ is given in Figure 3.3.9. The graph is a straight line. Every horizontal line intersects this graph exactly once. Thus, $f$ is a one-to-one function.

A complete graph of $g(x) = x^3 - 4x$ is given in Figure 3.3.10. Each horizontal line between the local minimum of $g$ and the local maximum of $g$ intersects the graph of $g$ exactly three times. Therefore, $g$ is *not* one-to-one. •

The graph of any function of the form $f(x) = ax + b$, where $a \neq 0$, is a straight line that is neither horizontal nor vertical. Such functions are one-to-one, so their inverses are also functions. In the Exercises you will be asked to show that, as suggested by Example 2, the inverse of a linear function is also a linear function. The graph of a quadratic function $f(x) = ax^2 + bx + c$ is a parabola that opens upward or downward. Quadratic functions are never one-to-one. Therefore, the inverse relation of a quadratic function is *never* a function.

The function $y = x^3$ is increasing on $(-\infty, \infty)$, and the function $y = -x^3$ is decreasing on $(-\infty, \infty)$. Every horizontal line will intersect the graph of either function exactly once. Thus, $y = x^3$ and $y = -x^3$ are one-to-one and so have inverses. These two functions provide a special case of the following theorem we give without proof.

• **THEOREM 7**   Let $f$ be a function with domain an interval $I$.

(a) If $f$ is increasing on $I$, then it is a one-to-one function and its inverse is a function.
(b) If $f$ is decreasing on $I$, then it is a one-to-one function and its inverse is a function.

Notice that the domain of the function $g$ of Example 4 is the interval $[0, \infty)$, and $g$ is increasing on this interval. By Theorem 7, $g$ is a one-to-one function and $g^{-1}$ is a function. Theorem 7 does not apply to the next example because the domain of that function is *not* an interval. However, the function is one-to-one so that its inverse is a function.

• **EXAMPLE 6:** Show that $f(x) = \frac{2x+7}{x+3}$ is one-to-one and find a rule for its inverse.

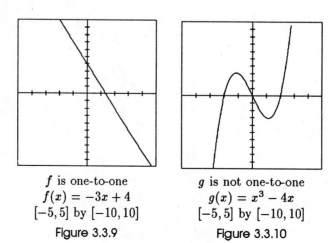

$f$ is one-to-one
$f(x) = -3x + 4$
$[-5, 5]$ by $[-10, 10]$
Figure 3.3.9

$g$ is not one-to-one
$g(x) = x^3 - 4x$
$[-5, 5]$ by $[-10, 10]$
Figure 3.3.10

## 3.3 Inverse Relations

**SOLUTION:** We can use long division to write $f$ in the form $f(x) = 2 + \frac{1}{x+3}$. A complete graph of the rational function $f$ has a vertical asymptote at $x = -3$ and has $y = 2$ as a horizontal asymptote (Figure 3.3.11). Notice that $f$ is one-to-one and its inverse must be a function.

A complete graph of the inverse can be obtained by reflecting the graph of $f$ through the line $y = x$ (Figure 3.3.12). Thus, the inverse will have $x = 2$ as a vertical asymptote and $y = -3$ as a horizontal asymptote. This follows because reflecting through the line $y = x$ sends horizontal lines to vertical lines and vice versa.

We can determine an equation for $f^{-1}$ by setting $y = f(x)$, interchanging $x$ and $y$, and then solving for $y$:

$$y = \frac{2x+7}{x+3},$$
$$x = \frac{2y+7}{y+3}.$$

If we solve this equation for $y$, we find that $y = \frac{-3x+7}{x-2}$. Thus, $f^{-1}(x) = \frac{-3x+7}{x-2}$. You can check that the computer generated graph of $f^{-1}$ in the $[-10, 10]$ by $[-10, 10]$ viewing rectangle coincides with the graph in Figure 3.3.12. •

We must be very careful when we find rules for inverses that are functions. Suppose $y = f(x)$ and $f^{-1}$ is a function. When we solve the equation $x = f(y)$ for $y$ we may introduce extraneous solutions. This can lead to an incorrect rule for $f^{-1}$. Example 7 illustrates one such possibility.

• **EXAMPLE 7:** Let $f(x) = \sqrt{x+3}$. Show that $f^{-1}$ is a function, find a rule for $f^{-1}$, and determine the domain and range of $f^{-1}$.

**SOLUTION:** The graph of $f$ in Figure 3.3.13 shows that the domain of $f$ is the interval $[-3, \infty)$, and $f$ is increasing on $[-3, \infty)$, so by Theorem 7 $f$ is one-to-one and $f^{-1}$ is a function.

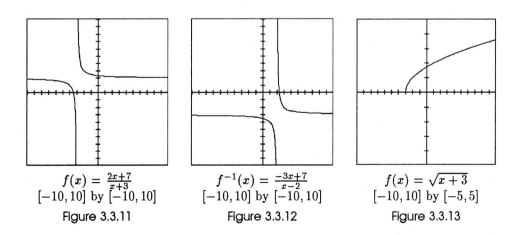

$f(x) = \frac{2x+7}{x+3}$
$[-10, 10]$ by $[-10, 10]$
Figure 3.3.11

$f^{-1}(x) = \frac{-3x+7}{x-2}$
$[-10, 10]$ by $[-10, 10]$
Figure 3.3.12

$f(x) = \sqrt{x+3}$
$[-10, 10]$ by $[-5, 5]$
Figure 3.3.13

To find a rule for $f^{-1}$ we set $y = f(x)$, interchange $x$ and $y$, and then solve for $y$:

$$y = \sqrt{x+3},$$
$$x = \sqrt{y+3},$$
$$x^2 = y + 3,$$
$$y = x^2 - 3.$$

Thus, $f^{-1}(x) = x^2 - 3$ is a *rule* for $f$. However, because we squared each side of the equation $x = \sqrt{y+3}$, we may have introduced extraneous solutions and must be careful *not* to conclude that $f^{-1}(x) = x^2 - 3$ with no restriction on $x$ without further investigation. Figure 3.3.13 shows that the domain of $f^{-1}$ is $[0, \infty)$ because the domain of $f^{-1}$ is the same as the range of $f$. Thus, $f^{-1}(x) = x^2 - 3$ *with the restriction* $x \geq 0$ correctly describes the inverse function. Simply stating that $f^{-1}(x) = x^2 - 3$ without restricting the domain is not correct. A complete graph of $f^{-1}(x) = x^2 - 3$, $x \geq 0$ coincides with the graph in Figure 3.3.14. •

The next example illustrates that a function can be one-to-one and have an inverse without satisfying the conditions of Theorem 7.

• **EXAMPLE 8:** Determine the domain, the range, where the function is increasing and decreasing, and show that it is one-to-one:

$$f(x) = \begin{cases} \frac{x}{x-1} & \text{for } x < 1 \\ x^2 + 1 & \text{for } x \geq 1 \end{cases}.$$

SOLUTION: A complete graph of $f$ is given in Figure 3.3.15. The domain of $f$ is $(-\infty, \infty)$ and the range of $f$ is $(-\infty, 1) \cup [2, \infty)$. We can see that $f$ is decreasing in $(-\infty, 1)$ and increasing in $[1, \infty)$. Notice that each horizontal line intersects the graph in Figure 3.3.15 in either zero or one point. The horizontal line test shows that $f$ is one-to-one. •

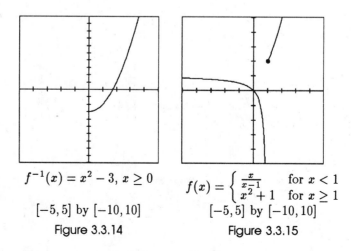

$f^{-1}(x) = x^2 - 3$, $x \geq 0$

$[-5, 5]$ by $[-10, 10]$

Figure 3.3.14

$f(x) = \begin{cases} \frac{x}{x-1} & \text{for } x < 1 \\ x^2 + 1 & \text{for } x \geq 1 \end{cases}$

$[-5, 5]$ by $[-10, 10]$

Figure 3.3.15

## 3.3 Inverse Relations

- **Rule for $f^{-1}$** Let $f$ be a function with domain $S$, range $T$, and for which $f^{-1}$ is also a function. If we have a rule for $f$, then a rule for $f^{-1}$ can be determined in the following way:
  1. Set $y = f(x)$.
  2. Interchange $x$ and $y$ in the equation in 1.
  3. Solve for $y$ in 2.

  The rule $y = f^{-1}(x)$ with $y$ given by 3 is a rule for $f^{-1}$. The domain of $f^{-1}$ is $T$ and its range is $S$.

### EXERCISES 3-3

Let $(-2, 3)$ and $(4, 2)$ be two points on the graph of the inverse of the relation given by a line $l$.

1. Determine an equation for the inverse of the line $l$.
2. Determine an equation for the line $l$.

Consider the relation $y = 2x + 4$.

3. Is the point $(0, 4)$ on the graph of the inverse of the relation?
4. Is the point $(4, 0)$ on the graph of the inverse of the relation?
5. Find a point $(a, b)$ that is on both the graph of the relation and the graph of its inverse.

Draw a complete graph of the inverse of each relation.

6.

7.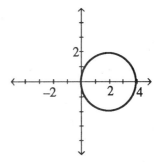

8. Let the domain of $f$ be $(-\infty, 0]$ and let $f(x) = x^2$ when $x \le 0$. Find a rule for the inverse of $f$. Is $f^{-1}$ a function?
9. Let $f(x) = \sqrt{x - 2}$. Find a rule for the inverse of $f$. Is $f^{-1}$ a function? Determine the domain and range of $f$. Determine the domain and range of $f^{-1}$.
10. Explain geometrically the relationship between the domain and range of a function and the domain and range of its inverse.

Draw a complete graph of each relation and its inverse. Do not use a graphing utility. Determine an equation for the inverse. Specify which inverses are functions.

11. $y = 3 - 2x$
12. $y = \dfrac{4 + 7x}{3}$
13. $y = x^2 + 1$
14. $y^2 = x$
15. $x^2 + y^2 = 4$
16. $y = \sqrt{x - 3}$

Determine whether each function is one-to-one. Determine an equation for the inverse of each function. Is the inverse a function? For those functions with *inverses* that are functions, find an explicit rule for the inverse. Draw a graph of the function and its inverse relation without using a graphing utility. Check your answer with a graphing utility.

17. $f(x) = 3x - 6$
18. $g(x) = (x - 2)^2$
19. $h(x) = \sqrt{x + 2}$
20. $g(x) = x^3$
21. $f(x) = \dfrac{x + 3}{x - 2}$
22. $t(x) = \dfrac{2x - 3}{x + 1}$
23. Determine the domain and range of both the function and its inverse in Exercise 18.
24. Determine the domain and range of both the function and its inverse in Exercise 19.

Use a graphing utility to determine whether each function is one-to-one. Draw a complete graph of the function. Draw its inverse without the aid of a graphing utility. Identify the functions with inverses that are also functions.

25. $f(x) = x^3 - 8x$
26. $f(x) = x^4 - 2x + 3$
27. $f(x) = \dfrac{x^2 - 2x + 3}{x + 2}$
28. $f(x) = x^3 - 2x - 6$
29. $f(x) = x^3 + 2x + 2$
30. $f(x) = \sin x$
31. Determine the domain and range of both the function and its inverse in Exercise 25.
32. Determine the domain and range of both the function and its inverse in Exercise 26.
*33. Show that if $f(x) = f(y)$ implies that $x = y$ for all $x$ and $y$ in the domain of $f$, then $f$ is one-to-one.
*34. Show that if $x \neq y$, but $f(x) = f(y)$ for some $x$ and $y$ in the domain of $f$, then $f$ is *not* one-to-one.
35. Prove that $f(x) = x^3 - 8x$ is *not* a one-to-one function.
36. Show, by example, that a relation that is not a function can have an inverse that is a function.
37. Find at least three functions such that $f^{-1} = f$.
38. If $f^{-1}(x) = f(x)$ for each $x$ in the domain of $f$, then determine $f \circ f(x)$.
39. Let $f$ be the function of Example 8. Determine the domain and range of $f$ and $f^{-1}$, a rule for $f^{-1}$, and draw a complete graph of $f^{-1}$.

## 3.4 Inverse Trigonometric Functions

In the previous sections we defined the inverse relation of a relation. The inverse relation of a given function is a function if and only if the given function is one-to-one. We will see that the inverse relations of the six basic trig functions are not functions. However, in this section we see how to restrict the domains of the basic trig functions so that the inverses of the new functions are functions. These inverses are referred to as the **inverse trig functions**. We determine complete graphs of the inverse trig functions.

Figure 3.4.1 contains a complete graph of $y = \sin x$, and Figure 3.4.2 contains a complete graph of the inverse of this function. The inverse is *not* a function because it fails the vertical line test. Each vertical line should intersect the graph of the inverse of $y = \sin x$ in Figure

## 3.4 Inverse Trigonometric Functions

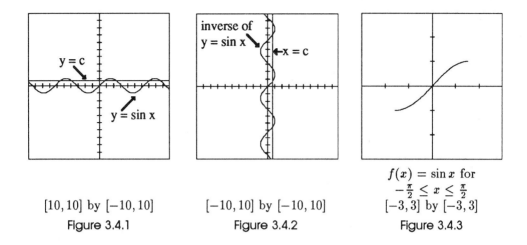

[10, 10] by [−10, 10]
Figure 3.4.1

[−10, 10] by [−10, 10]
Figure 3.4.2

$f(x) = \sin x$ for $-\frac{\pi}{2} \leq x \leq \frac{\pi}{2}$
[−3, 3] by [−3, 3]
Figure 3.4.3

3.4.2 at most once if the inverse is to be a function. However, vertical lines $x = c$ with $-1 \leq c \leq 1$ intersect the graph of the inverse of $y = \sin x$ infinitely many times.

In Section 3.3 we used the horizontal line test to decide if the inverse of a function is itself a function. This test states that the inverse of a function $f$ is also a function if and only if every horizontal line intersects the graph of $f$ at most once. Notice that the horizontal line $y = c$ with $-1 \leq c \leq 1$ intersects the graph of $y = \sin x$ infinitely many times (Figure 3.4.1). A function is said to be one-to-one if and only if every horizontal line intersects the graph of $f$ in at most one point. Thus, the inverse of a function $f$ is also a function if and only if $f$ is one-to-one.

Notice in Figure 3.4.1 that $y = \sin x$ takes on every value in $[-1, 1]$ *exactly* once if the $x$-values are restricted to the interval $\left[-\frac{\pi}{2}, \frac{\pi}{2}\right]$. Thus, the function $f(x) = \sin x$ with domain $\left[-\frac{\pi}{2}, \frac{\pi}{2}\right]$ is a one-to-one function. Notice that every horizontal line intersects the graph of $f$ in at most one point (Figure 3.4.3). Therefore, the inverse of $f$ is also a function for this specified domain (Figure 3.4.4).

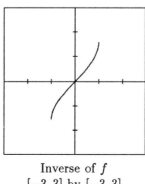

Inverse of $f$
[−3, 3] by [−3, 3]
Figure 3.4.4

Traditionally, the inverse notation $y = \sin^{-1} x$ has been reserved for the *function* with the graph in Figure 3.4.4, that is, the inverse of $f(x) = \sin x$ with domain $-\frac{\pi}{2} \leq x \leq \frac{\pi}{2}$. We will hold to this custom. Thus, even though the relation with the graph in Figure 3.4.2 is the inverse of the sine function, the notation $y = \sin^{-1} x$ will be reserved for the function with the graph in Figure 3.4.4. That is, $y = \sin^{-1} x$ is, by definition, the inverse of the function with the graph in Figure 3.4.3. More precisely, we have the following definition.

- **DEFINITION** The **inverse sine function**, denoted by $y = \sin^{-1} x$, is the function with domain $-1 \leq x \leq 1$ and range $-\frac{\pi}{2} \leq y \leq \frac{\pi}{2}$ that satisfies $\sin y = x$.

  It is also customary to use the notation $y = \arcsin x$ interchangeably with $y = \sin^{-1} x$. We give the following theorem without proof.

- **THEOREM 1** Let $f(x) = \sin x$ with $-\frac{\pi}{2} \leq x \leq \frac{\pi}{2}$. Then $f^{-1}(x) = \sin^{-1} x$ and
  (a) $f^{-1} \circ f(x) = x$ for all $x$ in $\left[-\frac{\pi}{2}, \frac{\pi}{2}\right]$.
  (b) $f \circ f^{-1}(x) = x$ for all $x$ in $[-1, 1]$.

  The domain and range of the composite function $f^{-1} \circ f$ of Theorem 1 is $\left[-\frac{\pi}{2}, \frac{\pi}{2}\right]$, and the domain and range of the composite function $f \circ f^{-1}$ is $[-1, 1]$. Therefore, $f^{-1} \circ f \neq f \circ f^{-1}$ because the domains of the two functions are different. Many times in the previous chapter we used the calculator key INV, 2nd F, or SHIFT together with the SIN key to solve an equation of the form $\sin x = 0.21$. The unique number produced by a calculator in this case is $\sin^{-1}(0.21)$. Thus, we have implicitly used the idea of inverse functions when we solved trig equations.

- **EXAMPLE 1**: Find each of the following.
  (a) $\sin^{-1}(0.5)$   (b) $\sin^{-1}(-0.7)$   (c) $\sin^{-1}(1.2)$

  SOLUTION:

  (a) By definition, $y = \sin^{-1}(0.5)$ if and only if $\sin y = 0.5$ and $-\frac{\pi}{2} \leq y \leq \frac{\pi}{2}$. If you happen to remember that $\sin \frac{\pi}{6} = \sin 30° = 0.5$, then $\sin^{-1}(0.5) = \frac{\pi}{6}$. Otherwise, we can use a calculator to determine $\sin^{-1}(0.5)$. With a calculator set in radian mode you will find that $\sin^{-1}(0.5)$ is $0.5235987756$. This is a decimal approximation to $\frac{\pi}{6}$. If you put your calculator in degree mode you will find $\sin^{-1}(0.5) = 30°$.
  (b) You can use a calculator to show that $\sin^{-1}(-0.7) = -0.7753975$ radians or $\sin^{-1}(-0.7) = -44.427004°$.
  (c) If you use a calculator you get an error message for $\sin^{-1}(1.2)$. This happens because there is *no* angle with sine equal to 1.2. The sine of any angle or real number is a number in the interval $[-1, 1]$.

  In Example 1(a), we wrote $y = \sin^{-1}(0.5)$ in the form $\sin y = 0.5$. There is an important subtle distinction between solving the equation $\sin y = 0.5$ and finding $\sin^{-1}(0.5)$. The equation $\sin y = 0.5$ has infinitely many solutions. That is, there are infinitely many real

## 3.4 Inverse Trigonometric Functions

numbers whose sine equals 0.5. However, the notation $\sin^{-1}(0.5)$ stands for that *single* number in $\left[-\frac{\pi}{2}, \frac{\pi}{2}\right]$ with sine equal to 0.5.

- **EXAMPLE 2:** Find $\cos(\sin^{-1}\mu)$ and $\tan(\sin^{-1}\mu)$.

  **SOLUTION:** $\sin^{-1}\mu$ stands for an angle between $-\frac{\pi}{2}$ and $\frac{\pi}{2}$ with sine equal to $\mu$. If we call this angle $\theta$, we can put $\theta$ in standard form as illustrated in Figure 3.4.5. If $\mu > 0$, $\theta$ is in the first quadrant; and if $\mu < 0$, $\theta$ is in the fourth quadrant. We can use the Pythagorean theorem to find that the length of the side adjacent to $\theta$ is $\sqrt{1-\mu^2}$. Thus,

  $$\cos\left(\sin^{-1}\mu\right) = \sqrt{1-\mu^2}.$$

  Notice that the cosine of the angle $\theta$ is positive whether $\mu$ is positive or negative.
  Next,

  $$\tan\left(\sin^{-1}\mu\right) = \frac{\mu}{\sqrt{1-\mu^2}}.$$

  If $\mu < 0$, then $\tan\theta < 0$; and if $\mu > 0$, then $\tan\theta > 0$. This agrees with the fact that if $\mu < 0$, then $\theta$ is in quadrant IV; and if $\mu > 0$, then $\theta$ is in quadrant I. •

- **EXAMPLE 3:** Determine the domain and the range of each function, and draw a complete graph.
  
  (a) $f(x) = \sin^{-1}(2x)$        (b) $g(x) = \sin^{-1}(\frac{1}{3}x)$

  **SOLUTION:**

  (a) In Section 2.6, we found that the effect of the coefficient 2 of $x$ in $y = h(2x)$ is to shrink horizontally the graph of $y = h(x)$ by a factor of $\frac{1}{2}$. Thus, the graph of $f(x) = \sin^{-1}(2x)$ can be obtained by applying a horizontal shrink of $\frac{1}{2}$ to the graph of $y = \sin^{-1}x$. This is confirmed by the computer drawn complete graph of $f(x) = \sin^{-1}(2x)$ in Figure 3.4.6. The domain of $f(x) = \sin^{-1}(2x)$ is $\left[-\frac{1}{2}, \frac{1}{2}\right]$ and its range is $\left[-\frac{\pi}{2}, \frac{\pi}{2}\right]$. Alternatively, $-1 \leq 2x \leq 1$ implies $-\frac{1}{2} \leq x \leq \frac{1}{2}$.

Figure 3.4.5

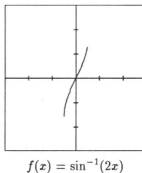

$f(x) = \sin^{-1}(2x)$
$[-3, 3]$ by $[-3, 3]$

Figure 3.4.6

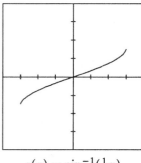

$g(x) = \sin^{-1}(\frac{1}{3}x)$
$[-4, 4]$ by $[-4, 4]$

Figure 3.4.7

(b) The graph of $g(x) = \sin^{-1}\left(\frac{1}{3}x\right)$ can be obtained by applying a horizontal stretch of 3 to the graph of $y = \sin^{-1} x$ (Figure 3.4.7). The domain of $g(x) = \sin^{-1}\left(\frac{1}{3}x\right)$ is $[-3, 3]$, and its range is $\left[-\frac{\pi}{2}, \frac{\pi}{2}\right]$. •

• **EXAMPLE 4:** Find the domain and the range of $f(x) = 3\sin^{-1}\left(\frac{1}{4}x + 2\right) - 5$, and draw a complete graph.

**SOLUTION:** We rewrite the expression for $f$ as follows:

$$f(x) = 3\sin^{-1}\left(\frac{1}{4}(x+8)\right) - 5.$$

The graph of $f$ can be obtained from the graph of $y = \sin^{-1} x$ by applying, in order, the following transformations.

1. A horizontal stretch by a factor of 4.
2. A horizontal shift of $-8$ units.
3. A vertical stretch by a factor of 3.
4. A vertical shift of $-5$ units.

This is confirmed by the computer drawn complete graph of $f$ in Figure 3.4.8. The domain of $f$ is $[-12, -4]$, and its range is $\left[-\frac{3\pi}{2} - 5, \frac{3\pi}{2} - 5\right]$, or $[-9.71, -0.29]$ accurate to hundredths. •

Complete graphs of $y = \cos x$ and $y = \tan x$ are given in Figures 3.4.9 and 3.4.10, respectively. Every possible value of each function is obtained for some value of $x$ in the domain represented by the $x$ values in the unshaded portions of the two graphs.

Restricting the domain of $y = \cos x$ to $[0, \pi]$ and the domain of $y = \tan x$ to $-\frac{\pi}{2}, \frac{\pi}{2}$ produces one-to-one functions. We use these restrictions to define the inverse cosine and inverse tangent functions.

$f(x) = 3\sin^{-1}(\frac{1}{4}x+2)-5$
$[-15, 0]$ by $[-12, 2]$
**Figure 3.4.8**

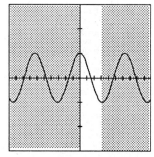

$y = \cos x$
$[-10, 10]$ by $[-3, 3]$
**Figure 3.4.9**

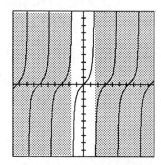

$y = \tan x$
$[-10, 10]$ by $[-10, 10]$
**Figure 3.4.10**

## 3.4 Inverse Trigonometric Functions

- **DEFINITION**   The **inverse cosine function**, denoted by $y = \cos^{-1} x$ or $y = \arccos x$, is the function with domain $[-1, 1]$, range $[0, \pi]$ that satisfies $\cos y = x$.

- **DEFINITION**   The **inverse tangent function**, denoted by $y = \tan^{-1} x$ or $y = \arctan x$, is the function with domain $(-\infty, \infty)$, range $\left(-\frac{\pi}{2}, \frac{\pi}{2}\right)$ that satisfies $\tan y = x$.

The graphs of $y = \cos^{-1} x = \arccos x$ and $y = \tan^{-1} x = \arctan x$ are given in Figures 3.4.11 and 3.4.12, respectively. The domain and range of $y = \cos^{-1} x$ are $[-1, 1]$ and $[0, \pi]$, respectively. The domain and range of $y = \tan^{-1} x$ are the set of all real numbers and $\left(-\frac{\pi}{2}, \frac{\pi}{2}\right)$, respectively. Notice that $y = -\frac{\pi}{2}$ and $y = \frac{\pi}{2}$ are horizontal asymptotes of $y = \tan^{-1} x$.

The inverse cotangent, secant, and cosecant functions are so rarely used that we will not give definitions of these three functions.

- **EXAMPLE 5:** Without using a calculator, find the value of each expression.
  (a) $\sin^{-1}\left(\tan \frac{3\pi}{4}\right)$
  (b) $\cos\left(\tan^{-1} \frac{1}{2}\right)$
  (c) $\sin(\arccos(-\frac{2}{3}) - \arctan \frac{3}{5})$

SOLUTION:

(a) We know from Section 2.1 that the point $(-1, 1)$ is on the terminal side of $\frac{3\pi}{4}$ (Figure 3.4.13). Thus, $\tan \frac{3\pi}{4} = -1$ and $\sin^{-1}\left(\tan \frac{3\pi}{4}\right) = \sin^{-1}(-1)$. Because $-\frac{\pi}{2} \leq \sin^{-1}(-1) \leq \frac{\pi}{2}$ and $\sin\left(-\frac{\pi}{2}\right) = -1$, we have $\sin^{-1}(-1) = -\frac{\pi}{2}$. Therefore, $\sin^{-1}\left(\tan \frac{3\pi}{4}\right) = -\frac{\pi}{2}$.

(b) Let $\theta = \tan^{-1} \frac{1}{2}$. Then, $\theta$ is an acute angle because $-\frac{\pi}{2} \leq \tan^{-1} \frac{1}{2} \leq \frac{\pi}{2}$ and $\tan \theta = \frac{1}{2} > 0$. Figure 3.4.14 shows $\theta$ in standard position. Thus, $\cos\left(\tan^{-1} \frac{1}{2}\right) = \cos \theta = 2/\sqrt{5}$.

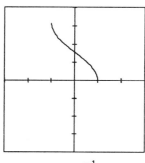

$y = \cos^{-1} x$
$[-3, 3]$ by $[-4, 4]$
Figure 3.4.11

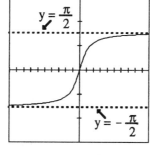

$y = \tan^{-1} x$
$[-10, 10]$ by $[-3, 3]$
Figure 3.4.12

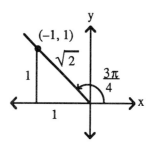

Figure 3.4.13

(c) Let $\alpha = \arccos\left(-\frac{2}{3}\right)$ and $\beta = \arctan\frac{3}{5}$. Figures 3.4.15 and 3.4.16 give the standard positions of the angles $\alpha$ and $\beta$.

Because $0 \leq \alpha \leq \pi$ and $\cos\alpha = -\frac{2}{3} < 0$, $\alpha$ must be in the second quadrant. $\beta$ must be in the first quadrant. (Why?) We can use the addition formulas from Section 3.2 to help make the computation:

$$\sin\left(\arccos\left(-\frac{2}{3}\right) - \arctan\frac{3}{5}\right) = \sin(\alpha - \beta)$$

$$= \sin\alpha\cos\beta - \cos\alpha\sin\beta$$

$$= \left(\frac{\sqrt{5}}{3}\right)\left(\frac{5}{\sqrt{34}}\right) - \left(-\frac{2}{3}\right)\left(\frac{3}{\sqrt{34}}\right)$$

$$= \frac{5\sqrt{5} + 6}{3\sqrt{34}}.$$

Calculators can be used to check our solutions to the above example. For example, we can use a calculator to compute directly the number in (b) and then compare this value with a decimal approximation to $2/\sqrt{5}$.

- **EXAMPLE 6**: Solve for $x$.
  (a) $\sin x = 0.6$    (b) $\cot x = 2.5$

SOLUTION:

(a) We use a calculator to determine that $\sin^{-1} 0.6 = 0.6435011$. Now $x$ must be an angle in the first or second quadrant because $\sin x > 0$. Therefore, $x$ can be any angle whose terminal side is the same as the terminal side of $0.6435011$ in the first quadrant, or whose terminal side is the same as $\pi - 0.6435011 = 2.4980915$ in the second quadrant. Recall that angles with the same terminal side are said to be coterminal. Thus, the solutions to $\sin x = 0.6$ are $0.6435011 + 2k\pi$ or $2.4980915 + 2k\pi$, where $k$ is any integer.

(b) The period of $\cot x$ is $\pi$. Because $\cot x > 0$, $x$ is in the first quadrant. Calculators do not have $\cot^{-1} x$ as a built-in function. However, if $\cot x = 2.5$, then $\tan x =$

Figure 3.4.14    Figure 3.4.15    Figure 3.4.16

0.4. Now, we use a calculator to find that $\tan^{-1} 0.4 = 0.3805064$. Thus, $x$ is $0.3805064 + k\pi$ where $k$ is any integer. ●

In the last example of this section we establish an identity involving inverse trig functions.

● **EXAMPLE 7:** $\sin^{-1} x + \cos^{-1} x = \frac{\pi}{2}$ for all $x$ in $[-1, 1]$.

**SOLUTION:** Let $\theta = \sin^{-1} x$. Then, $x = \sin\theta$ and $-\frac{\pi}{2} \leq \theta \leq \frac{\pi}{2}$ so that $-\frac{\pi}{2} \leq -\theta \leq \frac{\pi}{2}$ is also true. Adding $\frac{\pi}{2}$ to each portion of the last inequality gives $0 \leq \frac{\pi}{2} - \theta \leq \pi$. Now $\cos\left(\frac{\pi}{2} - \theta\right) = \sin\theta = x$. Because $0 \leq \frac{\pi}{2} - \theta \leq \pi$, it follows that $\frac{\pi}{2} - \theta = \cos^{-1} x$. Replacing $\theta$ by $\sin^{-1} x$ gives

$$\frac{\pi}{2} - \theta = \cos^{-1} x$$

$$\frac{\pi}{2} - \sin^{-1} x = \cos^{-1} x$$

$$\frac{\pi}{2} = \sin^{-1} x + \cos^{-1} x.$$

●

## EXERCISES 3-4

Use a calculator to evaluate. Express your answer in degrees.
1. $\sin^{-1}(0.362)$
2. $\arcsin(-1.67)$
3. $\tan^{-1}(0.125)$
4. $\tan^{-1}(-2.8)$

Use a calculator to evaluate. Express your answer as a real number.
5. $\sin^{-1}(0.46)$
6. $\cos^{-1}(-0.853)$
7. $\tan^{-1}(2.37)$
8. $\tan^{-1}(-22.8)$

Compute the exact value without using a calculator. Express your answer as a real number.
9. $\sin^{-1} 1$
10. $\tan^{-1} \sqrt{3}$
11. $\sin^{-1} \frac{\sqrt{2}}{2}$
12. $\cos^{-1}\left(-\frac{\sqrt{3}}{2}\right)$
13. $\tan^{-1}(-\sqrt{3})$
14. $\sin^{-1}(-1)$

Use a calculator to evaluate. Express your answer as a real number.
15. $\sin(\sin^{-1}(0.36))$
16. $\sin(\arccos(0.568))$
17. $\sin^{-1}(\cos 20)$
18. $\cos(\sin^{-1}(-0.125))$
19. $\sin(\sin^{-1} 1.2)$
20. $\sin^{-1}(\sin 1.2)$
21. $\sin(\sin^{-1} 2)$
22. $\sin^{-1}(\sin 2)$
23. $\tan^{-1}(\sin 2)$
24. $\tan(\arctan 3)$
25. $\sin^{-1}(\sin(\cos^{-1} 0.23))$
26. $\cos(\tan^{-1} 3.5 - \sin^{-1} 0.35)$

Compute the exact value without using a calculator.

27. $\cos\left(\sin^{-1}\frac{1}{2}\right)$

28. $\sin(\tan^{-1} 1)$

29. $\cos\left(2\sin^{-1}\frac{1}{2}\right)$

30. $\cos(\tan^{-1}\sqrt{3}) - \sin(\tan^{-1} 0)$

31. $\cos(\sin^{-1} 0.6)$

32. $\sin(\tan^{-1} 2)$

Draw a complete graph. Determine the domain and range.

33. $y = 1 - \arcsin x$

34. $y = 3 + \cos^{-1}(x-2)$

35. $y = -0.25\tan^{-1}(x - \pi)$

36. $g(x) = 3\arccos(\frac{1}{2}x - \pi)$

37. $y = \sin(\sin^{-1} x)$

38. $y = \sin^{-1}(\sin x)$

Solve for $x$. Find the exact solution(s).

39. $\sin(\sin^{-1} x) = x$

40. $\cos^{-1}(\cos x) = x$

41. $2\sin^{-1} x = 1$

42. $\tan^{-1} x = -1$

Find an equivalent algebraic expression not involving trig functions.

43. $\sin(\tan^{-1} x)$

44. $\cos(\tan^{-1} x)$

45. $\tan(\sin^{-1} x)$

46. $\cot(\cos^{-1} x)$

Solve for $x$. Find *all* solutions.

47. $\tan x = 2.3$

48. $\sin x = -0.75$

49. $\sec x = 3$

50. $\cot x = -5$

Describe what happens to the values of $f(x)$.

51. $f(x) = \dfrac{\sin^{-1} x}{x}$ as $x \to 0$

52. $f(x) = 2\tan^{-1} x$ as $|x| \to \infty$

Solve the inequality in $[-2\pi, 2\pi]$.

53. $(\sin x)(\tan^{-1} x) \geq 0$

54. $\dfrac{\sin^{-1} 2x}{\sin x} \geq 2$

Verify the following identities.

55. $\sin^{-1}(-x) = -\sin^{-1} x$ for $|x| \leq 1$

56. $\sin^{-1} x = \tan^{-1}\dfrac{x}{\sqrt{1-x^2}}$ for $|x| < 1$

57. $\arccos\theta + \arcsin\theta = 90°$

58. $\cos(\sin^{-1} x) = \sqrt{1-x^2}$ for $|x| \leq 1$

59. The length $L$ of the shadow cast by a tower 50 feet tall depends on $\theta$, the angle of elevation of the sun (measured from the horizontal). Express $\theta$ as a function of $L$. Draw a complete graph of the function. Which portion of the graph represents the problem situation?

60. A revolving light beacon $L$ stands 3 miles from the closest point $P$ along a straight shore. Let $\theta$ be the angle $PLQ$ (Figure 3.4.17). Express this angle as a function of the distance from

### 3.4 Inverse Trigonometric Functions • 241

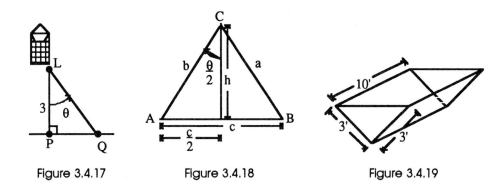

Figure 3.4.17   Figure 3.4.18   Figure 3.4.19

$P$ to $Q$. Draw a complete graph of the function. Which portion of the graph represents the problem situation?

*61. Let triangle $ABC$ be isosceles with $a = b$, and let $\angle C = \theta$ (Figure 3.4.18). Show that the area of the triangle is $\frac{1}{2}a^2 \sin \theta$. Hint: Let $h$ be the altitude of the triangle. Notice that the area is $\frac{1}{2}ch$. Find $\sin \frac{\theta}{2}$ and $\cos \frac{\theta}{2}$ and use the half-angle formulas.

The ends of a feeding trough 10 feet long are isosceles triangles with 3-foot sides, as shown in Figure 3.4.19. Let $\theta$ be the angle between the two equal sides.

62. Use the area result of Exercise 61 to write a *function* that gives the angle $\theta$ as a function of the *volume* $V$ of the trough.

63. Draw a complete graph of the function in Exercise 62. Explain why the graph represents only a portion of the problem situation.

64. What values of $\theta$ make sense in the problem situation?

65. Write an *equation* that relates angle $\theta$ and $V$ for all values of $\theta$ and $V$ that make sense in the problem situation.

66. Is there a maximum volume? What is the associated angle $\theta$?

67. For what values of $x$ is $\sin^{-1}\left(\dfrac{2x}{x^2+1}\right) = 2\tan^{-1} x$?

68. How would you define $\csc^{-1} x$?

69. How could you compute $\sin^{-1} x$ for all $|x| < 1$ on a calculator that had only a $\tan^{-1}$ key?

70. Show that for $A > 0$, $A\cos(Bx - C) = a\cos Bx + b\sin Bx$ where $A = \sqrt{a^2 + b^2}$ and $C = \tan^{-1}\left(\frac{b}{a}\right)$.

71. Use Exercise 70 to write $\sin x + \cos x$ in the form $A\cos(Bx - C)$. Check your answer with a graphing utility.

72. Use Exercise 70 to write $20\sin(10\pi t) + 30\cos(10\pi t)$ in the form $A\cos(Bt - C)$. Check your answer with a graphing utility.

73. Use Exercise 70 to write $3\sin 2x + 4\cos 2x$ in the form $A\cos(Bx - C)$. Check your answer with a graphing utility.

74. Generalize Exercise 70 to write $4\sin x - 2\cos x$ in the form $A\cos(Bx - C)$. Check your answer with a graphing utility.

## 3.5 Solving Equations and Inequalities

In this section we solve equations and inequalities that involve trig functions both algebraically and graphically. You will again see that a graphing method is more general and powerful. The identities established in previous sections are used to transform equations into equivalent forms to which algebraic techniques can be applied. Periodicity of functions is used to facilitate finding solutions regardless of the method used. We also determine local maximum and minimum values of trig functions.

The circular property of the trig functions gives $\sin(x + 2k\pi) = \sin x$ for any integer $k$. This property is used to solve the equation in Example 1.

• **EXAMPLE 1:** Solve $\sin x = 0.8$.

SOLUTION: Figure 3.5.1 shows that the equation $\sin x = 0.8$ has infinitely many solutions. The circular property of $\sin x$ suggests that we first solve the equation $\sin x = 0.8$ in $[0, 2\pi]$. There are two solutions in this interval. We can estimate from Figure 3.5.1, that the two solutions in $[0, 2\pi]$ are about 0.9 and 2.2. The solution near 0.9 represents an angle in the first quadrant and the solution near 2.2 an angle in the second quadrant. This is not to be confused with the first quadrant portion of the graphs in Figure 3.5.1. The first quadrant solution is $\sin^{-1} 0.8 = 0.927$, and the second quadrant solution is $\pi - \sin^{-1} 0.8 = 2.214$. (Why?) The complete solution of $\sin x = 0.8$ consists of all real numbers of the form $0.927 + 2k\pi$ or $2.214 + 2k\pi$, where $k$ is any integer. •

We must be careful when we use a complete period of periodic functions to solve equations, inequalities, and so forth. For example, Figure 3.5.1 suggests that the equation $\sin x = 0$ has three solutions in $[0, 2\pi]$. However, we can ignore the zero at $2\pi$, because it is obtained when we add the period $2\pi$ to the zero at 0. In some cases, we may need to look at the interval $[0, 2\pi)$ instead of $[0, 2\pi]$ to simplify observations.

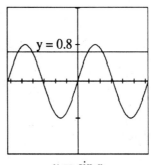

$y = \sin x$
$[-2\pi, 2\pi]$ by $[-2, 2]$
Figure 3.5.1

## 3.5 Solving Equations and Inequalities

Any function, no matter how complicated, that involves only the six basic trig functions is periodic. Periodicity is useful when solving equations.

- **EXAMPLE 2:** Solve $2\cos^2 t + \sin t + 1 = 0$.

**SOLUTION:** It can be shown that the left-hand side of this equation is a periodic function with period $2\pi$ (Figure 3.5.2). If we find the solutions to the equation that are in the interval $[0, 2\pi]$, then we can easily find all the rest of the solutions of the equation. There appears to be one solution in $[0, 2\pi]$ at about 4.7. The identity $\sin^2 t + \cos^2 t = 1$ can be used to replace $\cos^2 t$ by $1 - \sin^2 t$. In this way we obtain an equivalent equation that involves only one trig function:

$$2\cos^2 t + \sin t + 1 = 0$$
$$2(1 - \sin^2 t) + \sin t + 1 = 0$$
$$2 - 2\sin^2 t + \sin t + 1 = 0$$
$$-2\sin^2 t + \sin t + 3 = 0$$
$$2\sin^2 t - \sin t - 3 = 0.$$

We have transformed the original equation into an equivalent equation that is quadratic in $\sin t$. Notice that this quadratic equation can be factored. We could, of course, use the quadratic formula:

$$2\sin^2 t - \sin t - 3 = 0$$
$$(2\sin t - 3)(\sin t + 1) = 0.$$

Now, $t$ is a solution to the equation $2\cos^2 t + \sin t + 1 = 0$ if and only if $t$ is a solution to one of the equations

$$2\sin t - 3 = 0 \quad \text{or} \quad \sin t + 1 = 0.$$

The first of these equations can be rewritten in the form $\sin t = \frac{3}{2}$ to see that it has *no* solutions because $|\sin t| \leq 1$ for all $t$. If we rewrite $\sin t + 1 = 0$ in the form $\sin t = -1$,

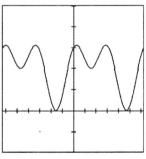

$f(t) = 2\cos^2 t + \sin t + 1$
$[-2\pi, 2\pi]$ by $[-2, 5]$

Figure 3.5.2

we can see that $\frac{3\pi}{2}$ (about 4.71) is the only solution to this equation in the interval $[0, 2\pi]$. Using the circular property of the trig functions we can conclude that the solutions to the original equation consist of the numbers of the form $\frac{3\pi}{2} + 2k\pi$, where $k$ is any integer. These are precisely the angles that are coterminal with $\frac{3\pi}{2}$. The graph of $f(t) = 2\cos^2 t + \sin t + 1$ in $[0, 2\pi]$ by $[-5, 5]$ confirms that there is only one solution of $2\cos^2 t + \sin t + 1 = 0$ in the interval $[0, 2\pi]$ (Figure 3.5.3). You may need to zoom-in around $t = \frac{3\pi}{2}$ to convince yourself that there is only one $x$-intercept. •

Once we observed in Example 2 that the function $f(t) = 2\cos^2 t + \sin t + 1$ was periodic of period $2\pi$ we were able to focus our attention on the interval $[0, 2\pi]$. This considerably simplifies both the algebraic and graphical methods of solution.

From now on, answers given will either be exact or accurate to hundredths unless otherwise stated. When you work the example yourself you will be able to tell whether the answers we give are exact or approximations.

• **EXAMPLE 3:** Solve $\tan x = 3 \cos x$.

**SOLUTION:** Again, we need to find the solutions to this equation only in the interval $[0, 2\pi]$ because the function $f(x) = \tan x - 3\cos x$ is periodic with period $2\pi$. The graph of $f$ in Figure 3.5.4 shows that the equation $\tan x = 3\cos x$ has two solutions in $[0, 2\pi]$. Notice that $f$ has $x = \frac{\pi}{2}$ and $x = \frac{3\pi}{2}$ as vertical asymptotes. (Why?) Zoom-in could be used to determine the solutions in the interval $[0, 2\pi]$. We give an algebraic solution to this equation.

Often it helps to rewrite trig equations so that the only trig functions involved are the *sine* and *cosine* functions. We are able to do this because $\tan x = \frac{\sin x}{\cos x}$, $\cot x = \frac{\cos x}{\sin x}$, $\sec x = \frac{1}{\cos x}$, and $\csc x = \frac{1}{\sin x}$. If we replace $\tan x$ by $\frac{\sin x}{\cos x}$ in $\tan x = 3\cos x$ we obtain $\frac{\sin x}{\cos x} = 3\cos x$. It seems reasonable to multiply each side of the last equation by $\cos x$. However, this process can introduce extraneous solutions. This means we must check all

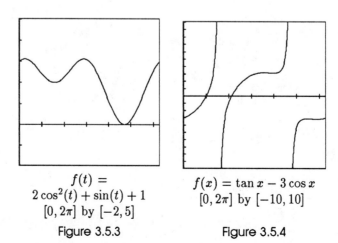

$f(t) = 2\cos^2(t) + \sin(t) + 1$
$[0, 2\pi]$ by $[-2, 5]$

Figure 3.5.3

$f(x) = \tan x - 3\cos x$
$[0, 2\pi]$ by $[-10, 10]$

Figure 3.5.4

solutions in the original equation. We can proceed as follows:

$$\tan x = 3\cos x$$
$$\frac{\sin x}{\cos x} = 3\cos x$$
$$\sin x = 3\cos^2 x$$
$$\sin x = 3(1 - \sin^2 x)$$
$$3\sin^2 x + \sin x - 3 = 0.$$

The quadratic formula gives $\sin x = (-1 \pm \sqrt{37})/6$. (Why?) We can use a calculator to determine that the number $(-1 - \sqrt{37})/6$ is strictly less than $-1$. Thus, the equation $\sin x = (-1 - \sqrt{37})/6$ has no solutions. However, the equation $\sin x = (-1 + \sqrt{37})/6$ has two solutions in the interval $[0, 2\pi]$. One is in the first quadrant and the other is in the second quadrant because $(-1 + \sqrt{37})/6 > 0$.

We can use a calculator to determine that $x = \sin^{-1}[(-1 + \sqrt{37})/6] = 1.01$ is the solution in the first quadrant. The solution in the second quadrant is $\pi - 1.01 = 2.13$. You can check that the values are also solutions to the original equation. Thus, the solutions to the original equation are $1.01 + 2k\pi$ or $2.13 + 2k\pi$, where $k$ is any integer. •

From now on we will use the following convention.

---

• **DEFINITION**   If $f$ is a periodic function with period $h$, then we will say that the solutions of the equation $f(x) = 0$ are **periodic** with period $h$. •

- **Zeros of Periodic Functions**  Let $f$ be a periodic function with period $h$, and let $x_1, x_2, \ldots, x_n$ be all the solutions to $f(x) = 0$ in an interval of length the period $h$. Then, $x_i + kh$ for $i = 1, 2, \ldots, n$ and $k$ any integer are all the solutions to $f(x) = 0$.

If one of the zeros occurs at the left-hand endpoint of an interval of length $h$, then there is also a zero at the right-hand endpoint. You should use only one of these two zeros in the above description of the zeros of a periodic function.

- **EXAMPLE 4:** Solve $3\tan^4 x = 1 + \sec^2 x$.

SOLUTION: First, we show that the solutions to this equation are periodic with period $\pi$ by showing that the function $f(x) = 3\tan^4 x - 1 - \sec^2 x$ is periodic with period $\pi$ (Figure 3.5.5). We have shown in Section 2.6 that $\tan x$ is periodic with period $\pi$. Thus, we need to show only that $\sec^2 x$ is periodic with period $\pi$. (Why?) Notice that

$$\sec^2(x + \pi) = (-\sec x)^2 = \sec^2 x.$$

Now, we can be sure that $f$ has period $\pi$. The graph of $f$ in Figure 3.5.6 suggests that this equation has two solutions in the interval $[-\frac{\pi}{2}, \frac{\pi}{2}]$. The graph of $f$ has $x = -\frac{\pi}{2}$ and $x = \frac{\pi}{2}$ as vertical asymptotes. (Why?)

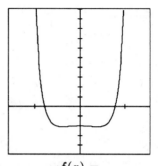

$f(x) =$
$3\tan^4 x - 1 - \sec^2 x$
$[-2\pi, 2\pi]$ by $[-5, 10]$
Figure 3.5.5

$f(x) =$
$3\tan^4 x - 1 - \sec^2 x$
$[-\frac{\pi}{2}, \frac{\pi}{2}]$ by $[-5, 10]$
Figure 3.5.6

We find the solutions algebraically by using the identity $1 + \tan^2 x = \sec^2 x$ to rewrite the equation:

$$3\tan^4 x = 1 + \sec^2 x$$
$$3\tan^4 x = 1 + 1 + \tan^2 x$$
$$3\tan^4 x - \tan^2 x - 2 = 0$$
$$(3\tan^2 x + 2)(\tan^2 x - 1) = 0.$$

Thus, we need to solve the following equations:

$$3\tan^2 x + 2 = 0 \qquad \tan^2 x - 1 = 0$$
$$\tan^2 x = -\frac{2}{3} \qquad \tan^2 x = 1$$
$$\tan x = \pm 1$$

The equation $\tan^2 x = -\frac{2}{3}$ has no solutions. One solution to $\tan x = 1$ is $\tan^{-1} 1$. Similarly, one solution to $\tan x = -1$ is $\tan^{-1}(-1)$. Now, $\tan^{-1} 1 = \frac{\pi}{4}$ and $\tan^{-1} 1 = -\frac{\pi}{4}$. Thus, the two solutions to $f(x) = 0$ in $\left[-\frac{\pi}{2}, \frac{\pi}{2}\right]$ are $\frac{\pi}{4}$ or $-\frac{\pi}{4}$. The periodicity of the solutions implies that the solutions to the original equation are $-\frac{\pi}{4} + k\pi$ or $\frac{\pi}{4} + k\pi$, where $k$ is any integer. These solutions can also be rewritten in the form $\frac{\pi}{4} + k\frac{\pi}{2}$, where $k$ is any integer. (Why?)
•

In Example 5 we factor the left side of the equation by grouping. We may not always include a graph with each example, but it is a good idea for you to start by viewing a graph to obtain preliminary information about the functions involved in an equation or inequality.

• **EXAMPLE 5:** Solve $2\cot x \cos x - 3\cos x + 6\cot x - 9 = 0$.

## 3.5 Solving Equations and Inequalities

SOLUTION: The function $f(x) = 2\cot x \cos x - 3\cos x + 6\cot x - 9$ can be shown to be periodic with period $2\pi$ and to have two zeros in any interval of length $2\pi$. We can group terms and factor as follows:

$$(2\cot x \cos x - 3\cos x) + (6\cot x - 9) = 0$$
$$(\cos x)(2\cot x - 3) + 3(2\cot x - 3) = 0$$
$$(\cos x + 3)(2\cot x - 3) = 0.$$

The factor $\cos x + 3$ can never be zero. (Why?) To solve the equation $\cot x = \frac{3}{2}$ with a calculator you will need to solve the equivalent equation $\tan x = \frac{2}{3}$. (We used $\tan x = \frac{1}{\cot x}$.) The solutions to the original equation are periodic with period $\pi$ because they are also solutions to the equation $\tan x = \frac{2}{3}$. The equation $\tan x = \frac{2}{3}$ has exactly one solution in any interval of length $\pi$ and $\tan^{-1}\left(\frac{2}{3}\right) = 0.59$. Thus, the solutions to the original equation are $0.59 + k\pi$, where $k$ is any integer. •

The same observations about periodicity for equations can be used to help solve inequalities involving trig functions.

• EXAMPLE 6: Solve $\sin x < \cot x$.

SOLUTION: $\sin x$ is periodic with period $2\pi$ and $\cot x$ is periodic with period $\pi$. Thus, we can solve this inequality by first finding the solutions in the interval $[0, 2\pi]$. We need to determine where the graph of $f(x) = \sin x - \cot x$ is *below* the $x$-axis (Figure 3.5.7). Notice that $f$ has $x = 0$, $x = \pi$, and $x = 2\pi$ as vertical asymptotes. We can see from this figure that $f$ has two zeros in the interval $[0, 2\pi]$. These zeros can be found graphically using zoom-in or algebraically.

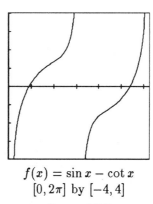

$f(x) = \sin x - \cot x$
$[0, 2\pi]$ by $[-4, 4]$
Figure 3.5.7

We will determine the zeros algebraically:

$$\sin x - \cot x = 0$$
$$\sin x - \frac{\cos x}{\sin x} = 0$$
$$\sin^2 x - \cos x = 0$$
$$1 - \cos^2 x - \cos x = 0$$
$$\cos^2 x + \cos x - 1 = 0.$$

Using the quadratic formula we find that $\cos x = (-1 \pm \sqrt{5})/2$. Now, $\cos x = (-1-\sqrt{5})/2$ has no real solutions. (Why?) The solutions to $\cos x = (-1+\sqrt{5})/2$ in $[0, 2\pi]$ are 0.90 and 5.38. The solution to the original inequality in $[0, 2\pi]$ is $(0, 0.90) \cup (\pi, 5.38)$. Furthermore, the solutions to the original inequality are periodic with period $2\pi$. •

We were able to use algebraic techniques in all of the previous examples to obtain solutions. Generally, it is difficult or even impossible to find zeros of trig functions algebraically. The periodicity of the functions in the previous examples helped simplify finding solutions. However, we should not expect all equations and inequalities to involve only periodic functions. The next example involves a function that is not periodic.

• **EXAMPLE 7:** Solve $\sin 2x \leq x$.

**SOLUTION:** Figure 3.5.8 gives a complete graph of $f(x) = \sin 2x$ and $g(x) = x$. Notice that neither $g$ nor $f - g$ is periodic. We need to determine those values of $x$ for which the graph of $f(x) = \sin 2x$ is *below* or intersects the graph of $g(x) = x$. We can see from this figure that the graphs of $f$ and $g$ have three points in common. One of these is $(0, 0)$. The graphs of both $f$ and $g$ are symmetric with respect to the origin because $f(-x) = -f(x)$ and $g(-x) = -g(x)$. We can use zoom-in to determine that the point of intersection in the first quadrant is $(0.948, 0.948)$ with error at most 0.01. By symmetry, $(-0.948, -0.948)$ is the point of intersection in the third quadrant. Thus, the graph of $f$ is below the graph of $g$ in $[-0.948, 0] \cup [0.948, \infty)$. Therefore, $\sin 2x \leq x$ in $[-0.948, 0] \cup [0.948, \infty)$. •

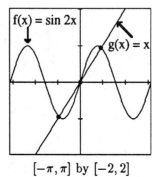

$[-\pi, \pi]$ by $[-2, 2]$
Figure 3.5.8

$\sin \frac{x}{2} + \sin \frac{x}{3}$
$[0, 12\pi]$ by $[-3, 3]$
Figure 3.5.9

$f(x) = \sin^3 x - 2\sin x + 1$
$[0, 2\pi]$ by $[-1, 3]$
Figure 3.5.10

In Section 3.1 we showed that the function of the next example is periodic with period $12\pi$.

- **EXAMPLE 8:** Determine all the zeros of $f(x) = \sin \frac{x}{2} + \sin \frac{x}{3}$ and the local maximum and minimum values of $f$ in $[0, 12\pi]$.

SOLUTION: Because $f$ is periodic with period $12\pi$, we can determine the zeros of $f$ in $[0, 12\pi]$ and then add $12k\pi$, $k$ any integer, to each of them. The graph of $f$ in Figure 3.5.9 is a complete graph of $f$ and shows that $f$ has seven zeros in $[0, 12\pi]$. However, the zero at $12\pi$ is obtained by adding $12\pi$ to the zero at 0. Thus, we need to determine only the other six zeros of $f$ in $[0, 12\pi]$.

Two of the zeros are 0 and $6\pi$ because both $\sin \frac{x}{2}$ and $\sin \frac{x}{3}$ are zero for these values. We can use zoom-in to determine that the solutions, excluding $12\pi$, to the equation $\sin \frac{x}{2} + \sin \frac{x}{3} = 0$ in the interval $[0, 12\pi]$ are 0, 7.54, 15.08, 18.85, 22.62, or 30.16 with error at most 0.01. Now, adding $12k\pi$, $k$ any integer, to these six values gives all solutions to the equation.

In Example 6 of Section 3.1 we found that $f$ had a local maximum value of 1.91 at 3.62. The other two local maximum values in $[0, 2\pi]$ are 0.22 and 1.22 and occur at 16.79 and 26.97, respectively. The three local minimum values of $f$ in $[0, 12\pi]$ are $-1.22$, $-0.22$, and $-1.91$, and they occur at 10.72, 20.91, and 34.08, respectively. •

- **EXAMPLE 9:** Find all solutions to $\sin^3 x - 2\sin x + 1 = 0$ in the interval $[0, 2\pi]$.

SOLUTION: The function $f(x) = \sin^3 x - 2\sin x + 1$ is periodic with period $2\pi$. The graph of $f$ in Figure 3.5.10 is a complete graph and shows that $f$ has three zeros in $[0, 2\pi]$. We can use zoom-in to determine that these zeros are 0.666, 1.571 or 2.475 with error at most 0.01. •

## EXERCISES 3-5

Solve the equation algebraically. Check with a graphing utility.

1. $\sin x = 0.7$
2. $\cos x = 0.9$
3. $\sin x = 1.3$
4. $\tan x = 2.75$

Solve the equation algebraically in the interval $[0, 2\pi]$. Check with a graphing utility.

5. $\sin 2x = 1$
6. $\sin 3t = 1$
7. $2 \sin x = 1$
8. $2 \sin 3x = 1$
9. $3 \cos 2x = 1$
10. $\sin 2t = \sin t$
11. $\sin^2 x = 0$
12. $\sin x \tan x + \sin x = 0$
13. $\cos x (\sin x - 1) = 0$
14. $\sin^2 x - 1 = 0$

Solve the equation algebraically. Check with a graphing utility.

15. $\sin^2 \theta - 2 \sin \theta = 0$
16. $2 \cos 2t = 1$
17. $2 \sin^2 x + 3 \sin x - 2 = 0$
18. $\cos (\sin x) = 1$
19. $3 \sin t = 2 \cos^2 t$
20. $2 \cos^2 x + \cos x - 1 = 0$
21. $\cos 2x + \cos x = 0$
22. $\sin 2t = \sin t$
23. $1 = \csc x - \cot x$
24. $\tan^2 x \cos x + 5 \cos x = 0$

Solve the inequality algebraically. Check with a graphing utility.

25. $\sin x < \dfrac{1}{2}$ for $0 \le x \le 2\pi$
26. $|\sin x| < \dfrac{1}{2}$ for $0 \le x \le 2\pi$
27. $\sin 2x < \cos 2x$
28. $\tan x > 0$

Solve the equation or inequality.

29. $3 \sin x = x$
30. $\tan x = x$ for $-\pi \le x \le \pi$
31. $3 \sin 2x - x^2 = 0$
32. $\sin x < \cos(x - \pi)$
33. $5 \sin 2x < 3 \cos x$
34. $x \sin x \ge 1$ for $-10 \le x \le 10$
35. $\sin 3x + \cos x = 0$
36. $3 \sin 2x = \cos x$

Draw the graph in the indicated interval. Determine all local maximum and minimum values in the interval.

37. $g(x) = 2 - 3 \sin(\frac{1}{2}x - \frac{\pi}{2})$ for $0 \le x \le 4\pi$
38. $f(x) = x^2 \sin x$ for $-\pi \le x \le \pi$
39. $g(x) = -1 + 2 \cos(\pi x - 2)$ for $0 \le x \le 3$
40. $f(x) = e^{-x/2} \sin 2x$ for $0 \le x \le 8$
41. Determine all extrema for the function in Example 8.

### 3.5 Solving Equations and Inequalities

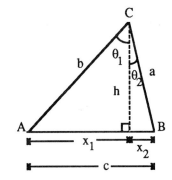

Figure 3.5.11

Figure 3.5.12

42. Determine all extrema for the function in Example 9.

## Applications Involving Sinusoids

43. Show that $g(x) = 2 - 3\cos(2x - \pi)$ is a sinusoid by finding $A$, $B$, $C$, and $D$ so that $A + B\sin(Cx + D) = g(x)$.

The height of a tidal wave has a sinusoidal representation as a function of time $t$ with period $p$ minutes and *maximum* height $h$ feet. (See Figure 3.5.11.) Assume the height is 0 at time $t = 0$.

44. Show that $y = \frac{h}{2} + \frac{h}{2}\sin(\frac{2\pi}{p}(t - \frac{p}{4}))$ is a sinusoid representation of the tidal wave.
45. Draw a complete graph of the sinusoid representation if the period is 40 minutes and the *maximum* height is 27 feet.

## The Cannon Problem

A cannon shell is fired with initial velocity $V$ feet per second. If the cannon barrel makes an angle of $\theta$ with the ground, then the horizontal distance the shell travels before it hits the ground is given by $d = \frac{V^2}{16}\sin\theta\cos\theta$.

46. Let $V = 500$ feet per second. Draw a complete graph of $d$ as a function of $\theta$. What portion of the graph represents the problem situation?
47. At what angle should the cannon be aimed to hit a target 2350 feet away if the initial velocity is 500 feet per second?
48. What is the maximum horizontal distance the shell travels if the initial velocity is 500 feet per second?
49. What is the maximum horizontal distance the shell travels if the initial velocity is 1200 feet per second?

## Area of a Triangle

The area $A$ of a triangle $ABC$ is given by $A = \frac{1}{2}ab\sin C$. In Figure 3.5.12, $x_1 + x_2 = c$ and $h$ is the altitude of the triangle.

50. Derive the area formula. Hints:
   i) Note that $A = \frac{1}{2}x_1 h + \frac{1}{2}x_2 h$,
   ii) $\frac{1}{2}ab \sin C = \frac{1}{2}ab \sin(\theta_1 + \theta_2)$
   $= \frac{1}{2}ab(\sin \theta_1 \cos \theta_2 + \sin \theta_2 \cos \theta_1)$,
   iii) Compute $\sin \theta_1$, $\cos \theta_2$, $\sin \theta_2$ and $\cos \theta_1$.

51. A single cell in a beehive is a regular hexagonal prism open at the front with a trihedral top at the back. It can be shown that the surface area of a cell is given by

$$S(\theta) = 6ab + \frac{3}{2}b^2\left(-\cot \theta + \frac{\sqrt{3}}{\sin \theta}\right)$$

where $\theta$ is the trihedral angle, $a$ is the depth of the cell, and $2b$ is the length of the line segment through the center connecting opposite vertices of the hexagonal front. Assume $a = 1.75$ (inches) and $b = 0.65$ (inches).

(a) Draw a complete graph of the trigonometric representation for the surface area.
(b) What values of $\theta$ make sense in the problem situation?
(c) What value of $\theta$ gives the minimum surface area? (*Remark:* This answer is remarkably close to the observed angle in nature.)
(d) What is the minimum surface area?

52. The range, $D$, of a projectile shot at an angle of elevation, $\theta$, and with an initial velocity, $v$, is given by

$$D(\theta) = \left(\frac{v^2}{g}\right) \sin 2\theta$$

where $g$ is the acceleration due to gravity (32 feet per second per second). Assume $v = 85$ feet per second. (See Figure 3.5.13.)

(a) Draw a complete graph of the trigonometric representation.
(b) What values of $\theta$ make sense in this problem situation?
(c) What value of $\theta$ gives the maximum range?
(d) A target is 215 feet down range. At what angle should the projectile be aimed in order to hit the target?
(e) Derive a formula for the trigonometric representation of the problem situation.
(f) Compare with the cannon problem (46–49).

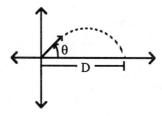

Figure 3.5.13

53. Determine the exact solutions to Example 9.
54. For what values of $x$ is $\log(\tan x) = \log(\sin x) - \log(\cos x)$?
55. For what values of $x$ is $\log(\cot x) = \log(\cos x) - \log(\sin x)$?

## 3.6 The Law of Sines

In Section 2.2 we solved right triangles. Recall that to solve a triangle means to find the remaining parts of a triangle when some of its angles and sides are specified. In this and the next section we solve triangles that do not necessarily contain a right angle. A triangle that does not contain a right angle is called an **oblique triangle**. In this section we solve such triangles using a formula called the *law of sines* that is valid for any triangle. We also investigate applications whose models involve a triangle. Solving the application will require the use of the law of sines to find the missing parts of the triangle.

We use $A$, $B$, and $C$ to denote the vertices of a triangle. The angles at $A$, $B$, and $C$ are denoted by $\alpha$, $\beta$, and $\gamma$, respectively. The measures of the sides opposite $A$, $B$, and $C$ are denoted by $a$, $b$, and $c$, respectively (Figure 3.6.1). Again, $\alpha$, $\beta$, and $\gamma$ stand for the name of the angle as well as the measure of the angle. At most one angle in an oblique triangle can have measure greater than $90°$. (Why?)

If three of the six parts $\alpha$, $\beta$, $\gamma$, $a$, $b$, or $c$ of a triangle are specified, then often a *unique* triangle is determined. In some cases, no triangle is determined or two triangles are determined. If three angles are specified and the sum of their measures is $180°$, then infinitely many triangles are possible. Recall that two triangles are *similar* if they have corresponding angles equal. There are infinitely many triangles similar to a given triangle. Figure 3.6.2 suggests one way to find additional triangles similar to a given triangle $ABC$. Any pair of sides of triangle $ABC$ can be extended or shrunk as suggested by Figure 3.6.2 to form additional triangles similar to $ABC$.

Figure 3.6.1    Figure 3.6.2

Figure 3.6.3

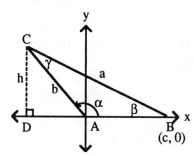
Figure 3.6.4

In this section we determine the triangle(s) when two angles and a side are specified, or, in some instances, when two sides and an angle are specified. The remaining possibilities are covered in the next section. First, we need the following important theorem.

• **THEOREM 1**     **(The Law of Sines)** In any triangle $ABC$ (Figure 3.6.3) we have

$$\frac{\sin\alpha}{a} = \frac{\sin\beta}{b} = \frac{\sin\gamma}{c}.$$

**PROOF** Without any loss of generality we can put the vertex $A$ at the origin and the vertex $B$ on the positive $x$-axis at $(c, 0)$ (Figure 3.6.4). We have drawn the triangle $ABC$ to suggest $\alpha$ is greater than $90°$. The same proof is valid if $\alpha$ is acute.

We choose $D$ on the $x$-axis so that $CD$ is perpendicular to the $x$-axis. Then, the distance $h$ from $C$ to the $x$-axis is a height, or altitude, of the triangle $ABC$. Because $\alpha$ is in standard position we have $\sin\alpha = \frac{h}{b}$. From the right triangle $BCD$ we have $\sin\beta = \frac{h}{a}$. Thus, $h = b\sin\alpha$ and $h = a\sin\beta$. It follows that $b\sin\alpha = a\sin\beta$, or $\frac{\sin\alpha}{a} = \frac{\sin\beta}{b}$. If we put the vertex $C$ on the positive $x$-axis we obtain Figure 3.6.5. Then, the same argument as above gives $\frac{\sin\alpha}{a} = \frac{\sin\gamma}{c}$. Therefore, $\frac{\sin\alpha}{a} = \frac{\sin\beta}{b} = \frac{\sin\gamma}{c}$.

Figure 3.6.5

## Two Angles and a Side Specified

If the measures of any two angles of a triangle are given, then the measure of the third angle can be determined. (Why?) If the measure of one side is also given, the law of sines can be used to find the measure of the remaining two sides as illustrated in Example 1.

- **EXAMPLE 1:** Solve triangle $ABC$ if $\alpha = 36°$, $\beta = 48°$, and $a = 8$.

    SOLUTION: First, $\gamma = 180° - (36° + 48°) = 96°$. Next, we use the law of sines:

    $$\frac{\sin \alpha}{a} = \frac{\sin \beta}{b} \qquad\qquad \frac{\sin \alpha}{a} = \frac{\sin \gamma}{c}$$

    $$\frac{\sin 36°}{8} = \frac{\sin 48°}{b} \qquad\qquad \frac{\sin 36°}{8} = \frac{\sin 96°}{c}$$

    $$b = \frac{8 \sin 48°}{\sin 36°} \qquad\qquad c = \frac{8 \sin 96°}{\sin 36°}$$

    $$b = 10.11 \qquad\qquad c = 13.54$$

Notice that the procedure of Example 1 works no matter which pair of angles is given or which side is given. Thus, if the measure of two angles and a side are given, then a unique triangle is determined, provided we choose a pair of angles whose sum is less than $180°$.

## Two Sides and an Angle Specified

We use the law of cosines in Section 3.7 to determine the triangle(s) if two sides and the *included* angle are given (Figure 3.6.6). This situation can be handled directly using the definitions of the trig functions. However, the method of Section 3.7 is considerably simpler. In this section, we determine the triangle(s) if two sides and an angle *opposite* one of the sides are given. This is sometimes called the ambiguous case because there may be *no* triangle, exactly *one* triangle, or *two* triangles in this situation.

Without any loss of generality we may assume that $a$, $b$, and $\beta$ are given. We first assume that $\beta$ is acute and put $\beta$ in standard position with $C$ in the first quadrant. Then, $A$ must be on the positive $x$-axis (Figure 3.6.7).

Figure 3.6.6          Figure 3.6.7

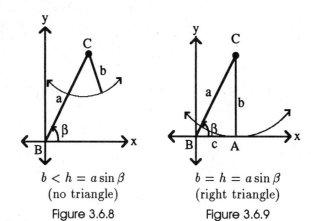

$b < h = a \sin \beta$
(no triangle)
Figure 3.6.8

$b = h = a \sin \beta$
(right triangle)
Figure 3.6.9

If $h$ is the distance from $C$ to the $x$-axis, then $h = a \sin \beta$ and gives the altitude of any such triangle. (Why?) We can think of $b$ as swinging like a pendulum. If $b < h = a \sin \beta$, then *no* triangle is determined (Figure 3.6.8). If $b = h = a \sin \beta$, then a *right* triangle, as shown in Figure 3.6.9, is determined. If $b > h = a \sin \beta$ and $b < a$, then *two* triangles are determined (Figure 3.6.10). If $b > h = a \sin \beta$ and $b \geq a$, then *one* triangle is determined (Figure 3.6.11).

It may seem that we should get two triangles in this last case; the second triangle formed using the origin, $C$, and the point where the arc in Figure 3.6.10 intersects the negative $x$-axis. This cannot happen because we have assumed that $\beta$ is acute and $A$ is in the first quadrant. (Why?) We illustrate the four situations above with $\beta$ acute in Example 2.

- **EXAMPLE 2**: Solve triangle $ABC$ if $\beta = 30°$, $a = 6$, and

  (a) $b = 2$.      (b) $b = 3$.      (c) $b = 5$.      (d) $b = 7$.

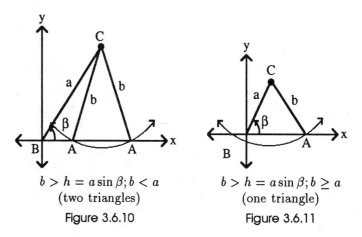

$b > h = a \sin \beta; b < a$
(two triangles)
Figure 3.6.10

$b > h = a \sin \beta; b \geq a$
(one triangle)
Figure 3.6.11

# 3.6 The Law of Sines

SOLUTION: Any such triangle must have height $h = a \sin \beta = 6 \sin 30° = 3$.

(a) Because $b = 2 < a \sin \beta = 3$ there should be no solution. We use the law of sines:

$$\frac{\sin \alpha}{a} = \frac{\sin \beta}{b}$$

$$\frac{\sin \alpha}{6} = \frac{\sin 30°}{2}$$

$$\sin \alpha = 1.5.$$

There is no solution in this case because $|\sin \alpha| \leq 1$ for any $\alpha$.

(b) There should be exactly one solution in this case because $b = 3 = a \sin \beta$. We use the law of sines:

$$\frac{\sin \alpha}{a} = \frac{\sin \beta}{b}$$

$$\frac{\sin \alpha}{6} = \frac{\sin 30°}{3}$$

$$\sin \alpha = 1.$$

The only solution to $\sin \alpha = 1$ with $0° < \alpha < 180°$ is $\alpha = 90°$. Thus, $\gamma = 180° - (30° + 90°) = 60°$. Next, we solve for $c$:

$$\frac{\sin \alpha}{a} = \frac{\sin \gamma}{c}$$

$$\frac{\sin 90°}{6} = \frac{\sin 60°}{c}$$

$$c = 6 \sin 60°$$

$$c = 3\sqrt{3}.$$

We have used the fact that the exact value of $\sin 60°$ is $\frac{\sqrt{3}}{2}$. If a calculator is used, we would find $c = 5.20$ accurate to hundredths. In this case we get the triangle in Figure 3.6.12.

Figure 3.6.12

(c) There are two solutions in this case. First, we find $\alpha$:

$$\frac{\sin \alpha}{a} = \frac{\sin \beta}{b}$$

$$\frac{\sin \alpha}{6} = \frac{\sin 30°}{5}$$

$$\sin \alpha = \frac{6 \sin 30°}{5}$$

$$\sin \alpha = \frac{3}{5}.$$

There are two solutions to $\sin \alpha = \frac{3}{5}$ with $0° < \alpha < 180°$, namely $36.87°$ and $143.13°$. If $\alpha = 36.87°$, then $\gamma = 113.13°$ and the law of sines determines $c$:

$$\frac{\sin \beta}{b} = \frac{\sin \gamma}{c}$$

$$\frac{\sin 30°}{5} = \frac{\sin 113.13°}{c}$$

$$c = 9.20.$$

Thus, $\alpha = 36.87°$, $\gamma = 113.13°$, and $c = 9.20$ (Figure 3.6.13). If $\alpha = 143.13°$, then $\gamma = 6.87°$ and the law of sines determines $c$:

$$\frac{\sin \beta}{b} = \frac{\sin \gamma}{c}$$

$$\frac{\sin 30°}{5} = \frac{\sin 6.87°}{c}$$

$$c = 1.20.$$

Thus, $\alpha = 143.13°$, $\gamma = 6.87°$, and $c = 1.20$ (Figure 3.6.14).

Figure 3.6.13

Figure 3.6.14

## 3.6 The Law of Sines

Figure 3.6.15

(d) There is one solution in this case. We use the law of sines to find $\alpha$:

$$\frac{\sin \alpha}{a} = \frac{\sin \beta}{b}$$

$$\frac{\sin \alpha}{6} = \frac{\sin 30°}{7}$$

$$\sin \alpha = \frac{6 \sin 30°}{7}$$

$$\sin \alpha = \frac{3}{7}.$$

There are two solutions to $\sin \alpha = \frac{3}{7}$ with $0 < \alpha < 180°$, namely $25.38°$ and $154.62°$. However, if $\alpha$ is $154.62°$, then $\alpha + \beta$ is $184.62°$, which is not possible. Therefore, $\alpha = 25.38°$ and $\gamma = 124.62°$:

$$\frac{\sin \beta}{b} = \frac{\sin \gamma}{c}$$

$$\frac{\sin 30°}{7} = \frac{\sin 124.62°}{c}$$

$$c = 11.52.$$

We get the triangle in Figure 3.6.15. •

Figure 3.6.16 illustrates the corresponding four cases if $\beta$ is assumed to be obtuse. Notice that there is either exactly one solution or no solution if $\beta$ is obtuse because $A$ must be on the positive $x$-axis by the initial placement of the proposed triangle.

Next, we consider applications that involve the use of the law of sines.

• **EXAMPLE 3**: A forest ranger at an observation point $A$ along a straight road sights a fire in the direction $32°$ east of north. Another ranger at a second observation point $B$ on the road, 10 miles due east of $A$, sights the same fire at $48°$ west of north. Find the distance from each observation point to the fire, and find the shortest distance from the road to the fire.

SOLUTION: Assume the fire is located at point $C$ (Figure 3.6.17). Because the fire is $32°$ east of north, $\alpha$ must be $90° - 32°$ or $58°$. Similarly, $\beta$ must be $42°$. Let $h$ be the length of the line segment from $C$ perpendicular to the road. Then, $h$ is the shortest distance from the road to the fire. We must determine $a$, $b$, and $h$ in Figure 3.6.17.

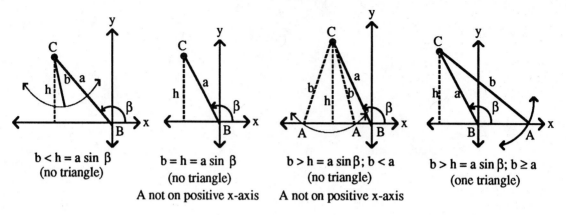

Figure 3.6.16

In this triangle we know that $\alpha = 58°$, $\beta = 42°$, $\gamma = 80°$, and $c = 10$. We use the law of sines:

$$\frac{\sin \alpha}{a} = \frac{\sin \gamma}{c} \qquad\qquad \frac{\sin \beta}{b} = \frac{\sin \gamma}{c}$$

$$\frac{\sin 58°}{a} = \frac{\sin 80°}{10} \qquad\qquad \frac{\sin 42°}{b} = \frac{\sin 80°}{10}$$

$$a = \frac{10 \sin 58°}{\sin 80°} \qquad\qquad b = \frac{10 \sin 42°}{\sin 80°}$$

$$a = 8.61 \qquad\qquad b = 6.79$$

Thus, the fire is 6.79 miles from $A$ and 8.61 miles from $B$. Further, the perpendicular distance $h$ from the fire to the road can be found from either

$$\sin 58° = \frac{h}{b} \qquad \text{or} \qquad \sin 42° = \frac{h}{a}.$$

We find that the shortest distance from the road to the fire is 5.76 miles. •

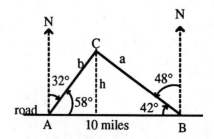

Figure 3.6.17

## 3.6 The Law of Sines

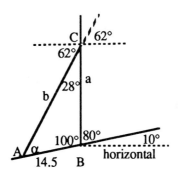

Figure 3.6.18

- **EXAMPLE 4:** A vertical telephone pole stands by the side of a road that slopes at an angle of 10° with the horizontal. When the angle of elevation of the sun is 62°, the telephone pole casts a 14.5-foot shadow downhill parallel to the road. Find the height of the telephone pole.

**SOLUTION:** Let $C$ and $B$ denote the top and bottom of the telephone pole, respectively. Let $A$ be the end of the shadow along the road. Refer to Figure 3.6.18. Because the angle of elevation of the sun is 62°, $\gamma$ must be 28°. Because the road makes a 10° angle with the horizontal, $\beta$ must be 100°. Thus, $\alpha$ is 52°. The length of the shadow is $c = 14.5$, and the height of the pole is $a$. We use the law of sines to determine $a$.

$$\frac{\sin \alpha}{a} = \frac{\sin \gamma}{c}$$

$$\frac{\sin 52°}{a} = \frac{\sin 28°}{14.5}$$

$$a = \frac{14.5 \sin 52°}{\sin 28°}$$

$$a = 24.34.$$

Thus, the height above the ground of the telephone pole is 24.34 feet.

---

## EXERCISES 3-6

1. Give the side lengths $a$, $b$, and $c$ for *two different* triangles $ABC$ whose interior angles are $\alpha = 32°$, $\beta = 75°$, and $\gamma = 73°$.

2. Give the side lengths $a$, $b$, and $c$ of *two different* triangles $ABC$ whose interior angles are $\alpha = 54°$, and $\gamma = 16°$.

Solve the triangle.

3. $\alpha = 40°$, $\beta = 30°$, $b = 10$
4. $\alpha = 60°$, $a = 3$, $b = 4$
5. $\beta = 30°$, $a = 12$, $b = 6$
6. $\alpha = 50°$, $\beta = 62°$, $a = 4$
7. $\alpha = 33°$, $\beta = 79°$, $b = 7$
8. $\beta = 85°$, $a = 4$, $b = 6$

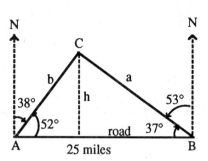

Figure 3.6.19    Figure 3.6.20

9. $\alpha = 50°$, $a = 4$, $b = 5$
10. $\beta = 38°$, $a = 16$, $b = 20$
11. $\beta = 38°$, $a = 16$, $b = 12$
12. $\beta = 116°$, $a = 11$, $b = 13$
13. $\beta = 116°$, $a = 11$, $b = 10$
14. $\beta = 116°$, $a = 11$, $b = 8$
15. $\beta = 152°$, $a = 8$, $b = 10$
16. $\gamma = 103°$, $\beta = 16°$, $c = 12$
17. If $a = 10$ and $\beta = 42°$, determine the values of $b$ for which $\alpha$ has
    (a) two values.    (b) one value.    (c) no value.
18. Two markers $A$ and $B$ on the same side of a canyon rim are 56 feet apart (Figure 3.6.19). A hiker is located across the rim at point $C$. A surveyor determines that $\angle BAC = 72°$ and $\angle ABC = 53°$.
    (a) What is the distance between the hiker and point $A$?
    (b) What is the distance between the two canyon rims? (Assume they are parallel.)
19. A forest ranger at an observation point $A$ along a straight road sights a fire in the direction 38° east of north. Another ranger at an observation point $B$ on the road, 25 miles due east of $A$, sights the same fire at 53° west of north (Figure 3.6.20). Find the distance from each observation point to the fire, and find the shortest distance from the road to the fire.
20. A vertical telephone pole stands by the side of a road that slopes at an angle of 15° with the horizontal. When the angle of elevation of the sun is 62°, the telephone pole casts a 16-foot shadow downhill parallel to the road (Figure 3.6.21). Find the height of the telephone pole.

Figure 3.6.21

## 3.6 The Law of Sines

21. A hot air balloon is seen over Park City, Utah, simultaneously by two observers at points $A$ and $B$. Assume the balloon and the observers are in the same vertical plane and the angles of elevation and the distances are as shown in Figure 3.6.22. How high above ground is the balloon?

22. A 4-foot airfoil is attached to the cab of a truck as shown in Figure 3.6.23. The angle between the airfoil and the cab top is 18°. What is the length of a *vertical* brace positioned as shown in Figure 3.6.23 if angle $\beta$ is 10°?

23. Solve the triangle $ABC$ in Figure 3.6.24, given that $a = 5$, $b = 8$, and $\gamma = 22°$. *Hint*: Erect a perpendicular from $A$ to the line through $B$ and $C$.

## King of the Mountains

"Just a few feet, it seems, can make a difference. Last March, George Wallerstein, an astronomer at the University of Washington, stunned mountaineers and geologists by declaring that the Himalayan mountain known as K-2 might be 36 ft. taller than Mount Everest, long thought to be the world's highest peak. This month, however, an eight-man Italian expedition, led by Geologist Ardito Desio, 90, refuted that claim. Using satellite signals and surveying techniques, they found that Everest towers 29,108 ft. above sea level—80 ft. taller than previously believed and 840 ft. higher than K-2.

To accomplish their lofty task, the Italians carried computerized radio receivers to stations on each mountain. The instruments used timed signals from U.S. Navstar satellites to calculate the exact longitude, latitude and altitude of each receiver. Armed with these coordinates, the researchers then measured the angles formed by the peaks and the receiver stations with a surveyor's theodolite, as well as the distance between the stations. Since the length of one side of each triangle and two angles were known, the peaks' heights could be accurately determined by simple geometry.

Wallerstein readily concedes that his measurement, taken while climbing K-2 may have been inaccurate. 'Mine was done with a secondhand receiver, while theirs was made with first-rate equipment,' he says. 'I guess it proves that in this business there's no place for amateurs.'" (From *Time* magazine, November 2, 1987.)

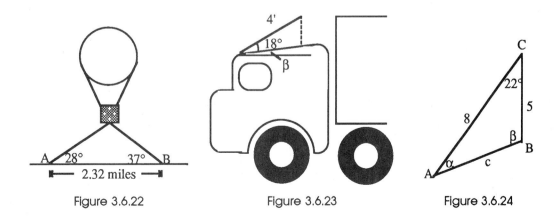

Figure 3.6.22  Figure 3.6.23  Figure 3.6.24

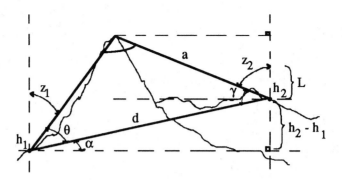

Figure 3.6.25

24. Using the (oversimplified) diagram in Figure 3.6.25 and "simple geometry" determine the height of the mountain peak above one of the radio receiver stations. The following information is known:

$$h_2 - h_1 = 895 \quad \text{feet}, \quad d = 2.63 \quad \text{miles}, \quad \angle z_1 = 41°, \quad \text{and} \quad \angle z_2 = 63°.$$

## 3.7 The Law of Cosines and Area of Triangles

In this section we obtain a formula called the *law of cosines* that is valid for any triangle. Applications whose solutions require finding missing parts of a triangle using the law of cosines will be investigated. We obtain formulas that give the area of a triangle in terms of the measure of its sides and angles. We use the law of cosines to obtain a formula for the angle between two lines.

- **THEOREM 1** (**The Law of Cosines**) Let $ABC$ be a triangle labeled in the usual way (Figure 3.7.1). Then,

$$a^2 = b^2 + c^2 - 2bc \cos \alpha,$$
$$b^2 = a^2 + c^2 - 2ac \cos \beta, \quad \text{and}$$
$$c^2 = a^2 + b^2 - 2ab \cos \gamma.$$

**PROOF** Without any loss of generality we can put the vertex $A$ at the origin and the vertex $B$ on the positive $x$-axis at $(c, 0)$ (Figure 3.7.2). We have drawn the triangle $ABC$ as if angle $\alpha$ is obtuse. The proof will be valid even if $\alpha$ is acute. We choose $D$ on the $x$-axis so that $CD$ is perpendicular to the $x$-axis. Then, the distance $h$ from $C$ to the $x$-axis is an altitude of the triangle $ABC$. Let the coordinates of $D$ be $(d, 0)$. Because $\alpha$ is in standard position we have

$$\cos \alpha = \frac{d}{b} \quad \text{and} \quad \sin \alpha = \frac{h}{b}.$$

Thus, $d = b \cos \alpha$ and $h = b \sin \alpha$. Notice that $d(B, C) = a$. Using the distance formula we have

$$(d(B, C))^2 = (c - d)^2 + (0 - h)^2$$
$$a^2 = (c - d)^2 + h^2.$$

# 3.7 The Law of Cosines and Area of Triangles

Substituting the above values of $d$ and $h$ we obtain
$$\begin{aligned} a^2 &= (c-d)^2 + h^2 \\ &= (c - b\cos\alpha)^2 + (b\sin\alpha)^2 \\ &= c^2 - 2bc\cos\alpha + b^2\cos^2\alpha + b^2\sin^2\alpha \\ &= c^2 - 2bc\cos\alpha + b^2(\cos^2\alpha + \sin^2\alpha) \\ &= b^2 + c^2 - 2bc\cos\alpha. \end{aligned}$$

This is the first of the three formulas to be obtained. Now, put the angle $\beta$ in standard position with $A$ on the positive $x$-axis (Figure 3.7.3). If we proceed in a manner similar to the above derivation, we will obtain $b^2 = a^2 + c^2 - 2ac\cos\beta$. This is the second of the three formulas. If we put the angle $\gamma$ in standard position, we will obtain the last of the three formulas.

There are two situations remaining to complete the investigation of the possible triangles determined when three of its six parts are specified.

## Three Sides Specified

Suppose that the measures of the three sides of a triangle are specified. In order that there actually be such a triangle, the sum of the measures of any two sides must exceed the measure of the third side. (Why?)

- **EXAMPLE 1:** Solve triangle $ABC$ if $a = 9$, $b = 7$, and $c = 5$.

SOLUTION: We use the law of cosines to find angle $\alpha$ (Figure 3.7.4):
$$\begin{aligned} a^2 &= b^2 + c^2 - 2bc\cos\alpha \\ 9^2 &= 7^2 + 5^2 - 2(7)(5)\cos\alpha \\ \cos\alpha &= \frac{7^2 + 5^2 - 9^2}{(2)(7)(5)} \\ \cos\alpha &= -\frac{1}{10}. \end{aligned}$$

Figure 3.7.1

Figure 3.7.2

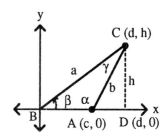

Figure 3.7.3

The only solution to $\cos \alpha = -\frac{1}{10}$ with $0° < \alpha < 180°$ is $95.74°$.

We use the law of cosines to find angle $\beta$:

$$b^2 = a^2 + c^2 - 2ac \cos \beta$$
$$7^2 = 9^2 + 5^2 - 2(9)(5) \cos \beta$$
$$\cos \beta = \frac{9^2 + 5^2 - 7^2}{(2)(9)(5)}$$
$$\cos \beta = \frac{19}{30}.$$

Thus, $\beta = 50.70°$. It follows that $\gamma = 33.56°$. •

## Two Sides and the Included Angle Specified

If two sides and an angle opposite one of those sides are given, we found in Section 3.1 that zero, one, or two triangles were possible. If two sides and an included angle of measure between $0°$ and $180°$ are specified, then precisely one triangle is determined as illustrated in Example 2.

• **EXAMPLE 2:** Solve triangle $ABC$ if $a = 4$, $b = 7$, and $\gamma = 42°$.

**SOLUTION:** We use the law of cosines to find the third side:

$$c^2 = a^2 + b^2 - 2ab \cos \gamma$$
$$c^2 = 4^2 + 7^2 - 2(4)(7) \cos 42°$$
$$c^2 = 23.383890$$
$$c = 4.8356892.$$

We did not round $c$ to hundredths because we need to use the value of $c$ again. Next, we use the law of sines to determine $\alpha$:

$$\frac{\sin \alpha}{a} = \frac{\sin \gamma}{c}$$
$$\frac{\sin \alpha}{4} = \frac{\sin 42°}{4.8356892}$$
$$\sin \alpha = 0.5534935$$
$$\alpha = 33.61°.$$

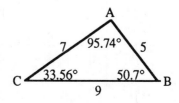

Figure 3.7.4

### 3.7 The Law of Cosines and Area of Triangles

Thus, $\alpha = 33.61°$, $\beta = 104.39°$, $c = 4.84$, and we have determined the triangle in Figure 3.7.5. •

Now we have covered all possibilities if three of the six parts of a triangle are specified. We have the following summary.

SUMMARY    Let three of the six parts of a triangle be given.

| Parts Given | Number of Possible Triangles |
| --- | --- |
| Three angles (sum equals 180°) | Infinitely many, they are all similar |
| Two angles (sum less than 180°), one side | One |
| One angle, two sides | Zero, one, or two |
| Three sides (sum of any two greater than the third) | One |

## Area of Triangles

• **THEOREM 2**    Let $ABC$ be a triangle labeled in the usual way (Figure 3.7.6). Then, the area $\mathcal{A}$ of the triangle is given by

$$\mathcal{A} = \frac{1}{2}bc\sin\alpha = \frac{1}{2}ac\sin\beta = \frac{1}{2}ab\sin\gamma.$$

PROOF    First, we put the angle $\alpha$ of triangle $ABC$ in standard position (Figure 3.7.6). We have drawn angle $\alpha$ as if it is acute. The same proof works if $\alpha$ is obtuse. Let $h$ be an altitude of the triangle $ABC$. Then, the area of the triangle is given by $\mathcal{A} = \frac{1}{2}ch$. Notice that $\sin\alpha = \frac{h}{b}$ so that $h = b\sin\alpha$. Substituting this

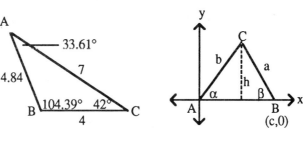

Figure 3.7.5          Figure 3.7.6

value for $h$ into the area formula gives

$$A = \frac{1}{2}ch$$
$$= \frac{1}{2}cb \sin \alpha$$
$$= \frac{1}{2}bc \sin \alpha.$$

If we interchange $b$ with $a$ and $\beta$ with $\alpha$ we obtain the second form for $A$. The third form follows in a similar way.

- **EXAMPLE 3:** Find the area of the triangle $ABC$ if $a = 8$, $b = 5$, and $\gamma = 52°$.

    SOLUTION: From Theorem 2 we know that the area is given by the following formula:

    $$A = \frac{1}{2}ab \sin \gamma$$
    $$= \frac{1}{2}(8)(5) \sin 52°$$
    $$= 15.76.$$

Thus, the area is 15.76 square units. •

The next theorem gives the area of a triangle in terms of the measure of its three sides.

- **THEOREM 3** (**Heron's Formula.**) Let $ABC$ be a triangle labeled in the usual way. Then, the area $A$ of the triangle is given by

    $$A = \sqrt{s(s-a)(s-b)(s-c)},$$

    where $s = \frac{1}{2}(a + b + c)$ is one-half of the perimeter.

    PROOF From Theorem 2 we have $A = \frac{1}{2}ab \sin \gamma$. We proceed as follows:

    $$A = \frac{1}{2}ab \sin \gamma$$
    $$2A = ab \sin \gamma$$
    $$4A^2 = a^2b^2 \sin^2 \gamma$$
    $$= a^2b^2(1 - \cos^2 \gamma)$$
    $$= a^2b^2 - a^2b^2 \cos^2 \gamma.$$

    Next, we multiply each side of the above equation by 4 and then use the law of cosines to replace $2ab \cos \gamma$ by $a^2 + b^2 - c^2$:

    $$4A^2 = a^2b^2 - a^2b^2 \cos^2 \gamma$$
    $$16A^2 = 4a^2b^2 - (2ab \cos \gamma)^2$$
    $$= 4a^2b^2 - (a^2 + b^2 - c^2)^2.$$

## 3.7 The Law of Cosines and Area of Triangles

It can be verified that the right-hand side of the above equation is the same as $(a+b+c)(a+b-c)(b+c-a)(c+a-b)$. Thus,

$$16A^2 = (a+b+c)(a+b-c)(b+c-a)(c+a-b).$$

Notice that $a+b+c = 2s$, $a+b-c = 2s-2c$, $b+c-a = 2s-2a$, and $a+c-b = 2s-2b$. Substituting these values we obtain

$$16A^2 = 2s(2s-2c)(2s-2a)(2s-2b)$$
$$16A^2 = 16s(s-c)(s-a)(s-b)$$
$$A = \sqrt{s(s-a)(s-b)(s-c)}.$$

- **EXAMPLE 4:** Find the area of the triangle $ABC$ if $a = 9$, $b = 7$, and $c = 5$.

  **SOLUTION:** First, $s = \frac{1}{2}(a+b+c) = 10.5$. Then, we compute the area using Heron's formula:

  $$\begin{aligned} a &= \sqrt{s(s-a)(s-b)(s-c)} \\ &= \sqrt{(10.5)(1.5)(3.5)(5.5)} \\ &= 17.41. \end{aligned}$$

  Thus, the area of the triangle is 17.41 square units.

Next, we consider applications that involve the use of the law of cosines.

- **EXAMPLE 5:** In order to determine the distance between two points $A$ and $B$ on opposite sides of a building, a surveyor chooses a point $C$ that is 110 feet from $A$ and 160 feet from $B$ (Figure 3.7.7). If the measure of the angle at $C$ is 54°, find the distance between $A$ and $B$.

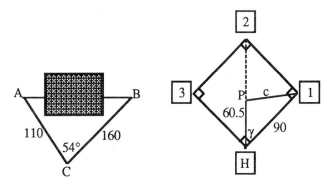

Figure 3.7.7　　　　　Figure 3.7.8

SOLUTION: The distance from $A$ to $B$ in the triangle $ABC$ is $c$. Now, we use the law of cosines:

$$c^2 = a^2 + b^2 - 2ab\cos\gamma$$
$$c^2 = 160^2 + 110^2 - 2(160)(110)\cos 54°$$
$$c = 130.42.$$

Thus, the distance between $A$ and $B$ is 130.42 feet. •

• **EXAMPLE 6:** In major league baseball, the four bases form a square with sides of length 90 feet. The front edge of the pitching rubber is 60.5 feet from home plate. Find the distance from the front edge of the pitching rubber to first base.

SOLUTION: Let $P$ be the front edge of the pitching rubber and $H$ be home plate (Figure 3.7.8). Let $c$ denote the distance from the front edge of the pitching rubber to first base. The measure of angle $\gamma$ is 45° because the angle at each base is 90° and a diagonal of a square bisects the angle at a vertex.

The law of cosines gives the following for the distance $c$:

$$c^2 = (60.5)^2 + (90)^2 - 2(60.5)(90)\cos 45°$$
$$c = 63.72.$$

Thus, the distance from the front edge of the pitching rubber to first base is 63.72 feet. •

## Angle Between Two Lines

In Section 2.4 we defined the angle between two lines to be the acute angle determined by the lines, or 90° if the two lines are perpendicular. In the next example we use the law of cosines to determine the angle between two lines.

• **EXAMPLE 7:** Let $\ell_1$ and $\ell_2$ be two lines with slopes $m_1$ and $m_2$, respectively. The angle between $\ell_1$ and $\ell_2$ is either

$$\alpha = \cos^{-1}\left(\frac{1+m_1 m_2}{\sqrt{(1+m_1^2)(1+m_2^2)}}\right) \quad \text{or} \quad 180° - \alpha \quad \text{if} \quad \alpha > 90°.$$

SOLUTION: Without loss of generality, we can assume that the line $\ell_2$ makes the larger counterclockwise angle with the positive $x$-axis. With this agreement, Figure 3.7.9 illustrates the three possibilities for $\ell_1$ and $\ell_2$ corresponding to $m_1$ and $m_2$ positive, $m_2$ negative and $m_1$ positive, or $m_1$ and $m_2$ negative. In each case, let $P$ and $Q$ be the intersection of the vertical line $x = 1$ with $\ell_1$ and $\ell_2$, respectively. Notice that the coordinates of $P$ are $(1, m_1)$ and the coordinates of $Q$ are $(1, m_2)$. (Why?)

Let $\alpha$ be the angle of the triangle $OPQ$ between $OP$ and $OQ$. Either $\alpha$ or $180° - \alpha$ is acute. Thus, either $\alpha$ or $180° - \alpha$ is the angle between $\ell_1$ and $\ell_2$.

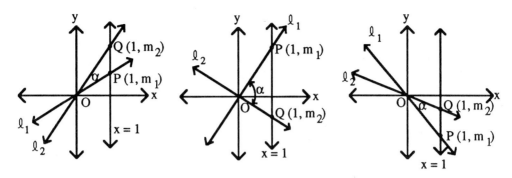

Figure 3.7.9

Now, $d(P,Q) = |m_1 - m_2|$, $d(O,P) = \sqrt{1+m_1^2}$, and $d(O,Q) = \sqrt{1+m_2^2}$. (Why?) We use the law of cosines to determine $\alpha$:

$$d^2(P,Q) = d^2(O,P) + d^2(O,Q) - 2d(O,P)d(O,Q)\cos\alpha$$

$$|m_1 - m_2|^2 = (\sqrt{1+m_1^2})^2 + (\sqrt{1+m_2^2})^2 - 2(\sqrt{1+m_1^2})(\sqrt{1+m_2^2})\cos\alpha$$

$$m_1^2 - 2m_1m_2 + m_2^2 = 1 + m_1^2 + 1 + m_2^2 - 2(\sqrt{(1+m_1^2)(1+m_2^2)})\cos\alpha$$

$$\cos\alpha = \frac{1 + m_1 m_2}{\sqrt{(1+m_1^2)(1+m_2^2)}}$$

$$\alpha = \cos^{-1}\left(\frac{1 + m_1 m_2}{\sqrt{(1+m_1^2)(1+m_2^2)}}\right).$$

If $\cos\alpha > 0$, then $\alpha$ is acute. If $\cos\alpha < 0$, then $90° < \alpha < 180°$ and $180° - \alpha$ is acute. If $\cos\alpha = 0$, then $\alpha = 90°$ and $\ell_1$ is perpendicular to $\ell_2$. •

## EXERCISES 3-7

Solve the triangle $ABC$.

1. $a = 1$, $b = 5$, $c = 4$
2. $a = 1$, $b = 5$, $c = 8$
3. $a = 3.2$, $b = 7.6$, $c = 6.4$
4. $\alpha = 21°$, $\beta = 17°$, $c = 15$
5. $\alpha = 55°$, $b = 12$, $c = 7$
6. $\beta = 125°$, $a = 25$, $c = 41$
7. $\beta = 103°$, $b = 13$, $a = 18$
8. $\alpha = 36°$, $b = 17$, $a = 14$
9. $\beta = 110°$, $b = 13$, $c = 15$
10. $a = 5$, $b = 7$, $c = 6$
11. If $a = 8$ and $\beta = 58°$, determine the values of $b$ for which $\alpha$ has
    (a) two values.   (b) one value.   (c) no value.
12. If $a = 12$ and $\beta = 32°$, determine the values of $b$ for which $\alpha$ has
    (a) two values.   (b) one value.   (c) no value.

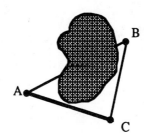

Figure 3.7.10

Find the area of the triangle $ABC$.

13. If $b = 6$, $c = 8$, and $\alpha = 47°$
14. If $a = 17$, $c = 14$, and $\beta = 103°$
15. If $\alpha = 15°$, $\beta = 65°$, and $a = 8$
16. If $\alpha = 10°$, $\gamma = 110°$, and $c = 12.3$
17. If $a = 2$, $b = 6$, and $c = 7$
18. If $a = 20$, $b = 36$, and $c = 50$

19. In order to determine the distance between two points $A$ and $B$ on opposite sides of a lake, a surveyor chooses a point $C$ that is 860 feet from $A$ and 175 feet from $B$ (Figure 7.3.10). If the measure of the angle at $C$ is $78°$, find the distance between $A$ and $B$.

20. Consider the ball diamond problem of Example 6 (Figure 3.7.8). Determine
    (a) the distance from the front edge of the pitching rubber to third base.
    (b) the distance between first and third bases.
    (c) the angle between home plate, the front edge of the pitching rubber, and first base.
    (d) the angle between the front edge of the pitching rubber, first base and home plate.

21. In women's softball, the four bases form a square with sides of length 60 feet. The front edge of the pitching rubber is 40 feet from home plate. Find the distance from the front edge of the pitching rubber to first base.

22. Find the radian measure of the largest angle of the triangle with sides of length 4, 5, and 6.

23. The sides of a parallelogram are of lengths 18 feet and 26 feet, and one angle is $39°$. Find the length of the longer diagonal.

24. Two observers are 600 feet apart on opposite sides of a flag pole. The angles of elevation from the observers to the top of the pole are $19°$ and $21°$. Find the height of the flag pole.

25. A blimp is sighted simultaneously by two observers: $A$ at the top of a 650 foot tower and $B$ at the base of the tower. Find the distance of the blimp from observer $A$ if the angle of elevation (from the horizontal) as viewed by $A$ is $32°$ and the angle of elevation as viewed by $B$ is $56°$. How high is the blimp?

26. In a parallelogram, two adjacent sides meet at an angle of $35°$ and are 3 feet and 8 feet in length. What is the length of the shorter diagonal of the parallelogram?

27. Two observers are 400 feet apart on opposite sides of a tree. If the angles of elevation from the observers to the top of the tree are 15° and 20°, how tall is the tree?

## • KEY TERMS

Listed below are the key terms, vocabulary, and concepts in this chapter. You should be familiar with their definitions and meanings as well as be able to illustrate each of them.

Absolute maximum
Absolute minimum
Addition formulas
Amplitude
Arccos
Arcsin
Arctan
Area of triangles
Circular properties of trig functions
Coterminal
Double-angle formulas
End behavior
Half-angle formulas
Heron's formula
Horizontal line test
Horizontal shifting
Horizontal shrinking
Identities for inverse trig functions
Initial side
Inverse relation

Inverse trigonometric functions
Law of cosines
Law of sines
Oblique triangle
One-to-one function
Periodic functions
Phase shift
Reference angle
Sinusoid
Solving oblique triangles
Standard position
Supplementary angle relationships
Terminal sides
Trigonometric equations
Trigonometric functions
Trigonometric identities
Trigonometric inequalities
Unit circle
Zeros of periodic functions

## • REVIEW EXERCISES

This section contains representative problems from each of the previous seven sections. You should use these problems to test your understanding of the material covered in this chapter.

Solve the triangle in Figure R.1.

1. $\alpha = 79°$, $\beta = 33°$, and $a = 7$
2. $a = 5$, $b = 8$, and $\beta = 110°$
3. $a = 14.7$, $\alpha = 29.3°$, and $\gamma = 37.2°$
4. $a = 8$, $b = 3$, and $\beta = 30°$
5. $\alpha = 34°$, $\beta = 74°$, and $c = 5$
6. $c = 41$, $\alpha = 22.9°$, and $\gamma = 55.1°$
7. $a = 5$, $b = 7$, and $c = 6$
8. $\alpha = 85°$, $a = 6$, and $b = 4$

Figure R.1　　　　　　　　　　　　　Figure R.2

9. If $a = 12$ and $\beta = 28°$, determine the values of $b$ for which $\alpha$ has (a) two values, (b) one value, and (c) no value.

Find the area of the triangle $ABC$.

10. $a = 3$, $b = 5$, and $c = 6$
11. $a = 10$, $b = 6$, and $\gamma = 50°$

Write as a sinusoid $y = A\sin(x + \alpha)$ and confirm by graphing.

12. $y = 2\sin x - 5\cos x$
13. $y = -4\sin x + 3\cos x$

14. Draw a complete graph of the inverse of the relation given by the graph of Figure R.2.

Draw a complete graph of each relation and its inverse. Do not use a graphing utility. Determine an equation for the inverse. Specify which inverses are functions.

15. $y = 5 - 3x$
16. $y = x^2 - 2$

Determine if each function is one-to-one. Determine an equation for the inverse of each function. Is the inverse a function? For those functions whose inverses are *functions* find an explicit rule $y = f^{-1}(x)$ for the inverse. Draw a graph of the function and its inverse relation without using a graphing utility. Check your answer with a graphing utility.

17. $f(x) = (x + 2)^2$
18. $h(x) = 2\sqrt{x - 4}$

19. Determine the domain and range of both the function and its inverse in Exercise 17.

Use a graphing utility to determine if each function is one-to-one. Sketch a complete graph of the inverse of each function without the aid of a graphing utility. Identify the functions whose inverses are also functions.

20. $f(x) = x^3 - 8x$
21. $f(x) = \dfrac{x^2 - 2x + 3}{x + 2}$

22. Determine the domain and range of both the function and its inverse in Exercise 21.

Verify the following identities by graphing each side of the equation and comparing the graphs. Then verify the identity algebraically.

23. $\tan^2 x - \sin^2 x = \sin^2 x \tan^2 x$
24. $2\sin\theta \cos^3\theta + 2\sin^3\theta \cos\theta = \sin 2\theta$

Draw a complete graph. Determine the domain and range.

25. $y = 3x\cos x + \sin 2x$
26. $y = 2\sin 3x - 5\cos 2x$
27. $y = \dfrac{2x - 1}{\tan x}$
28. $f(x) = x^2 \sin x$

# Review Exercises

Draw a complete graph and determine the end behavior of the function.

29. $y = \dfrac{1}{x}\sin(\dfrac{1}{x})$  

30. $y = x^3 \cos x$

Investigate graphically what happens to the values of $f(x)$ as $x \to 0$.

31. $f(x) = \dfrac{1}{x}\sin(\dfrac{1}{x})$  

32. $f(x) = \dfrac{\sin x}{x^2}$

33. Determine the domain, range, period, and draw a complete graph of the function $h(x) = 2\sin x - \cos 2x$.

34. Determine the zeros in the interval $[0, 2\pi]$ of the function in Exercise 33.

35. Determine all local and absolute maximum and minimum values in the interval $[0, 2\pi]$ of the function in Exercise 33.

36. Use an algebraic method to determine all real solutions of $0.25 - \cos^2 x = 0$ in the interval $[0, \pi]$. Check your answer with a graphing utility.

Use algebraic means to verify the following identities. Confirm your analysis by graphing.

37. $\csc x - \cos x \cot x = \sin x$

38. $\dfrac{\tan \theta + \sin \theta}{2 \tan \theta} = \cos^2 \dfrac{\theta}{2}$

39. $\dfrac{1 + \tan \theta}{1 - \tan \theta} + \dfrac{1 + \cot \theta}{1 - \cot \theta} = 0$

40. $\sin 3\theta = 3\cos^2 \theta \sin \theta - \sin^3 \theta$

Use a graphing utility to investigate if the equation is an identity. If not, determine a *counterexample*.

41. $\sec x - \sin x \tan x = \cos x$

42. $(\sin^2 \alpha - \cos^2 \alpha)(\tan^2 \alpha + 1) = \tan^2 \alpha - 1$

Determine the *exact* value. Check your answer with a calculator.

43. $\cos \dfrac{\pi}{8}$

44. $\sin \dfrac{7\pi}{12}$

If $\sin x = \dfrac{2}{3}$ and $\dfrac{\pi}{2} < x < \pi$, then determine the trig value.

45. $\sin \dfrac{x}{2}$

46. $\tan x$

47. Find the domain and range of the function $g(x) = 5\cos^2 2x$, and draw a complete graph.

48. Write $\sin 3\theta + \cos 3\theta$ as an expression involving only $\sin \theta$ and $\cos \theta$.

Solve the equation for $x$ in the interval $[0, 2\pi]$. Find exact answers when possible. Check your answer using a graphing utility.

49. $2\cos^2 x + \cos x - 1 = 0$

50. $\sin x = \dfrac{\sqrt{2}}{2}$

51. Use a calculator to evaluate $\sin^{-1}(0.766)$. Express your answer in degrees.

52. Use a calculator to evaluate $\sin^{-1}(0.479)$. Express your answer as a real number.

Compute the exact value without using a calculator.

53. $\sin^{-1}(\dfrac{\sqrt{3}}{2})$

54. $\tan^{-1} 1$

Use a calculator to evaluate. Express your answer as a real number.

55. $\sin(\sin^{-1}(0.25))$

56. $\cos^{-1}(\tan(0.2))$

57. $\tan(\sin^{-1} 2)$

58. $\sin^{-1}(\sin(\sqrt{3}\cos 3))$

Figure R.3  Figure R.4

59. Compute the exact value of $\tan(\sin^{-1} 0.5)$ without using a calculator.

Draw a complete graph. Do not use a graphing utility. Determine the domain and range. Check with a graphing utility.

60. $y = \arccos x - 1$
61. $f(x) = (\sin^{-1} x)^2$
62. Solve $2\sin^{-1} x = \sqrt{2}$ for $x$. Find the exact solution(s).
63. Find an equivalent algebraic expression *not involving trig functions* for $\sin(\cos^{-1} x)$.
64. Solve $\cos x = -0.707$ for $x$. Find *all* solutions.
65. Solve the inequality $\sin 2x / [\cos(x/2)] \geq 1$ in $[-2\pi, 2\pi]$.

Solve the equation algebraically in the interval $[0, 2\pi]$. Check with a graphing utility.

66. $2\sin 2x = 1$
67. $2\cos x = 1$

Solve the equation algebraically. Find *all* real solutions. Check with a graphing utility.

68. $\sin 3x = \sin x$
69. $\sin^2 x - 2\sin x - 3 = 0$
70. $\cos 2t = \cos t$
71. $\sin(\cos x) = 1$

Solve the inequality algebraically. Check with a graphing utility.

72. $2\cos 2x > 1$ for $0 \leq x \leq 2\pi$
73. $\sin 2x > 2\cos x$ for $0 \leq x \leq 2\pi$
74. Suppose $\cos x = 0.2$ with $0 \leq x \leq \frac{\pi}{2}$. Compute $\sin(2x)$ exactly.
75. Draw the graph of $g(x) = 5 + 2x\sin(2x + \pi)$ for $-2\pi \leq x \leq 2\pi$. Determine all local maximum and minimum values in the interval.
76. Two markers $A$ and $B$ on the same side of a canyon rim are 80 feet apart (Figure R.3). A hiker is located across the rim at point $C$. A surveyor determines that $\angle BAC = 70°$ and $\angle ABC = 65°$.
    (a) What is the distance between the hiker and point $A$?
    (b) What is the distance between the two canyon rims? (Assume they are parallel.)
77. A hot air balloon is seen over Hanover, Indiana simultaneously by two observers at points $A$ and $B$. Assume the balloon and the observers are in the same vertical plane and the angles of elevation and the distances are as shown in Figure R.4. How high above ground is the balloon?

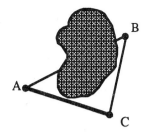

Figure R.5

78. In order to determine the distance between two points $A$ and $B$ on opposite sides of a lake, a surveyor chooses a point $C$ that is 900 feet from $A$ and 225 feet from $B$ (Figure R.5). If the measure of the angle at $C$ is 70°, find the distance between $A$ and $B$.

79. Find the radian measure of the largest angle of the triangle with sides of length 9, 8, and 10.

A 6-foot fence stands 3.5 feet from a high wall (Figure R.6). Let $L$ be the length of a ladder extending from some point $C$ on the ground to some point $A$ on the high wall. The shortest such ladder must just touch the fence. A *shorter* ladder would reach from $C$ to a point slightly below $A$ on the wall. Let $\theta$ be the angle of inclination shown in Figure R.6.

(a) Determine a trigonometric representation of the length of such a ladder (as a function of $\theta$).

(b) Draw a complete graph of the trigonometric representation.

(c) What values of $\theta$ make sense in the problem situation?

(d) Find the angle of inclination of the shortest such ladder.

81. Sarah is in a rowboat at a point $P$ 1.5 miles from a straight shore. Her home is on the shore at a point $C$ that is 6 miles from the point on the shore ($A$) closest to Sarah. Assume that she can walk twice as fast as she can row and that she walks 3 miles per hour. Assume she rows to a point $B$ and then walks to her home at $C$. Let $\theta$ be the angle shown in Figure R.7.

(a) Determine a trigonometric representation of the time it takes Sarah to reach home.

Figure R.6

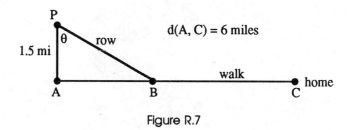

Figure R.7

(b) Draw a complete graph of the trigonometric representation.
(c) What values of $\theta$ make sense in this problem situation?
(d) What is her quickest way home?

# 4  Logarithmic and Exponential Functions

## • Introduction

In this chapter we introduce operations on functions and continue the study of piecewise-defined functions started in Section 1.3. A special procedure for entering piecewise-defined functions on a graphing utility is given because it is not usually possible to enter them directly. Exponential and logarithmic functions are introduced as inverses of each other. We see that the graphs of a relation and its inverse are symmetric with respect to the line $y = x$. Although the inverse of a function need not be a function, we show that inverses of one-to-one functions are functions.

Important applications about growth and decay are studied. For example, we study compound interest, annuities, population change, bacterial growth, and decay of radioactive substances. These problem situations have exponential functions as models, and algebraic solutions of such problems involve logarithmic functions. A special technique is needed to obtain a complete graph of most logarithmic functions with a graphing utility. Logarithm functions are used to help analyze data. Equations and inequalities that involve exponential and logarithmic functions are solved algebraically and graphically.

We assume that you can now quickly sketch a graph of $y = af(x + c) + d$ given the graph of $y = f(x)$. We make frequent use of this information to obtain graphs in this and subsequent chapters.

## 4.1 Operations on Functions

A function consists of a *domain* and a *rule* that assigns a unique element of the *range* to each element of the domain. Sometimes a separate rule is needed for various portions of the domain. Such functions are said to be *piecewise-defined* and are studied in this section. We obtain a computer drawn graph of some piecewise defined functions by expressing the function as a sum of products of other functions. The sum, difference, product, and quotient of two functions are defined in this section. We have been implicitly using these definitions in this textbook.

- **DEFINITION** Let $f$ and $g$ be two functions. We define the **sum** $f+g$, the **difference** $f-g$, and the **product** $fg$ to be the functions with domains the set of all numbers common to the domains of $f$ and $g$ and with rules given by the following formulas:

$$(f+g)(x) = f(x) + g(x),$$
$$(f-g)(x) = f(x) - g(x),$$
$$(fg)(x) = f(x)g(x).$$

- **EXAMPLE 1:** Let $f(x) = \sqrt{2-x}$ and $g(x) = \frac{1}{x}$. Find the domains and rules of $f+g$, $f-g$, and $fg$.

SOLUTION: The domain of $f$ is the set of all real numbers for which $2 - x \geq 0$. (Why?) Thus, the domain of $f$ is $(-\infty, 2]$. The domain of $g$ is the set of all nonzero real numbers. The set of elements common to the domains of $f$ and $g$ is $(-\infty, 0) \cup (0, 2]$. Therefore, the domains of $f+g$, $f-g$, and $fg$ are $(-\infty, 0) \cup (0, 2]$. The rules for these functions are given by the following formulas:

$$(f+g)(x) = f(x) + g(x) = \sqrt{2-x} + \frac{1}{x},$$
$$(f-g)(x) = f(x) - g(x) = \sqrt{2-x} - \frac{1}{x},$$
$$(fg)(x) = f(x)g(x) = \frac{1}{x}\sqrt{2-x}.$$

- **EXAMPLE 2:** Use complete graphs of $f(x) = x$ and $g(x) = \sin x$ to obtain a complete graph of $h(x) = x + \sin x$.

SOLUTION: Complete graphs of $f$ and $g$ are given in Figure 4.1.1. For each value of $x$ we add the corresponding values of $f(x)$ and $g(x)$ to obtain the value of $h(x)$. Thus, if $(a, b)$ and $(a, c)$ are points on the graphs of $f$ and $g$, respectively, then $(a, b+c)$ is a point on the graph of $h$ (Figure 4.1.2). Each point on the graph of $h$ can be obtained this way.

## 4.1 Operations on Functions

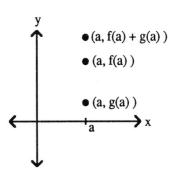

Figure 4.1.1

Figure 4.1.2

Notice that the values of $g(x)$ vary between $-1$ and $1$ for all $x$, but that $f(x) \to \infty$ as $x \to \infty$ and $f(x) \to -\infty$ as $x \to -\infty$. For values of $x$ far away from 0, the values of $h(x) = x + \sin x$ are dominated by the values of $f(x) = x$. In fact, for $|x|$ large, $f(x) = x$ is a very good approximation to $h(x) = x + \sin x$ because the values of $g(x) = \sin x$ are relatively small. For each value of $x$, the value of $h(x)$ is within one unit of the value of $f(x)$. Figure 4.1.3 gives a complete graph of $h$. Compare the graphs of $h(x) = x + \sin x$ and $f(x) = x$ that are shown together in Figure 4.1.4. •

An additional restriction is needed to define the quotient of two functions.

• **DEFINITION**   Let $f$ and $g$ be two functions. We define the **quotient** $\frac{f}{g}$ to be the function with domain the set of all numbers $x$ common to the domains of $f$ and $g$ for which $g(x) \neq 0$, and with rule given by the following formula:

$$\frac{f}{g}(x) = \frac{f(x)}{g(x)}, \quad \text{provided} \quad g(x) \neq 0.$$ •

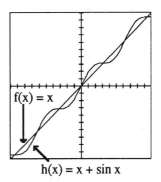

$h(x) = x + \sin x$

$h(x) = x + \sin x$

$[-10, 10]$ by $[-10, 10]$

$[-10, 10]$ by $[-10, 10]$

Figure 4.1.3

Figure 4.1.4

- **EXAMPLE 3:** Let $f(x) = x^2 - 1$ and $g(x) = x^2 + 1$. Find the domain and rule of $\frac{f}{g}$ and $\frac{g}{f}$.

  **SOLUTION:** The domains of $f$ and $g$ are the set of all real numbers. Thus, the domain of $\frac{f}{g}$ is the set of all real numbers $x$ for which $g(x) \neq 0$. Because $g$ has no real zeros, the domain of $\frac{f}{g}$ is the set of all real numbers, and its rule is given by

  $$\frac{f}{g}(x) = \frac{f(x)}{g(x)} = \frac{x^2 - 1}{x^2 + 1}.$$

  The domain of $\frac{g}{f}$ is the set of all real numbers $x$ for which $f(x) \neq 0$. The real zeros of $f$ are $-1$ and $1$. Therefore, the domain of $\frac{g}{f}$ is $(-\infty, -1) \cup (-1, 1) \cup (1, \infty)$ and its rule is given by

  $$\frac{g}{f}(x) = \frac{g(x)}{f(x)} = \frac{x^2 + 1}{x^2 - 1}.$$

Example 3 suggests that every rational function is a quotient of functions. Recall that $f$ is a rational function if it can be expressed in the form $\frac{p(x)}{h(x)}$ where $p$ and $h$ are polynomial functions. Thus, $f$ is the quotient of the polynomial functions $p$ and $h$, that is, $f = \frac{p}{h}$.

## Graphing Piecewise-Defined Functions

Suppose that $f$ is the following piecewise-defined function:

$$f(x) = \begin{cases} x^2 & \text{for } x < 2 \\ 7 - x & \text{for } x \geq 2 \end{cases}.$$

The domain of $f$ is the set of all real numbers. A complete graph of $f$ is given in Figure 4.1.5.

One goal of this section is to show how a graph of $f$ can be obtained directly using a graphing calculator or computer graphing utility.

$$f(x) = \begin{cases} x^2 & \text{for } x < 2 \\ 7 - x & \text{for } x \geq 2 \end{cases}$$

Figure 4.1.5

## 4.1 Operations on Functions

• **EXAMPLE 4:** Express

$$f(x) = \begin{cases} x^2 & \text{for } x < 2 \\ 7 - x & \text{for } x \geq 2 \end{cases}$$

as a sum of two functions with domains that are the set of all real numbers.

**SOLUTION:** We define $f_1$ and $f_2$ as follows:

$$f_1(x) = \begin{cases} x^2 & \text{for } x < 2 \\ 0 & \text{for } x \geq 2 \end{cases} \qquad f_2(x) = \begin{cases} 0 & \text{for } x < 2 \\ 7 - x & \text{for } x \geq 2 \end{cases}.$$

The domains of $f_1$ and $f_2$ are the set of all real numbers. Next, we show that $f = f_1 + f_2$. If $x < 2$, then

$$(f_1 + f_2)(x) = f_1(x) + f_2(x) = x^2 + 0 = x^2.$$

If $x \geq 2$, then

$$(f_1 + f_2)(x) = f_1(x) + f_2(x) = 0 + (7 - x) = 7 - x.$$

Thus, $f = f_1 + f_2$. •

The next step is to find a rule for each of the two functions $f_1$ and $f_2$, defined in Example 4, that can be entered in a graphing utility. Then, the sum $f = f_1 + f_2$ can be entered in the graphing utility. To accomplish this task we make use of the signum function that is available on most computers. Loosely speaking, the signum function assigns to each real number its sign. To be more precise, we give the following definition.

---

• **DEFINITION**   The **signum function**, denoted by **sgn** $(x)$, is defined by

$$\text{sgn}(x) = \begin{cases} -1 & \text{for } x < 0 \\ 0 & \text{for } x = 0 \\ 1 & \text{for } x > 0 \end{cases}.$$

•

The domain of sgn $(x)$ is $(-\infty, \infty)$, and its range is $\{-1, 0, 1\}$. Notice that sgn$(x)$ is discontinuous at $x = 0$. A complete graph of $y = \text{sgn}(x)$ is given in Figure 4.1.6. *The three closed and open circles have been added by hand.*

The signum function is built in on most computers, but is not generally a built-in function on graphing calculators. The next example gives a way to emulate this function on a graphing utility that does not have the signum function built in.

• **EXAMPLE 5:** Determine the domain, the range, and draw a complete graph of $f(x) = \frac{|x|}{x}$.

**SOLUTION:** This function can usually be entered in a graphing utility as $f(x) = \frac{\text{ABS}(x)}{x}$, where $\text{ABS}(x) = |x|$. A computer drawn graph of $f$ is given in Figure 4.1.7.

The function $f$ is the quotient of the functions $y = |x|$ and $y = x$. Thus, the domain of $f$ is the set of all nonzero real numbers. Notice that for each $x \neq 0$, $f(x)$ is either 1 or $-1$. Therefore, the range of $f$ is $\{-1, 1\}$, and the graph of $f$ in Figure 4.1.7 is complete. •

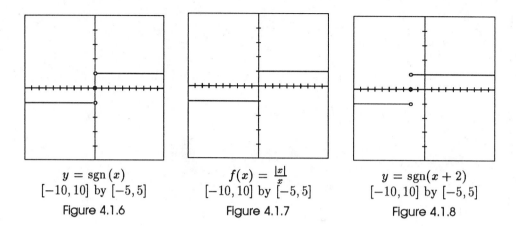

$y = \text{sgn}(x)$
$[-10, 10]$ by $[-5, 5]$
Figure 4.1.6

$f(x) = \frac{|x|}{x}$
$[-10, 10]$ by $[-5, 5]$
Figure 4.1.7

$y = \text{sgn}(x + 2)$
$[-10, 10]$ by $[-5, 5]$
Figure 4.1.8

The signum function and the function $f(x) = \frac{|x|}{x}$ are discontinuous at $x = 0$. Example 5 shows that

$$f(x) = \frac{|x|}{x} = \begin{cases} -1 & \text{for } x < 0 \\ 1 & \text{for } x > 0 \end{cases}.$$

Notice that $f(x) = \frac{|x|}{x}$ agrees with $\text{sgn}(x)$ for $x \neq 0$. At $x = 0$, $f$ is not defined and $\text{sgn}(0) = 0$. If the graphing utility you use does not have $\text{sgn}(x)$ as a built-in function, then you can obtain a graph of $y = \text{sgn}(x)$ that looks correct by graphing the function $f(x) = \frac{|x|}{x}$. The two *computer drawn* graphs appear to be identical because they are identical for $x \neq 0$, and the point $(0, 0)$ that is on the graph of $y = \text{sgn}(x)$ and not on the graph of $f(x) = \frac{|x|}{x}$ appears to be on the graph of $f$ because it is the origin.

- **EXAMPLE 6:** Determine the domain, the range, and draw a complete graph of each function.
    (a) $f(x) = \text{sgn}(x + 2)$ \qquad (b) $g(x) = 1 + \text{sgn}(x - 1)$

SOLUTION: The graph of $f(x) = \text{sgn}(x + 2)$ can be obtained by shifting the graph of $y = \text{sgn}(x)$ horizontally 2 units to the left (Figure 4.1.8). Similarly, the graph of $g(x) = 1 + \text{sgn}(x - 1)$ can be obtained by shifting the graph of $y = \text{sgn}(x)$ horizontally 1 unit to the right and then shifting the resulting graph vertically up 1 unit (Figure 4.1.9).

The domain of each function is the set of all real numbers. The range of $f$ is $\{-1, 0, 1\}$ and the range of $g$ is $\{0, 1, 2\}$. •

Example 6 can be extended to obtain the following, which you will be asked to prove in the exercises.

- **THEOREM 1** \qquad The domain of $\text{sgn}(x - a)$ is the set of all real numbers, and

$$\text{sgn}(x - a) = \begin{cases} -1 & \text{for } x < a \\ 0 & \text{for } x = a \\ 1 & \text{for } x > a \end{cases}.$$

## 4.1 Operations on Functions

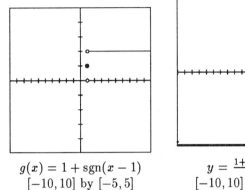

$g(x) = 1 + \text{sgn}(x-1)$
$[-10, 10]$ by $[-5, 5]$
Figure 4.1.9

$y = \frac{1+\text{sgn}(x-2)}{2}$
$[-10, 10]$ by $[-2, 2]$
Figure 4.1.10

$y = \frac{1-\text{sgn}(x-2)}{2}$
$[-10, 10]$ by $[-2, 2]$
Figure 4.1.11

The functions in the next example are the type of function needed to enter piecewise functions directly in graphing utilities.

- **EXAMPLE 7:** Determine the domain, the range, and draw a complete graph of each function.

  (a) $y = \dfrac{1 + \text{sgn}(x-2)}{2}$  (b) $y = \dfrac{1 - \text{sgn}(x-2)}{2}$

SOLUTION: The domain of each function is the set of all real numbers. For $x > 2$,

$$\frac{1 + \text{sgn}(x-2)}{2} = \frac{1+1}{2} = 1 \quad \text{and} \quad \frac{1 - \text{sgn}(x-2)}{2} = \frac{1-1}{2} = 0.$$

For $x < 2$,

$$\frac{1 + \text{sgn}(x-2)}{2} = \frac{1+(-1)}{2} = 0 \quad \text{and} \quad \frac{1 - \text{sgn}(x-2)}{2} = \frac{1-(-1)}{2} = 1.$$

For $x = 2$, each function is equal to $\frac{1}{2}$ because $\text{sgn}(0) = 0$. Summarizing, we have

$$\frac{1 + \text{sgn}(x-2)}{2} = \begin{cases} 0 & \text{for } x < 2 \\ \frac{1}{2} & \text{for } x = 2 \\ 1 & \text{for } x > 2 \end{cases} \quad \text{and} \quad \frac{1 - \text{sgn}(x-2)}{2} = \begin{cases} 1 & \text{for } x < 2 \\ \frac{1}{2} & \text{for } x = 2 \\ 0 & \text{for } x > 2 \end{cases}.$$

Complete graphs of the two functions are given in Figures 4.1.10 and 4.1.11, respectively. •

To save space we make the following definitions.

- **DEFINITION** Let $a$ be any real number. We define the functions $C_{(a,\infty)}$ and $C_{(-\infty,a)}$ by

$$C_{(a,\infty)}(x) = \frac{1 + \text{sgn}(x-a)}{2} \quad \text{and} \quad C_{(-\infty,a)}(x) = \frac{1 - \text{sgn}(x-a)}{2}.$$

- **THEOREM 2**   Let $a$ be any real number. Then,

$$C_{(a,\infty)}(x) = \begin{cases} 0 & \text{for } x < a \\ \frac{1}{2} & \text{for } x = a \\ 1 & \text{for } x > a \end{cases} \quad \text{and} \quad C_{(-\infty,a)}(x) = \begin{cases} 1 & \text{for } x < a \\ \frac{1}{2} & \text{for } x = a \\ 0 & \text{for } x > a \end{cases}.$$

In advanced mathematics courses, mathematicians define the *characteristic function of a set $S$*. This function is equal to 1 at values in $S$ and equal to 0 at values outside of $S$. The functions $C_{(a,\infty)}$ and $C_{(-\infty,a)}$ agree with the characteristic functions of the sets $(a,\infty)$ and $(-\infty,a)$, respectively, except at $x = a$.

- **EXAMPLE 8:** Find a single rule that when entered in a graphing utility gives a complete graph of

$$f(x) = \begin{cases} x^2 & \text{for } x < 2 \\ 7 - x & \text{for } x \geq 2 \end{cases}$$

except possibly at $x = 2$.

**SOLUTION:** Let $g(x) = x^2$ and $h(x) = 7 - x$ for all real numbers $x$. We show that, for $x \neq 2$, $f = C_{(-\infty,2)}g + C_{(2,\infty)}h$. For $x < 2$, we have

$$(C_{(-\infty,2)}g + C_{(2,\infty)}h)(x) = C_{(-\infty,2)}(x)g(x) + C_{(2,\infty)}(x)h(x)$$
$$= 1 \cdot x^2 + 0 \cdot (7 - x)$$
$$= x^2.$$

For $x > 2$ we have

$$(C_{(-\infty,2)}g + C_{(2,\infty)}h)(x) = C_{(-\infty,2)}(x)g(x) + C_{(2,\infty)}(x)h(x)$$
$$= 0 \cdot x^2 + 1 \cdot (7 - x)$$
$$= 7 - x.$$

Therefore, for $x \neq 2$, $f = C_{(-\infty,2)}g + C_{(2,\infty)}h = \frac{1-\text{sgn}(x-2)}{2}x^2 + \frac{1+\text{sgn}(x-2)}{2}(7-x)$.

For $x = 2$, $C_{(-\infty,2)}g + C_{(2,\infty)}h$ is equal to 4.5 and $f(2) = 5$. If we replace the point $(2, 4.5)$ of the computer drawn graph of $C_{(-\infty,2)}g + C_{(2,\infty)}h$ by $(2, 5)$, we obtain the complete graph of $f$ given in Figure 4.1.12. Thus, $C_{(-\infty,2)}g + C_{(2,\infty)}h$ is a *single rule* that when entered in a graphing utility gives a complete graph of $f$ for $x \neq 2$. You often will not be able to see the point $(2, 4.5)$. Thus, except for a single point the computer drawn graph of $y = C_{(-\infty,2)}g + C_{(2,\infty)}h$ appears to be a complete graph of $f$.   ●

If your graphing utility does *not* have the signum function built in, then be sure to replace $\text{sgn}(x-a)$ by $\frac{|x-a|}{x-a}$ in Example 8. If you use this replacement, the function entered on the graphing utility agrees with $f$ except at $x = 2$ and is not defined at $x = 2$.

In general, we are able to obtain a single rule that when entered in a graphing utility gives a graph of a piecewise-defined function except at the values of $x$ at which the function rule changes. Then, we add, by hand, open or closed circles at the end points where the function rule changes to complete the graph.

# 4.1 Operations on Functions

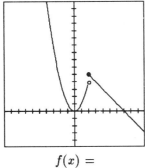

$f(x) = \begin{cases} x^2 & \text{for } x < 2 \\ 7 - x & \text{for } x \geq 2 \end{cases}$
$[-10, 10]$ by $[-5, 15]$
Figure 4.1.12

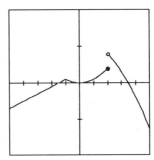

$f(x) = \begin{cases} 2x + 3 & \text{for } x < -1 \\ x^2 & \text{for } -1 \leq x \leq 2 \\ 12 - x^2 & \text{for } x > 2 \end{cases}$
$[-5, 5]$ by $[-20, 20]$
Figure 4.1.13

It is possible to extend Example 8 to piecewise-defined functions where the domain is partitioned into three or more parts. However, it seems much easier to do this by hand. If you are not familiar with a particular rule used for a portion of the domain of a piecewise-defined function you can use a graphing utility to investigate that rule in the appropriate domain.

- **EXAMPLE 9**: Sketch a complete graph of

$$f(x) = \begin{cases} 2x + 3 & \text{for } x < -1 \\ x^2 & \text{for } -1 \leq x \leq 2 \\ 12 - x^2 & \text{for } x > 2 \end{cases}.$$

SOLUTION: Here the rule on each of the three parts of the domain is a familiar one. The graph in Figure 4.1.13 is a complete graph of $f$. Notice the function is discontinuous at $x = 2$ and continuous on the intervals $(-\infty, 2]$ and $(2, \infty)$. (Why?)

## EXERCISES 4-1

Determine the domain and a rule for each of $f + g$, $f - g$, $fg$, and $f/g$.

1. $f(x) = 2x - 1$; $g(x) = x^2$
2. $f(x) = (x - 1)^2$; $g(x) = 3 - x$
3. $f(x) = x^2$; $g(x) = 2x$
4. $f(x) = \sqrt{x}$; $g(x) = x - 2$
5. $f(x) = x + 3$; $g(x) = \dfrac{2x - 1}{3}$
6. $f(x) = \dfrac{1}{x - 1}$; $g(x) = (x + 2)^2$

7–8. Find the range of the functions $f + g$, $f - g$, $fg$, and $f/g$ with $f$ and $g$ defined as in Exercises 5 and 6. You may need to use a graphing utility.

9. Explain why $fg$ and $f \circ g$ are different in general. Give an example.

10. Determine if *addition* of polynomial functions is commutative, associative, has an identity, or has an inverse.

11. Determine if *multiplication* of polynomial functions is commutative, associative, has an identity, or has an inverse.

Draw complete graphs of $f$, $g$, and $f + g$ in the same viewing rectangle.

12. $f(x) = -x$; $g(x) = \cos x$
13. $f(x) = \sin x$; $g(x) = \cos x$
14. $f(x) = \dfrac{1}{x}$; $g(x) = \sin x$
15. $f(x) = -x$; $g(x) = \dfrac{1}{x}$.

16. Which function in Exercise 12 will most closely approximate the sum function $f + g$ for $x$ large in absolute value? Why?

Determine the domain and the range of each function, and sketch a complete graph without using a graphing utility.

17. $f(x) = \begin{cases} 2x & \text{for } x > 1 \\ x - 3 & \text{for } x \leq 1 \end{cases}$
18. $f(x) = \begin{cases} \sqrt{x+3} & \text{for } x \geq 1 \\ -x & \text{for } x < 0 \end{cases}$

19. $f(x) = \begin{cases} x^2 & \text{for } x > -3 \\ 2 - x & \text{for } x \leq -3 \end{cases}$
20. $g(x) = \begin{cases} \dfrac{1}{x} & \text{for } x \geq -1 \\ x^2 - x & \text{for } x < -1 \end{cases}$

21. $f(x) = \begin{cases} 2 & \text{for } x \geq 5 \\ -2x & \text{for } -2 \leq x < 5 \\ 2 - x^2 & \text{for } x < -2 \end{cases}$
22. $f(x) = \begin{cases} \sqrt{x+1} & \text{for } x \geq 8 \\ x - 5 & \text{for } 0 < x < 8 \\ x^2 - 5 & \text{for } x \leq 0 \end{cases}$

23. Where is the function in Exercise 21 discontinuous?
24. Where is the function in Exercise 22 discontinuous?

Determine the domain, the range, and draw a complete graph.

25. $f(x) = \operatorname{sgn}(x - 5)$
26. $g(x) = \operatorname{sgn}(x + 5)$
27. $f(x) = \dfrac{|x - 5|}{x - 5}$
28. $g(x) = \dfrac{|x + 5|}{x + 5}$
29. $f(x) = \dfrac{1 - \operatorname{sgn}(x - 5)}{2}$
30. $f(x) = \dfrac{1 + \operatorname{sgn}(x - 5)}{2}$
31. $f(x) = \dfrac{1 - \frac{|x-5|}{x-5}}{2}$
32. $f(x) = \dfrac{1 + \frac{|x+5|}{x+5}}{2}$

Determine the domain, the range, and find a single rule that when entered in a graphing utility gives a complete graph of each function, except possibly at the $x$ values at which the rule changes.

33. $f(x) = \begin{cases} x^3 + x + 1 & \text{for } x \geq 2 \\ x^4 - x^2 & \text{for } x < 2 \end{cases}$
34. $g(x) = \begin{cases} 2x + x^5 & \text{for } x \geq 1 \\ 5x^3 - x - 1 & \text{for } x < 1 \end{cases}$

35. Where is the function in Exercise 33 discontinuous?
36. Where is the function in Exercise 34 discontinuous?
37. Prove Theorem 1.
38. Prove Theorem 2.

## Income Tax
In 1988, Federal tax laws required that income be taxed at the following rates.

| Taxable Income | Tax |
|---|---|
| $0 to $17,850 | 15% of taxable income |
| over $17,850 to $43,150 | $2677.50 plus 28% of amount over $17,850 |
| over $43,150 to $100,480 | $9761.50 plus 33% of amount over $43,150 |
| over $100,480 | $28,680.40 plus 28% of amount over $100,480 |

39. Explain how the $9761.50 figure was determined for a taxable income of $43,150.
40. Determine a piecewise-defined function that is an algebraic representation of this problem situation.
41. Determine the domain and range of this function.
42. Sketch a graph of this problem situation. Do not use a graphing utility.

## Long Distance Charges
A certain long distance phone company charges the following amounts for calls after 6 P.M. and before 8 A.M. from Columbus, Ohio to San Francisco, California. For calls up to 5 minutes in length the charge is $0.72 per minute or fraction of a minute. For calls greater than 5 and up to 15 minutes in length the charge is $0.63 per minute or fraction of a minute. For calls greater than 15 minutes in length, the charge is $0.51 per minute or fraction of a minute.

43. Determine a piecewise-defined function that is an algebraic representation of the problem situation.
44. Determine the domain and range of this function.
45. Sketch a graph of this problem situation. Do not use a graphing utility.

A rocket launched from ground level is $d$ miles above the earth $t$ minutes after launch. While the rocket engine is on, $d$ is given by
$$d = 6t^2 + 185t.$$
When the engines are off, $d$ is given by
$$d = 2250 + 30t - t^2.$$

46. Assume the rocket engine is on for only the first 10 minutes. Determine a piecewise-defined function that is a model for this problem situation.
47. Draw a complete graph of the problem situation.
48. When does the rocket hit the earth?

49. What is the maximum height reached by the rocket? When does it reach its maximum height?

## Depreciation and Appreciation

The following notation and definitions will be used in connection with depreciation and appreciation of an asset.

$C$ = original value (cost) of the asset

$n$ = length of term (useful life under consideration)

$S$ = value of the asset at end of term (in case of depreciation, $S$ is called the *salvage value*)

$D = C - S$ is the *total depreciation* if $S < C$

and is the *total appreciation* if $S > C$

$B$ = *book value*, the value for accounting or tax purposes at any time $t$

$\dfrac{C - S}{n}$ = the annual depreciation or appreciation amount

A continuous model for *book value* at any time $t$ using the *straight line method* is

$$B = C - t\left(\frac{C - S}{n}\right) = C - \frac{Dt}{n}$$

where $0 \leq t \leq n$.

50. A machine costs $17,000, has a useful life of 6 years, and has salvage value of $1200. Draw the graph of the "book value" model $y = B(t)$. Why is the model called straight line depreciation?
51. What is the book value of the machine in 4 years, 3 months?
52. Assume a home valued today at $78,000 has a "useful life" of 25 years and *appreciates* at $5000 per year using the straight line method. Draw a complete graph of this problem situation.
53. What is the value of the home in Exercise 52 after 10 years? 15 years? 25 years?

A precision computer-assisted design (CAD) machine purchased from Japan costs $22,600. It appreciates in value for 3 years to $26,400 (due to the fall in the value of the dollar) and then depreciates in value for 4 years (when the age of the machine overshadows the fall in the value of the dollar) and has a salvage value of $4700. Assume straight line appreciation and depreciation for computing book value.

54. Write a piecewise-defined function of time that is a model of the book value for this problem situation.
55. Draw a complete graph of the model of the book value of the machine.
56. What is the book value of the machine after 5 years, 6 months?
57. Prove that for any real number $x$, $x = \text{sgn}(x)\,\text{abs}(x)$; that is, every real number is the product of its signum and its absolute value.

## 4.2 Exponential Functions

In this section we begin the study of exponential functions. We establish important properties of these functions as well as complete graphs. Problem situations about growth or decay of bacteria, radioactive substances, and population have exponential functions as models. Applications of this type are studied in this section.

In calculus you will be given a careful definition of an exponential function. In this textbook we assume that $a^x$ is a real number for $a > 0$ and use a calculator or computer to compute the values of an exponential function. We further assume that in exponential functions $y = a^x$, the base is positive. The usual properties of exponents are given in the following theorem without proof.

- **THEOREM 1**     Let $a, x, y$ be real numbers with $a > 0$.

    (a) $a^0 = 1$        (b) $a^x a^y = a^{x+y}$        (c) $\dfrac{a^x}{a^y} = a^{x-y}$

    (d) $a^{-x} = \dfrac{1}{a^x}$        (e) $(a^x)^y = a^{xy}$

- **DEFINITION**     Let $a$ be a positive real number. The function $f(x) = a^x$ with domain the set of all real numbers is called the **exponential function with base $a$**.

- **EXAMPLE 1:** Draw and compare complete graphs of $f(x) = 2^x$ and $g(x) = 0.5^x$. Find the domain and range of each function.

    SOLUTION: We can use a calculator or computer to determine the following end behavior of $f$:

    $$f(x) \to \infty \quad \text{as} \quad x \to \infty,$$
    $$f(x) \to 0 \quad \text{as} \quad x \to -\infty.$$

    The $y$-intercept of the graph of $f$ is 1 because $2^0 = 1$. A complete graph of $f$ is given in Figure 4.2.1. The function $f$ is increasing in $(-\infty, \infty)$.

    We find the opposite end behavior for $g$:

    $$g(x) \to 0 \quad \text{as} \quad x \to \infty,$$
    $$g(x) \to \infty \quad \text{as} \quad x \to -\infty.$$

    Notice that $g(x) = 0.5^x = \left(\frac{1}{2}\right)^x = 2^{-x}$ so that $g(x) = f(-x)$. Thus, a complete graph of $g$ can be obtained by reflecting a complete graph of $f$ through the $y$-axis. A complete graph of $g$ is given in Figure 4.2.2. The function $g$ is decreasing in $(-\infty, \infty)$.

    The domain of each function is $(-\infty, \infty)$, the range of each function is $(0, \infty)$, and the $x$-axis is a horizontal asymptote for each function.

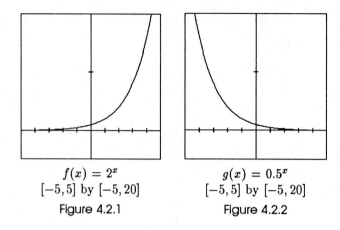

$f(x) = 2^x$
$[-5, 5]$ by $[-5, 20]$
Figure 4.2.1

$g(x) = 0.5^x$
$[-5, 5]$ by $[-5, 20]$
Figure 4.2.2

If $a > 1$, the graph of $y = a^x$ looks essentially like the graph of $y = 2^x$. If $0 < a < 1$, the graph of $y = a^x$ looks essentially like the graph of $y = 0.5^x$.

- **EXAMPLE 2:** Draw and compare complete graphs of $y = 2^x$ and $y = 3^x$.

  **SOLUTION:** A complete graph of each function is given in Figure 4.2.3. The end behavior of each function is the same, and each has $y$-intercept 1. The graphs in Figure 4.2.3 suggest that $2^x < 3^x$ in $(0, \infty)$ because the graph of $y = 2^x$ appears to be *below* the graph of $y = 3^x$. Indeed, this is true, but you may need to view the two graphs in several additional viewing rectangles to convince yourself of this fact. For $x < 0$, the graph of $y = 2^x$ is *above* the graph of $y = 3^x$. The graphs in Figure 4.2.4 illustrate this. Thus, $2^x > 3^x$ in $(-\infty, 0)$.  •

If $a > b > 1$, the graphs of $y = a^x$ and $y = b^x$ are related as suggested in Example 2. In the Exercises you will compare these graphs for $0 < a < b < 1$.

A rough sketch of the graph of $y = a(b^{cx+d}) + k$ can be obtained by applying standard geometric transformations to the graph of $y = b^x$. First we show that $c$ can be assumed to be equal to 1. Notice that we can use properties of exponents to rewrite the exponential

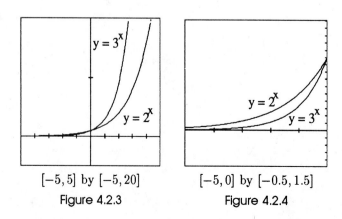

$[-5, 5]$ by $[-5, 20]$
Figure 4.2.3

$[-5, 0]$ by $[-0.5, 1.5]$
Figure 4.2.4

expression $2^{3x+4}$ with base 2 as an exponential expression with base $2^3 = 8$.
$$2^{3x+4} = 2^{3(x+4/3)} = (2^3)^{x+4/3} = 8^{x+4/3}$$
Similarly, $y = a(b^{cx+d}) + k = a(b^c)^{x+d/c} + k$.

- **EXAMPLE 3**: Determine the domain, the range, and draw a complete graph of
  (a) $y = 2(3^{x+1}) - 5$.
  (b) $y = 2^{3-2x}$.

**SOLUTION:**

(a) A complete graph of $y = 2(3^{x+1}) - 5$ can be obtained by applying, in order, the following transformation to the graph of $y = 3^x$.

  1. Vertical stretch by a factor of 2 to obtain $y = 2(3^x)$.
  2. Horizontal shift left 1 unit to obtain $y = 2(3^{x+1})$.
  3. Vertical shift down 5 units to obtain $y = 2(3^{x+1}) - 5$.

  The computer drawn graph in Figure 4.2.5 confirms this observation. The domain is $(-\infty, \infty)$, and the range is $(-5, \infty)$.

  Alternatively, we can obtain a rough sketch in the following way. The function $y = 2(3^{x+1}) - 5$ must increase in $(-\infty, \infty)$ like $y = 3^x$ because the coefficient 2 of $3^{x+1}$ is positive. Notice that the function $y = -2(3^{x+1}) - 5$ is a decreasing function in $(-\infty, \infty)$. The $-5$ in $y = 2(3^{x+1}) - 5$ tells us that the line $y = -5$ is the horizontal asymptote of the function. We can see that the point $(0, 1)$ of $y = 3^x$ is moved to the point $(-1, -3)$ under the transformations that produce $y = 2(3^{x+1}) - 5$ from $y = 3^x$. This point, the horizontal asymptote, the y-intercept and the increasing behavior allow us to produce the rough sketch in Figure 4.2.6.

(b) First we rewrite $y = 2^{3-2x}$ in the form $y = a^{x-h}$ for appropriate $a$ and $h$:
$$2^{3-2x} = 2^{-2(x-\frac{3}{2})} = (2^{-2})^{x-1.5} = 0.25^{x-1.5}.$$

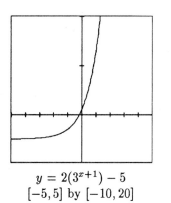

$y = 2(3^{x+1}) - 5$
$[-5, 5]$ by $[-10, 20]$

Figure 4.2.5

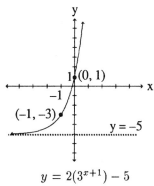

$y = 2(3^{x+1}) - 5$

Figure 4.2.6

The graph of $y = 2^{3-2x} = 0.25^{x-1.5}$ must look like the graph of $y = 0.25^x$ except that it is shifted to the right 1.5 units. Thus, the function $y = 2^{3-2x}$ is a decreasing function in $(-\infty, \infty)$, and the $x$-axis is its horizontal asymptote. The point $(1.5, 1)$ on the graph of $y = 2^{3-2x}$ corresponds to the point $(0, 1)$ on the graph of $y = 0.25^x$, and $(0, 8)$ is the $y$-intercept of $y = 2^{3-2x}$. Figure 4.2.7 gives a rough sketch of $y = 2^{3-2x}$ using this information. •

Exponential functions of the form $y = a(b^{cx+d}) + k$ are either always increasing or always decreasing. Thus, such functions are one-to-one and their inverses are functions.

• **EXAMPLE 4:** Let $f(x) = 3^x$. Draw a complete graph of $f^{-1}$. Show that $f^{-1}$ is a function.

**SOLUTION:** In Section 3.3 we showed that a complete graph of $f^{-1}$ can be obtained by reflecting a complete graph of $f$ through the line $y = x$. Complete graphs of $f$ and $f^{-1}$ are given in Figure 4.2.8.
We can see from the graph in Figure 4.2.8 that $f$ is one-to-one. Thus, $f^{-1}$ is a function. Notice also that the graph of $f^{-1}$ satisfies the vertical line test for a function. •

The horizontal asymptote of the function in the next example involves a number that is important in the study of mathematics and applications.

• **EXAMPLE 5:** Find the domain, the range, an end behavior model, and draw a complete graph of $f(x) = \left(1 + \frac{1}{x}\right)^x$.

**SOLUTION:** The graph of $f$ in the $[-10, 10]$ by $[-10, 10]$ viewing rectangle is actually a complete graph (Figure 4.2.9). We must have $1 + \frac{1}{x} > 0$ because, by definition, the base of an exponential function must be positive. This means that $x > 0$ or $x < -1$. (Why?) Thus, the domain of $f$ is $(-\infty, -1) \cup (0, \infty)$ as suggested by Figure 4.2.9. From this graph we can see that $f(x) \to \infty$ as $x \to -1^-$ and $f(x) \to 1$ as $x \to 0^+$. We can confirm this

$y = 2^{3-2x}$

Figure 4.2.7

Figure 4.2.8

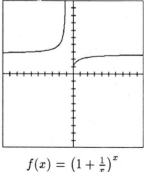

$f(x) = \left(1 + \frac{1}{x}\right)^x$
$[-10, 10]$ by $[-10, 10]$

Figure 4.2.9

information using a calculator. The graph of $f$ has a horizontal asymptote. The horizontal asymptote is $y = e$, where $e$ is an irrational number that is equal to 2.718281828 accurate to 9 decimal places. The importance of the number $e$ was first observed by Euler in the 1700's. Thus, $y = e$ is an end behavior model for $f$. You can approximate the $y$-coordinate of the horizontal asymptote graphically by a careful selection of viewing rectangles. The graph of $f$ in $[0, 100]$ by $[1, 3]$ is a good starting point (Figure 4.2.10). From Figure 4.2.10, it appears that $e$ is a bit larger than 2.7. It is possible to determine $e$ graphically with accuracy up to the limits of machine precision. The range of $f$ is $(1, e) \cup (e, \infty)$.

In the Exercises we ask you to show that $y = e^r$ is an end behavior model of $f(x) = \left(1 + \frac{r}{x}\right)^x$. This means $y = e^r$ is the horizontal asymptote and thus $\left(1 + \frac{r}{x}\right)^x \to e^r$ as $x \to \infty$.

## Population

Suppose that the population of a certain town is $P$ and is increasing at constant rate $r$ each year, where $r$ is in decimal form. Then, the population of the town one year later is $P + Pr = P(1 + r)$. Two years later the population is $P(1 + r) + P(1 + r)r = P(1 + r)^2$. In general, the population $t$ years later is $P(1 + r)^t$. If the population of a town is decreasing at constant rate $r$ each year, then the population one year later is $P - Pr = P(1 - r)$. Two years later the population is $P(1 - r) - P(1 - r)r = P(1 - r)^2$. In general, the population $t$ years later is $P(1 - r)^t$. These are examples of **exponential growth or decay**.

• **EXAMPLE 6:** The population of a town is 50,000 and is increasing at the rate of 2.5% each year.

(a) Write the population $P$ as a function of time.
(b) Draw a complete graph of the function $P$ in (a). What portion of the graph represents the problem situation?
(c) Determine when the population of the town will be 100,000.

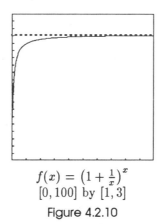

$f(x) = \left(1 + \frac{1}{x}\right)^x$
$[0, 100]$ by $[1, 3]$
Figure 4.2.10

SOLUTION:

(a) From the discussion preceding the example, we know that $P(t) = (50,000)(1 + 0.025)^t$, or $P(t) = (50,000)1.025^t$ where $t$ represents the numbers of years after the population is 50,000.

(b) A complete graph of $P$ is given in Figure 4.2.11. According to the statement of the example, only the portion of the graph in Figure 4.2.11 that is to the right of the $y$-axis represents the problem situation. However, if we also know that the population of the town had been increasing at the rate of 2.5% in prior years, then a portion of the graph to the left of the $y$-axis could also represent the problem situation. (Why?)

(c) We need to determine $t$ so that $P(t) = (50,000)1.025^t = 100,000$. We can use the graph in Figure 4.2.11 to see that $t$ is a little less than 30. Zoom-in can be used to determine that the value of $t$ that makes $P(t) = 100,000$ is 28.07 with error at most 0.01. Thus, the population of the town will be 100,000 about 28.07 years later. ●

We can generalize the population example to general exponential growth and exponential decay problems. Suppose that there are $P$ units of a substance present initially. If the substance $P$ is growing at constant rate $r$, $r$ in decimal form, then the amount of substance present $t$ years later is $P(1 + r)^t$. If the substance is decaying at constant rate $r$, then the amount of substance present $t$ years later is $P(1 - r)^t$. In general, $P \cdot a^t$ is a model for exponential growth if $a > 1$ and is a model for exponential decay if $0 < a < 1$. Finally, the unit of time need not be years. For example, $t$ could be measured in months, days, or hours.

In the next example we see how to determine an exponential growth model when the doubling time for a substance is known.

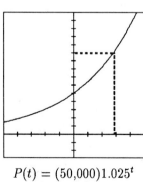

$P(t) = (50,000)1.025^t$
$[-50, 50]$ by
$[-30,000, 150,000]$
Figure 4.2.11

## Biological Growth

Assume that the number of bacteria in a certain bacterial culture doubles every hour and that there are 100 present initially. The number of bacteria after one hour is $200 = 2(100)$, after two hours is $400 = 2^2(100)$, and after three hours is $800 = 2^3(100)$. After $t$ hours the number of bacteria present in the culture is given by $f(t) = (100)2^t$.

- **EXAMPLE 7:** Assume that the number of bacteria in a certain bacterial culture doubles every hour and that there are 100 present initially.

  (a) How many bacteria are present after 7 hours and 15 minutes?
  (b) Draw a graph that shows the number of bacteria present during the first 12 hours.
  (c) Determine when the number of bacteria will be 350,000.

**SOLUTION:**

(a) From the discussion preceding this example, $f(t) = (100)2^t$ represents the number of bacteria present after $t$ hours. Notice that $t = 7.25$ is equivalent to 7 hours and 15 minutes. Thus, the number of bacteria present after 7 hours and 15 minutes is $f(7.25) = (100)2^{7.25}$ which is about 15,222.

(b) The graph of $f$ in the $[0, 13]$ by $[0, 450,000]$ viewing rectangle shows the number of bacteria present during the first 12-hour period (Figure 4.2.12). Notice how the values of $f$ appear to be 0 during the first 5 hours. The number of bacteria is so large near the end of the 12-hour period that the number present during the first part of the 12-hour period, by comparison, seems to be 0. We need a second graph to better illustrate the values of $f$ during the first six hours (Figure 4.2.13). Even this figure fails to illustrate clearly the values during the first hour.

(c) We need to determine $t$ so that $f(t) = (100)2^t = 350,000$. The graph of $f$ and $y = 350,000$ in Figure 4.2.14 shows that $t$ is approximately 11.7. We can zoom in and find that $t$ is 11.77 with error at most 0.01. Thus, after 11 hours and 46.2 minutes, the number of bacteria is approximately 350,000. ●

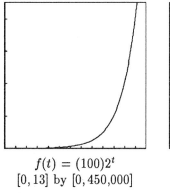

$f(t) = (100)2^t$
$[0, 13]$ by $[0, 450,000]$
Figure 4.2.12

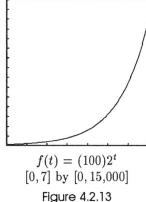

$f(t) = (100)2^t$
$[0, 7]$ by $[0, 15,000]$
Figure 4.2.13

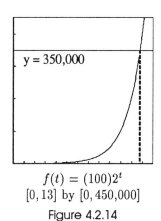

$f(t) = (100)2^t$
$[0, 13]$ by $[0, 450,000]$
Figure 4.2.14

## Radioactive Decay

One example of exponential decay is that associated with radioactive substances. The rate of such decay is based on the half-life of the substance. The *half-life* of a radioactive substance is the amount of time it takes for half of the substance to decay. Suppose that the half-life of a certain radioactive substance is 20 days, and there are 5 grams present initially. Then, after 20 days there are $5(\frac{1}{2})$ grams remaining. After 40 days there are $5(\frac{1}{2})^2$ grams remaining. In general, after $t$ days there will be $5(\frac{1}{2})^{t/20}$ grams remaining. Notice the exponent on $\frac{1}{2}$ is 1 when $t$ is 20, and 2 when $t$ is 40.

- **EXAMPLE 8:** The half-life of a certain radioactive substance is 20 days, and there are 5 grams present initially.

   (a) Write the amount $A$ of substance remaining as a function of time.
   (b) Draw a complete graph of the function in (a).
   (c) When will there be less than 1 gram of the substance remaining?

SOLUTION:

   (a) From the discussion preceding the example, we can write $A(t) = 5(0.5)^{t/20}$, where $t$ is the number of days after the amount of radioactive substance is 5 grams.
   (b) A complete graph of $A$ is given in Figure 4.2.15.
   (c) $A$ is a decreasing function. If we find when $A(t) = 1$, then the solution we seek consists of all $t$ greater than this number. We could graph $A$ and $y = 1$ in the same viewing rectangle and use zoom-in to determine that the $t$ coordinate of the point of intersection is 46.44. Thus, there will be less than 1 gram of the radioactive substance left after 46.44 days. •

In Section 4.4 we will learn how to solve problems like Example 8 algebraically using logarithms.

The exponential function $f(x) = a \cdot b^t$ is an extremely important model for compound growth or decay problems. The function is increasing and represents growth if $a > 0$ and

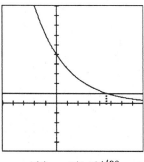

$A(t) = 5(0.5)^{t/20}$
$[-50, 80]$ by $[-5, 10]$

Figure 4.2.15

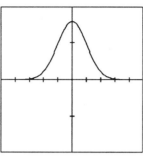

$\phi(x) = \frac{1}{2\pi}e^{-x^2/2}$
$[-5, 5]$ by $[-0.2, 0.2]$
Figure 4.2.16

$b > 1$. Similarly, the function is decreasing and represents decay if $a > 0$ and $0 < b < 1$. In Section 4.3, we will use this model in economic applications.

We close this section by investigating an important graph that occurs in the study of statistics; the so called *bell-shaped curve* or *normal distribution curve*.

- **EXAMPLE 9:** Let $\phi(x) = \frac{1}{2\pi}e^{-x^2/2}$.

  (a) Draw a complete graph of $\phi$.
  (b) Find any maximum and minimum values of $\phi$.

SOLUTION: A complete graph of $\phi$ is given in Figure 4.2.16. Notice that $\phi(x) \to 0$ as $x \to \infty$ and as $x \to -\infty$. Thus, the $x$-axis is a horizontal asymptote of $\phi$. The figure suggests the maximum value of $\phi$ occurs at $x = 0$. Thus, $\phi(0) = \frac{1}{2\pi}$ is the maximum value of $\phi$. (Why?) We can use zoom-in to show that the maximum value of $\phi$ is 0.159 with error at most 0.01. Check that $\frac{1}{2\pi}$, accurate to thousandths, is 0.159. •

## EXERCISES 4-2

1. Graph $y = (\frac{1}{4})^x$ and $y = 4^{-x}$ in the same viewing rectangle.

2. Draw the graphs of $y = (\frac{1}{4})^x$, $y = (\frac{1}{3})^x$, $y = (\frac{1}{2})^x$ in the same $[-2, 4]$ by $[-1, 2]$ viewing rectangle. Solve each inequality.

   (a) $(\frac{1}{4})^x > (\frac{1}{3})^x$
   
   (b) $(\frac{1}{3})^x > (\frac{1}{2})^x$
   
   (c) $(\frac{1}{4})^x < (\frac{1}{3})^x$
   
   (d) $(\frac{1}{3})^x < (\frac{1}{2})^x$

Without using a graphing utility, draw a complete graph of each function. Check your answer with a graphing utility.

3. $f(x) = 1 + 2^x$
4. $f(x) = 1 - 3^x$
5. $f(x) = 2^{x-3}$
6. $y = 3 \cdot 2^{x+2}$
7. $g(x) = -2 - 2^{x+3}$
8. $D(x) = -2 + (\frac{1}{2})^{x+1}$

9. Determine the domain and range of the function in Exercise 5.
10. Determine the domain and range of the function in Exercise 7.

Use rules of exponents to solve each equation. Check with a graphing utility.

11. $2^x = 4^2$
12. $x^4 = 16$
13. $8^{x/2} = 4^{x+1}$
14. $(-8)^{5/3} = 2(4^{x/2})$
15. $(1 - 5x)^{1/2} = 20$
16. $x^{1/2} - (x - 1.25)^{1/2} = 0.5$
*17. $e^{4x} - 2e^{2x} - 3 = 0$
*18. $3(2^{2x}) + 7(2^x) - 6 = 0$

Draw a complete graph of the function. Determine the domain of each function.

19. $f(x) = e^{2x-1}$
20. $g(x) = 2^{x^2-1}$
21. $f(x) = x3^x$
22. $f(x) = xe^{-x}$

Determine where each function is increasing and decreasing. Determine all local maximum and minimum values, and the range of each function.

23. $f(x) = e^{2x-1}$
24. $g(x) = 2^{x^2-1}$
25. $f(x) = x3^x$
26. $f(x) = xe^{-x}$
27. $f(x) = x2^{-x^2}$
28. $g(x) = -x10^{x^2/50}$

29. Let $f(x) = 2^x$. Show that $f$ is one-to-one and sketch a graph of $f$ and $f^{-1}$ on the same coordinate system.

30. Let $f(x) = (\frac{1}{2})^x$. Show that $f$ is one-to-one and sketch a graph of $f$ and $f^{-1}$ on the same coordinate system.

31. Investigate graphically the end behavior of $y = (1+2/x)^x$ by drawing the graph of $y = (1+2/x)^x$ and the line $y = e^2$ in the $[0, 100]$ by $[0, 10]$ viewing rectangle. Repeat in the $[100, 1000]$ by $[7, 8]$ viewing rectangle.

32. Investigate graphically the end behavior of $y = (1+3/x)^x$ by drawing the graph of $y = (1+3/x)^x$ and the line $y = e^3$ in the $[0, 100]$ by $[0, 25]$ viewing rectangle. Repeat in the $[100, 1000]$ by $[19, 21]$ viewing rectangle.

33. Investigate graphically the end behavior of $y = (1 + 0.1/x)^x$ by drawing the graph of $y = (1 + 0.1/x)^x$ and the line $y = e^{0.1}$ in the $[0, 0.1]$ by $[0.75, 1.25]$ viewing rectangle. Repeat in the $[1, 10]$ by $[1.08, 1.12]$ viewing rectangle.

34. Investigate graphically the behavior of $f(x) = (1 + x)^{1/x}$ as $x \to 0$. What is the domain of $f$? Notice $f(0)$ is *not* defined, so $f$ is *not* continuous at $x = 0$. How would you define the value of $f$ at $x = 0$ in order to make $f$ continuous in the interval $(-1, \infty)$? *Hint:* Use zoom-in. Does this suggest an alternate definition for $e$?

Draw a complete graph of each inequality.

35. $y < e^x - 1$
36. $y \geq 2e^{x-3}$

## 4.2 Exponential Functions

37. Find the simultaneous solutions to the system
$$\begin{cases} y = x^2 \\ y = 2^x \end{cases},$$
that is, solve $x^2 = 2^x$. For what values of $x$ is $2^x > x^2$? For what values of $x$ is $x^2 > 2^x$?

38. Find the simultaneous solution of the system
$$\begin{cases} y = x^3 \\ y = 3^x \end{cases},$$
that is, solve $x^3 = 3^x$. For what values of $x$ is $3^x > x^3$? For what values of $x$ is $x^3 > 3^x$?

39. Assume that the number of bacteria in a certain bacterial culture doubles every three hours and that there are 2500 present initially.
    (a) How many bacteria are present after 6 hours?
    (b) Draw a graph that shows the number of bacteria present during the first 24-hour period.
    (c) Determine when the number of bacteria will be 100,000.

40. Assume that the number of rabbits in a certain population doubles every month and that there are 20 rabbits present initially.
    (a) How many rabbits are present after one year? After 5 years?
    (b) Draw a graph that shows the number of rabbits present during the first year.
    (c) Determine when the number of rabbits will be 10,000.
    (d) Explain why this exponential growth model is not a good model for rabbit population growth over a long period of time. What factors influence population growth?

41. The population of a town is 475,000 and is increasing at the rate of 3.75% each year.
    (a) Write the population $P$ as a function of time.
    (b) Draw a complete graph of the function $P$ in (a). What portion of the graph represents the problem situation?
    (c) Determine when the population of the town will be one million.

42. The population of a town is 123,000 and is decreasing at the rate of 2.375% each year.
    (a) Write the population $P$ as a function of time.
    (b) Draw a complete graph of the function $P$ in (a). What portion of the graph represents the problem situation?
    (c) Determine when the population of the town will be 50,000.

43. The population of a small town in the year 1890 was 6250. Assume the population increased at the rate of 3.75% each year.
    (a) Write the population $P$ as a function of time.
    (b) Draw a complete graph of the function $P$ in (a). What portion of the graph represents the problem situation?
    (c) Determine the population of the town in 1915 and in 1940.

44. The half-life of a certain radioactive substance is 14 days, and there are 6.58 grams present initially.
    (a) Write the amount $A$ of substance remaining as a function of time.
    (b) Draw a complete graph of the function in (a).
    (c) When will there be less than 1 gram of the substance remaining?

45. The half-life of a certain radioactive substance is 65 days, and there are 3.5 grams present initially.
    (a) Write the amount $A$ of substance remaining as a function of time.
    (b) Draw a complete graph of the function in (a).
    (c) When will there be less than 1 gram of the substance remaining?

46. The half-life of a certain radioactive substance is 1.5 seconds. Let $S$ be the amount of the substance initially (in grams).
    (a) Determine an algebraic representation for $S$.
    (b) Draw a complete graph of the algebraic representation in (a) if there are 2 grams of the substance initially.
    (c) Draw a graph of the problem situation in (b).
    (d) What is the initial amount of the substance needed if there is to be 1 gram left after 1 minute?
    *(e) Discuss the stability of this substance.

## 4.3 Economic Applications

In this section we continue the study of exponential functions. Simple interest, compound interest, and annuities are introduced. Economic applications are studied in this section.

### Simple Interest

Suppose that $200 is invested at 7% simple interest for one year. The *interest earned* at the end of one year is $200 (0.07), or $14. We say that $200 is invested at *simple annual interest rate* 0.07 or at simple annual interest rate of 7%.

• **DEFINITION**     Suppose that $P$ dollars are invested at simple interest rate $r$, where $r$ is a decimal. Then, $P$ is called the **principal**, $r$ the **rate of interest**, and $Pr$ the **interest received at the end of one interest period.**  •

Suppose that $P$ dollars are invested at simple interest rate $r$. Then the total value of the investment at the end of one period is $P + Pr$, or $P(1 + r)$, that is, the principal *plus* the interest. At the end of the second period the value of the investment is $P + 2Pr$, or

## 4.3 Economic Applications

$P(1+2r)$. In general, the total value of the investment at the end of $n$ periods using the simple interest rate model is given by the formula

**Simple Interest:** $S = P(1+nr)$.

- **EXAMPLE 1:** Five hundred dollars is invested at 7% simple annual interest. Determine the value of the investment 10 years later.

  **SOLUTION:** The value of the investment is given by $S = P(1+nr)$ where $P = 500$, $n = 10$, and $r = 0.07$. Thus, $S = 850$, and the value of the investment after 10 years is \$850.

## Compound Interest

Usually, financial institutions allow interest to compound; that is, they pay interest *on the interest*. In this case, the total value of the investment at the end of the first period is *also* $P + Pr = P(1+r)$, the same as simple interest. This amount is considered the new principal for the second investment period. The interest on this investment for the second period is $P(1+r)r$, and the value of the investment at the end of the second period is $P(1+r)+P(1+r)r$, or $P(1+r)^2$. At the end of the third period the value of the investment is $P(1+r)^3$. At the end of the $n$th period the value of the investment is $P(1+r)^n$. Interest accumulated in this way is called **compound interest**, and the value of the investment at the end of $n$ periods is given by the formula

**Compound Interest:** $S = P(1+r)^n$.

- **EXAMPLE 2:** Five hundred dollars is invested at 7% interest compounded annually. Determine the value of the investment 10 years later.

  **SOLUTION:** The value of the investment is given by $S = P(1+r)^n$ where $P = 500$, $n = 10$, and $r = 0.07$. Thus, $S = 983.58$ and the value of the investment after 10 years is \$983.58.

Suppose that money is invested at 9% per year *compounded quarterly*. Usually we say that money is invested at 9% compounded quarterly. There are four interest periods per year, the time period is 3 months and the interest rate per quarter is $\frac{9}{4}$%, or 0.0225. For example, if \$200 is invested at 9% compounded quarterly, then at the end of $n$ quarters ($3n$ months) the value of the investment is $200(1+0.0225)^n$, or $200(1.0225)^n$. We say that 9% is the *annual percentage rate* and $\frac{9}{4}$% is the *interest rate per quarter*.

- **DEFINITION** Suppose that a sum of money is invested at $r$% per year compounded $k$ times a year. Then $r$% is called the **annual percentage rate (APR)** and $\frac{r}{k}$% the **interest rate per period**.

**Compound Interest Formula** Suppose that $P$ dollars are invested at annual interest rate $r$ ($r$ in decimal form), and that interest is compounded $k$ times per year. The value of the investment at the end of $n$ periods is given by the formula

$$S = P\left(1 + \frac{r}{k}\right)^n.$$

- **EXAMPLE 3:** Suppose that $500 is invested at 9% APR compounded monthly.
  
  (a) Find the value of the investment after 5 years.
  (b) Draw a graph that shows the value of the investment during the first 30 years.
  (c) Determine when the value of the investment is $3000.

SOLUTION:

(a) The compound interest formula for this investment is $S = 500\left(1 + \frac{0.09}{12}\right)^n = 500(1.0075)^n$. We need to find the value $S$ of the investment when $n = 60$. Now, $S = 500(1.0075)^{60} = 782.84$. The value of the investment after 5 years is $782.84.

(b) We need a graph of $S = 500(1.0075)^n$ that shows the values of $S$ for $0 \leq n \leq 360$. The graph of $S$ in the $[0, 360]$ by $[0, 8000]$ viewing rectangle is given in Figure 4.3.1.

(c) We need to find a value of $n$ for which $S = 500(1.0075)^n$ is 3000. In the next section we solve this equation algebraically. Now, we find a solution graphically by determining the $x$-coordinate of the point of intersection of the graph of $A = 500(1.0075)^x$ with the graph of $y = 3000$ (Figure 4.3.2). The answer appears to be about 240 months or 20 years. We can use zoom-in to find that $x$ is 239.796 with error at most 0.01. Thus, the value of the investment will be $3000 in about 239.796 months or 19.983 years. Depending on the way interest is paid, a better answer may be 20 years. ●

- **EXAMPLE 4:** How much must be invested at 8% APR compounded daily so that the value of the investment 4 years later is $5000?

SOLUTION: In the compound interest formula $S = P\left(1 + \frac{r}{k}\right)^n$ we have $S = 5000$ and $r = 0.08$. (Why?) The length of a compounding period is one day so that $k$ is 365 and $n$ is

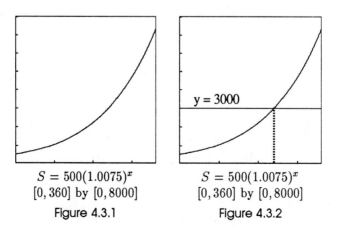

$S = 500(1.0075)^x$
$[0, 360]$ by $[0, 8000]$
Figure 4.3.1

$S = 500(1.0075)^x$
$[0, 360]$ by $[0, 8000]$
Figure 4.3.2

## 4.3 Economic Applications

5(365) or 1825. We ignore the one or two leap years that could occur in the 5-year period. Substituting these values in the compound interest formula we obtain

$$5000 = P\left(1 + \frac{0.08}{365}\right)^{1825}.$$

We can determine from this equation that $P = 3351.75$. Thus, if we invest \$3,351.75 at 8% APR compound daily, the value after 5 years will be \$5,000. •

• **EXAMPLE 5:** How long will it take to accumulate \$10,000 if \$1,000 is invested at 7.5% APR compounded quarterly?

**SOLUTION:** In the compound interest formula $S = P\left(1 + \frac{r}{k}\right)^n$ we have $S = 10,000$, $P = 1,000$, $k = 4$, $r = 0.075$, and we need to determine $n$, the number of quarters required for the value of the investment to be \$10,000. With these substitutions we obtain

$$10,000 = 1000\left(1 + \frac{0.075}{4}\right)^n.$$

We can rewrite the above equation in the following way:

$$10 = 1.01875^n.$$

We can determine $n$ graphically by finding the $x$-coordinate of the point of intersection of the graphs of $y = 10$ and $y = 1.01875^x$ (Figure 4.3.3). We can estimate from this graph that $n(x)$ is about 125. If we use zoom-in, we will find that $n$ is 123.95. Thus, after 123.95 quarters the value of the investment will be \$10,000. Depending on the way interest is paid, a better answer may be 124 quarters or 31 years. •

In the next section we will see how to solve equations like those in Examples 3 and 5 algebraically using logarithms.

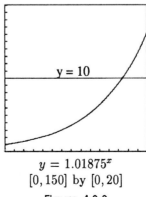

$y = 1.01875^x$
$[0, 150]$ by $[0, 20]$

Figure 4.3.3

• **EXAMPLE 6:** What APR rate compounded monthly is required for a $2,000 investment to accumulate to $4,000 in 7 years? Alternatively, we could solve $2 = (1+x)^{84}$ for $x$ obtaining $x = 2^{1/84} - 1$, and use a calculator to determine $x$.

**SOLUTION:** The compound interest formula for this problem situation is

$$4000 = 2000\left(1 + \frac{r}{12}\right)^{84}.$$

Let $x = \frac{r}{12}$. The above equation can be rewritten as

$$2 = (1+x)^{84}.$$

We can use the graphs of $y = 2$ and $y = (1+x)^{84}$ to determine $x$ (Figure 4.3.4). Notice that we must choose a viewing rectangle with very small horizontal width because $x$ represents a monthly interest rate. We can use zoom-in to determine that $x$ is 0.008285892. The APR is $12x$ or 0.0994296. Thus, investing $2,000 at an 9.943% APR compounded monthly will produce $4,000 in 7 years. Alternatively, we could solve $2 = (1+x)^{84}$ for $x$ obtaining $x = 2^{1/84} - 1$, and use a calculator to determine $x$. •

• **EXAMPLE 7:** Suppose that $1000 is invested at 8% APR compound interest. Determine the value of the investment after one year if the interest is compounded annually, quarterly, monthly, weekly, or daily.

**SOLUTION:** In the compound interest formula we set $P = 1000$ and $r = 0.08$. We use a calculator to complete Table 1. •

You might be surprised to see so little difference in the value of the investment in the table in Example 7. Even if we compound more often, the value of the investment will not increase very much. For example, compounding *hourly* gives

$$S = 1000\left(1 + \frac{0.08}{8760}\right)^{8760} = \$1083.29$$

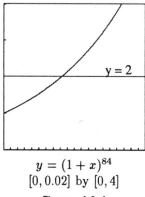

$y = (1+x)^{84}$
$[0, 0.02]$ by $[0, 4]$
Figure 4.3.4

## 4.3 Economic Applications

| Compounding Period | Value of Investment After One Year |
|---|---|
| annually | $S = 1000(1 + 0.08) = \$1080.00$ |
| quarterly | $S = 1000\left(1 + \frac{0.08}{4}\right)^4 = \$1082.43$ |
| monthly | $S = 1000\left(1 + \frac{0.08}{12}\right)^{12} = \$1083.00$ |
| weekly | $S = 1000\left(1 + \frac{0.08}{52}\right)^{52} = \$1083.22$ |
| daily | $S = 1000\left(1 + \frac{0.08}{365}\right)^{365} = \$1083.28$ |

Table 1

where $8760 = (24)(365)$ is the number of hours in a 365-day year. It appears that $S$ is approaching a value less than \$1084 as the number of compounding periods increases. From the Exercises of the previous section you can conclude that $\left(1 + \frac{0.08}{n}\right)^n \to e^{0.08}$ as $n \to \infty$. Thus, $S = 1000\left(1 + \frac{0.08}{n}\right)^n \to 1000e^{0.08} = 1083.29$ as $n \to \infty$. In general, if $P$ dollars are invested at compound interest rate $r$ for $t$ years and the number of compounded periods approaches infinity, then the value of the investment approaches $Pe^{rt}$.

- **DEFINITION**  If $P$ dollars are invested at APR $r$ (in decimal form) **compounded continuously**, then the value of the investment after $t$ years is given by $S = Pe^{rt}$, where $e$ is approximately 2.718281828.

- **EXAMPLE 8:** Suppose that \$1000 is invested at 8% compounded continuously. Find the value of the investment after 1 year and after 5 years.

  SOLUTION: The value of the investment after 1 year at 8% compounded continuously is

  $$S = 1000e^{0.08} = \$1083.29.$$

  Most calculators have $e$ built in so that this value can be computed easily with a calculator. We can also enter 2.718281828 for $e$ and compute the value of $S$. Compare this value with the values found in Table 1 of this section.

  The value of the investment after 5 years is

  $$S = 1000e^{0.08(5)} = \$1491.82.$$

## Annuities

- **DEFINITION** — An **annuity** consists of a sequence of payments or deposits made at various times in the future.

We will assume that the payments or deposits are equal and that the periods between payments or deposits are equal. We further assume that the payments or deposits are made at the *end* of each period. Such annuities are usually called **ordinary annuities** but we will simply refer to them as *annuities* in this textbook.

Suppose that $n$ payments or deposits of $R$ dollars each are made and that the interest rate per payment period is $i$. The following time line represents this problem situation:

Notice that the first payment or deposit of $R$ dollars occurs at the end of the first period. Usually the length of a period is a month.

- **DEFINITION** — The **future value** $S$ of an annuity consisting of $n$ equal payments or deposits of $R$ dollars each with interest rate $i$ per period is given by

$$S = R\frac{(1+i)^n - 1}{i}.$$

Let $r$ be the APR that gives interest rate $i$ for the period length of the above annuity, and let $k$ be the number of such periods in one year. Then, $i = \frac{r}{k}$ or $r = ki$. Let $A$ be the amount that must be invested today at APR $r$ compounded $k$ times a year so that the value of the investment will be the same as the future value of the above annuity consisting of $n$ equal payments or deposits of $R$ dollars each. Then, we have the following:

$$A(1+i)^n = R\frac{(1+i)^n - 1}{i}$$

$$A = R\frac{1 - (1+i)^{-n}}{i}$$

- **DEFINITION** — The **present value** $A$ of an annuity consisting of $n$ equal payments or deposits of $R$ dollars each with interest rate $i$ per period is given by

$$A = R\frac{1 - (1+i)^{-n}}{i}.$$

- **EXAMPLE 9:** What monthly payments are required for a 4-year $9,000 car loan at 12.5% APR compounded monthly?

SOLUTION: This is an example of a 4-year annuity with present value $9,000. (Why?) There are to be 60 equal monthly payments of $R$ dollars each with monthly interest rate of $\frac{0.125}{12}$. We can determine $R$ from the following equation:

$$9000 = R \frac{1 - \left(1 + \frac{0.125}{12}\right)^{-60}}{\frac{0.125}{12}}.$$

We can use a calculator to show that $R = 202.48$. Thus, 60 monthly payments of $202.48 are required to pay for a $9,000 car at 12.5% APR compound monthly. •

• EXAMPLE 10: Greg invests $75 per month into an IRA annuity for 20 years at 11.8% APR. Determine the value of Greg's IRA investment after 20 years.

SOLUTION: By the assumptions of this section, the first $75 deposit occurs at the end of the first month of the 20-year investment period. The monthly interest rate is $\frac{0.118}{12}$, and the number of deposits is 240. We use the future value formula to compute the value $S$ of the investment after 20 years:

$$S = R \frac{(1+i)^n - 1}{i}$$

$$= 75 \frac{\left(1 + \frac{0.118}{12}\right)^{240} - 1}{\frac{0.118}{12}}$$

$$= 72{,}225.47.$$

Thus, the value of Greg's IRA investment after 20 years is $72,225.47. •

Often annuities requiring *monthly* payments are given with APR rates which have no compounding frequency stated. In such cases, the APR rate is assumed to be compounded monthly.

# EXERCISES 4-3

1. Draw a complete graph of each function in the same viewing rectangle for $0 \leq x \leq 10$.
   (a) $y = 1000(1 + .05x)$
   (b) $y = 1000(1.05)^x$
2. Draw a complete graph of each function in the same viewing rectangle for $0 \leq x \leq 10$.
   (a) $y = 1000(1 + .08x)$
   (b) $y = 1000(1.08)^x$
3. Draw a complete graph of each function in the same viewing rectangle for $0 \leq x \leq 10$.
   (a) $y = 1000(1 + .06x)$
   (b) $y = 1000(1 + .09x)$
4. Draw a complete graph of each function in the same viewing rectangle for $0 \leq x \leq 10$.
   (a) $y = 1000(1.06)^x$
   (b) $y = 1000(1.09)^x$
5. Explain how the graphs in Exercises 1 and 2 are related to investing $1000 in different kinds of interest bearing savings accounts.

6. Explain how the graphs in Exercises 3 and 4 are related to investing $1000 in an interest bearing savings account of the same type.
7. Determine $r$ if $2500 = 1000(1 + 12r)$.
8. Determine $t$ if $2500 = 1000(1 + .07t)$.
9. Determine $P$ if $2500 = P(1 + .08(16))$

Consider the equation $S = P(1 + rt)$. Solve the equation for the variable specified in terms of the other three variables.

10. $r$            11. $t$            12. $P$

13. Determine $r$ if $2500 = 1000(1 + r)^{12}$.
14. Determine $t$ if $2500 = 1000(1.07)^n$.
15. Determine $P$ if $2500 = P(1.08)^{16}$.

Consider the equation $S = P(1 + r)^t$. Solve the equation for the variable specified in terms of the other three variables.

16. $r$            17. $P$

18. What is the value of an initial investment of $2575.00 at $8% compounded continuously for 6 years?
19. Determine when an investment doubles in value at 6% compounded continuously.
20. Determine $r$ if $2300 = 1500\, e^{10r}$. Make up a problem situation for which the equation is a model.
21. Determine when an investment of $2300 accumulates to a value of $4150 if the investment earns interest at the rate of 9% compounded quarterly.
22. Determine when an investment of $1500 accumulates to a value of $3750 if the investment earns interest at the rate of 8% compounded monthly.
23. A $1580 investment earns interest compounded annually. Determine the annual interest rate if the value of the investment is $3000 after 8 years.
24. A $22,000 investment earns interest compounded monthly. Determine the annual interest rate if the value of the investment is $36,500 after 5 years.
25. Determine how much time is required for an investment to *double* in value if interest is earned at the rate of 5.75% compounded quarterly.
26. Determine how much time is required for an investment to *triple* in value if interest is earned at the rate of 6.25% compounded monthly.
27. Sally deposits $500 in a bank that pays 6% annual interest. Assume that she makes no other deposits or withdrawals. How much has she accumulated after 5 years if the bank pays interest compounded
    (a) annually?        (b) quarterly?        (c) monthly?
    (d) daily?            (e) continuously?
28. Pete deposits $4500 in a bank that pays 7% annual interest. Assume that he makes no other deposits or withdrawals. How much has he accumulated after 8 years if the bank pays interest compounded
    (a) annually?        (b) quarterly?        (c) monthly?
    (d) daily?            (e) continuously?

## 4.3 Economic Applications

29. Draw a graph that shows the value during the first 10 years of a $1000 investment that earns 8% simple interest. In the same viewing rectangle, draw the graph of the model for a $1000 investment that earns 8% interest compounded monthly.

30. Compare the two graphs in Exercise 29 in the interval $[0, 1]$ and in the interval $[1, 10]$. Describe this comparison in terms of the problem situations given in Exercise 29.

### Effective Annual Rate

The effective annual rate $i_{eff}$ of APR $r$ compounded $k$ times per year is given by

$$i_{eff} = \left(1 + \frac{r}{k}\right)^k - 1.$$

Effective annual rates can be used to compare interest rates with compounding periods of different lengths as illustrated in Exercises 31 to 34. Effective annual rate is often called "effective yield" in business circles. It is not to be confused with annual percentage rate (APR).

31. What investment yields the greatest return: 6% compounded quarterly, or 5.75% compounded daily?

32. What investment yields the greatest return: 8.25% compounded monthly, or 8% compounded daily?

33. What investment yields the greatest return: 7% compounded quarterly, or 7.20% compounded daily?

34. What investment yields the greatest return: 8.5% compounded quarterly, or 8.40% compounded monthly?

35. Draw a complete graph of each function in the same viewing rectangle for $0 \leq x \leq 240$.
    (a) $y = 100 \dfrac{(1.005)^x - 1}{0.005}$
    (b) $y = 100 \dfrac{\left(1 + \frac{0.08}{12}\right)^x - 1}{\frac{0.08}{12}}$

36. Draw a complete graph of each function in the same viewing rectangle for $0 \leq x \leq 60$.
    (a) $y = 200 \dfrac{1 - \left(1 + \frac{0.08}{12}\right)^{-x}}{\frac{0.08}{12}}$
    (b) $y = 200 \dfrac{1 - (1.01)^{-x}}{0.01}$

37. Explain how the graphs in Exercises 35 are related to a retirement annuity requiring payments of $100 per month.

38. Explain how the graphs in Exercises 36 are related to financing a car requiring monthly payments of $200.

39. What are the monthly payments of a car loan of $8,250 for 5 years at 13% annual interest? *Note:* Remember that the 13% is an APR rate, so the actual *monthly* interest rate is $\frac{0.13}{12}$.

40. What are the monthly payments of a home mortgage of $81,500 for 30 years at 10.25% annual interest?

41. Amy contributes $50 per month into an IRA annuity for 25 years. Assuming that the IRA earns 6.25% annual interest, what is the value of Amy's IRA account after 25 years?

42. Frank contributes $50 per month into an IRA annuity for 15 years. Assuming that the IRA earns 5.5% annual interest, what is the value of Frank's IRA account after 15 years?

43. An $86,000 mortgage for 30 years at 12% APR requires monthly payments of $884.61. Suppose you decided to make monthly payments of 1050.00. When would the mortgage loan be completely paid?

44. Suppose you make payments of $884.61 for the $86,000 mortgage in Exercise 43 for 10 years and then make payments of $1050 until the loan is paid. When will the mortgage loan be completely paid under these circumstances?

45. Consider an $86,000 mortgage loan at 12% APR requiring monthly payments. Determine the required monthly payment if the loan has a term of

    (a) 30 years.　　(b) 25 years.　　(c) 20 years.　　(d) 15 years.

46. Explain why the formula

$$B(n) = R\frac{1-(1+i)^{-(360-n)}}{i}$$

gives the outstanding loan balance of a 30-year mortgage loan requiring monthly payments of $R$ dollars with APR $12i$ as a function of the number of payments ($n$) made.

47. (a) Draw a complete graph of

$$y = \$884.61\frac{1-(1.01)^{-(360-x)}}{0.01}.$$

   (b) Explain how this graph relates to the outstanding loan balance of an $86,000 mortgage loan with a 30-year term requiring monthly payments of $884.61.

48. Some mortgage loans are available that require payments every *two weeks* (26 times each year). Consider a mortgage loan of $80,000 for a 30-year term at 10% APR requiring monthly payments.

    (a) Determine the monthly payment.

    (b) Suppose 1/2 of the monthly payment was made every two weeks (26 times each year). When would the mortgage loan be completely paid? Assume the interest rate per two week interval is 1/2 of the monthly interest rate.

49. (a) Solve

$$86{,}000 = R\frac{1-(1.01)^{-x}}{0.01}$$

   for $R$.

   (b) Draw a complete graph of $R = f(x)$ for $0 \leq x \leq 360$.

   (c) Explain how the graph in (b) relates to a $86,000 mortgage loan at 12% APR.

50. Solve for $t$ using a graphing utility:

$$10{,}000 = 200\frac{1-(1.01)^{-t}}{0.01}.$$

Explain how this problem relates to a car loan of $10,000 requiring monthly payments of $200.00.

## 4.3 Economic Applications

51. Solve for $i$ using a graphing utility:
$$10{,}000 = 238 \frac{1-(1+i)^{-60}}{i}.$$
Explain how this problem relates to a car loan of $10,000 requiring monthly payments of $238.00.

*52. Explain why $A(1+i)^n = S$ where $S$ and $A$ are the future value and present value, respectively, of the same annuity.

*53. Explain why $S = R + R(1+i) + R(1+i)^2 + R(1+i)^3 + \cdots + R(1+i)^{n-1}$ is the future value of an annuity of $n$ payments of $R$ dollars each made at equal intervals of time where the interest rate $i$ is the interest rate per payment interval.

*54. Use the sum formula for a finite geometric series to show that
$$S = R \frac{(1+i)^n - 1}{i}.$$

*55. Use Exercises 53 and 54 to show that
$$A = R \frac{1-(1+i)^{-n}}{i}.$$

## The Constant Percentage Depreciation Method

The following notation and definitions will be used in connection with depreciation and appreciation of an asset.

$C$ = original value (cost) of the asset

$n$ = length of term (useful life under consideration)

$S$ = value of the asset at end of term (in case of depreciation, $S$ is called the *salvage value*)

$D = C - S$ is the *total depreciation* if $S < C$ and is the *total appreciation* if $S > C$

$B$ = book value, the value for accounting or tax purposes at any time $t$

$\dfrac{C-S}{n}$ = the annual depreciation or appreciation amount

A continuous model for *book value* at any time $t$ using the *constant percentage* method is
$$B = C(1-r)^t \quad \text{where} \quad S = C(1-r)^n \quad \text{and} \quad 0 \le t \le n.$$

Consider a machine costing $17,000 with a useful life of 6 years and that has a salvage value of $1200.

56. Assume that the machine is depreciated using the constant percentage method. Use the fact that $S = C(1-r)^n$ to determine $r$.

57. Draw a complete graph of the *book value* model $y = B(t)$.

58. What values of $t$ make sense in this problem situation?

59. What is the book value in 4 years, 3 months?
60. Draw the graph of *book value* of the machine using both the straight line method (see Exercise 50 from Section 4.1) and the constant percentage method in the same viewing rectangle for $0 \leq t \leq 6$.

## Appreciation

The constant percentage depreciation method is closely related to a very common method of determining appreciation (inflation or growth).

61. Show that if $r < 0$, then $C(1-r)^n = S$ becomes $C(1+r)^n = S$ with $r > 0$. Further show that $S > C$, which means that the asset has *appreciated* over time.
62. Let $C = \$55,000$ and $r = -0.08$. Draw a complete graph of $y = S(n) = c(1-r)^n$.

## Inflation

The formula $S = C(1+r)^n$ is frequently used to model inflation. In such a case, $r$ is the annual inflation rate, $C$ is the value today, and $S$ is the *inflated* value $n$ years from now. The *purchasing power* of one dollar $n$ years from now (assuming an annual inflation rate of $r$) is $C = 1/(1+r)^n$.

63. Assume an inflation rate of 8%. Use a graph to determine the value of a \$55,000 house in 7 years. (Assume no other factors affect the value of the house.)
64. What is the purchasing power of one dollar in 10 years if the annual inflation rate is 3%?
65. What is the purchasing power of one dollar in 10 years if the annual inflation rate is 8%?
66. What is the purchasing power of one dollar in 10 years if the annual inflation rate is 15%?

## 4.4 Logarithmic Functions

In this section we introduce logarithmic functions as inverses of exponential functions. In the past, logarithms were necessary for computation. Today calculators make computational use of logarithms obsolete. However, logarithmic functions are still very important in mathematics. For example, if $x$ can be expressed as an exponential function of $y$, then we will see that $y$ can be expressed as a logarithmic function of $x$. We give some of the important algebraic properties of logarithms and see how to use them to solve equations involving exponents. In the next section we see how to obtain their graphs with graphing utilities.

- **THEOREM 1**   Let $a > 0$ and $a \neq 1$. The inverse of the exponential function $y = a^x$ is a function.

    **PROOF**   In Section 4.2 we found that the function $y = a^x$ is increasing on $(-\infty, \infty)$ if $a > 1$ and is decreasing on $(-\infty, \infty)$ if $0 < a < 1$. Thus, the inverse of the one-to-one function $y = a^x$ is a function.

## 4.4 Logarithmic Functions

- **DEFINITION**    The inverse of the exponential function $y = a^x$ is called the **logarithmic function with base $a$** and is denoted by $y = \log_a x$. The value of $\log_a x$ is called the **logarithm of $x$ with base $a$**. The function $y = \log_{10} x$ is called the **common logarithmic function**, and the function $y = \log_e x$ is called the **natural logarithmic function**.

Whenever the base of the logarithm is omitted, it is assumed to be 10. Thus, we write $\log_{10} x = \log x$. If the base is $e$, we write $\log_e x = \ln x$.

Most calculators have a key to evaluate the functions $y = \log x$ and $y = \ln x$. In the next section we see how to use a calculator to compute the values of $y = \log_a x$ for other values of $a$. Computers generally only have $y = \ln x$ built in. Unfortunately, $y = \ln x$ is often accessed on a computer by entering $y = \log x$. You should always check to see which logarithm function is built in on the computer you use. In the next section we use graphing utilities to graph $y = \log_a x$ for $a$ different from 10 or $e$. In this section we graph logarithmic functions by using the fact that they are the inverses of exponential functions.

- **EXAMPLE 1:** Specify the domain, the range and draw a complete graph of each function.

    (a)   $y = \log_2 x$            (b)   $y = \log x$

    (c)   $y = \ln x$               (d)   $y = \log_{0.5} x$

**SOLUTION:** The domain of each of these functions is $(0, \infty)$, and the range is $(-\infty, \infty)$. This is because the corresponding exponential functions have domain $(-\infty, \infty)$ and range $(0, \infty)$. The function $y = \log_2 x$ is the inverse of $y = 2^x$, $y = \log x$ is the inverse of $y = 10^x$, $y = \ln x$ is the inverse of $y = e^x$, and $y = \log_{0.5} x$ is the inverse of $y = 0.5^x$. In Section 3.3, we showed that the graph of the inverse of a relation can be obtained by reflecting the graph of the relation through the line $y = x$. We use this property to draw the complete graphs of $y = \log_2 x$, $y = \log x$, $y = \ln x$, and $y = \log_{0.5} x$ that are given in Figures 4.4.1 through 4.4.4, respectively.

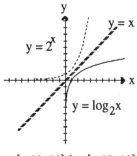

$[-10, 10]$ by $[-10, 10]$

Figure 4.4.1

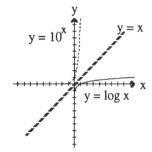

$[-10, 10]$ by $[-10, 10]$

Figure 4.4.2

We can compute the values of a logarithmic function by computing the values of the corresponding exponential function.

- **EXAMPLE 2:** Use the definition of logarithms to show that $\log_2 8 = 3$.

SOLUTION: To show that $\log_2 8 = 3$ we show that $(8, 3)$ is a solution to $y = \log_2 x$. Notice that $(3, 8)$ is a solution to $y = 2^x$, that is, $8 = 2^3$. Thus, $(8, 3)$ is a solution to $y = \log_2 x$ and $3 = \log_2 8$. Use Figure 4.4.1 to check this computation. •

We can generalize Example 2 in the following way.

SUMMARY      If $a > 0$ and $a \neq 1$, then $y = \log_a x$ if and only if $x = a^y$.

- **EXAMPLE 3:** Use exponents to compute $\log_5 \frac{1}{25}$.

SOLUTION: If we let $x = \log_5 \frac{1}{25}$, then $5^x = \frac{1}{25}$. Because $\frac{1}{25} = 5^{-2}$, we have $5^x = \frac{1}{25} = 5^{-2}$ so that $x = -2$. Therefore, $\log_5 \frac{1}{25} = -2$. •

- **EXAMPLE 4:** Use exponents to solve for $x$.

     (a) $\log_5 x = 1.5$          (b) $\log_2(x - 3) = 4$

     (c) $\log_x 3 = 2$          (d) $\log_x \frac{1}{8} = -3$

SOLUTION:

(a) We can write $\log_5 x = 1.5$ in exponential form as $x = 5^{1.5}$. Thus, accurate to hundredths, the solution is 11.18.
(b) $\log_2(x - 3) = 4$ is equivalent to $x - 3 = 2^4$. Thus, $x = 19$.
(c) $\log_x 3 = 2$ is equivalent to $x^2 = 3$. Thus, $x = \sqrt{3}$ because the base of a logarithm must be positive.
(d) $\log_x \frac{1}{8} = -3$ is equivalent to $x^{-3} = \frac{1}{8} = 2^{-3}$. Thus, $x = 2$. •

$[-10, 10]$ by $[-10, 10]$

Figure 4.4.3

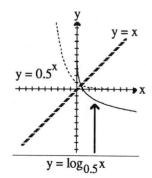

$[-10, 10]$ by $[-10, 10]$

Figure 4.4.4

## Properties of Logarithms

Some properties of logarithms are given in Theorems 2 and 3.

- **THEOREM 2**  Let $a > 0$ and $a \neq 1$. Then,
  - (a) $a^{\log_a x} = x$ for every positive real number $x$.
  - (b) $\log_a a = 1$.
  - (c) $\log_a 1 = 0$.

  **PROOF**
  - (a) We know that $y = \log_a x$ is equivalent to $x = a^y$. Thus, $x = a^y = a^{\log_a x}$.
  - (b) The logarithmic form of $a^1 = a$ is $\log_a a = 1$.
  - (c) The logarithmic form of $a^0 = 1$ is $0 = \log_a 1$.

- **THEOREM 3**  Let $a$, $r$, and $s$ be positive real numbers with $a \neq 1$. Then,
  - (a) $\log_a rs = \log_a r + \log_a s$.
  - (b) $\log_a \dfrac{r}{s} = \log_a r - \log_a s$.
  - (c) $\log_a r^c = c \log_a r$ for every real number $c$.
  - (d) $\log_a a^x = x$ for every real number $x$.

  **PROOF** Let $\log_a r = u$ and $\log_a s = v$. Then, in exponential form, we have $a^u = r$ and $a^v = s$.

  (a) $rs = a^u a^v = a^{u+v}$. The logarithmic form of $rs = a^{u+v}$ is $\log_a rs = u + v$. Substituting for $u$ and $v$ we find
  $$\log_a rs = u + v = \log_a r + \log_a s.$$

  (b) $\dfrac{r}{s} = \dfrac{a^u}{a^v} = a^{u-v}$. The exponential form of $\dfrac{r}{s} = a^{u-v}$ is $\log_a \dfrac{r}{s} = u - v$. Substituting for $u$ and $v$ we find
  $$\log_a \dfrac{r}{s} = \log_a r - \log_a s.$$

  (c) $r^c = (a^u)^c = a^{uc}$. Thus, $\log_a r^c = uc$. Substituting for $u$ we find
  $$\log_a r^c = c \log_a r.$$

  (d) Let $r = a$ and $c = x$ in (c) and we obtain $\log_a a^x = x \log_a a$. By Theorem 2(b), we know that $\log_a a = 1$. Thus, $\log_a a^x = x$.

If we let $f(x) = \log_a x$, then the inverse of $f$ is the function $f^{-1}(x) = a^x$. It follows from Theorem 2(a) that $f^{-1} \circ f(x) = x$ for every positive real number $x$. Similarly, Theorem 3(d) gives $f \circ f^{-1}(x) = x$ for every real number $x$. Notice that the domain of $f^{-1} \circ f$ is $(0, \infty)$ and the domain of $f \circ f^{-1}$ is $(-\infty, \infty)$.

Before the easy availability of calculators, the formulas in Theorem 3 were used to compute the value of complicated arithmetic expressions. This was possible because a logarithmic function is one-to-one. This means that if the number is known, its logarithm is also known and vice versa. This was accomplished by first using Theorem 3 to express the logarithm of the complicated expression in terms of the logarithms of its simpler components. Then, logarithm tables were used to finish the computation. The basic idea is illustrated in the next example.

- **EXAMPLE 5:** Express $\log_a(x^3y^{3/2})/\sqrt{z}$ in terms of $\log_a x$, $\log_a y$, and $\log_a z$.

  **SOLUTION:** We use Theorem 3 to obtain the following:

  $$\log_a \frac{x^3 y^{3/2}}{\sqrt{z}} = \log_a(x^3 y^{3/2}) - \log_a \sqrt{z}$$
  $$= \log_a x^3 + \log_a y^{3/2} - \log_a z^{1/2}$$
  $$= 3\log_a x + \frac{3}{2}\log_a y - \frac{1}{2}\log_a z$$

  Notice that the logarithm of an expression that involves products, quotients and exponents can be expressed as sums and differences of products of logarithms and numbers. On the other hand, Example 6 shows that an exponential function with an exponent that involves sums and differences can be expressed as a product and quotient of exponential expressions.

- **EXAMPLE 6:** Express $a^{2x+3y-z}$ in terms of $a^x$, $a^y$, and $a^z$.

  **SOLUTION:** We use properties of exponents to obtain the following:

  $$a^{2x+3y-z} = a^{2x} \cdot a^{3y} \cdot a^{-z}$$
  $$= \frac{(a^x)^2 \cdot (a^y)^3}{a^z}.$$

  In the previous two sections we used graphs to solve equations involving exponents. Now we can obtain an algebraic solution using logarithms.

- **EXAMPLE 7:** The population of a town is 50,000 and is increasing at the rate of 2.5% per year. Determine when the population of the town is 100,000.

  **SOLUTION:** This was (c) of Example 6 of Section 4.2. In that section we answered this question using a graph and zoom-in. Now we use logarithms. The population of the town at any time $t$ is $P(t) = 50{,}000(1.025)^t$. Thus, we need to solve the following equation for $t$:

  $$50{,}000(1.025)^t = 100{,}000.$$

  If we divide each side of the above equation by 50,000 we obtain the following equivalent equation:

  $$1.025^t = 2.$$

  At this point we use the important *well-defined property* of a function $f$ that states that if $a = b$, then $f(a) = f(b)$. In our case, this means we can take the logarithm of each side of an equation to obtain a new equation. Because $1.025^t = 2$, we can use the well-defined property of the logarithm base 10 function to continue the algebraic procedure

## 4.4 Logarithmic Functions

in the following way:

$$1.025^t = 2$$
$$\log 1.025^t = \log 2$$
$$t \log 1.025 = \log 2 \text{ (Why?)}$$
$$t = \frac{\log 2}{\log 1.025}.$$

Thus, accurate to hundredths, $t$ is 28.07. Check that this is the solution obtained in Example 6 of Section 4.2. •

### Data Analysis

Scientists often try to determine the relationship between variables empirically. They collect enough experimental data to establish that a given rule is plausible. Once a rule is conjectured, additional evidence can be used to verify or support the rule. We illustrate this idea using logarithms.

Suppose $y = ax^b$ is a power function of $x$. We further assume $a$, $x$, and $y$ are positive. The following equations are equivalent:

$$y = ax^b$$
$$\ln y = \ln ax^b$$
$$\ln y = \ln a + \ln x^b$$
$$\ln y = \ln a + b \ln x.$$

Notice that $\ln y$ is a *linear* function of $\ln x$.

---

• **THEOREM 4**   Let $a$, $x$, and $y$ be positive. Then $y = ax^m$ if and only if $\ln y$ is a linear function of $\ln x$.

**PROOF**   We have already showed that if $y = ax^m$, then $\ln y = \ln a + m \ln x$. Thus, $\ln y$ is a linear function of $\ln x$.

Conversely, suppose $\ln y$ is a linear function of $\ln x$. Then there are real numbers $m$ and $c$ so that $\ln y = m \ln x + c$. The natural logarithm function is one-to-one, and has range $(-\infty, \infty)$. Thus, there is a unique positive real number $a$ such that $\ln a = c$. We substitute this form for $c$ in the expression for $\ln y$:

$$\ln y = m \ln x + c$$
$$\ln y = m \ln x + \ln a$$
$$\ln y = \ln x^m + \ln a$$
$$\ln y = \ln ax^m$$
$$y = ax^m.$$

320 • Chapter 4                                            Logarithmic and Exponential Functions

| $x$ | $y$ |
|---|---|
| 1 | 5 |
| 3 | 31.2 |
| 5 | 73.1 |
| 8 | 160 |

Table 1

- **EXAMPLE 8:** Use the information in Table 1 and logarithms to determine a power rule model for $y$ in terms of $x$.

**SOLUTION:** According to Theorem 4 we can show that $y$ is a power function of $x$ by showing that $\ln y$ is a linear function of $\ln x$. Figure 4.4.5 shows the four points $(\ln x, \ln y)$ for $x$ and $y$ as given by the rows of Table 1. Notice that the four points appear to be on a straight line. We could draw a line through the four points and estimate the slope $m$ and $y$-intercept $b$ to write $\ln y = m \ln x + b$. In fact, even if the points did not lie on a line we could draw a "best" fitting line through the data. This line would give rise to a "best" fitting power function through the data points $(x, y)$.

Alternatively, we could use two of the four pairs of values $(\ln x, \ln y)$ to determine $m$ and $b$ in $\ln y = m \ln x + b$. First we use $(\ln 1, \ln 5)$:

$$\ln 5 = m \ln 1 + b$$
$$\ln 5 = m \cdot 0 + b$$
$$\ln 5 = b.$$

Thus, $b = \ln 5$. Next, we use $(\ln 8, \ln 160)$:

$$\ln 160 = m \ln 8 + \ln 5.$$

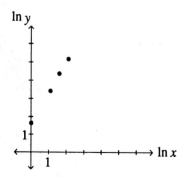

Figure 4.4.5

Solving for $m$ we find $m = 1.666666667$, which suggests that $m = \frac{5}{3}$. In fact, $m = \frac{5}{3}$ is an exact solution to $\ln 160 = m \ln 8 + \ln 5$:

$$\ln 160 = m \ln 8 + \ln 5$$
$$\ln 160 = \ln(8^m)(5)$$
$$160 = 5(8^m)$$
$$32 = 8^m$$
$$2^5 = 2^{3m}$$
$$m = \frac{5}{3}.$$

Generally, we would simply approximate the value of $m$.

We can use the values of $m$ and $b$ to write $y$ as a power function of $x$:

$$\ln y = m \ln x + b$$
$$\ln y = \frac{5}{3} \ln x + \ln 5$$
$$\ln y = \ln(5x^{5/3})$$
$$y = 5x^{5/3}.$$

The power function model determined in Example 8 can be used to predict the value of $y$ given a new value of $x$.

## EXERCISES 4-4

Compute without using a calculator.

1. $\log_4 16$
2. $\log_{10} 10$
3. $\log_{1/2} 16$
4. $\ln 1$
5. $\log_2(-3)$
6. $\log_4(-2)^4$

Solve each equation. Do not use a calculator.

7. $\log_9 x = 2$
8. $\log_2 x = 5$
9. $\log_x \frac{1}{125} = -3$
10. $\log_3(x+1) = 2$
11. $\log_x 81 = 9$
12. $\log_2 x^2 = -2$
13. $\log_3 |x| = 1$
14. $\log_6(x^2 - 2x + 1) = 0$

Sketch a complete graph. Use the inverse property. Do not use a graphing utility.

15. $f(x) = \log_3 x$
16. $g(x) = \log_5(x)$
17. $y = \log_{1/4} x$
18. $y = \log_2(-x)$

Express in logarithmic form.

19. $x + y = 2^8$
20. $(1 + r)^n = P$

Express in exponential form.

21. $\log_3 \frac{x}{y} = -2$
22. $\frac{\ln P}{\ln(1+r)} = n$

Use properties of logarithms to write the expression as a sum, difference, or product of simple logarithms (logarithms of expressions that do not involve sums, products, quotients, or exponents).

23. $\log[5000x(1+r)^{360}]$

24. $\log(\sqrt[5]{216z^3})$

25. Explain why the functions $f(x) = \log_2 x^2$ and $g(x) = 2\log_2 x$ are different even though Theorem 3(c) implies that $\log_2 x^2 = 2\log_2 x$.

Use a calculator to investigate numerically the end behavior of the function. That is, find what $f(x)$ approaches as $|x| \to \infty$.

26. $f(x) = a^{1/x}$

27. $f(x) = \dfrac{\ln x}{x}$

28. $f(x) = x^{1/x}$

29. $f(x) = \dfrac{x}{\ln x}$

In Exercises 30–34, use logarithms to solve the problem algebraically.

30. The population of a town is 475,000 and is increasing at the rate of 3.75% each year. Determine when the population of the town will be one million.

31. The population of a town is 123,000 and is decreasing at the rate of 2.375% each year. Determine when the population of the town will be 50,000.

32. The population of a small town in the year 1890 was 6250. Assume the population increased at the rate of 3.75% each year. Determine the population of the town in 1915 and in 1940.

33. The half-life of a certain radioactive substance is 14 days, and there are 6.58 grams present initially. When will there be less than 1 gram of the substance remaining?

34. The half-life of a certain radioactive substance is 65 days, and there are 3.5 grams present initially. When will there be less than 1 gram of the substance remaining?

35. Consider the data in Tables 2, 3, and 4. Which data have a power rule for an algebraic representation? Why?

36. For Tables 2, 3, and 4 with a power rule algebraic representation, determine the explicit power rule.

It is established in the Exercises in Section 4.5 that if the points $(x, \ln y)$ are on the same line with slope $m$ and $y$-intercept $c$ then an algebraic representation of $y$ in terms of $x$ is the *exponential function* $y = ab^x$ where $\ln a = c$ and $\ln b = m$.

37. Which data in Tables 2, 3, and 4 have an exponential algebraic representation?

| $x$ | $y$ |
|---|---|
| 2 | 7.48 |
| 3 | 7.14 |
| 7.5 | 1.94 |
| 7.7 | 0.84 |

Table 2

| $x$ | $y$ |
|---|---|
| 4 | 2 816 |
| 6.5 | 31,908 |
| 8.5 | 122,019 |
| 10 | 275,000 |

Table 3

| $x$ | $y$ |
|---|---|
| 8 | 23.84 |
| 12 | 58.2 |
| 15 | 113.69 |
| 40 | 30,092.66 |

Table 4

## 4.4 Logarithmic Functions

38. For the tables in Exercise 37 that have an exponential model, determine the explicit exponential rule.

Use an algebraic method to solve the equation for $t$. Check with a graphing utility.

39. $2500 = 1000(1.08)^t$
40. $6000 = 4600(1.05)^t$
41. $10{,}000 = 3500(1.07)^t$
42. Explain how Exercises 39–41 relate to interest bearing saving accounts.
43. Solve for $t$: $S = P(1+r)^t$
44. Use an algebraic method to determine how long it will take for an initial deposit of $1250 to double in value at 7% compounded monthly.
45. Use an algebraic method to determine how long it will take for an initial deposit of $1250 to triple in value at 7% compounded monthly.

Use an algebraic method to solve the equation for $t$. Check with a graphing utility.

46. $10{,}000 = 225 \dfrac{1 - (1.01)^{-t}}{0.01}$
47. $10{,}000 = 300 \dfrac{1 - (1.01)^{-t}}{0.01}$
48. Explain how Exercises 46 and 47 relate to a car loan of $10,000.
49. Use a graphing utility to solve $10{,}000 = 250[1 - (1+i)^{-60}]/i$ for $i$. *Note:* Think about solving this equation algebraically. There is no known method of expicitly solving this equation for $i$!
50. Explain how Exercise 49 relates to a car loan of $10,000 requiring monthly payments of $250.

## Planetary Motion

Table 5 gives the period of one complete revolution about the sun for the listed planets along with the length of the semi-major orbit axes.

51. Let $P$ be the orbit period and $x$ be the length of the major axis. Plot $\ln P$ against $\ln x$. That is, plot the pairs $(\ln x, \ln P)$. Verify the relationship is linear and estimate the slope $m$ and $y$-intercept.

| Planet | Period (days) | Length of Semi-Major Axes (miles) |
|---|---|---|
| Earth | 365 | 92,600,000 |
| Mercury | 88 | 36,000,000 |
| Venus | 225 | 67,100,000 |
| Mars | 687 | 141,700,000 |
| Jupiter | 4330 | 483,400,000 |
| Saturn | 10,750 | 886,100,000 |

Table 5

| Weight $x$ (lbs.) | Blood Pressure $P$ |
|---|---|
| 20 | 106 |
| 50 | 133 |
| 80 | 150 |
| 110 | 162 |
| 125 | 167 |
| 140 | 172 |

Table 6

52. Use Exercise 51 to show that $P = ax^m$ for constants $a$ and $m$. This result was discovered by Kepler in the early 1700's and is known as *Kepler's third law of planetary motion*.

53. Predict the period of Pluto's orbit if its semi-major orbit axis length is 2,200,000,000 miles.

## Blood Pressure

The *at rest blood pressure* $P$ and weight $x$ of various primates were measured as shown in Table 6.

54. Plot $\ln P$ against $\ln x$. That is, plot the pairs $(\ln x, \ln P)$. Verify the relationship is linear. Determine the slope $m$ and $y$-intercept.

55. Use Exercise 54 to find a formula expressing the blood pressure as a function of weight.

## 4.5 More on Logarithms

In this section we obtain a formula that gives the logarithms of numbers in one base in terms of the logarithms of numbers in another base. This formula, called the change of base formula, can be used to compute logarithms with any base on a calculator and to obtain a complete graph of any logarithmic function with a graphing utility.

- **THEOREM 1** (Change of base formula) Let $a$ and $b$ be positive real numbers. Then,

   (a) $\log_b x = \dfrac{\log_a x}{\log_a b}$,  (b) $\log_b a = \dfrac{1}{\log_a b}$.

## 4.5 More on Logarithms

**PROOF**

(a) The exponential form of $y = \log_b x$ is $x = b^y$. Next, we apply the logarithm base $a$ to both sides of the equation $x = b^y$ and solve for $y$:

$$x = b^y$$
$$\log_a x = \log_a b^y$$
$$\log_a x = y \log_a b$$
$$y = \frac{\log_a x}{\log_a b}.$$

Thus, $y = \log_b x = \frac{\log_a x}{\log_a b}$.

(b) Let $x = a$ in the formula in (a). Then we have

$$\log_b x = \frac{\log_a x}{\log_a b}$$
$$\log_b a = \frac{\log_a a}{\log_a b}.$$

Because $\log_a a = 1$, we have

$$\log_b a = \frac{1}{\log_a b}.$$

- **EXAMPLE 1:** Use a calculator to compute each value.

  (a) $\log_5 4$  (b) $\log_2 9$

  **SOLUTION:** Most calculators have both logarithms base 10, [log], and logarithms base $e$, [ln], as built-in functions.

  (a) We can use logarithm base 10 and enter $\log_5 4$ as $\frac{\log 4}{\log 5}$. Thus, $\log_5 4 = 0.86$ accurate to hundredths. Because $5^{\log_5 4} = 4$, the calculator answer for $\log_5 4$ can be checked by raising 5 to that power to see if you get approximately 4. That is, we can use a calculator to check that $5^{0.86}$ is 3.9912984, which is approximately 4.

  (b) We can use logarithm base $e$ and enter $\log_2 9$ as $\log_2 9 = \frac{\ln 9}{\ln 2}$ to find that $\log_2 9$ is 3.17 accurate to hundredths. •

Theorem 1(a) is the relationship we need to graph logarithmic functions with a graphing utility. Most calculators have both $\log x$ and $\ln x$ as built-in functions. Generally computers have $\ln x$ as a built-in function that is accessed by entering $\log x$. As long as a graphing utility has one logarithmic function built in, we can use this logarithmic function to graph *any* other logarithmic function.

- **EXAMPLE 2:** Express $\log_3 x$ in terms of $\log x$ and $\ln x$. Then, draw a complete graph of $y = \log_3 x$.

SOLUTION: From Theorem 1(a) we can write

$$\log_3 x = \frac{\log x}{\log 3} \quad \text{or} \quad \log_3 x = \frac{\ln x}{\ln 3}.$$

Now, depending on which logarithmic function is built in on your graphing utility, a complete graph of $y = \log_3 x$ can be obtained by entering $y = \frac{\log x}{\log 3}$ or $y = \frac{\ln x}{\ln 3}$ (Figure 4.5.1).

A rough sketch of the graph of $y = a \log_b(cx + d) + k$ can be obtained by applying standard geometric transformations to the graph of $y = \log_b x$.

- **EXAMPLE 3**: Determine the domain, the range, and draw a complete graph of $y = \log_3(-x)$.

SOLUTION: A complete graph of $y = \log_3(-x)$ can be obtained by reflecting the graph of $y = \log_3 x$ through the $y$-axis (Figure 4.5.2). We can also obtain a complete graph of $y = \log_3(-x)$ with a graphing utility by entering $y = \frac{\ln(-x)}{\ln 3}$. The domain of $y = \log_3(-x)$ is $(-\infty, 0)$, and the range is $(-\infty, \infty)$.

- **EXAMPLE 4**: Determine the domain, the range, the asymptotes, and draw a complete graph of each function.

    (a) $y = \log_5(x + 2)$      (b) $y = \log_5(3 - x)$

SOLUTION:

(a) A complete graph of $y = \log_5(x+2)$ can be obtained by shifting a complete graph of $y = \log_5 x$ horizontally to the left 2 units (Figure 4.5.3). We can also obtain a complete graph with a graphing utility by writing $y = \log_5(x+2) = \frac{\log(x+2)}{\log 5}$. The domain of $y = \log_5(x+2)$ is $(-2, \infty)$ and the range is $(-\infty, \infty)$. The line $x = -2$ is a vertical asymptote of $y = \log_5(x+2)$ because the $y$-axis ($x = 0$) is a vertical asymptote of $y = \log_5 x$.

(b) Notice that $y = \log_5(3 - x) = \log_5((-1)(x - 3))$. Thus, a complete graph of $y = \log_5(3 - x)$ can be obtained by reflecting the graph of $y = \log_5 x$ through the

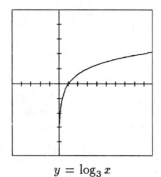

$y = \log_3 x$
$[-5, 10]$ by $-5, 5]$

Figure 4.5.1

Figure 4.5.2

Figure 4.5.3

## 4.5 More on Logarithms

$y$-axis and then shifting the resulting graph 3 units to the right (Figure 4.5.4). You can check this result with a graphing utility. The domain of $y = \log_5(3-x)$ is $(-\infty, 3)$ and the range is $(-\infty, \infty)$. The line $x = 3$ is a vertical asymptote of $y = \log_5(3-x)$. (Why?)

In the next example, we determine the effect of the coefficient $a$ in $y = \log_b(ax+c)$.

• **EXAMPLE 5:** Determine the domain, the range, the asymptotes, and draw a complete graph of $y = \log_2(3x+5)$.

SOLUTION: We rewrite $y$ as follows:

$$y = \log_2(3x+5)$$
$$= \log_2\left((3)\left(x+\frac{5}{3}\right)\right)$$
$$= \log_2(3) + \log_2\left(x+\frac{5}{3}\right).$$

Now, we can see that a graph of $y = \log_2(3x+5)$ can be obtained by shifting the graph of $y = \log_2 x$ horizontally to the left $\frac{5}{3}$ units, and then shifting the resulting graph vertically $\log_2 3$ units which is approximately 1.58 units. The computer drawn graph of $y = \log_2(3x+5)$ given in Figure 4.5.5 confirms this observation. The domain of $y = \log_2(3x+5)$ is $(-\frac{5}{3}, \infty)$, and the range is $(-\infty, \infty)$. The line $x = -\frac{5}{3}$ is a vertical asymptote.

Alternatively, we can obtain a rough sketch of a complete graph of $y = \log_2(3x+5)$ in the following way. We can see that the domain of $y = \log_2(3x+5)$ is $(-\frac{5}{3}, \infty)$ because we must have $3x+5 > 0$. Therefore, the line $x = -\frac{5}{3}$ must be a vertical asymptote. The graph of $y = \log_2(3x+5)$ must be like $y = \log_2 x$. Thus, we can find two points of the graph of $y = \log_2(3x+5)$ and use them to obtain a rough sketch of the graph. The $x$- and $y$-intercepts are a natural pair of points to find. If $\log_2(3x+5) = 0$, then $3x+5 = 2^0$

Figure 4.5.4

$f(x) = 3x^5 - 2x^4 + 6x^3 - 4x^2 - 24x + 16$
$[-5, 5]$ by $[-20, 50]$

Figure 4.5.5

so that $x = -\frac{4}{3}$. The $x$-intercept is the point $(-\frac{4}{3}, 0)$. The $y$-intercept is $\log_2 5$, which is approximately 2.3. Now, we can use the points $(-\frac{4}{3}, 0)$ and $(0, \log_2 5)$ together with the fact that the graph of $y = \log_2(3x + 5)$ is like the graph of $y = \log_2 x$ to draw a rough sketch of the graph of $y = \log_2(3x + 5)$. •

• **EXAMPLE 6:** Determine the domain, the range, and sketch a complete graph of $y = -3\log_4(2x - 5) + 1$.

**SOLUTION:** To obtain a computer drawn graph we enter $y$ in the form $y = -3\frac{\log(2x-5)}{\log 4} + 1$. A complete graph is given in Figure 4.5.6. We can see from this graph that the domain is $(\frac{5}{2}, \infty)$ and the range is $(-\infty, \infty)$. The domain is $(\frac{5}{2}, \infty)$ because we must have $2x - 5 > 0$. Thus, $x = \frac{5}{2}$ is a vertical asymptote.

To obtain a rough sketch of $y = -3\log_4(2x - 5) + 1$ we use any convenient pair of points of the graph. If $x = 3$, then $y = -3\log_4(1) + 1 = 1$ because $\log_4 1 = 0$. If $x = \frac{9}{2}$, then $y = -3\log_4(9 - 5) + 1 = -2$ because $\log_4 4 = 1$. Because we know the general shape of $y = -3\log_4(2x - 5) + 1$, we can use the points $(3, 1)$ and $(\frac{9}{2}, -2)$ to draw a rough sketch (Figure 4.5.6). •

In the next example we look at graphs of more complicated functions involving logarithmic functions.

• **EXAMPLE 7:** Determine the domain, the range, the asymptotes, and draw a complete graph of each function.

(a) $y = \log|x + 2|$  (b) $y = \log_3\sqrt{x - 1}$  (c) $y = \log\dfrac{2x + 1}{x - 3}$

**SOLUTION:**

(a) The graph of $y = \log|x + 2|$ in the $[-10, 10]$ by $[-5, 5]$ viewing rectangle is given in Figure 4.5.7. Notice that the graph is symmetric with respect to the vertical line $x = -2$ and has this line for a vertical asymptote. Because $|x + 2| \geq 0$ for all $x$, the

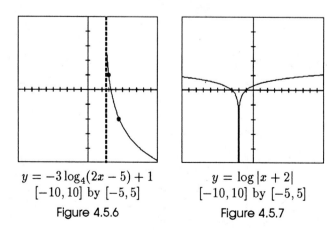

$y = -3\log_4(2x - 5) + 1$
$[-10, 10]$ by $[-5, 5]$
Figure 4.5.6

$y = \log|x + 2|$
$[-10, 10]$ by $[-5, 5]$
Figure 4.5.7

domain is the set of all real numbers different from $-2$, and the range is $(-\infty,\infty)$. The right branch of $y = \log|x+2|$ is the graph of $y = \log x$ shifted horizontally left 2 units. The left branch is the reflection of the right branch through the line $x = -2$. We can write $y$ as a piecewise-defined function:

$$y = \begin{cases} \log-(x+2) & \text{for } x < -2 \\ \log(x+2) & \text{for } x \geq -2 \end{cases}$$

(b) The graph of $y = \log_3 \sqrt{x-1}$ in Figure 4.5.8 is a complete graph. This graph looks very much like the graph of $y = \log_3(x-1)$. We rewrite $y$ to see why this is true:

$$y = \log_3 \sqrt{x-1}$$
$$= \log_3(x-1)^{1/2}$$
$$= \frac{1}{2}\log_3(x-1).$$

The domain is $(1,\infty)$, and the range is $(-\infty,\infty)$. The line $x = 1$ is a vertical asymptote.

(c) The domain of this function consists of those values of $x$ for which $\frac{2x+1}{x-3} > 0$. The solution to this inequality is $(-\infty,-\frac{1}{2}) \cup (3,\infty)$. (Why?) Thus, the domain is $(-\infty,-\frac{1}{2}) \cup (3,\infty)$. A complete graph is given in Figure 4.5.9. The range is $(-\infty,\infty)$. The lines $x = -\frac{1}{2}$ and $x = 3$ are vertical asymptotes. This graph appears to have a horizontal asymptote. If we write $\frac{2x+1}{x-3} = 2 + \frac{7}{x-3}$ we can see why there is a horizontal asymptote. Notice that $2 + \frac{7}{x-3} \to 2$ as $x \to \infty$ or as $x \to -\infty$. Therefore,

$$\log\frac{2x+1}{x-3} = \log\left(2 + \frac{7}{x-3}\right) \to \log 2$$

as $x \to \infty$ or as $x \to -\infty$. The line $y = \log 2 = 0.30$ is a horizontal asymptote. ●

We close this section with an example that examines the effect of the monthly payment on the term (length) of a loan. Suppose that an $86,000 loan at APR 12% compounded

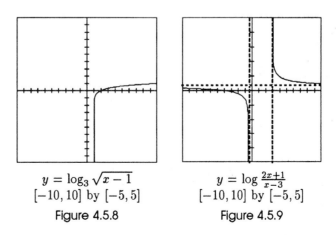

$y = \log_3 \sqrt{x-1}$
$[-10, 10]$ by $[-5, 5]$

Figure 4.5.8

$y = \log \frac{2x+1}{x-3}$
$[-10, 10]$ by $[-5, 5]$

Figure 4.5.9

monthly is required to purchase a certain home. The interest payment required the first month is $(0.01)(86{,}000) = \$860$. Thus, monthly payments must exceed $860 in order to pay off the loan. We assume the interest rate and payment are fixed.

- **EXAMPLE 8:** Consider an $86,000 loan at APR 12% compounded monthly.
    (a) Draw a graph that shows how the term of the loan depends on the amount of the monthly payment.
    (b) What monthly payments are required to pay off the loan in 25 years?
    (c) If the amount in (b) is increased by $50, how long will it take to pay off the loan?

    SOLUTION:
    (a) This loan is an example of an annuity with present value $86,000 and monthly interest rate 0.01. Let $x$ (in dollars) be the amount of the monthly payment and $t$ (in months) be the number of months required to pay off the loan. The present value formula of Section 4.3 gives the following algebraic representation:
    $$86{,}000 = x \frac{1 - 1.01^{-t}}{0.01}.$$
    Next, we solve the above equation for $t$:
    $$86{,}000 = x \frac{1 - 1.01^{-t}}{0.01}$$
    $$860 = x(1 - 1.01^{-t})$$
    $$\frac{860}{x} = 1 - 1.01^{-t}$$
    $$1.01^{-t} = 1 - \frac{860}{x}$$
    $$1.01^{-t} = \frac{x - 860}{x}$$
    $$1.01^{t} = \frac{x}{x - 860}$$
    $$\ln(1.01^{t}) = \ln\left(\frac{x}{x - 860}\right)$$
    $$t = \frac{1}{\ln 1.01} \ln\left(\frac{x}{x - 860}\right).$$

    Figure 4.5.10 gives the graph of $t$ as a function of $x$ in the $[0, 2000]$ by $[0, 360]$ viewing rectangle. Notice that $t$ is a decreasing function in the interval $(0, 2000)$. This is reasonable because as the monthly payment $x$ increases we would expect the term $t$ (length) of the loan to decrease. However, notice how sharply the graph decreases. This means that small increases in the monthly payments dramatically reduce the term of the loan.

    (b) Twenty five years is the same as 300 months. We can use zoom-in to determine that the $x$-coordinate of the point of intersection of the graph in Figure 4.5.10

$$t = \frac{1}{\ln 1.01} \ln \frac{x}{x-860}$$
$[0, 2000]$ by $[0, 360]$
Figure 4.5.10

with the horizontal line $t = 300$ is 905.77. This means that monthly payments of $905.77 are required to pay off the loan in 25 years.

(c) We can use the equation

$$t = \frac{1}{\ln 1.01} \ln \left( \frac{x}{x - 860} \right)$$

to determine that the number of months required to pay off the loan with monthly payments of $955.77 is 231.21. This is equivalent to 19.27 years. Thus, the term of the loan is decreased by nearly $5\frac{3}{4}$ years (about 23%) if monthly payments are increased by $50 (about 6%). •

## • EXERCISES 4-5

1. Use a graphing utility to draw complete graphs of $y = \log_2 x$, $y = \ln x$, $y = \log_3 x$, $y = \log_5 x$, and $y = \log x$ in the same viewing rectangle.

Solve each inequality graphically.

2. $\log_2 x < \log_3 x$
3. $\log_2 x > \log_3 x$
4. $\ln x < \log_3 x$
5. $\ln x > \log_3 x$

Without using a graphing utility, sketch a complete graph of each function. Check your answer with a graphing utility.

6. $y = 2\ln(x + 3)$
7. $y = -1 - \ln(x - 1)$
8. $y = 1 + \log_3(x - 2)$
9. $y = 2 - \log_2(2x + 6)$
10. Determine the domain and range of the function in Exercise 7.
11. Determine the domain and range of the function in Exercise 8.

Draw a complete graph and determine the domain of each function.

12. $y = \log_5 \sqrt{x - 3}$
13. $y = \log_2(x + 3)^2$
14. $y = \ln(5 - 2x)$

15. $y = \dfrac{\ln \sqrt{x}}{x-1}$   16. $y = \dfrac{1}{2} \ln \left(\dfrac{1+x}{1-x}\right)$   17. $y = \ln(x + \sqrt{x^2 + 1})$

Determine the end behavior of each function. That is, determine what $f(x)$ approaches as $|x| \to \infty$.

18. $y = x \ln x$   19. $y = x^2 \ln x$

20. $y = x^2 \ln |x|$   21. $y = \dfrac{\ln x}{x}$

Draw a complete graph of each function. Determine where the function is increasing and decreasing. Determine all local maximum and minimum values.

22. $f(x) = x \ln x$   23. $g(x) = x^2 \ln x$

24. $f(x) = x^2 \ln |x|$   25. $g(x) = \dfrac{\ln x}{x}$

Sketch the graph of each inequality. Do not use a graphing utility.

26. $y < \ln(3 - x)$   27. $y \geq 2 - \ln(x + 2)$

Solve each system of equations using a graphing utility.

28. $x + y = 6$   29. $y = x^2 - 2$
    $y = \ln(x - 2)$       $y - 1 = \ln|x + 5|$

30. Show that $y = \dfrac{1}{2}(e^x - e^{-x})$ and $y = \ln(x + \sqrt{x^2 + 1})$ are inverses of each other. Give a graphical argument.

For each pair of functions $f$ and $g$, find the solutions to $f(x) > g(x)$, $f(x) < g(x)$, and $f(x) = g(x)$ graphically.

31. $f(x) = 3^x$; $g(x) = x^3$   32. $f(x) = e^x$; $g(x) = x^e$

33. $f(x) = 10^x$; $g(x) = x^x$   34. $f(x) = e^x$; $g(x) = (\ln x)^x$

35. $f(x) = \ln x$; $g(x) = \ln(\ln x)$   36. $f(x) = \ln x$; $g(x) = x^{1/3}$

37. How would you graph $y = \log_x 4$ using a graphing utility?

Use the results of Exercise 37 to graph the equation.

38. $y = \log_x 4$   39. $y = \log_x(x - 1)$

40. Determine the domain and range of the function in Exercise 38.

41. Determine the domain and range of the function in Exercise 39.

## Earthquake Intensity

Earthquake intensity is reported using the Richter scale,

$$R = \log\left(\dfrac{a}{T}\right) + B,$$

where $R$ is the magnitude (of intensity), $a$ is the amplitude in microns of the vertical ground motion at the receiving station, $T$ is the period of the seismic wave (in seconds), and $B$ is a factor that accounts for the weakening of the seismic wave with increasing distance from the epicenter of the earthquake.

42. What is the magnitude on the Richter scale of an earthquake if $a = 250$, $T = 2$, and $B = 4.250$?

43. Explain why an earthquake of magnitude 6 on the Richter scale is 100 times more intense than one with the same epicenter of magnitude 4 on the Richter scale. Assume $T$ and $B$ are constant. The change in the vertical ground motion (amplitude) is directly related to the *intensity* of the earthquake.

44. Draw a complete graph of the Richter scale model (magnitude $R$ as a function of amplitude $a$). Assume $T = 2$ and $B = 4.250$. What are the values of $a$ that make sense in this problem situation?

## Light Absorption

The *Beer-Lambert law of light absorption* is given by

$$\log\left(\frac{I}{I_0}\right) = Kx,$$

where $I_0$ and $I$ denote the intensity of light of a particular type before and after passing through a body of material, respectively, and $x$ denotes the length of the path followed by the beam of light passing through the material. $K$ is a constant.

45. Let $I_0 = 12$ lumens and assume $K = -0.00235$. Write $I$ as a function of $x$.

46. Draw a complete graph of $I = f(x)$.

Suppose $I_0$ represents the intensity of light measured at the surface of a lake. The Beer-Lambert law can be used to determine the intensity $I$ of the light measured at a depth of $x$ feet from the surface.

47. Assume the constants given in Exercise 45 for Lake Erie. What is the intensity of the light at a depth of 30 feet?

48. Suppose for Lake Superior, $K = -0.0125$ and the surface intensity of the light is 12 lumens. What is the intensity at a depth of 30 feet?

## RL Circuit

It is known from the theory of electricity that in an $RL$ circuit the current $I$ in the circuit is given by

$$I = \frac{V}{R}(1 - e^{-(Rt)/L})$$

as a function of time $t$. Here $V$, $R$, and $L$ are constants for voltage, resistance, and self-inductance, respectively.

49. Assume $V = 10$ volts, $R = 3$ ohms, and $L = 0.03$ Henries. Draw a complete graph of $I = f(t)$.

50. Show that $I \to V/R = 10/3$ as $t \to \infty$. Give a graphical argument.

## Rule of 72

51. Graph $y = r$ and $y = \ln(1 + r)$ for $0 \le r \le 1$ in the same viewing rectangle. When is $\ln(1 + r)$ approximately equal to $r$?

52. Show how using 0.72 as an approximate value for $\ln 2$ you can derive an easy to remember "rule" that says money invested at annual rate $r$ compounded annually will double in value after

| $x$ | $y$ |
|---|---|
| 2.3 | 43.8 |
| 4 | 283.5 |
| 5.5 | 1473.1 |
| 7 | 7654.5 |
| 9 | 68,890.5 |

Table 1

$72/100r$ years. For example, at 6% interest compounded annually, you would expect money to double in value after $72/[100(0.06)] = 72/6 = 12$ years. Is this a good "rule of thumb?"

## Semilogarithmic Graphing

Exponential relations appear to be linear when plotted on *semilogarithmic* graph paper. Before calculators and computers were readily available, techniques for data analysis frequently involved semilogarithmic graph paper. The same analysis can be carried out using *ordinary* graph paper and a scientific or graphing calculator.

53. Plot the pairs $(x, \log y)$ on regular graph paper using the values $x$ and of $y$ from Table 1. Verify that the points are collinear (on the same line).

54. Compute the slope $m$ and the $y$-intercept $y_0$ of the line in Exercise 53. Verify that the data can be represented by the *exponential* relationship $y = ab^x$ where $m = \log b$ and $\log a = y_0$.

55. Prove that if $f$ is the exponential function $f(x) = ab^x$ then $\log f(x)$ varies *linearly* with $x$. That is, show that $\log f(x) = mx + y_0$ for some constants $m$ and $y_0$.

56. Determine the actual exponential function which provides a model of the data in Exercise 53. *Hint:* Use Exercise 55.

57. Let $f(x) = ab^x$. Show that

$$\log b = \frac{\log f(x_2) - \log f(x_1)}{x_2 - x_1}.$$

for any pair of real numbers $x_1$, $x_2$ with $x_1 \neq x_2$. Assume $a > 0$ and $b > 1$.

58. Consider a $75,000 mortgage loan at 10.50% APR compounded monthly.
    (a) Draw a graph that shows how the term of the loan depends on the amount of the monthly payment.
    (b) What monthly payments are required to pay off the loan in 25 years?
    (c) If the amount in (b) is increased by $50, how long will it take to pay off the loan?

59. Consider an $110,000 mortgage loan at 9% APR compounded monthly.
    (a) Draw a graph that shows how the term of the loan depends on the amount of the monthly payment.
    (b) What monthly payments are required to pay off the loan in 25 years?
    (c) If the amount in (b) is increased by $50, how long will it take to pay off the loan?

## 4.6 Equations, Inequalities, and Extreme-Value Problems

In this section we solve equations and inequalities that involve exponential or logarithmic functions both algebraically and graphically. We find local maximum and minimum values of functions that involve exponential and logarithmic functions. You will find that solving these equations and inequalities algebraically involves moving from the exponential form to the logarithmic form and vice versa, and applying the properties of logarithms.

- **EXAMPLE 1:** Solve $\log x = -2.5$.

  **SOLUTION:** One way to solve this equation is to write it in the following equivalent exponential form:
  $$x = 10^{-2.5}.$$
  Now, the exponentiation key on a calculator can be used to show that $x$ is approximately 0.0031623. •

When we solve equations and inequalities involving logarithms algebraically we often apply the properties of logarithms to algebraic expressions. However, we must be careful because using these properties need not lead to equivalent equations. For example, suppose we replace $\log \frac{x}{x-2}$ by $\log x - \log(x-2)$ or vice versa. The graphs of $y = \log \frac{x}{x-2}$ and $y = \log x - \log(x-2)$ in Figure 4.6.1 show that $\log \frac{x}{x-2} = -1$ has a solution but $\log x - \log(x-2) = -1$ does not. In fact, the domain $(2, \infty)$ of $y = \log x - \log(x-2)$ is a subset of the domain $(-\infty, 0) \cup (2, \infty)$ of $y = \log \frac{x}{x-2}$, but *not* vice versa. Thus, the domains are different.

We may introduce extraneous solutions, but we do not lose solutions if we replace $\log_a f(x) - \log_a g(x)$ by $\log_a(f(x)/g(x))$ because the domain of $y = \log_a f(x) - \log_a g(x)$ is contained in the domain of $y = \log_a(f(x)/g(x))$. Replacement the other way may lose

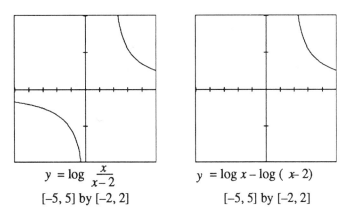

$y = \log \frac{x}{x-2}$       $y = \log x - \log(x-2)$
[−5, 5] by [−2, 2]      [−5, 5] by [−2, 2]

Figure 4.6.1

solutions because the domain of $y = \log_a(f(x)/g(x))$ generally contains numbers *not* in the domain of $y = \log_a f(x) - \log_a g(x)$. Similarly, we replace $\log_a f(x) + \log_a g(x)$ by $\log_a f(x)g(x)$ and not vice versa when solving algebraically for precisely the same reason. If $r > 0$, then both $\log_a r^s$ and $s \log_a r$ are defined and equal for all real numbers $s$. With these relationships in mind, we are ready to solve algebraically. To be safe, always check to see if a solution is extraneous.

- **EXAMPLE 2:** Solve $\frac{1}{2}\log_5(x+6) - \log_5 x = 0$.

**SOLUTION:** Our strategy is to rewrite the equation in the form $\log_5 f(x) = c$ and use the exponential form $f(x) = 5^c$ to solve for $x$. Remember, we replace $\log_a g(x) - \log_a h(x)$ by $\log_a(g(x)/h(x))$ and $\log_a g(x) + \log_a h(x)$ by $\log_a g(x)h(x)$, but not the other way, to avoid losing solutions. We proceed as follows:

$$\frac{1}{2}\log_5(x+6) - \log_5 x = 0$$
$$\log_5(x+6)^{1/2} - \log_5 x = 0$$
$$\log_5 \sqrt{x+6} - \log_5 x = 0$$
$$\log_5 \frac{\sqrt{x+6}}{x} = 0$$
$$\frac{\sqrt{x+6}}{x} = 5^0$$
$$\frac{\sqrt{x+6}}{x} = 1.$$

Next, we solve $\sqrt{x+6}/x = 1$ in the usual way:

$$\frac{\sqrt{x+6}}{x} = 1$$
$$\sqrt{x+6} = x$$
$$x + 6 = x^2$$
$$0 = x^2 - x - 6$$
$$0 = (x-3)(x+2).$$

We must check that $-2$ and $3$ are solutions to the original equation. (Why?) A calculator can be used to show that $\frac{1}{2}\log_5(3+6) - \log_5 3 = 0$. (Remember that $\log_5 a = \log a / \log 5$.) Thus, 3 is a solution to the original equation. Notice $-2$ is an extraneous solution to the original equation because $\log_5 x$ is not defined at $x = -2$. Therefore, only 3 is a solution to the original equation. The graph of $y = \frac{1}{2}\log_5(x+6) - \log_5 x$ in Figure 4.6.2 suggests that the domain of this function is $(0, \infty)$ and that 3 is a solution to the original equation. Because this graph is complete there can only be one solution to the original equation. (Why?) •

- **EXAMPLE 3:** Solve $\log_2(2x+1) - \log_2(x-3) = 2\log_2 3$.

## 4.6 Equations, Inequalities, and Extreme-Value Problems

SOLUTION: The following equations need not be equivalent. However, all solutions to the original equation are also solutions to each new equation. (Why?)

$$\log_2(2x+1) - \log_2(x-3) = 2\log_2 3$$
$$\log_2 \frac{2x+1}{x-3} = \log_2 9$$
$$\frac{2x+1}{x-3} = 9$$
$$9x - 27 = 2x + 1$$
$$7x = 28$$
$$x = 4.$$

To check that 4 is a solution to the original equation we need to show that $\log_2 9 - \log_2 1 = 2\log_2 3$. Now, $\log_2 9 = 2\log_2 3$ and $\log_2 1 = 0$, so this equation is true, and 4 is the solution to the original equation.

The graph of $y = \log_2(2x+1) - \log_2(x-3) - 2\log_2 3$ in Figure 4.6.3 is complete and suggests that the domain of this function is $(3, \infty)$ and that 4 is the only solution to the original equation. (Why?) •

The next two examples involve equations containing exponential expressions.

- **EXAMPLE 4:** Solve $1.08^x = 3$.

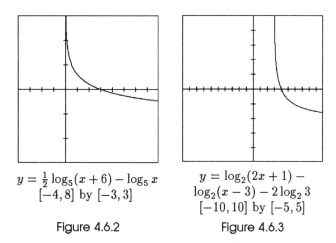

$y = \frac{1}{2}\log_5(x+6) - \log_5 x$
$[-4, 8]$ by $[-3, 3]$

$y = \log_2(2x+1) - \log_2(x-3) - 2\log_2 3$
$[-10, 10]$ by $[-5, 5]$

Figure 4.6.2         Figure 4.6.3

SOLUTION: We take the logarithm base 10 of each side of the equation:

$$1.08^x = 3$$
$$\log 1.08^x = \log 3$$
$$x \log 1.08 = \log 3$$
$$x = \frac{\log 3}{\log 1.08}.$$

We can check that this value of $x$ gives a solution to the original equation. This solution is approximately 14.274915. •

• **EXAMPLE 5:** Solve $\frac{3^x - 3^{-x}}{2} = 5$.

SOLUTION: It can be shown that the graphs of $y = \frac{3^x - 3^{-x}}{2}$ and $y = 5$ in Figure 4.6.4 are complete graphs. This figure suggests that there is only one real solution to this equation, and its value is approximately 2. We could use zoom-in to determine this solution more accurately.

Alternatively, we can find this solution algebraically. The following equations are equivalent:

$$\frac{3^x - 3^{-x}}{2} = 5$$
$$3^x - 3^{-x} = 10$$
$$(3^x)^2 - 1 = (10)3^x \quad \text{(Multiply each side by } 3^x.\text{)}$$
$$(3^x)^2 - (10)3^x - 1 = 0.$$

This equation is quadratic in $3^x$. The quadratic formula gives

$$3^x = \frac{10 \pm \sqrt{104}}{2} = 5 \pm \sqrt{26}.$$

$y = \frac{3^x - 3^{-x}}{2}$
$[-10, 10]$ by $[-10, 10]$
Figure 4.6.4

## 4.6 Equations, Inequalities, and Extreme-Value Problems

The equation $3^x = 5 - \sqrt{26}$ has no real solution because $5 - \sqrt{26}$ is negative. We can use logarithms to solve the equation $3^x = 5 + \sqrt{26}$ as we did in Example 4. You will find that $x$ is approximately 2.1048721. •

The algebraic technique needed to solve an inequality involving logarithms is similar to the technique used to solve equations. The resulting inequalities are not necessarily equivalent, so solutions must be checked in the original inequality.

• **EXAMPLE 6:** Solve $3 \log_4 x - 1 < 0$.

**SOLUTION:** Notice that we must have $x > 0$. (Why?) We can rewrite the inequality as $\log_4 x < \frac{1}{3}$. Figure 4.6.5 gives the graph of $y = \log_4 x$ and $y = \frac{1}{3}$. We can use zoom-in to show that the solution is $0 < x < 1.59$. •

In the next two examples we find local maximum and minimum values of functions involving exponential and logarithmic functions.

• **EXAMPLE 7:** Find the domain, the range, the local maximum and minimum values, and draw a complete graph of $y = \ln x(3 - x)$.

**SOLUTION:** The domain of this function consists of those values of $x$ for which $x(3-x) > 0$. It can be shown that the solution to $x(3-x) > 0$ is $(0, 3)$. The graph of $y = \ln x(3-x)$ in Figure 4.6.6 is a complete graph.

We can use zoom-in to determine that the local maximum value of this function is 0.81 with error at most 0.01. There are *no* local minimum values, and the range is $(-\infty, 0.81]$. •

• **EXAMPLE 8:** Find the domain, the range, the local maximum and minimum values, and draw a complete graph of $y = 10(e^{(x-100)/50} + e^{(100-x)/50})$.

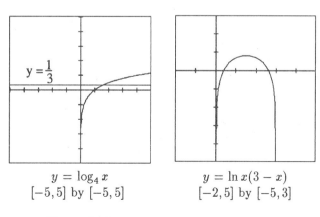

$y = \log_4 x$
$[-5, 5]$ by $[-5, 5]$

Figure 4.6.5

$y = \ln x(3 - x)$
$[-2, 5]$ by $[-5, 3]$

Figure 4.6.6

SOLUTION: The domain of this function is the set of all real numbers. The graph in Figure 4.6.7 is a complete graph.

We can use zoom-in to determine that the local minimum value is 20 with error at most 0.01. There is *no* local maximum value and the range is $[20, \infty)$. It can be shown that the coordinates of the lowest point are exactly $(100, 20)$. •

In the next example we introduce another application that has an exponential function for a model. Scientists have established that standard atmospheric pressure of 14.7 pounds per square inch is reduced by half for each 3.6 miles of vertical ascent. This rule for atmospheric pressure holds for altitudes up to 50 miles.

• **EXAMPLE 9**: Express the atmospheric pressure $P$ as a function of altitude and determine when the atmospheric pressure will be 4 pounds per square inch. What is the pressure at an altitude of 50 miles?

SOLUTION: The information above gives $P = 14.7(0.5)^{h/3.6}$ where $h$ is the altitude in miles. To determine when atmospheric pressure is 4 pounds per square inch, we must solve the following equation:

$$14.7(0.5)^{h/3.6} = 4.$$

We can solve this equation as follows:

$$14.7(0.5)^{h/3.6} = 4$$

$$(0.5)^{h/3.6} = \frac{4}{14.7}$$

$$\frac{h}{3.6} \log(0.5) = \log\left(\frac{4}{14.7}\right)$$

$$h = \frac{3.6 \log(4/14.7)}{\log(0.5)}.$$

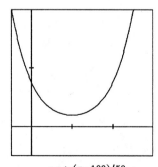

$y = 10(e^{(x-100)/50} + e^{(100-x)/50})$
$[-50, 300]$ by $[-50, 200]$

**Figure 4.6.7**

### 4.6 Equations, Inequalities, and Extreme-Value Problems

Using a calculator we find that $h$ is approximately 6.7598793. Thus, the atmospheric pressure will be 4 pounds per square inch at an altitude of approximately 6.7598793 miles.

To find the pressure at an altitude of 50 miles we need to find $P(50)$. Using a calculator we find that $P(50)$ is approximately 0.000969. Thus, the atmospheric pressure at an altitude of 50 miles is about 0.000969 pounds per square inch.

## EXERCISES 4-6

Solve each equation algebraically. Check your answer with a graphing utility.

1. $\log x = 4$
2. $\ln x = -1$
3. $\log_4(x - 5) = -1$
4. $\log_x(1 - x) = 1$
5. $\log_x(12 - x) = 2$
6. $3 \log_2 x = 4$
7. $\ln x + \ln 2 = 3$
8. $\frac{1}{2} \log_3(x + 1) = 2$
9. $\log_2 x + \log_2(x + 3) = 2$
10. $\log_4(x + 1) - \log_4 x = 1$

Compare the domain and range of each pair of functions.

11. $y = 2 \log x$; $y = \log x^2$
12. $y = \log x + \log(x + 1)$; $y = \log(x(x + 1))$
13. $y = \log_5 \sqrt{x + 6} - \log_5 x$; $y = \log_5 \frac{\sqrt{x + 6}}{x}$
14. $y = \log_2 x - \log_2(x + 1)$; $y = \log_2 \left( \frac{x}{x + 1} \right)$

Find the *exact* solutions to each equation.

15. $(1.06)^x = 4.1$
16. $(0.98)^x = 1.6$
17. $\frac{2^x - 2^{-x}}{3} = 4$
18. $\frac{2^x + 2^{-x}}{2} = 3$

19–22. Use a calculator to find a decimal approximation to each solution in Exercises 15–18. Then use a graphing utility to check the solution.

Draw a complete graph of each function. Determine the domain and the range, and find where the function is increasing or decreasing. Find all local maximum and minimum values.

23. $y = xe^x$
24. $y = x^2 e^{-x}$
25. $y = \frac{3^x - 3^{-x}}{3}$
26. $y = \frac{3^x + 3^{-x}}{3}$
27. $y = \ln(x^2 + 2x)$
28. $y = \ln \left( \frac{x}{2 + x} \right)$

Solve each inequality.

29. $2 \log_2 x - 4 \log_2 3 > 0$
30. $2 \log_3(x + 1) - 2 \log_3 6 < 0$

Solve each equation or inequality.

31. $e^x + x = 5$
32. $\ln |x| - e^{2x} \geq 3$
33. $e^x < 5 + \ln x$
34. $e^{2x} - 8x + 1 = 0$

Solve each equation.

35. $\frac{\ln x}{x} = \frac{1}{10} \ln 2$
36. $2^x = x^{10}$

37. Explain why Exercises 35 and 36 have the same *positive* solutions.
38. Show that $f(x) = (e^x + e^{-x})/2$ is an even function and that $g(x) = (e^x - e^{-x})/2$ is an odd function.

## Probability Density Function

The function $f(x) = (25/(r\sqrt{2\pi}))e^{-(x-u)^2/(2r^2)}$ is an adjusted normal probability density function with mean $u$ and variance $r^2$. Let $u = 11.6$ and $r^2 = 2.35$.

39. Draw a complete graph of $f$.
40. Determine the domain of $f$.
41. Show that the graph of $f$ is symmetric with respect to the line $x = u$.
42. Determine all $x$ where $f(x) = 2$.

## Approximations for $e^x$

Let

$$f_1(x) = x + 1$$

$$f_2(x) = \frac{x^2}{2!} + x + 1$$

$$f_3(x) = \frac{x^3}{3!} + \frac{x^2}{2!} + x + 1$$

$$f_4(x) = \frac{x^4}{4!} + \frac{x^3}{3!} + \frac{x^2}{2!} + x + 1$$

$$\vdots$$

$$f_n(x) = \frac{x^n}{n!} + \frac{x^{n-1}}{(n-1)!} + \ldots + x + 1$$

$$\vdots$$

Note: $n! = n(n-1)(n-2)\cdots 3 \cdot 2 \cdot 1$.

43. Show that $f_n(x) = \dfrac{x^n}{n!} + f_{n-1}(x)$.
44. Draw the graphs of $y = f_1(x)$, $y = f_2(x)$, $y = f_3(x)$, $y = f_4(x)$, $y = f_5(x)$ and $y = e^x$ in the $[-5, 5]$ by $[-25, 25]$ viewing rectangle.
45. For what values of $x$ does $f_3(x)$ approximate $e^x$ with an error less than 0.1?
46. For what values of $x$ does $f_5(x)$ approximate $e^x$ with an error less than 0.1?

## Potential Energy

In a certain molecular structure, the total potential energy $E$ between two ions is given by

$$E = \frac{-5.6}{r} + 10e^{-r/3},$$

where $r$ is the distance separating the nuclei.

47. Draw a complete graph of $E$ as a function of $r$. Determine the domain of $E$.
48. What values of $r$ make sense in the problem situation?
49. Find any relative maximum and minimum values of the total potential energy on $(0, \infty)$ and the number $r$ for which they occur.

## The Sinking Fund Depreciation Method

The following notation and definitions will be used in connection with depreciation of an asset:

$C =$ original value (cost) of the asset.

$n =$ length of term (useful life under consideration).

$S =$ value of the asset at the end of term; $S$ is called the *salvage value*.

$D = C - S$ is the *total depreciation*.

$B =$ *book value*, the value for accounting and/or tax purposes at any time $t$.

A continuous model for *book value* at any time $t$ using the *sinking fund* method is

$$B = C - \left(\frac{C - S}{((1+i)^n - 1)/i}\right)\left(\frac{(1+i)^t - 1}{i}\right)$$

where $0 \leq t \leq n$ and $i$ is an annual interest rate. In this case, the annual depreciation charge is assumed to earn interest at rate $i$ per year, and the annual depreciation charges accumulate to $D = C - S$, the total depreciation.

50. A machine costs $17,000, has a useful life of 6 years, and has salvage value of $1200. Draw a complete graph of the "book value" model $y = B(t)$. What values of $t$ make sense in the problem situation? Assume $i = 5\%$.
51. Consider the machine in Exercise 50. What is the book value of the machine in 4 years, 3 months?
52. Consider the machine in Exercise 50. Draw three graphs of the book value of the machine using
    (a) the straight line method (see Exercise 50, Section 6.1).
    (b) the constant percentage method (see Exercise 56, Section 6.4).
    (c) the sinking fund method.
    
    For each graph use the same viewing rectangle for $0 \leq t \leq 6$.
53. Use the graphs in Exercise 52 to explain which method allows the greatest depreciation in the first 2 years of the useful life of the machine.

## Mortgage Interest Rates

There is no known method to determine an exact, explicit solution for the following two problems. Use a graphing zoom-in method.

54. Amy and Steve can afford to make monthly mortgage payments of $800. They want to purchase a home requiring a 30-year, $92,000 mortgage. What APR interest rate will be required?

55. Jill and Benny can afford to make monthly mortgage payments of $1200. They want to purchase a home requiring a 30-year $130,000 mortgage. What APR interest rate will be required?

## • KEY TERMS

Listed below are the key terms, vocabulary, and concepts in this chapter. You should be familiar with their definitions and meanings as well as be able to illustrate each of them.

Annuity
$C_{(a,\infty)}$, $C_{(-\infty,a)}$, and $C_{(a,b)}$
Common logarithmic function
Compound interest
Compounded continuously
Data analysis
$e$
Exponential function
Exponential growth
Future value
Half-life
Interest

Logarithmic function
Natural logarithmic function
Piecewise-defined function
Present value
Principal
Radioactive decay
Rate of interest
Signum function
Sum, difference, product, and quotient of functions
Well-defined property of functions

## • REVIEW EXERCISES

This section contains representative problems from each of the previous six sections. You should use these problems to test your understanding of the material covered in this chapter.

1. Determine the domain and a rule for each of $f+g$, $f-g$, $fg$, and $f/g$ where $f(x) = 1/(x-1)$ and $g(x) = (x+2)^2$.

Determine the domain and the range of each function, and sketch a complete graph without using a graphing utility.

2. $f(x) = \begin{cases} -2x & \text{for } x < 1 \\ \sqrt{x+5} & \text{for } x \geq 1 \end{cases}$

3. $f(x) = \begin{cases} 3-x & \text{for } x < -3 \\ (x-2)^2 & \text{for } -3 \leq x < 2 \\ x^3 & \text{for } x \geq 2 \end{cases}$

4. Where are the functions in Exercises 2 and 3 discontinuous?

Determine the domain, the range, and draw a complete graph of each function.

5. $f(x) = -\text{sgn}(x-3)$

6. $f(x) = 2\dfrac{|x-3|}{x-3}$

7. Determine the domain and the range of the function

$$g(x) = \begin{cases} 2x + x^5 & \text{for } x < 2 \\ x^3 - x - 1 & \text{for } x \geq 2 \end{cases},$$

and find a single rule that when entered in a graphing utility gives a complete graph.

Without using a graphing utility, draw a complete graph of each function. Determine the domain and range. Check your answer with a graphing utility.

8. $f(x) = 2 + 3^x$
9. $f(x) = 3 - 2^{x+2}$

Without using a graphing utility, sketch a complete graph of each function. Check your answer with a graphing utility.

10. $y = 2 + \ln(x - 3)$
11. $y = \ln(x - 2) - 1$
12. $y = 3 \ln x$
13. $y = 2 \ln(x + 4)$
14. Use rules of exponents to solve $2^{x+1} = 4^3$.
15. Use a graphing utility to draw a complete graph of $f(x) = (8x)2^{-x}$. Determine the domain and range.

Determine where each function is increasing and decreasing. Determine all local maximum and minimum values.

16. $f(x) = e^{-2x}$
17. $g(x) = (8x)2^{-x}$

Compute without using a calculator.

18. $\log_5 125$
19. $\log_3(-9)$

Solve each equation. Do not use a calculator.

20. $\log_x \dfrac{1}{9} = -2$
21. $\log_4 x = 3$

22. Use properties of logarithms to write the expression $\log[(1 + r)^{12}/r]$ as a sum, difference, or product of simple logarithms.
23. Use properties of logarithms to solve $z = (23 + x)^M$ for $M$.
24. Use a calculator to investigate numerically the end behavior of the function $f(x) = x^2/\ln x$. That is, find what $f(x)$ approaches as $x \to \infty$.
25. Use a calculator to investigate the behavior of $g(x) = x/\ln(x - 2)$ as $x \to 2^+$.
26. Use a calculator to investigate the behavior of $f(x) = x/\ln x^2$ as $x \to 1^-$.
27. Solve $\ln x > \log_2 x$ graphically.

Draw a complete graph. Determine the domain and range.

28. $y = \dfrac{\ln \sqrt{x + 2}}{x - 2}$
29. $y = \log_3(x - 5)^2$

30. Determine the end behavior of $y = x^3 \ln x$. That is, determine what $f(x)$ approaches as $|x| \to \infty$.
31. Draw a complete graph of $y = x^3 \ln x$. Determine where the function is increasing and decreasing. Determine all local maximum and minimum values.

Solve each equation algebraically. Check your answer with a graphing utility.

32. $5 \log_3 x = 2$
33. $\log_5 x + \log_5(x - 4) = 1$
34. Compare the domain and range of the functions $y = \log_3 x - \log_3(x+2)$ and $y = \log_3[x/(x+2)]$.
35. Find the *exact* solution to the equation $(1.5)^x = 0.90$.

36. Use a calculator to find a decimal approximation to the solution in Exercise 35. Then use a graphing utility to check the solution.

Draw a complete graph of each function. Determine the domain and the range, and find where the function is increasing or decreasing. Find all local maximum and minimum values.

37. $y = \ln\left(\dfrac{x}{x-3}\right)$

38. $y = \dfrac{2^x + 2^{-x}}{3}$

Solve algebraically. Check with a graphing utility.

39. $3\log_5(x-1) - 2\log_5 4 > 0$

40. $\log_2(x-1) + \log_2(x+2) = 2$

41. $\log_3(x+5) + 2\log_3 x > 1$

Use a graphing utility to solve each equation or inequality.

42. $e^x + \ln x = 5$

43. $e^{2x} + 3e^x \leq 10$

A rocket launched from ground level is $d$ miles above the earth $t$ minutes after launch (Figure R.1). While the rocket engine is on, $d$ is given by

$$d = 6t^2 + 185t.$$

When the engines are off, $d$ is given by

$$d = 1728 + 25t - t^2.$$

44. Assume the rocket engine is on for only the first 8 minutes. Determine a piecewise-defined function that is a model for this problem situation.

45. Linda deposits $500 in a bank that pays 7% annual interest. Assume that she makes no other deposits or withdrawals. How much has she accumulated after 5 years if the bank pays interest compounded
   (a) annually?
   (b) quarterly?
   (c) monthly?
   (d) daily?
   (e) continuously?

46. Determine when an investment of $1500 accumulates to a value of $2280 if the investment earns interest at the rate of 7% compounded monthly.

47. Determine how much time is required for an investment to *double* in value if interest is earned at the rate of 6.25% compounded quarterly.

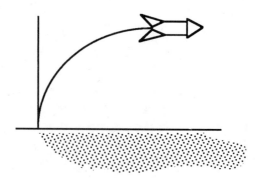

Figure R.1

48. Draw a graph that shows the value at any time $t$ of a $500 investment that earns 9% simple interest. In the same viewing rectangle, draw the graph that shows the value at any time $t$ of a $500 investment that earns 9% interest, compounded monthly.

49. The population of a town is 625,000 and is increasing at the rate of 4.05% each year.
    (a) Write the population $P$ as a function of time.
    (b) Draw a complete graph of the function $P$ in (a). What portion of the graph represents the problem situation?
    (c) Determine when the population of the town is one million.

50. The half-life of a certain radioactive substance is 21 days, and there are 4.62 grams present initially.
    (a) Write the amount $A$ of substance remaining as a function of time.
    (b) Draw a complete graph of the function in (a).
    (c) When will there be less than one gram of the substance remaining?

Earthquake intensity is reported using the Richter scale,

$$R = \log\left(\frac{a}{T}\right) + B,$$

where $R$ is the magnitude, $a$ is the amplitude in microns of the vertical ground motion at the receiving station, $T$ is the period of the seismic wave (in seconds), and $B$ is a factor that accounts for the weakening of the seismic wave with increasing distance from the epicenter of the earthquake.

51. What is the magnitude on the Richter scale of an earthquake if $a = 275$, $T = 2.5$, and $B = 4.250$?

52. Draw a complete graph of the Richter scale model (magnitude $R$ as a function of amplitude $a$). Assume $T = 2.5$ and $B = 4.250$. What are the values of $a$ that make sense in this problem situation?

# 5 Trigonometry of Complex Numbers, Vectors, Polar Equations, and Parametric Equations

## • Introduction

In this chapter we represent complex numbers geometrically and write them in trigonometric form. This form is used to do the arithmetic of complex numbers. We state De Moivre's theorem for complex numbers and use it to determine integer powers and roots of complex numbers.

Polar coordinates are introduced and their relationship with rectangular coordinates investigated. Complete graphs of equations involving the polar coordinates $r$ and $\theta$ are obtained.

The study of vectors is started in this chapter. We use both geometric methods and components of vectors to do the arithmetic of vectors. Applications that involve vectors are studied. Vectors are used to describe the path of objects moving in a plane.

We define curves in the plane using parametric equations. Parametric equations are used to obtain graphs of relations and to simulate motion. We use both a polar grapher and a parametric grapher to obtain graphs in this chapter. Many graphing utilities are capable of drawing graphs of polar or parametric equations. There are relatively simple programs that allow graphing calculators to obtain graphs of polar and parametric equations. See the *Graphing Calculator and Computer Graphing Laboratory Manual* for details.

# 5.1 Trigonometry and the Geometry of Complex Numbers

In this section we introduce complex numbers, represent complex numbers geometrically, and obtain trigonometric representations for complex numbers. Trigonometric representations are used to do the arithmetic of complex numbers.

## Complex Numbers

Let $i$ be a symbol with the property $i^2 + 1 = 0$. Formally, $i$ is a solution to the equation $x^2 + 1 = 0$. Thus, $i^2 + 1 = 0$ or $i^2 = -1$. We also write $i = \sqrt{-1}$. Both $i$ and $-i$ are solutions to the quadratic equation $x^2 + 1 = 0$.

---

• **DEFINITION**  An expression of the form $a + bi$, where $a$ and $b$ are real numbers is called a **complex number**. The set of all such expressions is called the set of **complex numbers**, or simply the **complex numbers**. The real number $a$ is called the **real part** of the complex number $a + bi$, and the real number $b$ the **imaginary part** of the complex number. •

---

We agree that $a + 0i = a$. Then, the subset of complex numbers with $b = 0$ is just the set of real numbers. Before finding complex number zeros of polynomials we need to understand the arithmetic of complex numbers. Equality and the operations of addition, subtraction, multiplication, and division for complex numbers are defined as follows.

---

• **DEFINITION**  Let $a + bi$ and $c + di$ be any two complex numbers.

1. Equality: $a + bi = c + di$ if and only if $a = b$ and $c = d$.
2. Addition: $(a + bi) + (c + di) = (a + c) + (b + d)i$
3. Subtraction: $(a + bi) - (c + di) = (a - c) + (b - d)i$
4. Multiplication: $(a + bi)(c + di) = (ac - bd) + (ad + bc)i$
5. Division: $\dfrac{a + bi}{c + di} = \dfrac{ac + bd}{c^2 + d^2} + \dfrac{bc - ad}{c^2 + d^2}i$, where $c^2 + d^2 \neq 0$. •

---

Thus, to add or subtract two complex numbers, we add or subtract their corresponding real and imaginary parts. To divide by $c + di$ we need $c^2 + d^2 \neq 0$. This means that not both $c$ and $d$ can be zero, that is, $c + di \neq 0 + 0i$. Addition and multiplication of complex numbers satisfies the commutative and associative properties. Furthermore, the distributive property holds for complex numbers.

• **EXAMPLE 1:** Write each of the following in the form $a + bi$, where $a$ and $b$ are real numbers.

(a) $(7 - 3i) + (4 + 5i)$   (b) $(7 - 3i) - (4 + 5i)$

(c) $(2 + 3i)(5 - i)$   (d) $\dfrac{4 - 3i}{2 + 5i}$

## 5.1 Trigonometry and the Geometry of Complex Numbers

**SOLUTION:**

(a) We use the definition to determine the sum $(7 - 3i) + (4 + 5i)$:
$$(7 - 3i) + (4 + 5i) = (7 + 4) + (-3 + 5)i$$
$$= 11 + 2i.$$

(b) We use the definition to determine the difference $(7 - 3i) - (4 + 5i)$:
$$(7 - 3i) - (4 + 5i) = (7 - 4) + (-3 - 5)i$$
$$= 3 - 8i.$$

(c) First, we use the definition to find the product $(2 + 3i)(5 - i)$:
$$(2 + 3i)(5 - i) = ((2)(5) - (3)(-1)) + ((2)(-1) + (3)(5))i$$
$$= 13 + 13i.$$

Alternatively, we can treat complex numbers as binomials and multiply in the expected way:
$$(2 + 3i)(5 - i) = (2)(5) + 2(-i) + (3i)5 + (3i)(-i)$$
$$= 10 - 2i + 15i - 3i^2$$
$$= 10 + 13i + 3$$
$$= 13 + 13i.$$

Notice we replaced $i^2$ by $-1$ above.

(d) We use the definition to determine the quotient $\dfrac{4 - 3i}{2 + 5i}$:
$$\frac{4 - 3i}{2 + 5i} = \frac{(4)(2) + (-3)(5)}{2^2 + 5^2} + \frac{(-3)(2) - (4)(5)}{2^2 + 5^2}i$$
$$= -\frac{7}{29} - \frac{26}{29}i$$

In order to divide complex numbers without memorizing the definition, we need the following.

---

- **DEFINITION**     The **complex conjugate** of $u = a + bi$ is $\bar{u} = a - bi$.

---

- **THEOREM 1**     Let $u = a + bi$ be any complex number.
  (a) $(a + bi) + (0 + 0i) = a + bi$     (b) $(a + bi)(1 + 0i) = a + bi$
  (c) $u\bar{u} = (a + bi)(a - bi) = a^2 + b^2$

  **PROOF**
  (a) $(a + bi) + (0 + 0i) = (a + 0) + (b + 0)i = a + bi$
  (b) $(a + bi)(1 + 0i) = ((a)(1) - (b)(0)) + ((a)(0) + (b)(1))i = a + bi$

(c) As in Example 1, we treat $a + bi$ and $a - bi$ as binomials and expand as follows:

$$u\bar{u} = (a+bi)(a-bi) = a^2 - abi + abi - b^2 i^2$$
$$= a^2 + b^2.$$

With our agreement to write $0 + 0i$ as 0 and $1 + 0i$ as 1, Theorem 1 shows that 0 and 1 are the additive and multiplicative identities, respectively, for the complex numbers. The third part of Theorem 1 shows that the product of a complex number and its complex conjugate is a real number. We use this fact to divide complex numbers.

• **EXAMPLE 2:** Write each of the following in the form $a + bi$, where $a$ and $b$ are real numbers.

(a) $\dfrac{5+i}{2-3i}$  (b) $\dfrac{7-i}{2i}$

SOLUTION:

(a) We multiply the numerator and denominator of $\dfrac{5+i}{2-3i}$ by the complex conjugate of the denominator and simplify:

$$\frac{5+i}{2-3i} = \frac{(5+i)(2+3i)}{(2-3i)(2+3i)} = \frac{10+2i+15i-3}{4+9} = \frac{7}{13} + \frac{17}{13}i.$$

(b) We use the same procedure as in (a). Because the complex conjugate of $2i$ is $-2i$, we multiply the numerator and denominator of $\dfrac{7-i}{2i}$ by $-2i$:

$$\frac{7-i}{2i} = \frac{(7-i)(-2i)}{(2i)(-2i)} = \frac{-2-14i}{4} = -\frac{1}{2} - \frac{7}{2}i.$$

Alternatively, we can put $\dfrac{7-i}{2i}$ in the form $a + bi$ by multiplying numerator and denominator by $i$ instead of the complex conjugate $-2i$ of the denominator:

$$\frac{7-i}{2i} = \frac{(7-i)(i)}{(2i)(i)} = \frac{1+7i}{-2} = -\frac{1}{2} - \frac{7}{2}i. \qquad \bullet$$

## Geometric Representation of Complex Numbers

There is a one-to-one correspondence between the set of all complex numbers and the points in the coordinate plane. The complex number $a + bi$ corresponds to the point $P(a, b)$ of the coordinate plane (Figure 5.1.1). When the coordinate plane is viewed as a representation of the complex numbers, it is called the **complex plane**, or **Gaussian plane**, after the famous mathematician K. F. Gauss. The $x$-axis of the complex plane is called the **real axis**, and the $y$-axis is called the **imaginary axis**. Figure 5.1.2 illustrates the correspondence for the complex numbers $2 + 3i$, $2 - 3i$, and $-3 - 2i$. We also say that Figure 5.1.2 gives the graph of the complex numbers $2 + 3i$, $2 - 3i$, and $-3 - 2i$.

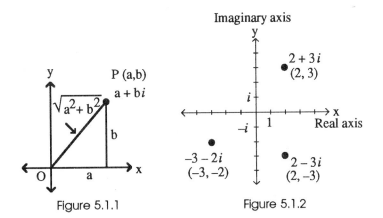

Figure 5.1.1  Figure 5.1.2

The distance between the complex number $a + bi$ and the origin $O$ is $\sqrt{a^2 + b^2}$ (Figure 5.1.1). We call $\sqrt{a^2 + b^2}$ the **absolute value** or **modulus** of the complex number $a + bi$. In general, the absolute value of a complex number $z$ is denoted by $|z|$. If $z = a + bi$ is real ($b = 0$), then $|z| = \sqrt{a^2} = |a|$.

We give a geometric interpretation of addition of complex numbers. Let $P = (a, b)$, $Q = (c, d)$, and $R = (a + c, b + d)$, the points corresponding to the complex numbers $a + bi$, $c + di$, and $(a + c) + (b + d)i$, respectively. We represent the complex numbers $a + bi$ and $c + di$ by the line segments $OP$ and $OQ$, respectively (Figure 5.1.3). It can be shown that the line segment $OR$ is a diagonal of the parallelogram determined by the line segments $OP$ and $OQ$.

Notice in Figure 5.1.3 that $PR$ is parallel to $OQ$ and has the same length as $OQ$. Thus, the point $R$ can be determined by drawing a line segment from $P$ to $R$ that is parallel to $OQ$ and has the same length as $OQ$ as shown in Figure 5.1.4.

- **EXAMPLE 3**: Use the line segment representation of complex numbers to determine the following complex numbers.

   (a) $(2 + i) + (1 + 3i)$    (b) $(3 + 5i) - (2 + i)$

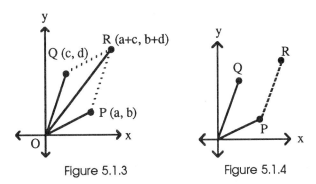

Figure 5.1.3  Figure 5.1.4

## SOLUTION:

(a) Let $P = (2, 1)$ and $Q = (1, 3)$ be the points corresponding to the complex numbers $2+i$ and $1+3i$, respectively. Draw a line segment parallel to $OQ$ with initial point $P$ and terminal point $R$ (Figure 5.1.5). Notice that the coordinates of $R$ are $(3, 4)$ and that by simple addition $(2 + i) + (1 + 3i) = 3 + 4i$.

(b) Let $R = (3, 5)$, and $P = (2, 1)$ be the points corresponding to the complex numbers $3 + 5i$ and $2 + i$, respectively. Let $z = (3 + 5i) - (2 + i)$, and let $Q$ be the point in the plane corresponding to $z$. Then $OR$ must be a diagonal of a parallelogram determined by $OP$ and $OQ$ because $3+5i = (2+i)+z$ (Figure 5.1.6). Furthermore, the line segment $PR$ must be parallel to $OQ$ and must have the same length as $OQ$. Thus, the coordinates of $Q$ must be $(1, 4)$. Thus, $z$ is the complex number $1 + 4i$ determined by the point $Q = (1, 4)$. Notice that by subtraction $(3+5i) - (2+i) = 1 + 4i$. •

## Trigonometric Form for Complex Numbers

Let $P(a, b)$ be the point in the complex plane that corresponds to the complex number $z = a + bi$. Let $\theta$ be any angle in standard position with the point $P$ on its terminal side. Let $r$ be the length of the segment $OP$. It follows that $r = \sqrt{a^2 + b^2}$, $\cos\theta = \frac{a}{r}$, and $\sin\theta = \frac{b}{r}$ (Figure 5.1.7). Thus, $a = r\cos\theta$ and $b = r\sin\theta$. We can write $z = a + bi = (r\cos\theta) + (r\sin\theta)i = r(\cos\theta + i\sin\theta)$. Often this form for $z$ is abbreviated to $z = r \operatorname{cis} \theta$. The angle $\theta$ can be given in degrees or radians.

• **DEFINITION**    Let $z = a + bi \neq 0$. A trigonometric form for $z$ is given by

$$z = r(\cos\theta + i\sin\theta) = r \operatorname{cis} \theta$$

where $r = |z| = \sqrt{a^2 + b^2}$, $\cos\theta = \frac{a}{r}$, and $\sin\theta = \frac{b}{r}$. •

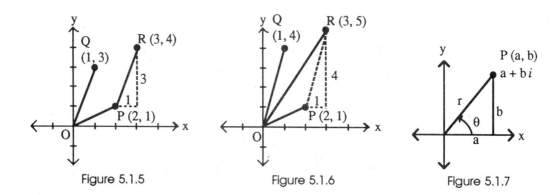

Figure 5.1.5         Figure 5.1.6         Figure 5.1.7

## 5.1 Trigonometry and the Geometry of Complex Numbers

The trig form for $z = a + bi$ is *not* unique because there are infinitely many choices for the angle $\theta$. In fact, we can write $z = r(\cos(\theta + 2k\pi) + i \sin(\theta + 2k\pi))$ for any integer $k$ if $\theta$ is measured in radians. If $\theta$ is measured in degrees then $z = r(\cos(\theta + k \cdot 360°) + i \sin(\theta + k \cdot 360°))$ for any integer $k$. For the complex number 0, we take $r = 0$ and $\theta$ can be any angle.

- **EXAMPLE 4**: Determine the trig form for the following complex numbers with $0 \leq \theta < 2\pi$. Then, give all possible trig forms. Also give the trig form in degrees for (a) and (b).
  (a) $-2 + 2i$          (b) $3 - 3\sqrt{3}\,i$          (c) $-3 - 4i$

SOLUTION:

(a) The distance from the origin to $-2 + 2i$ is $\sqrt{(-2)^2 + 2^2} = \sqrt{8} = 2\sqrt{2}$. The angle between 0 and $2\pi$ that $-2 + 2i$ makes with the positive $x$-axis is $3\pi/4$ (Figure 5.1.8). (Why?) Thus, we can write $z = -2 + 2i$ as $z = 2\sqrt{2}(\cos(3\pi/4) + i\sin(3\pi/4))$. Another trig form for $z$ can be obtained by using any other angle coterminal with $3\pi/4$. Therefore, all possible trig forms are given by $z = 2\sqrt{2}(\cos(3\pi/4 + 2k\pi) + i\sin(3\pi/4 + 2k\pi)) = 2\sqrt{2}\,\text{cis}(3\pi/4 + 2k\pi)$, where $k$ is any integer. In degrees, the trig form for $z$ is given by $z = 2\sqrt{2}(\cos(135° + k \cdot 360°) + i\sin(135° + k \cdot 360°))$, where $k$ is any integer.

(b) The distance from the origin to $3 - 3\sqrt{3}\,i$ is $\sqrt{(3)^2 + (-3\sqrt{3})^2} = \sqrt{36} = 6$. Thus, $3 - 3\sqrt{3}\,i = 6(\frac{1}{2} - i\frac{\sqrt{3}}{2})$. (Why?) Recall that $\cos\frac{\pi}{3} = \frac{1}{2}$ and $\sin\frac{\pi}{3} = \frac{\sqrt{3}}{2}$. Therefore, the angle that $3 - 3\sqrt{3}\,i$ makes with the positive $x$-axis is $\frac{5\pi}{3}$ (Figure 5.1.9). Thus, we can write $z = 3 - 3\sqrt{3}\,i$ as $z = 6\left(\cos\frac{5\pi}{3} + i\sin\frac{5\pi}{3}\right)$. All possible trig forms for $z$ can be obtained by using any other angle coterminal with $\frac{5\pi}{3}$. Therefore, all possible trig forms are given by $z = 6\,\text{cis}\left(\frac{5\pi}{3} + 2k\pi\right)$, where $k$ is any integer. In degrees, the trig form of $z$ is given by $z = 6\,\text{cis}(300° + k \cdot 360°)$, where $k$ is any integer.

(c) The distance from the origin to $-3 - 4i$ is $\sqrt{(-3)^2 + (-4)^2} = \sqrt{25} = 5$. Thus, $-3 - 4i = 5\left(-\frac{3}{5} - \frac{4}{5}i\right)$. Let $\theta$ be the angle $-3 - 4i$ makes with the positive $x$-axis (Figure 5.1.10). Then, $\theta$ is in the third quadrant and $\tan\theta = \frac{-4}{-3} = \frac{4}{3}$. Thus,

Figure 5.1.8          Figure 5.1.9

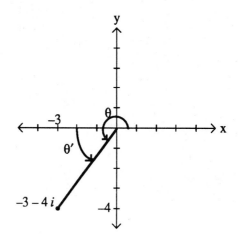

Figure 5.1.10

$\theta' = \tan^{-1}(\frac{4}{3})$ is the reference angle of $\theta$. We can use a calculator to show that $\theta'$ is about 0.93 radians. It follows that $\theta = \pi + 0.93 = 4.07$ radians. (Why?) Thus, $z = 5(\cos 4.07 + i \sin 4.07)$. All trig forms for $z$ are given by $z = 5\,\text{cis}(4.07 + 2k\pi)$, where $k$ is any integer. •

The trig forms for complex numbers can be used to multiply or divide complex numbers.

• **THEOREM 2**    Let $z_1 = r_1(\cos\theta_1 + i\sin\theta_1)$ and $z_2 = r_2(\cos\theta_2 + i\sin\theta_2)$. Then,

(a) $z_1 z_2 = r_1 r_2(\cos(\theta_1 + \theta_2) + i\sin(\theta_1 + \theta_2)) = r_1 r_2 \,\text{cis}(\theta_1 + \theta_2)$.

(b) $\frac{z_1}{z_2} = \frac{r_1}{r_2}(\cos(\theta_1 - \theta_2) + i\sin(\theta_1 - \theta_2)) = \frac{r_1}{r_2} \,\text{cis}(\theta_1 - \theta_2)$, if $z_2 \neq 0$.

**PROOF**

(a) We multiply $z_1$ by $z_2$ in the usual way:

$$z_1 z_2 = [r_1(\cos\theta_1 + i\sin\theta_1)][r_2(\cos\theta_2 + i\sin\theta_2)]$$
$$= r_1 r_2[(\cos\theta_1 \cos\theta_2 - \sin\theta_1 \sin\theta_2) + i(\sin\theta_1 \cos\theta_2 + \cos\theta_1 \sin\theta_2)]$$
$$= r_1 r_2(\cos(\theta_1 + \theta_2) + i\sin(\theta_1 + \theta_2))$$
$$= r_1 r_2 \,\text{cis}(\theta_1 + \theta_2).$$

The last step follows because

$$\cos(\theta_1 + \theta_2) = \cos\theta_1 \cos\theta_2 - \sin\theta_1 \sin\theta_2, \quad \text{and}$$
$$\sin(\theta_1 + \theta_2) = \sin\theta_1 \cos\theta_2 + \cos\theta_1 \sin\theta_2.$$

(b) We divide $z_1$ by $z_2$ by first multiplying numerator and denominator by the complex conjugate of the portion of the denominator between the parentheses:

$$\frac{z_1}{z_2} = \frac{r_1(\cos\theta_1 + i\sin\theta_1)}{r_2(\cos\theta_2 + i\sin\theta_2)}$$

$$= \frac{r_1}{r_2}\frac{(\cos\theta_1 + i\sin\theta_1)}{(\cos\theta_2 + i\sin\theta_2)} \cdot \frac{(\cos\theta_2 - i\sin\theta_2)}{(\cos\theta_2 - i\sin\theta_2)}$$

$$= \frac{r_1}{r_2}\frac{(\cos\theta_1\cos\theta_2 + \sin\theta_1\sin\theta_2) + i(\sin\theta_1\cos\theta_2 - \cos\theta_1\sin\theta_2)}{(\cos^2\theta_2 + \sin^2\theta_2)}.$$

$$= \frac{r_1}{r_2}(\cos(\theta_1 - \theta_2) + i\sin(\theta_1 - \theta_2))$$

$$= \frac{r_1}{r_2}\operatorname{cis}(\theta_1 - \theta_2)$$

- **EXAMPLE 5:** Use the trig forms to determine $z_1 z_2$ and $\frac{z_1}{z_2}$ if $z_1 = -2 + 2i$ and $z_2 = 3 - i3\sqrt{3}$.

**SOLUTION:** In Example 4 we found that $-2 + 2i = 2\sqrt{2}(\cos\frac{3\pi}{4} + i\sin\frac{3\pi}{4})$ and $3 - i3\sqrt{3} = 6(\cos\frac{5\pi}{3} + i\sin\frac{5\pi}{3})$.

We apply Theorem 2 to compute $z_1 z_2$:

$$z_1 z_2 = (-2 + 2i)(3 - i3\sqrt{3})$$

$$= \left[2\sqrt{2}\left(\cos\frac{3\pi}{4} + i\sin\frac{3\pi}{4}\right)\right]\left[6\left(\cos\frac{5\pi}{3} + i\sin\frac{5\pi}{3}\right)\right]$$

$$= 12\sqrt{2}\left[\cos\left(\frac{3\pi}{4} + \frac{5\pi}{3}\right) + i\sin\left(\frac{3\pi}{4} + \frac{5\pi}{3}\right)\right]$$

$$= 12\sqrt{2}\left(\cos\frac{29\pi}{12} + i\sin\frac{29\pi}{12}\right)$$

$$= 12\sqrt{2}\left(\cos\frac{5\pi}{12} + i\sin\frac{5\pi}{12}\right).$$

The last step follows because $\frac{29\pi}{12} = \frac{5\pi}{12} + 2\pi$. If we multiply $z_1$ and $z_2$ directly we obtain

$$(-2 + 2i)(3 - i3\sqrt{3}) = (-6 + 6\sqrt{3}) + (6 + 6\sqrt{3})i.$$

A calculator can be used to verify that $-6 + 6\sqrt{3} = 12\sqrt{2}\cos\frac{5\pi}{12}$ and $6 + 6\sqrt{3} = 12\sqrt{2}\sin\frac{5\pi}{12}$.

We apply Theorem 2 to compute $\frac{z_1}{z_2}$:

$$\frac{z_1}{z_2} = \frac{2\sqrt{2}\left(\cos\frac{3\pi}{4} + i\sin\frac{3\pi}{4}\right)}{6\left(\cos\frac{5\pi}{3} + i\sin\frac{5\pi}{3}\right)}$$

$$= \frac{\sqrt{2}}{3}\left[\cos\left(\frac{3\pi}{4} - \frac{5\pi}{3}\right) + i\sin\left(\frac{3\pi}{4} - \frac{5\pi}{3}\right)\right]$$

$$= \frac{\sqrt{2}}{3}\left[\cos\left(-\frac{11\pi}{12}\right) + i\sin\left(-\frac{11\pi}{12}\right)\right]$$

$$= \frac{\sqrt{2}}{3}\left(\cos\frac{11\pi}{12} - i\sin\frac{11\pi}{12}\right).$$

The last step follows because cosine is an even function and sine is an odd function. If we divide $z_1$ by $z_2$ directly we obtain

$$\frac{-2+2i}{3-i3\sqrt{3}} = \frac{-2+2i}{3-i3\sqrt{3}} \cdot \frac{3+i3\sqrt{3}}{3+i3\sqrt{3}}$$

$$= \frac{(-6-6\sqrt{3}) + i(6 - 6\sqrt{3})}{9+27}$$

$$= -\frac{1+\sqrt{3}}{6} + i\frac{1-\sqrt{3}}{6}.$$

Again a calculator can be used to show that $-\frac{1+\sqrt{3}}{6} = \frac{\sqrt{2}}{3}\cos\frac{11\pi}{12}$ and $\frac{1-\sqrt{3}}{6} = -\frac{\sqrt{2}}{3}\sin\frac{11\pi}{12}$. •

## EXERCISES 5-1

Compute. Express the answer in the form $a + bi$ where $a$ and $b$ are real numbers.

1. $(-1 + 2i) + (3 + 5i)$
2. $3(2 - 3i)$
3. $(2 - 4i)(3 + 2i)$
4. $(-1 + 2i)^2$
5. $\frac{1+i}{1-i}$
6. $\frac{2-3i}{4+5i}$

Graph in the complex plane.

7. $3$
8. $3i$
9. $4 - 4i$
10. $-4 + 3i$
11. $2(\cos 30° + i\sin 30°)$
12. $3(\cos\frac{\pi}{4} + i\sin\frac{\pi}{4})$

Consider complex numbers $z_1$ and $z_2$ as represented in Figure 5.1.11.

13. If $z_1 = a + bi$, determine $a$, $b$, and $|z_1|$.
14. If $z_2 = a + bi$, determine $a$, $b$, and $|z_2|$.
15. Determine the trig form of $z_1$.

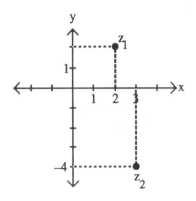

Figure 5.1.11

16. Determine the trig form of $z_2$.

Let $z_1 = -2 + 3i$, $z_2 = 3 + 4i$, and $z_3 = 2 - 5i$. Perform the indicated computation and write in the form $a + bi$.

17. $z_1 + z_2$
18. $4z_3$
19. $z_1 z_3$
20. $|z_1 - z_2|$
21. $|z_1 z_2|$
22. $\dfrac{z_2}{z_3}$
23. Determine $(1 + 3i) + (4 + 4i)$ using a geometric representation.
24. Determine $(-3 + i) - (4 + 2i)$ using a geometric representation.

Write the complex number in the form $a + bi$.

25. $2(\cos 60° + i \sin 60°)$
26. $5(\cos \frac{3\pi}{4} + i \sin \frac{3\pi}{4})$
27. $3.4(\cos 5 + i \sin 5)$
28. $10(\cos 6\pi + i \sin 6\pi)$

Write in trig form with $0 \leq \theta < 2\pi$, then give all possible trig forms.

29. $2 + 3i$
30. $-4 + 4i$
31. $\sqrt{2} - 8i$
32. $-1 - 2i$

33. Prove that $\bar{u} = r \operatorname{cis}(-\theta)$ if $u = r \operatorname{cis} \theta$. (Recall $\bar{u} = a - bi$ if $u = a + bi$ and $\operatorname{cis} \theta = \cos \theta + i \sin \theta$.)
34. Show that $|u \cdot v| = |u||v|$ for complex numbers $u$ and $v$.
35. Prove that for complex numbers $u$ and $v$, $u \cdot v = 0$ if and only if $u = 0$ or $v = 0$.

Use Theorem 2 to compute. (*Note:* $\operatorname{cis} \theta = \cos \theta + i \sin \theta$.)

36. $(2 \operatorname{cis} 30°)(3 \operatorname{cis} 60°)$
37. $\dfrac{2 \operatorname{cis} 30°}{3 \operatorname{cis} 60°}$
38. $(\sqrt{2} \operatorname{cis} 3)(\sqrt{3} \operatorname{cis} 5)$
39. $(2 - 3i)(-3 + 4i)$

**40-43** Write the two complex numbers in Exercises 36–39 in the form $a + bi$, perform the operations, and verify the results are the same as that obtained using the trig form and Theorem 2.

**44.** Use a calculator to confirm that $-6 + 6\sqrt{3} = 12\sqrt{2}\cos\frac{5\pi}{12}$ and $6 + 6\sqrt{3} = 12\sqrt{2}\sin\frac{5\pi}{12}$ (see Example 5).

## 5.2 De Moivre's Theorem, Roots of Complex Numbers, and Binomial Theorem

In this section we state the Binomial Theorem and a famous theorem of De Moivre. De Moivre's theorem is used to determine roots of complex numbers. We investigate the geometric relationship among the roots of a complex number.

### Binomial Theorem

De Moivre's theorem is about raising a complex number to an integer power. Raising a binomial to a power is usually tedious. For example, we can use multiplication to establish the following formulas:

$$(a+b)^2 = a^2 + 2ab + b^2,$$
$$(a+b)^3 = a^3 + 3a^2b + 3ab^2 + b^3,$$
$$(a+b)^4 = a^4 + 4a^3b + 6a^2b^2 + 4ab^3 + b^4.$$

In general, the following theorem is true.

• **THEOREM 1** (**The Binomial Theorem**). Let $n$ be a positive integer. Then,

$$(a+b)^n = a^n + na^{n-1}b + \frac{n(n-1)}{1 \cdot 2}a^{n-2}b^2 + \frac{n(n-1)(n-2)}{1 \cdot 2 \cdot 3}a^{n-3}b^3 + \cdots$$
$$+ \frac{n(n-1)\cdots(n-k+1)}{1 \cdot 2 \cdots k}a^{n-k}b^k + \cdots + nab^{n-1} + b^n.$$

If we use the pattern suggested by the coefficient of general term

$$\frac{n(n-1)\cdots(n-k+1)}{1 \cdot 2 \cdots k}a^{n-k}b^k$$

of the expansion of $(a+b)^n$ to compute the coefficient of $ab^{n-1}$, we obtain

$$\frac{n(n-1)\cdots 2}{1 \cdot 2 \cdots (n-1)}ab^{n-1}.$$

which simplifies to $nab^{n-1}$. (Why?)

• **EXAMPLE 1:** Use the Binomial Theorem to expand $(a+b)^5$.

SOLUTION: Applying the Binomial Theorem we obtain

$$(a+b)^5 = a^5 + 5a^4b + \frac{5\cdot 4}{1\cdot 2}a^3b^2 + \frac{5\cdot 4\cdot 3}{1\cdot 2\cdot 3}a^2b^3$$
$$+ \frac{5\cdot 4\cdot 3\cdot 2}{1\cdot 2\cdot 3\cdot 4}ab^4 + b^5.$$

We can simplify the coefficients and obtain the following form:

$$(a+b)^5 = a^5 + 5a^4b + 10a^3b^2 + 10a^2b^3 + 5ab^4 + b^5.$$

•

The binomial coefficients form an interesting triangular array of numbers called **Pascal's triangle**. The first row of the triangle consists of the single coefficient 1 of the expansion of $(a+b)^0$. The second row consists of the two coefficients 1 and 1 of the expansion of $(a+b)^1$. The third row consists of the coefficients 1, 2, and 1 of the expansion of $(a+b)^2 = a^2 + 2ab + b^2$. Continuing in this way we obtain the following triangular array.

$$\begin{array}{ccccccccccc}
 & & & & & 1 & & & & & \\
 & & & & 1 & & 1 & & & & \\
 & & & 1 & & 2 & & 1 & & & \\
 & & 1 & & 3 & & 3 & & 1 & & \\
 & 1 & & 4 & & 6 & & 4 & & 1 & \\
1 & & 5 & & 10 & & 10 & & 5 & & 1
\end{array}$$

$$\cdot \quad \cdot \quad \cdot \quad \cdot \quad \cdot \quad \cdot \quad \cdot$$
$$\cdot \quad \cdot \quad \cdot \quad \cdot \quad \cdot \quad \cdot \quad \cdot$$

Notice how each entry in a row, other than the first and the last, is the sum of the two closest entries of the previous row. For example, the entries in the seventh row above are 1, $6 = 1+5$, $15 = 5+10$, $20 = 10+10$, $15 = 10+5$, $6 = 5+1$, and 1.

The Binomial Theorem can be used to expand more complicated expressions.

• **EXAMPLE 2:** Use the Binomial Theorem to expand $(2x - 3y^2)^4$.

SOLUTION: We use the Binomial Theorem with $a = 2x$, $b = -3y^2$, and $n = 4$:

$$(2x - 3y^2)^4 = (2x)^4 + 4(2x)^3(-3y^2) + \frac{4\cdot 3}{1\cdot 2}(2x)^2(-3y^2)^2$$
$$+ \frac{4\cdot 3\cdot 2}{1\cdot 2\cdot 3}(2x)(-3y^2)^3 + (-3y^2)^4$$
$$= 16x^4 - 96x^3y^2 + 216x^2y^4 - 216xy^6 + 81y^8.$$

•

## De Moivre's Theorem

The Binomial theorem is fairly complicated. However, the formula for raising a complex number to an integer power is considerably simpler.

- **THEOREM 2** (De Moivre's Theorem) Let $n$ be any integer. Then,
$$[r(\cos\theta + i\sin\theta)]^n = r^n(\cos n\theta + i\sin n\theta),$$
$$(r\operatorname{cis}\theta)^n = r^n\operatorname{cis}(n\theta).$$

Let $a$ be a complex number and $n$ a positive integer. If $z$ is a complex number satisfying $z^n = a$, then $z$ is called an **nth root of a**. Every complex number $a$ has $n$ different $n$th roots. De Moivre's theorem can be used to find all the $n$th roots of any complex number. If $z$ is a complex number satisfying $z^n = 1$, then $z$ is called an **nth root of unity**. We begin by investigating the geometry of roots of unity.

- **EXAMPLE 3**: Determine the two square roots of unity, give their trig forms in degrees, and graph them.

SOLUTION: The square roots of unity are the two solutions of $z^2 = 1$ (Figure 5.2.1). They are $1$ and $-1$. The trig forms, in degrees, of the two square roots of unity are
$$1 = \cos 0° + i\sin 0°,$$
$$-1 = \cos 180° + i\sin 180°.$$

Notice that $1$ and $-1$ are $180°$ apart on the unit circle, and that $180° = \frac{360°}{2}$.

- **EXAMPLE 4**: Determine the four fourth roots of unity, give their trig forms in radians, and graph them.

SOLUTION: The fourth roots of unity are the four solutions of $z^4 = 1$. We can solve this equation algebraically:
$$z^4 = 1$$
$$z^4 - 1 = 0$$
$$(z^2 - 1)(z^2 + 1) = 0$$
$$(z - 1)(z + 1)(z^2 + 1) = 0.$$

Figure 5.2.1

## 5.2 De Moivre's Theorem, Roots of Complex Numbers, and Binomial Theorem

Now we can see that the four fourth roots of unity are $1, -1, -i$, and $i$ (Figure 5.2.2). Next, we give their trig forms in radians:

$$1 = \cos 0 + i \sin 0,$$
$$i = \cos \frac{\pi}{2} + i \sin \frac{\pi}{2},$$
$$-1 = \cos \pi + i \sin \pi,$$
$$-i = \cos \frac{3\pi}{2} + i \sin \frac{3\pi}{2}.$$

Notice that the fourth roots of unity are spaced $90°$ apart on the unit circle and that $90° = \frac{360°}{4}$.

Examples 3 and 4 are special cases of the following theorem, which we give without proof.

- **THEOREM 3**   The $n$th roots of a complex number $z$ are spaced $\left(\frac{360}{n}\right)°$ apart on a circle of radius $\sqrt[n]{|z|}$.

- **EXAMPLE 5:** Determine the cube roots of $-1$.

  **SOLUTION:** The cube roots of $-1$ are the three solutions of $z^3 = -1$. Notice that $-1$ is a cube root of $-1$ because $(-1)^3 = -1$. If the cube roots of $-1$ are indeed spaced $120°$ apart, then we can see from Figure 5.2.3 that the other two cube roots of $-1$ should be $\cos 60° + i \sin 60°$ and $\cos(-60°) + i \sin(-60°)$. We use De Moivre's theorem to show that

Figure 5.2.2      Figure 5.2.3

these complex numbers are cube roots of $-1$:

$$(\cos 60° + i \sin 60°)^3 = \cos 180° + i \sin 180°$$
$$= -1 + i(0)$$
$$= -1,$$
$$(\cos(-60°) + i \sin(-60°))^3 = \cos(-180°) + i \sin(-180°)$$
$$= -1 + i(0)$$
$$= -1.$$

Alternatively, we can write

$$\cos 60° + i \sin 60° = \frac{1}{2} + i\frac{\sqrt{3}}{2},$$
$$\cos(-60°) + i \sin(-60°) = \frac{1}{2} - i\frac{\sqrt{3}}{2},$$

and verify directly that $(\frac{1}{2} + i\frac{\sqrt{3}}{2})^3 = -1$ and $(\frac{1}{2} - i\frac{\sqrt{3}}{2})^3 = -1$.

Notice that $z^3 = -1$ is equivalent to $z^3 + 1 = 0$ and that $z^3 + 1 = (z+1)(z^2 - z + 1)$. The complex numbers $\frac{1}{2} + i\frac{\sqrt{3}}{2}$ and $\frac{1}{2} - i\frac{\sqrt{3}}{2}$ are the two solutions of $z^2 - z + 1 = 0$. •

• **EXAMPLE 6:** Use De Moivre's theorem to determine the value of $(-1+i)^{12}$.

SOLUTION: In trig form, $-1 + i = \sqrt{2}(\cos \frac{3\pi}{4} + i \sin \frac{3\pi}{4})$. Thus, we can compute as follows:

$$(-1+i)^{12} = \left[\sqrt{2}\left(\cos \frac{3\pi}{4} + i \sin \frac{3\pi}{4}\right)\right]^{12}$$
$$= 2^6 \left(\cos 9\pi + i \sin 9\pi\right)$$
$$= 64 \left(\cos \pi + i \sin \pi\right)$$
$$= -64.$$ •

We can conclude from Example 6 that $-1+i$ is a 12th root of $-64$. In the next example we illustrate a systematic procedure for determining the $n$th roots of a complex number.

• **EXAMPLE 7:** Determine the cube roots of $-8$.

SOLUTION: First, we put $-8$ in trig form:

$$-8 = 8(\cos \pi + i \sin \pi).$$

If $z = r(\cos \theta + i \sin \theta)$ is a cube root of $-8$, then $z^3 = -8$. We proceed as follows:

$$z^3 = -8$$
$$[r(\cos \theta + i \sin \theta)]^3 = 8(\cos \pi + i \sin \pi)$$
$$r^3(\cos 3\theta + i \sin 3\theta) = 8(\cos \pi + i \sin \pi).$$

Now, $r^3 = 8$ and $r$ is real, so $r = 2$. Next, $3\theta$ can be any angle coterminal with $\pi$. Thus,
$$3\theta = \pi + 2k\pi$$
where $k$ is any integer. It follows that
$$\theta = \frac{\pi}{3} + \frac{2k\pi}{3}.$$
Notice that
$$\theta = \begin{cases} \frac{\pi}{3} & \text{if } k = 0 \\ \pi & \text{if } k = 1 \\ \frac{5\pi}{3} & \text{if } k = 2 \end{cases}.$$

Any other integer value of $k$ produces an angle coterminal with one of $\frac{\pi}{3}$, $\pi$, or $\frac{5\pi}{3}$. The three distinct complex numbers $2(\cos\frac{\pi}{3} + i\sin\frac{\pi}{3})$, $2(\cos\pi + i\sin\pi)$, and $2(\cos\frac{5\pi}{3} + i\sin\frac{5\pi}{3})$ are the three cube roots of $-8$. Notice that

$$2\left(\cos\frac{\pi}{3} + i\sin\frac{\pi}{3}\right) = 2\left(\frac{1}{2} + i\frac{\sqrt{3}}{2}\right) = 1 + i\sqrt{3},$$

$$2(\cos\pi + i\sin\pi) = -2, \quad \text{and}$$

$$2\left(\cos\frac{5\pi}{3} + i\sin\frac{5\pi}{3}\right) = 2\left(\frac{1}{2} - i\frac{\sqrt{3}}{2}\right) = 1 - i\sqrt{3}.$$

You should check that the cube of each of these numbers is $-8$. ●

● **EXAMPLE 8:** Verify directly that the three cube roots of $-8$ lie on a circle of radius 2 centered at the origin and that they determine an equilateral triangle.

SOLUTION: We know from the trig form of the three complex numbers given in Example 7 that each has distance 2 from the origin. Thus, $1 + i\sqrt{3}$, $-2$, and $1 - i\sqrt{3}$ lie on the circle of radius 2 centered at the origin (Figure 5.2.4). To verify that the three cube roots of $-8$ form an equilateral triangle, we need to show that the distances between the points $P_1(1, \sqrt{3})$, $P_2(-2, 0)$, and $P_3(1, -\sqrt{3})$ are equal.

$$d(P_1, P_2) = \sqrt{3^2 + (\sqrt{3})^2} = \sqrt{12}$$

$$d(P_2, P_3) = \sqrt{(-3)^2 + (\sqrt{3})^2} = \sqrt{12}$$

$$d(P_3, P_1) = \sqrt{0^2 + (-2\sqrt{3})^2} = \sqrt{12}$$

Thus, the points $P_1$, $P_2$, and $P_3$ form an equilateral triangle. ●

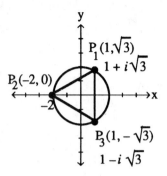

Figure 5.2.4

Examples 7 and 8 can be generalized in the following way.

- **THEOREM 4**  Let $c = r(\cos\theta + i\sin\theta)$ be any complex number with $r \neq 0$ and $n$ any positive integer. Then, $c$ has $n$ distinct $n$th roots $z_k$ given by

$$z_k = \sqrt[n]{r}\left(\cos\left(\frac{\theta + 2k\pi}{n}\right) + i\left(\sin\frac{\theta + 2k\pi}{n}\right)\right),$$

for $k = 0, 1, 2, \ldots, n - 1$. The $n$ distinct $n$th roots lie on the circle of radius $\sqrt[n]{r}$ centered at the origin and determine a *regular* polygon with $n$ equal sides.

If the measure of $\theta$ is given in degrees, then we must replace $2k\pi$ by $k \cdot 360°$ in Theorem 4.

- **DEFINITION**  A polygon with all sides equal is said to be **regular**. A polygon with $n$ equal sides and $n$ equal angles is called a **regular $n$-gon**.  •

  - **EXAMPLE 9:** Find the eight 8th roots of unity and draw the regular octagon (8-gon) determined by the roots.

    SOLUTION: We want to find the eight 8th roots of the complex number 1. In trig form, we have $1 = \cos 0 + i\sin 0$. By Theorem 4, the eight 8th roots of 1 are given by

    $$z_k = \cos\frac{0 + 2k\pi}{8} + i\sin\frac{0 + 2k\pi}{8} = \cos\frac{k\pi}{4} + i\sin\frac{k\pi}{4},$$

## 5.2 De Moivre's Theorem, Roots of Complex Numbers, and Binomial Theorem

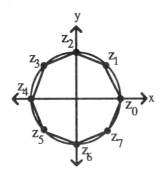

Figure 5.2.5

for $k = 0, 1, \ldots, 7$. We can write these eight complex numbers in the form $a + bi$ as follows:

$$z_0 = \cos 0 + i \sin 0 = 1,$$
$$z_1 = \cos \frac{\pi}{4} + i \sin \frac{\pi}{4} = \frac{1}{\sqrt{2}} + \frac{1}{\sqrt{2}} i,$$
$$z_2 = \cos \frac{\pi}{2} + i \sin \frac{\pi}{2} = i,$$
$$z_3 = \cos \frac{3\pi}{4} + i \sin \frac{3\pi}{4} = -\frac{1}{\sqrt{2}} + \frac{1}{\sqrt{2}} i,$$
$$z_4 = \cos \pi + i \sin \pi = -1,$$
$$z_5 = \cos \frac{5\pi}{4} + i \sin \frac{5\pi}{4} = -\frac{1}{\sqrt{2}} - \frac{1}{\sqrt{2}} i,$$
$$z_6 = \cos \frac{3\pi}{2} + i \sin \frac{3\pi}{2} = -i,$$
$$z_7 = \cos \frac{7\pi}{4} + i \sin \frac{7\pi}{4} = \frac{1}{\sqrt{2}} - \frac{1}{\sqrt{2}} i.$$

The above eight complex numbers all lie on the circle of radius 1 centered at the origin and determine the regular 8-gon illustrated in Figure 5.2.5. •

## EXERCISES 5-2

Write in trig form with $0 \leq \theta < 2\pi$.

1. $3 - 4i$
2. $-2 + 2i$
3. $5 + 3i$
4. $8 - 5i$
5. $-2i$
6. $-10$

Use Theorem 2 of Section 5.1 to compute. (*Note:* $\text{cis } \theta = \cos \theta + i \sin \theta$.)

7. $(2 \text{ cis } 30°)(3 \text{ cis } 60°)$
8. $\dfrac{2 \text{ cis } 45°}{3 \text{ cis } 30°}$

9. Write the two complex numbers in Exercise 7 in the form $a + bi$, perform the multiplication, and verify that the result is the same as that obtained using the trig form and Theorem 2 of Section 5.1.

10. Write the two complex numbers in Exercise 8 in the form $a+bi$, perform the division, and verify that the result is the same as that obtained using the trig form and Theorem 2 of Section 5.1.

Compute.

11. $\left(\text{cis}\,\dfrac{\pi}{4}\right)^5$

12. $(1+i)^5$

13. $(2\,\text{cis}\,\pi)^6$

14. $(3+4i)^{20}$

Find the $n$th roots of each complex number for the specified value of $n$. Graph each $n$th root in the complex plane.

15. $n = 4, \quad 1$

16. $n = 6, \quad 1$

17. $n = 4, \quad 2 + 2i$

18. $n = 4, \quad -2 + 2i$

19. $n = 6, \quad -2 + 2i$

20. $n = 5, \quad 32$

21. Determine $z$ and the three cube roots of $z$ if one cube root of $z$ is $1 + \sqrt{3}i$. Graph the roots.

22. Determine $z$ and the four fourth roots of $z$ if one fourth root of $z$ is $-2 - 2i$. Graph the roots.

23. Solve $z^4 = 5 - 5i$ and graph the solutions.

24. Solve $z^6 = 2i$ and graph the solutions.

25. In Example 6 we found that $-1 + i$ was a 12th root of $-64$. Determine the other eleven 12th roots of $-64$ and graph each in the complex plane.

26. Show geometrically that we can multiply two complex numbers by multiplying their absolute values and adding their arguments (Figure 5.2.6). That is, if $z_1 = r_1 \,\text{cis}\, \alpha$ and $z_2 = r_2 \,\text{cis}\, \beta$, then $z_1 z_2 = r_1 r_2 \,\text{cis}(\alpha + \beta)$. Graph $z_1$. Connect $z_1$ to the point $P = (1, 0)$. Graph $z_2$. Construct $\angle \alpha$ on $OQ$. Construct $\angle \gamma$ on $OQ$ with vertex at $Q$. Triangles $Oz_2 S$ and $OPz_1$ are similar. Thus, $\dfrac{d(0, S)}{d(0, z_1)} = \dfrac{d(0, z_2)}{d(0, P)} = \dfrac{d(0, z_2)}{1}$ so $d(0, S) = d(0, z_1) \cdot d(0, z_2) = r_1 \cdot r_2$.

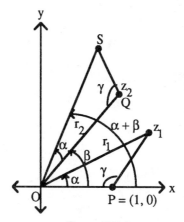

Figure 5.2.6

27. Construct a regular pentagon (5-gon) of radius 4; that is, each of the five vertices are at a distance 4 from the origin.
28. Construct a regular 10-gon of radius 6.

Expand.

29. $(x+y)^6$
30. $(a-2b^2)^4$
31. $(x-y)^6$
32. $(1+0.08)^5$
33. What is the coefficient of the $x^5y^3$ term in $(x+y)^8$?
34. What is the coefficient of the $x^3y^5$ term in $(x+y)^8$?

**Factorial and the binomial coefficient.** We use $n!$ to denote the product $n(n-1)(n-2)\cdots 2\cdot 1$ for a positive integer $n$. For example, $3! = 3\cdot 2\cdot 1 = 6$, $5! = 5\cdot 4\cdot 3\cdot 2\cdot 1 = 120$.

35. Show that the binomial coefficient of the general term $a^{n-k}b^k$ in the expansion of $(a+b)^n$ is $\dfrac{n!}{(n-k)!k!}$

36. Let $\binom{n}{k} = \dfrac{n!}{(n-k)!k!}$ for positive integers $k < n$. Compute

    (a) $\binom{5}{2}$     (b) $\binom{8}{5}$     (c) $\binom{8}{3}$     (d) $\binom{10}{1}$

37. Show that (a) $\binom{n}{k} = \binom{n}{n-k}$ and (b) $\binom{n}{1} = n$.

38. What would be sensible definitions for $0!$, $\binom{n}{0}$, and $\binom{n}{n}$?

## 5.3 Polar Coordinates and Graphs

In this section we see how to represent points in the coordinate plane with polar coordinates. This process is very closely related to finding the trig form for a complex number. We also graph polar equations, that is, equations in the two polar variables $r$ and $\theta$.

Let $P$ be a point in the coordinate plane with rectangular coordinates $(x, y)$ (Figure 5.3.1). The location of $P$ can be specified by giving its distance $r$ from the origin $O$ and an

Figure 5.3.1

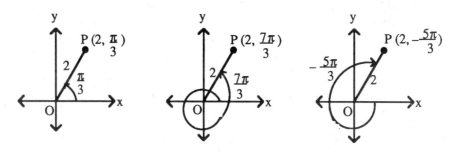

Figure 5.3.2

angle $\theta$ that the line through $O$ and $P$ makes with the positive $x$-axis. The measure of $\theta$ can be given in degrees or radians. We say that $(r, \theta)$ are **polar coordinates** of $P$. Notice that $r^2 = x^2 + y^2$ and $\tan \theta = \frac{y}{x}$ if $x \neq 0$. Polar coordinates of the point $(x, y)$ are *not unique* because $(r, \theta + 2n\pi)$, $n$ any integer, are also polar coordinates for the point $P$. For example, $(2, \frac{\pi}{3})$, $(2, \frac{7\pi}{3})$, and $(2, -\frac{5\pi}{3})$ are all polar coordinates for the same point $P$ (Figure 5.3.2).

We permit the variable $r$ to take on negative values with the following meaning. Let $r > 0$. Figure 5.3.3 illustrates how to graph the polar coordinates $(-r, \theta)$. The distance of the point $(-r, \theta)$ from the origin is $r$ but in the direction opposite to the direction given by the terminal side of $\theta$ in $(r, \theta)$. Notice that $(-r, \theta)$ also has polar coordinates $(r, \theta + \pi)$. For example, $(-2, \frac{\pi}{3})$ and $(2, \frac{4\pi}{3})$ are polar coordinates for the same point (Figure 5.3.4). Notice that $\frac{4\pi}{3} = \frac{\pi}{3} + \pi$. In fact, $(-2, \frac{\pi}{3} + 2n\pi)$ are also polar coordinates for the point $(-2, \frac{\pi}{3})$.

- **THEOREM 1**    Let $(r, \theta)$ be polar coordinates for a point $P$. Then, $P$ also has the following for polar coordinates:

$$(r, \theta + 2n\pi), \quad n \text{ any integer}$$
$$(-r, \theta + \pi + 2n\pi) = (-r, \theta + (2n+1)\pi), \quad n \text{ any integer}$$

The origin $O$ is assumed to have polar coordinates $(0, \theta)$, where $\theta$ can be any angle.

Theorem 2 gives the relationship between rectangular and polar coordinates of a point.

- **THEOREM 2**    Let a point $P$ have rectangular coordinates $(x, y)$ and polar coordinates $(r, \theta)$. Then, $x = r \cos \theta$, $y = r \sin \theta$, $r^2 = x^2 + y^2$, and $\tan \theta = \frac{y}{x}$ $(x \neq 0)$.

    **PROOF**    Let $P$ be a point with rectangular coordinates $(x, y)$ and polar coordinates $(r, \theta)$. Figure 5.3.5 gives one possibility for the point $P$ with $r > 0$ and $\theta$ in the second quadrant. We know that $r = \sqrt{x^2 + y^2}$ so that $r^2 = x^2 + y^2$. From the definitions of the trig functions at $\theta$, we have $\sin \theta = \frac{y}{r}$, $\cos \theta = \frac{x}{r}$, and $\tan \theta = \frac{y}{x}$. Thus, $y = r \sin \theta$ and $x = r \cos \theta$. This proof is valid if $\theta$ is in any of the other three quadrants.

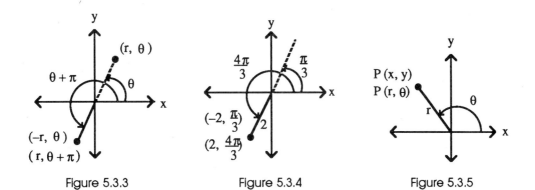

Figure 5.3.3    Figure 5.3.4    Figure 5.3.5

Figure 5.3.6 displays the relationship if $r < 0$ and $\theta$ in the second quadrant. In this case $P$ is in Quadrant IV, so $x > 0$ and $y < 0$. Again, this proof does not depend on the quadrant that contains $\theta$. Let $Q$ be the corresponding point in the second quadrant with polar coordinates $(-r, \theta)$. Then $Q$ has rectangular coordinates $(-x, -y)$. (Why?) Because $r < 0$ we have $|r| = -r$. We use the point $Q$ to determine the trig functions at $\theta$:

$$\sin\theta = \frac{-y}{|r|} = \frac{-y}{-r} = \frac{y}{r},$$

$$\cos\theta = \frac{-x}{|r|} = \frac{-x}{-r} = \frac{x}{r},$$

$$\tan\theta = \frac{-y}{-x} = \frac{y}{x}.$$

Thus, $y = r\sin\theta$, $x = r\cos\theta$, and $\tan\theta = \frac{y}{x}$. Also $r^2 = x^2 + y^2$ because $|r| = \sqrt{(-x)^2 + (-y)^2}$.

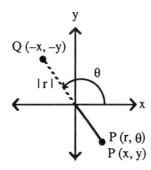

Figure 5.3.6

**372** • Chapter 5  Trigonometry of Complex Numbers, Vectors, Polar Equations, and Parametric Equations

• **EXAMPLE 1:** Find the rectangular coordinates of the points with the following polar coordinates.

(a) $(3, \frac{5\pi}{6})$ \hspace{2cm} (b) $(2, -200°)$

**SOLUTION:**

(a) Refer to Figure 5.3.7. Applying Theorem 2, we find that the rectangular coordinates of the point with polar coordinates $(3, \frac{5\pi}{6})$ are $(3\cos\frac{5\pi}{6}, 3\sin\frac{5\pi}{6}) = (-\frac{3\sqrt{3}}{2}, \frac{3}{2})$.

(b) Refer to Figure 5.3.8. Applying Theorem 2, we find that the rectangular coordinates of the point are $(2\cos(-200°), 2\sin(-200°)) = (-1.88, 0.68)$. •

Suppose that we are given the rectangular coordinates $(x, y)$ of a point $P$ and want to find polar coordinates $(r, \theta)$ for the point $P$. We can choose $r = \sqrt{x^2 + y^2}$. However, we must be careful if we use the inverse tangent keys on a calculator to determine $\theta$ from $\tan\theta = \frac{y}{x}$. If $P$ is in Quadrant I or IV, then $\tan^{-1}\frac{y}{x}$ is a correct angle to pair with $r = \sqrt{x^2 + y^2}$ to obtain polar coordinates for $P$. This is not true if $\theta$ is in Quadrant II or III, as the next example illustrates.

• **EXAMPLE 2:** Find polar coordinates for the points with the following rectangular coordinates.

(a) $(3, -5)$ \hspace{1.5cm} (b) $(-1, 1)$ \hspace{1.5cm} (c) $(-2, -3)$

**SOLUTION:**

(a) The point $P$ with rectangular coordinates $(3, -5)$ is in the fourth quadrant (Figure 5.3.9). If $(r, \theta)$ are polar coordinates of $P$ with $r > 0$, then $r = \sqrt{3^2 + (-5)^2} = \sqrt{34}$. Also, $\tan\theta = -\frac{5}{3}$. Now, $\tan^{-1}(-\frac{5}{3}) = -1.03$ is in the fourth quadrant, so we can choose $\theta = \tan^{-1}(-\frac{5}{3}) = -1.03$. Thus, $(\sqrt{34}, -1.03)$ or $(5.83, -1.03)$ give polar coordinates of $P$. This can be confusing because $(5.83, -1.03)$ look like

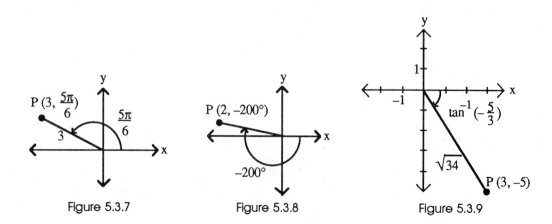

Figure 5.3.7 \hspace{2cm} Figure 5.3.8 \hspace{2cm} Figure 5.3.9

rectangular coordinates. We have to depend on the context to determine the type of coordinates being specified.

(b) The point $P$ with rectangular coordinates $(-1, 1)$ is in the second quadrant (Figure 5.3.10). Again, we can choose $r = \sqrt{(-1)^2 + 1^2} = \sqrt{2}$. Now, $\tan^{-1}(-1) = -\frac{\pi}{4}$ is not in the second quadrant. However, $\pi + \tan^{-1}(-1) = \frac{3\pi}{4}$ is in the second quadrant and can be paired with $r = \sqrt{2}$ to give polar coordinates for $P$. Thus, $(\sqrt{2}, \frac{3\pi}{4})$ are polar coordinates for $P$.

(c) The point $P$ with rectangular coordinates $(-2, -3)$ is in the third quadrant (Figure 5.3.11). Again, we can choose $r = \sqrt{(-2)^2 + (-3)^2} = \sqrt{13}$. Now, $\tan^{-1}(\frac{-3}{-2}) = 0.98$ is in the first quadrant. However, $\pi + \tan^{-1}(\frac{-3}{-2}) = 4.12$ is in the third quadrant. Thus, $(\sqrt{13}, 4.12)$, or $(3.61, 4.12)$ are polar coordinates for $P$. •

Example 2 can be generalized as follows.

- **THEOREM 3**    Let $P$ have rectangular coordinates $(x, y)$. Then, the following are polar coordinates for $P$ where $r = \sqrt{x^2 + y^2}$.

  (a) $(r, \tan^{-1} \frac{y}{x})$, if $(x, y)$ is in the first or fourth quadrant.
  (b) $(r, \pi + \tan^{-1} \frac{y}{x})$ if $(x, y)$ is in the second or third quadrant.

## Polar Graphs

- **DEFINITION**    The **graph of an equation** in the two polar variables $r$ and $\theta$ is the set of all points $(r, \theta)$ in the coordinate plane where $(r, \theta)$ is a solution to the equation. •

Distinct solutions may produce the same point in the plane. For example, if $(r, \theta)$ and $(r, \theta + 2\pi)$ are solutions to a given equation, they correspond to the same point in the coordinate plane.

Figure 5.3.10

Figure 5.3.11

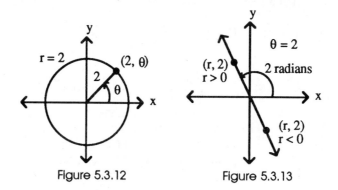

Figure 5.3.12    Figure 5.3.13

- **EXAMPLE 3**: Draw a complete graph of each polar equation.
  (a) $r = 2$   (b) $\theta = 2$

SOLUTION:

(a) The solutions to the polar equation $r = 2$ consist of all points $(2, \theta)$ when $\theta$ is any angle. The set of all such points is the *circle* with center at the origin and radius 2 (Figure 5.3.12).

(b) The solutions to the equation $\theta = 2$ consist of all points $(r, 2)$ where $r$ is any real number. The set of all such points is a *straight line* through the origin (Figure 5.3.13). •

If $r$ is a periodic function of $\theta$, then we can use the periodicity to help draw a complete graph. Example 4 illustrates this situation.

- **EXAMPLE 4**: Draw a complete graph of the polar equation $r = 3 + 3\sin\theta$.

SOLUTION: The function $r$ is periodic with period $2\pi$. (Why?) Thus, we will obtain a complete graph by restricting $\theta$ to the interval $[0, 2\pi]$. In fact, if we draw the graph of $r = 3 + 3\sin\theta$ for $2\pi \leq \theta \leq 4\pi$ we will completely retrace the graph of $r$ for $\theta$ in the interval $0 \leq \theta \leq 2\pi$. For $0 \leq \theta \leq \frac{\pi}{2}$, we have $0 \leq \sin\theta \leq 1$ and $3 \leq 3 + 3\sin\theta \leq 6$. Thus, as $\theta$ increases from 0 to $\frac{\pi}{2}$, $\sin\theta$ increases from 0 to 1, and $r$ increases from 3 to 6. Similarly, we can determine the remaining information given in Table 1.

You may need to compute several specific values to convince yourself that the graph in Figure 5.3.14 is complete. Notice that the graph is traced in a counterclockwise fashion starting with $(3, 0)$ and ending with $(3, 2\pi)$ as $\theta$ varies from 0 to $2\pi$. •

The graph in Figure 5.3.14 is called a **cardioid**, which refers to its heartlike shape. In fact, the graph of any function of the form $r = a \pm a\cos\theta$ or $r = a \pm a\sin\theta$ is called a cardioid.

In addition to specifying the viewing rectangle $L \leq x \leq R$, $B \leq y \leq T$ we usually need to specify the range of values of $\theta$ used to obtain a graph. For example, we could have

# 5.3 Polar Coordinates and Graphs

| $\theta$ | $\sin \theta$ | $r = 3 + 3\sin\theta$ |
|---|---|---|
| $0 \leq \theta \leq \frac{\pi}{2}$ | increases from 0 to 1 | increases from 3 to 6 |
| $\frac{\pi}{2} \leq \theta \leq \pi$ | decreases from 1 to 0 | decreases from 6 to 3 |
| $\pi \leq \theta \leq \frac{3\pi}{2}$ | decreases from 0 to $-1$ | decreases from 3 to 0 |
| $\frac{3\pi}{2} \leq \theta \leq 2\pi$ | increases from $-1$ to 0 | increases from 0 to 3 |

Table 1

indicated that the graph in Figure 5.3.14 was drawn for $\theta$ in $[0, 2\pi]$. However, because of the periodicity of the function $r$ this was not necessary. That is, the graph in Figure 5.3.14 is also the graph of $r = 3 + 3\sin\theta$ for $\theta$ in $(-\infty, \infty)$ or for $\theta$ in $[a, a + 2n\pi]$, $n$ any integer.

The polar equation $r = 3 + 3\sin\theta$ of Example 4 defines $r$ as a function of $\theta$. That is, each value of $\theta$ determines a unique value of $r$. However, the graph in Figure 5.3.14 fails the vertical line test for a function. This is not a contradiction because the vertical line test is used to determine if $y$ is a function of $x$ in an equation involving $x$ and $y$ where $x$ and $y$ are the usual rectangular coordinates. If we convert the polar equation to rectangular form, we can see directly that the equation does *not* define $y$ as a function of $x$:

$$r = 3 + 3\sin\theta$$
$$r = 3 + 3\frac{y}{r}$$
$$r^2 = 3r + 3y$$
$$x^2 + y^2 = 3\sqrt{x^2 + y^2} + 3y.$$

Now, we can see that $y$ is *not* a function of $x$ because each of the rectangular pairs $(0,0)$ and $(0,6)$ is a solution to the equation.

Most graphing utilities are able to obtain graphs of polar equations. If you are using a programmable graphing calculator, you probably will be able to write or obtain a short program that will draw graphs of polar equations. The *Graphing Calculator and Computer*

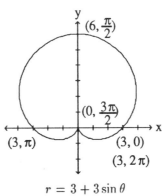

$r = 3 + 3\sin\theta$
Figure 5.3.14

*Graphing Laboratory Manual* accompanying this textbook contains a simple program that permits graphs of curves defined by polar equations to be obtained with a graphing calculator. The graphing utility *Master Grapher* used to draw graphs in this textbook has a polar equation option.

If a polar graphing utility is used, then the information given in Table 1 can be determined graphically by observing the drawing of the graph with $\theta$ restricted to the four intervals in the table. That is, setting the minimum value of $\theta$ to 0 and the maximum value of $\theta$ to $\frac{\pi}{2}$ will produce the first quadrant portion of the graph in Figure 5.3.14 drawn in counterclockwise fashion starting with $(3,0)$ and ending with $(6, \frac{\pi}{2})$.

- **EXAMPLE 5**: Draw a complete graph of the polar function $r = 5\cos 2\theta$.

SOLUTION: $r$ is a periodic function of $\theta$ with period $\pi$. (Why?) Even though the values of $r$ for $\pi \leq \theta \leq 2\pi$ are the same as the values of $r$ for $0 \leq \theta \leq \pi$, we obtain distinct points on the graph because the angles are different. For example, $r = 4.70$ for $\theta = 0.175$ or $3.316$, but the polar coordinates $(4.7, 0.175)$ and $(4.7, 3.316)$ name different points. However, we

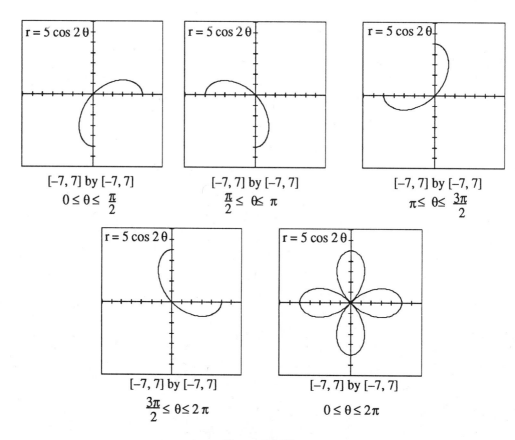

Figure 5.3.15

## 5.3 Polar Coordinates and Graphs

will get a complete graph if we let $\theta$ vary from 0 to $2\pi$. Again, the graph of $r = 5\cos 2\theta$ for $2\pi \leq \theta \leq 4\pi$ retraces the graph of $r$ for $0 \leq \theta \leq 2\pi$. We can determine the following information about $r$.

Figure 5.3.15 shows the graph of $r = 5\cos 2\theta$ for $0 \leq \theta \leq \frac{\pi}{2}$, $\frac{\pi}{2} \leq \theta \leq \pi$, $\pi \leq \theta \leq \frac{3\pi}{2}$, $\frac{3\pi}{2} \leq \theta \leq 2\pi$, and $0 \leq \theta \leq 2\pi$. As $\theta$ varies from 0 to $\frac{\pi}{4}$, the first quadrant portion of the graph in $0 \leq \theta \leq \frac{\pi}{2}$ is produced in counterclockwise fashion. The third quadrant portion of the graph in $0 \leq \theta \leq \frac{\pi}{2}$ is produced counterclockwise as $\theta$ varies from $\frac{\pi}{4}$ to $\frac{\pi}{2}$. The graphs for the ranges $\frac{\pi}{2} \leq \theta \leq \pi$, $\pi \leq \theta \leq \frac{3\pi}{2}$, and $\frac{3\pi}{2} \leq \theta \leq 2\pi$ can be similarly analyzed. The graph in $0 \leq \theta \leq 2\pi$ is complete and is called a **four-leafed rose**. To obtain a cumulative effect of the graph of $r = 5\cos 2\theta$ in the above $\theta$-intervals, try drawing the graph for $0 \leq \theta \leq \frac{\pi}{2}$, $0 \leq \theta \leq \pi$, $0 \leq \theta \leq \frac{3\pi}{2}$, and $0 \leq \theta \leq 2\pi$. •

Notice that the length of the petals of the graph in $0 \leq \theta \leq 2\pi$ of Figure 5.3.15 is 5 units. The graph of $r = \cos 2\theta$ is also a four-leafed rose but the length of each petal is only one unit. If $a > 1$, the effect of $a$ in $r = af(\theta)$ is to **stretch radially** the graph of

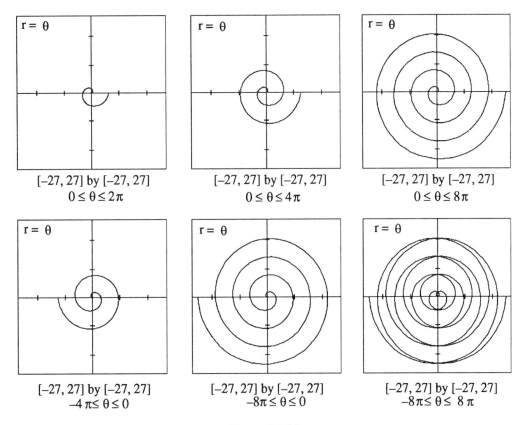

Figure 5.3.16

$r = f(\theta)$ by the factor $a$. If $0 < a < 1$, the effect of $a$ in $r = af(\theta)$ is to **shrink radially** the graph of $r = f(\theta)$ by the factor $a$.

In the Exercises, you will investigate the graphs of $r = a\sin n\theta$ and $r = a\cos n\theta$ for $a$ any real number and $n$ any integer. If $n$ is an odd integer greater than one, the graphs are $n$-leafed roses. If $n$ is an even positive integer, then the graphs are $2n$-leafed roses. The above statements are also true if $n$ is negative, except we must say $|n|$-leafed or $2|n|$-leafed rose in this case.

The graph in the next example is called a **spiral of Archimedes**.

- **EXAMPLE 6:** Draw the graph of $r = \theta$ for $0 \le \theta \le 2\pi$, $0 \le \theta \le 4\pi$, $0 \le \theta \le 8\pi$, $-4\pi \le \theta \le 0$, $-8\pi \le \theta \le 0$ and $-8\pi \le \theta \le 8\pi$.

SOLUTION: The six computer drawn graphs are given in Figure 5.3.16. The graph of $r = \theta$ goes around and around getting further from the origin as $|\theta|$ increases. •

## EXERCISES 5-3

Compute the polar coordinates $(r, \theta)$ with $0 \le \theta < 360°$ for the point with given rectangular coordinates.

1. $(1, 1)$
2. $(-10, 0)$
3. $(-3, 4)$
4. $(-2, 5)$

Compute *all* polar coordinate pairs for the point with given rectangular coordinates.

5. $(2, 2)$
6. $(-2, 5)$

Compute the rectangular coordinates for the point with given polar coordinates.

7. $(0, \pi)$
8. $(2, 30°)$
9. $(-3, 135°)$
10. $(-2, -120°)$

Compute the distance between the two points.

11. $(2.5, 30°), (4, 90°)$
12. $(-3, 120°), (2, 60°)$

Sketch a complete graph of each equation without using a polar graphing utility. If possible, check your answer with a polar graphing utility.

13. $r = 3$
14. $\theta = \frac{\pi}{4}$
15. $\theta = -\frac{\pi}{2}$
16. $r = -1$
17. $r = 3\cos\theta$
18. $r = 2\sin\theta$
19. $r = 2 + 2\cos\theta$
20. $r = 4 + 4\sin\theta$
21. $r = -1 + 2\cos\theta$
22. $r = \sin 2\theta$

Use a polar graphing utility to draw the polar graph for $\theta$ in the given intervals.

23. $r = 3\theta$; $0 \le \theta \le \frac{\pi}{2}$, $0 \le \theta \le \pi$, $0 \le \theta \le 2\pi$, $-\pi \le \theta \le \pi$, and $0 \le \theta \le 4\pi$.
24. $r = 5\sin\theta$, $0 \le \theta \le \frac{\pi}{2}$, $0 \le \theta \le \pi$, $0 \le \theta \le 2\pi$, $-\pi \le \theta \le \pi$, and $0 \le \theta \le 4\pi$.
25. $r = 5\cos 3\theta$, $0 \le \theta \le \frac{\pi}{2}$, $0 \le \theta \le \pi$, $0 \le \theta \le 2\pi$, $-\pi \le \theta \le \pi$, and $0 \le \theta \le 4\pi$.

26. $r = 5\sin 2\theta$, $0 \leq \theta \leq \frac{\pi}{2}$, $0 \leq \theta \leq \pi$, $0 \leq \theta \leq 2\pi$, $-\pi \leq \theta \leq \pi$, and $0 \leq \theta \leq 4\pi$.

Use a polar graphing utility to draw a complete graph of the polar equation. Specify an interval $a \leq \theta \leq b$ of smallest length that gives a complete graph.

27. $r = 2\theta$
28. $r = 0.275\theta$
29. $r = 2 - 3\cos\theta$
30. $r = 2\sin 3\theta$
31. $r = 3\sin^2\theta$
32. $r = 2 - 2\sin\theta$

33. Use a polar graphing utility to draw a complete graph of $r = 5\sin n\theta$ for $n = 1, 2, 3, 4, 5,$ and 6. Explain how the number of rose "petals" is related to $n$.

34. Use a polar graphing utility to draw a complete graph of $r = 5\cos n\theta$ for $n = 1, 2, 3, 4, 5,$ and 6. Explain how the number of rose "petals" is related to $n$.

35. Use a polar graphing utility to draw a complete graph of $r = a\sin 2\theta$ for $a = 1, 2, 3, 4, 5,$ and 10. What is the effect of the parameter $a$ on the graph?

36. Use a polar graphing utility to draw a complete graph of $r = a\cos 3\theta$ for $a = 1, 2, 3, 4, 5,$ and 10. What is the effect of the parameter $a$ on the graph?

Determine the length of *one* rose "petal."

37. $r = 5\sin 3\theta$
38. $r = 5\sin 2\theta$
39. $r = 8\sin 5\theta$
40. $r = 8\sin 4\theta$

Draw the graph of $r = 5\sin 2\theta$ and $r = 5\sin 3\theta$ in the standard viewing rectangle.

41. Predict what the graph of $r = 5\sin(2.5\theta)$ looks like, then draw a complete graph of $r = 5\sin(2.5\theta)$ with a polar graphing utility.

42. Give an algebraic argument for the appearance of the graph in Exercise 41.

43. Predict the graph of $r = 5\sin(3.5\theta)$. Confirm your graph by using a polar graphing utility.

44. Predict the graph of $r = a\sin\left(\frac{m}{n}\theta\right)$ for $m$ and $n$ relatively prime positive integers ($m$ and $n$ have no common factors).

## Polar Coordinate System

Let $\ell$ be the half line consisting of the positive $x$-axis. In the *polar coordinate system*, this half line is called the *polar axis* and the end $O$ (origin) is called the *pole* (Figure 5.3.17). If $r > 0$, the point $P$ with *polar coordinates* $(r, \theta)$ is the point $r$ units from the pole and on the terminal side of angle $\theta$. If $r < 0$, the point $P$ is the point $|r|$ units from the pole and on the half-line which is obtained by extending the terminal side of angle $\theta$ through the pole.

Figure 5.3.17

Figure 5.3.18

45. Find the polar coordinates of the vertices of a square with side length $a$ whose center is at the pole and with two sides parallel to the polar axis.
46. Find the polar coordinates of the vertices of a regular pentagon if the center is at the pole, one vertex is on the polar axis, and the distance from the center to a vertex is $a$.

Before computers, polar graphs were often sketched using polar graph paper as shown in Figure 5.3.18.

47. Use polar graph paper (you can make it yourself using ruler, protractor, and compass) and sketch a complete graph of $r = 3\sin 2\theta$ for $0 \leq \theta \leq 360°$. Plot points for $\theta = 0 + k \cdot 30°$, $k = 0, 1, 2, \ldots, 12$.

Solve the polar system by using a polar graphing utility.

48. $r = 5\sin\theta$
    $r = 5\cos\theta$

49. $r = 5\sin\theta$
    $r = 5\sin 3\theta$

50. $r = \dfrac{\theta}{\pi}$
    $r = 3\sin\theta$

51. $r = 2\cos\theta - 1$
    $r = 3\sin 2\theta$

52. Show that if $P_1$ and $P_2$ have polar coordinates $(r_1, \theta_1)$ and $(r_2, \theta_2)$ respectively, then the distance between $P_1$ and $P_2$ is $d(P_1, P_2) = \sqrt{r_1^2 + r_2^2 - 2r_1 r_2 \cos(\theta_2 - \theta_1)}$.

## 5.4 Vectors

In this section we begin the study of vectors. We investigate the arithmetic of vectors and use them with applications about motion. Many physical concepts can be represented by vectors. For example, the initial idea to use vectors grew out of the study of forces in physics. A **force** is a quantity that has both magnitude and direction. Forces are vectors that are often represented by arrows. The arrow points in the direction that the force is applied, and the length of the arrow represents the magnitude of the force.

## 5.4 Vectors

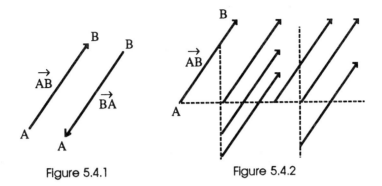

Figure 5.4.1     Figure 5.4.2

A **vector** is a **directed line segment**. For example, $\overrightarrow{AB}$ denotes the vector determined by the line segment $AB$ with direction assigned from $A$ to $B$ (Figure 5.4.1). $A$ is called the **initial point** and $B$ the **terminal point** of the vector $\overrightarrow{AB}$. $\overrightarrow{BA}$ is the vector determined by the line segment $AB$ with direction assigned from $B$ to $A$. $B$ is the initial point of $\overrightarrow{BA}$ and $A$ the terminal point. The **magnitude** of the vector $\overrightarrow{AB}$ is the length of the line segment $AB$. We denote the magnitude of $\overrightarrow{AB}$ by $|\overrightarrow{AB}|$. Two **vectors are equal** if and only if they have the same magnitude and direction. There are infinitely many vectors equal to a given vector $\overrightarrow{AB}$. The collection of vectors equal to $\overrightarrow{AB}$ consists of the vectors obtained by applying horizontal and/or vertical shifts to $\overrightarrow{AB}$; Figure 5.4.2 gives some of the possibilities. If a vector $\overrightarrow{CD}$ has been obtained by applying a combination of horizontal and vertical shifts to $\overrightarrow{AB}$, then $\overrightarrow{CD}$ is called a **parallel shift** of $\overrightarrow{AB}$.

In a coordinate system, we can be more precise about the meaning of the vector $\overrightarrow{AB}$. Consider the two points $A = (1, 2)$ and $B = (5, 4)$ (Figure 5.4.3). The magnitude of $\overrightarrow{AB}$ is $\sqrt{(5-1)^2 + (4-2)^2} = \sqrt{20}$, the distance from $A$ to $B$. The direction of $\overrightarrow{AB}$ is the angle $\theta$

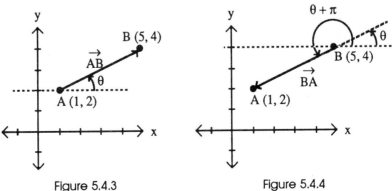

Figure 5.4.3     Figure 5.4.4

that $\overrightarrow{AB}$ makes with the positive $x$-axis. The magnitude of $\overrightarrow{BA}$ is also $\sqrt{20}$, but its direction with the positive $x$-axis is $\pi + \theta$ (Figure 5.4.4).

In a coordinate plane, we have the following theorem about equality of vectors.

• **THEOREM 1**  The **vectors $\overrightarrow{AB}$ and $\overrightarrow{CD}$ are equal** if and only if $\overrightarrow{AB}$ can be obtained from $\overrightarrow{CD}$ by a *parallel shift*. In other words, $\overrightarrow{AB}$ can be obtained from $\overrightarrow{CD}$ by applying appropriate horizontal or vertical shifts.

It follows from Theorem 1 that if $\overrightarrow{AB}$ and $\overrightarrow{CD}$ are equal, then $d(A,B) = d(C,D)$. Furthermore, vectors are not changed under parallel shifting.

• **EXAMPLE 1:** Let $A = (-4, 2)$, $B = (-1, 6)$, $O = (0, 0)$, and $P = (3, 4)$. Show that $\overrightarrow{AB} = \overrightarrow{OP}$.

SOLUTION: The length of each vector is $\sqrt{3^2 + 4^2} = 5$. If we shift $\overrightarrow{AB}$ horizontally to the right 4 units, and then shift the resulting vector vertically down 2 units we obtain the vector $\overrightarrow{OP}$ (Figure 5.4.5).  •

Notice that the coordinates of $A$ in $\overrightarrow{AB} = \overrightarrow{OP}$ give the shift factors needed to slide $\overrightarrow{OP}$ to $\overrightarrow{AB}$. In Example 1, shifting $\overrightarrow{OP}$ horizontally left 4 units and up 2 units produces $\overrightarrow{AB}$. The opposites of the coordinates of $A$ give the shift factors for obtaining $\overrightarrow{OP}$ from $\overrightarrow{AB}$.

Theorem 1 and Example 1 can be generalized as follows. Let $\overrightarrow{AB}$ be any vector. There is a *unique* point $P$ such that $\overrightarrow{AB} = \overrightarrow{OP}$, where $O = (0,0)$ is the origin. Thus, every vector can be represented by a directed line segment with initial point the origin of a coordinate system. In this case, the vector is completely determined by its terminal point.

Choosing the representation of a vector with initial point $(0, 0)$ from among a set of equal vectors is similar to choosing a reduced form of a set of equal fractions. For example, $\frac{1}{2}$ is the reduced form of $\frac{2}{4}, \frac{3}{6}, \ldots$, and is the representation most often used. In future mathematics courses, you will learn that equality of vectors and equality of rational numbers are each examples of **equivalence relations**.

Figure 5.4.5

## 5.4 Vectors

From now on we will use letters in **bold** type to represent vectors. If $(x, y)$ is the terminal point of the vector $v$ with initial point $(0,0)$, we will write $v = (x, y)$ because $v$ is completely determined by its terminal point. We call $x$ the **$x$-component** and $y$ the **$y$-component** of the vector $v$. The magnitude of $v$, is the length $\sqrt{x^2 + y^2}$ of $v$, and is denoted by $|v|$.

- **EXAMPLE 2:** Let $A = (2,5), B = (4,6), C = (-3,4)$, and $D = (-5,-2)$ be four points. Determine the vector $v = (x, y)$ that is equal to the vector

    (a) $\overrightarrow{AB}$.  (b) $\overrightarrow{CD}$.

SOLUTION:

(a) Let $P = (x, y)$ be the terminal point of $v$ (Figure 5.4.6). Then $d(O, P) = d(A, B)$ and $OP$ and $AB$ have the same slope. (Why?) The slope of $AB$ is $\frac{1}{2}$. (Why?) Thus, the slope $\frac{y}{x}$ is also equal to $\frac{1}{2}$. It follows that $\frac{y}{x} = \frac{1}{2}$ or $2y = x$. Now $d(A, B) = \sqrt{2^2 + 1^2} = \sqrt{5}$ so that $d(O, P) = \sqrt{x^2 + y^2} = \sqrt{5}$, or $x^2 + y^2 = 5$. Substituting $2y$ for $x$ in $x^2 + y^2 = 5$ we obtain

$$x^2 + y^2 = 5$$
$$4y^2 + y^2 = 5$$
$$y^2 = 1.$$

It follows that $y = 1$ because $P$ is in the first quadrant. (Why?) Thus, $x = 2$ and $v = (2, 1)$.

Alternatively, the terminal point $B$ of $\overrightarrow{AB}$ is 2 units right and 1 unit up from the initial point $A$. Therefore, the terminal point $P$ of $\overrightarrow{OP}$ must be 2 units right and 1 unit up from the initial point $(0, 0)$. Thus, $P = (2, 1)$ and $v = (2, 1)$.

(b) Let $P = (x, y)$ be the terminal point of $v$ (Figure 5.4.7). Both $x$ and $y$ must be negative. (Why?) Notice $P$ must be 2 units left and 6 units down from the origin so that $P = (-2, -6)$. We can use the fact that $OP$ and $CD$ have the same slope

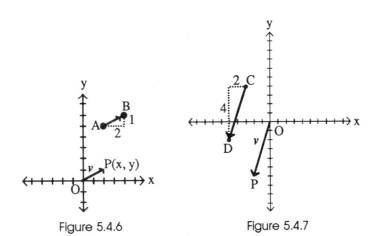

Figure 5.4.6    Figure 5.4.7

to obtain $y = 3x$, and $d(O,P) = d(C,D) = \sqrt{40}$ to obtain $x^2 + y^2 = 40$. Again $P = (-2,-6)$ so that $v = (-2,-6)$. •

Notice in Example 2(a) that the coordinates of the vector $v = (2,1)$ can be obtained by subtracting the coordinates of the point $A = (2,5)$ from the coordinates of the point $B = (4,6)$. That is, $v = (2,1) = (4-2, 6-5)$. Similarly, for the vector of Example 2(b) we have $v = (-2,-6) = (-5-(-3), -2-4)$. These are special cases of the following theorem.

• **THEOREM 2**  Let $A = (a_1, a_2)$ and $B = (b_1, b_2)$ be two points, and let $v = (x,y)$ be the vector with initial point the origin that is equal to $\overrightarrow{AB}$. Then $x = b_1 - a_1$ and $y = b_2 - a_2$.

Just as we say $\frac{2}{4} = \frac{1}{2}$ for rational numbers, we will say that $\overrightarrow{AB} = (b_1 - a_1, b_2 - a_2)$ for the vector $\overrightarrow{AB}$ in Theorem 2.

## Addition of Vectors

Let $u$ and $v$ be any two vectors. We define the **sum** $u + v$ as follows. Draw a vector equal to $v$ with initial point the terminal point of $u$ and terminal point $P$ (Figure 5.4.8). We may use $v$ for the name of this vector because it is equal to $v$. The vector $u + v$ is the vector with initial point $O$ and terminal point $P$.

If we close the parallelogram in Figure 5.4.8 with a vector equal to $u$, then the diagonal would represent $v + u$ (Figure 5.4.9). Thus, $u + v = v + u$ so that addition of vectors is *commutative*.

Notice that the addition of vectors is essentially the same as the geometric addition of complex numbers described in Section 5.1. If $(a,b)$ is the terminal point of $u$ and $(c,d)$ is the terminal point of $v$, then the following can be shown to be valid:

$$u + v = (a,b) + (c,d)$$
$$= (a+c, b+d).$$

Figure 5.4.8

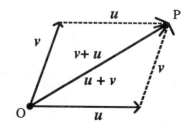

Figure 5.4.9

## 5.4 Vectors

In other words, if we know the *components* of $u$ and $v$, then the *components* of the sum $u + v$ are the sum of the *components* of $u$ and of $v$. In fact, complex numbers can be represented by vectors. For example, the vector $u$ with initial point $(0,0)$ and terminal point $(2,3)$ is a vector representation of the complex number $2+3i$ (Figure 5.4.10). It can be shown that addition of complex numbers is preserved by this representation. That is, if $a+bi$ corresponds to $v$ and $c + di$ corresponds to $u$, then $(a + c) + (b + d)i$ corresponds to $v + u$.

- **EXAMPLE 3**: Use the geometric definition of addition of vectors to find $u + v$, where $u = (2, 5)$ and $v = (4, 3)$. Compare with componentwise addition.

  SOLUTION: The vector equal to $v = (4, 3)$ with initial point $(2, 5)$ has terminal point $(6, 8)$ (Figure 5.4.11). (Why?) Thus, $u + v = (6, 8)$. Componentwise addition gives

  $$u + v = (2, 5) + (4, 3)$$
  $$= (2 + 4, 5 + 3)$$
  $$= (6, 8).$$

- **DEFINITION**    The vector with $(0, 0)$ as both the initial and terminal point is denoted by **0** and is called the **zero vector**.

  Notice that $u + 0 = 0 + u = u$ for any vector $u$.

- **DEFINITION**    Let $u$ be any vector. The vector with the same magnitude as $u$ but with opposite direction is denoted by $-u$ and is called the **additive inverse** or **negative of** $u$.

  Notice that $u+(-u) = (-u)+u = 0$ for any vector $u$. Some of the important properties about vectors are summarized in Theorem 3, which we give without proof.

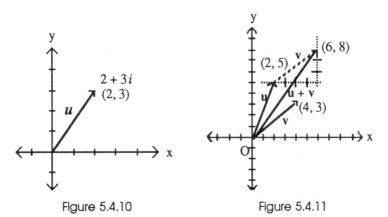

Figure 5.4.10      Figure 5.4.11

• **THEOREM 3**  Let $u$, $v$, and $w$ be arbitrary vectors. Then,

(1) $u + v = v + u$  (Commutative Property)
(2) $(u + v) + w = u + (v + w)$  (Associative Property)
(3) $u + 0 = u$  (Additive Identity Property)
(4) $u + (-u) = 0$  (Additive Inverse Property)

## Subtraction of Vectors

The difference of two vectors $u$ and $v$ is denoted by $u - v$ and is that vector which, when added to $v$ gives $u$ (Figure 5.4.12). That is, $v + (u - v) = u$. Notice that the vector $u - v$ has initial point the terminal point of $v$, and terminal point the terminal point of $u$. If $u = (a, b)$ and $v = (c, d)$ then subtraction can be accomplished *componentwise* because the following can be shown to be true:

$$u - v = (a, b) - (c, d)$$
$$= (a - c, b - d).$$

## Multiplication by a Scalar

We use a vector and a real number to form a new vector using the following definition.

• **DEFINITION**  Let $a$ be any real number and $u$ any vector. The **scalar multiple** of $u$ by $a$ is defined as follows.

(1) If $a = 0$, then $au = 0$.
(2) If $a > 0$, then the vector $au$ has the same direction as $u$ and has magnitude $a|u|$.
(3) If $a < 0$, then the vector $au$ has direction opposite to $u$ and has magnitude $|a||u|$.

Figure 5.4.12   Figure 5.4.13

## 5.4 Vectors

Figure 5.4.13 gives some examples of scalar multiples of a vector $u$. Notice that a scalar multiple of a vector $u$ is a stretch or shrink of $u$ in the same or opposite direction of $u$.

Some of the important properties about scalar multiplication are summarized in Theorem 4, which we give without proof.

- **THEOREM 4**  Let $u$ and $v$ be any two vectors and $a$ and $b$ any two scalars. Then,
  (1) $a(u+v) = au + av$
  (2) $(a+b)u = au + bu$
  (3) $(ab)u = a(bu)$
  (4) $(1)u = u$
  (5) $(-1)u = -u$
  (6) $a0 = 0,\ 0u = 0$

If $u = (a, b)$, then scalar multiplication can be accomplished *componentwise* because the following can be shown to be true:

$$cu = c(a, b)$$
$$= (ca, cb).$$

### Linear Combinations of Vectors

- **DEFINITION**  The vector $w$ is said to be a **linear combination** of the vectors $u$ and $v$ if there are scalars $a$ and $b$ such that $w = au + bv$.

Let $i = (1, 0)$ and $j = (0, 1)$ (Figure 5.4.14). Every vector can be expressed as a linear combination of $i$ and $j$. This can be determined as follows. Let $u = (a, b)$ be any vector:

$$u = (a, b)$$
$$= (a, 0) + (0, b)$$
$$= a(1, 0) + b(0, 1)$$
$$= ai + bj.$$

Notice that the scalars in this case are the $x$- and $y$-components of the vector $u = (a, b)$ (Figure 5.4.15). The magnitude of each of the vectors $i$ and $j$ is 1. Vectors with magnitude 1 are called **unit vectors**. Hence $i$ and $j$ are unit vectors.

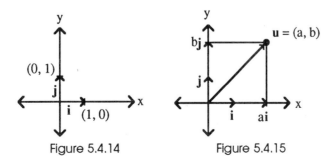

Figure 5.4.14  Figure 5.4.15

- **EXAMPLE 4:** Let $A = (-3, 2)$ and $B = (2, 6)$. Express the vector $\overrightarrow{AB}$ as a linear combination of $\boldsymbol{i}$ and $\boldsymbol{j}$.

  SOLUTION: We use Theorem 2 and compute as follows:
  $$\begin{aligned}\overrightarrow{AB} &= (2 - (-3), 6 - 2) \\ &= (5, 4) \\ &= 5(1, 0) + 4(0, 1) \\ &= 5\boldsymbol{i} + 4\boldsymbol{j}.\end{aligned}$$
  •

  If we know an angle that gives the direction of a vector, then we can determine its $x$- and $y$-components. This is very similar to the procedure used in Section 5.1 to find the trig form of a complex number.

- **THEOREM 5**    Let $\boldsymbol{u} = (a, b)$ be any vector and $P$ the point with coordinates $(a, b)$ (Figure 5.4.16). If $\theta$ is an angle in standard position with terminal side $OP$, then $a = |\boldsymbol{u}|\cos\theta$ and $b = |\boldsymbol{u}|\sin\theta$.

  PROOF   We know that $r = \sqrt{a^2 + b^2} = |\boldsymbol{u}|$. From the definition of the trig functions at $\theta$ we have $\cos\theta = \frac{a}{r}$ and $\sin\theta = \frac{b}{r}$. Thus, $a = r\cos\theta = |\boldsymbol{u}|\cos\theta$ and $b = r\sin\theta = |\boldsymbol{u}|\sin\theta$.

- **EXAMPLE 5:** Determine the $x$- and $y$-components of the vector $\boldsymbol{u}$ that makes an angle of $115°$ with the positive $x$-axis and has length 6 (Figure 5.4.17).

  SOLUTION: By Theorem 5, the $x$-coordinate of $\boldsymbol{u}$ is $6\cos 115° = -2.54$ and the $y$-coordinate is $6\sin 115° = 5.44$. Thus, $\boldsymbol{u} = (-2.54, 5.44)$.
  •

- **EXAMPLE 6:** Let $\boldsymbol{u} = (3, 2)$ and $\boldsymbol{v} = (-2, 5)$. Determine
  - (a) $2\boldsymbol{u} - 3\boldsymbol{v}$.
  - (b) $|\boldsymbol{u}|$ and $|\boldsymbol{v}|$.
  - (c) the angle that each vector makes with the positive $x$-axis.
  - (d) the angle with initial side $\boldsymbol{u}$ and terminal side $\boldsymbol{v}$.

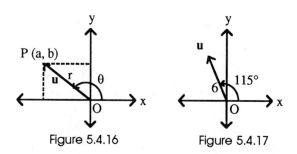

Figure 5.4.16      Figure 5.4.17

## 5.4 Vectors

**SOLUTION:**

(a) $2\mathbf{u} - 3\mathbf{v} = 2(3,2) - 3(-2,5)$
$= (6,4) - (-6,15)$
$= (12,-11)$

(b) $|\mathbf{u}| = \sqrt{3^2 + 2^2} = \sqrt{13}$
$|\mathbf{v}| = \sqrt{(-2)^2 + 5^2} = \sqrt{29}$

(c) The angle $\alpha$ that $\mathbf{u}$ makes with the positive $x$-axis is acute (Figure 5.4.18). Thus, $\alpha = \tan^{-1}(\frac{2}{3}) = 33.69°$. The angle $\beta$ that $\mathbf{v}$ makes with the positive $x$-axis is in the *second* quadrant. Now, $\tan^{-1}(-\frac{5}{2}) = -68.20°$ is *not* in the second quadrant. Thus, $\beta = 180° + \tan^{-1}(-\frac{5}{2}) = 111.80°$.

(d) We can see from Figure 5.4.18 that the angle $\theta$ with initial side $\mathbf{u}$ and terminal side $\mathbf{v}$ is $\beta - \alpha$, or $78.11°$. •

Example 6(d) can be solved directly without finding the angles that $\mathbf{u}$ and $\mathbf{v}$ make with the positive $x$-axis. Let $A = (3,2)$ and $B = (-2,5)$ be the terminal points of the vector $\mathbf{u}$ and $\mathbf{v}$ in Figure 5.4.18. Then from the Law of Cosines we have

$$d^2(A,B) = |\mathbf{u}|^2 + |\mathbf{v}|^2 - 2|\mathbf{u}||\mathbf{v}|\cos(\beta - \alpha)$$
$$34 = 12 + 29 - 2\sqrt{13}\sqrt{29}\cos(\beta - \alpha)$$
$$\beta - \alpha = \cos^{-1}\left(\frac{13 + 29 - 34}{2\sqrt{13}\sqrt{29}}\right)$$
$$\beta - \alpha = 78.11°$$

We can generalize this approach in the following way.

- **Angle between Vectors** Let $\mathbf{u}$ and $\mathbf{v}$ be vectors (Figure 5.4.19). The angle $\beta$ between $\mathbf{u}$ and $\mathbf{v}$ is given by

$$\beta = \cos^{-1}\left(\frac{|\mathbf{u}|^2 + |\mathbf{v}|^2 - d^2}{2|\mathbf{u}||\mathbf{v}|}\right)$$

where $d$ is the distance between the terminal points of $\mathbf{u}$ and $\mathbf{v}$.

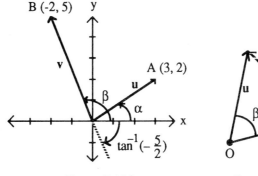

Figure 5.4.18        Figure 5.4.19

The **velocity** of a moving object is a vector because it has both magnitude and direction. Usually, the magnitude of velocity is called **speed**.

- **EXAMPLE 7:** A plane is flying on a bearing of 65° east of north at 500 mph. Express the velocity of the plane as a vector.

  SOLUTION: Let $v$ be the velocity vector (Figure 5.4.20). The angle that $v$ makes with the positive $x$-axis is 25°. The magnitude of $v$ is 500 because the plane is flying at 500 mph. It follows from Theorem 5 that

  $$v = (500\cos 25°, 500\sin 25°)$$
  $$= (453.15, 211.31).$$

- **EXAMPLE 8:** A plane is flying on a bearing of 65° east of north at 500 mph. A tail wind that adds to the plane's velocity is blowing in the direction 35° east of north at 80 mph. Express the actual velocity of the plane as a vector. Determine the actual speed and direction of the plane.

  SOLUTION: In Example 7 we found that the velocity of the plane in still air was $v = (453.15, 211.31)$. The velocity $u$ of the wind can be expressed as a vector as follows (Figure 5.4.21). The angle $u$ makes with the positive $x$-axis is 55° because $u$ is 35° east of north. Thus, $u = (80\cos 55°, 80\sin 55°)$, or $u = (45.89, 65.53)$. The actual speed and direction of the plane is given by the vector $u + v$:

  $$u + v = (45.89, 65.53) + (453.15, 211.31)$$
  $$= (499.04, 276.84).$$

The actual speed of the plane is $|u+v| = \sqrt{(499.04)^2 + (276.84)^2} = 570.68$ mph, in the direction $\tan^{-1}(\frac{276.84}{499.04}) = 29.02°$ north of east. We could also say that the direction of the plane is 60.98° east of north.

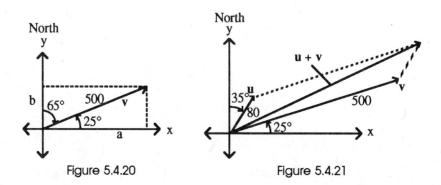

Figure 5.4.20

Figure 5.4.21

# 5.4 Vectors

Usually pilots face a slightly different problem. The actual speed and direction ($u+v$) are known as well as the tail or head wind vector ($u$). The flight bearing must be established. This is illustrated in Exercises 38 and 39.

## EXERCISES 5-4

Let $u = (-1, 3)$, $v = (2, 4)$, and $w = (2, -5)$ be vectors. Determine the following.

1. $u + v$
2. $u - v$
3. $u - w$
4. $3v$
5. $2u + 3w$
6. $2u - 4v$
7. $|u + v|$
8. $|u - v|$

Use the geometric definitions of addition and subtraction of vectors to find the sum $u + v$, the difference $u - v$, and then compare with the results obtained using the respective componentwise definitions.

9. $u = (1, 3)$, $v = (3, 6)$
10. $u = (-1, 2)$, $v = (4, -2)$

Let $A = (-1, 2)$, $B = (3, 4)$, $C = (-2, 5)$, and $D = (2, -8)$. Write each as a vector with *initial point at the origin*.

11. $\overrightarrow{AB}$
12. $3\overrightarrow{AB}$
13. $2\overrightarrow{AB} - \overrightarrow{BA}$
14. $\overrightarrow{AB} + \overrightarrow{CD}$
15. $\overrightarrow{AB} - \overrightarrow{CD}$
16. $2\overrightarrow{AB} + 3\overrightarrow{CD}$

Compute.

17. $|\overrightarrow{AB}|$
18. $|2\overrightarrow{AB}|$
19. $|\overrightarrow{AB} + \overrightarrow{BA}|$

Show geometrically that $\overrightarrow{AB} = \overrightarrow{OB} - \overrightarrow{OA}$.

20. $A = (2, 3)$, $B = (5, 2)$
21. $A = (-2, 4)$, $B = (2, 6)$

Express the vector $\overrightarrow{AB}$ as linear combination of $i = (1, 0)$ and $j = (0, 1)$.

22. $A = (3, 0)$, $B = (0, 6)$
23. $A = (2, 1)$, $B = (5, 0)$
24. $A = (3, -2)$, $B = (2, 6)$
25. $A = (-2, 1)$, $B = (-3, -5)$

Find the angle each vector makes with the positive $x$-axis. Then determine the angle between the vectors.

26. $u = (3, 4)$, $v = (1, 0)$
27. $u = (-1, 2)$, $v = (3, 2)$
28. $u = (-1, 2)$, $v = (3, -4)$
29. $u = (2, -3)$, $v = (-3, -5)$

30. Let $r = (1, 2)$ and $s = (2, -1)$. Show that the vector $v = (5, 7)$ can be expressed as linear combination of $r$ and $s$. Explain what this means geometrically.

31. Let $r = (1, 2)$ and $s = (2, -1)$. Show that *any* vector $v = (x, y)$ can be expressed as linear combination of $r$ and $s$.

32. Let $r = (a, b)$ and $s = (c, d)$. Show that any vector $v = (x, y)$ can be expressed as linear combination of $r$ and $s$ for almost all values of $a$, $b$, $c$, and $d$. For what values of $a$, $b$, $c$, and $d$ is it *impossible* to express $v$ as a linear combination of $r$ and $s$?

33. A plane is flying on a bearing of 25° west of north at 530 mph. Express the velocity of the plane as a vector.

392 • Chapter 5 Trigonometry of Complex Numbers, Vectors, Polar Equations, and Parametric Equations

34. A plane is flying on a bearing of 10° east of south at 460 mph. Express the velocity of the plane as a vector.

35. A plane is flying on a bearing of 20° west of north at 325 mph. A tail wind is blowing in the direction 40° west of north at 40 mph. Express the actual velocity of the plane as a vector. Determine the actual speed and direction of the plane.

36. A plane is flying on a bearing of 10° east of south at 460 mph. A tail wind is blowing in the direction 20° west of south at 80 mph. Express the actual velocity of the plane as a vector. Determine the actual speed and direction of the plane.

37. A sailboat under auxiliary power is proceeding on a bearing of 25° north of west at 6.25 mph in still water. Then a tail wind blowing 15 mph in the direction 35° south of west alters the course of the sailboat. Express the actual velocity of the sailboat as a vector. Determine the actual speed and direction of the boat.

38. A pilot must actually fly due west at a constant speed of 382 mph against a head wind of 55 mph blowing in the direction of 22° south of east. What direction and speed must the pilot maintain to keep on course (due west)?

39. A jet fighter pilot must actually fly due north at a constant speed of 680 mph against a head wind of 80 mph blowing in the direction of 10° west of south. What direction and speed must the pilot maintain to keep on course (due north)?

## Motion Problems

40. A ball is tossed up in the air at an angle of 70° with the horizontal and with initial velocity 36 feet per second (Figure 5.4.22). Neglecting air resistance *and* gravity, specify the position of the ball as a vector 1 second, 2.5 seconds, and 4 seconds after the ball is released. Express the position of the ball at time $t$, where $t$ is the number of seconds after the ball is released. Explain why this is not a realistic model. (When does the ball return to the ground?)

Consider an object dropped with initial velocity *zero* (a so-called "freely falling body") from a tower at point $A$ (Figure 5.4.23). Let point $B$ be the position of the object $t$ seconds after it is released (neglecting air resistance and wind effects). It is a fact from elementary physics that on Earth the

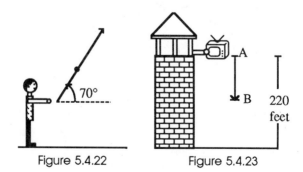

Figure 5.4.22   Figure 5.4.23

vector $\overrightarrow{AB}$ is equal to the "effect of gravity" vector $g = (0, -16t^2)$. Assume point $A$ is 220 above the ground.

41. Determine a vector that describes the speed and position of the object after 2 seconds, 3 seconds, and 4 seconds. Specify the coordinates of point $B$.
42. When will the object hit the ground? At what speed?
43. Rework Exercise 40 accounting for the **effect of gravity**. That is, the ball is tossed up and it eventually falls back to the ground (due to gravity). *Hint: Add* the two vectors $u$ and $g$ where $u$ is the vector in Exercise 40 (neglecting the effect of gravity) and $g$ is the "effect of gravity" vector.
44. When will the ball of Exercise 40 hit the ground? At what speed?
45. How high above the ground will the ball of Exercise 40 reach? How far will the ball travel in the horizontal direction?
46. Let $\ell_1$ and $\ell_2$ be perpendicular lines intersecting at the origin with $\ell_1$ in the first quadrant and third quadrant. Let the slope of $\ell_1$ be given by $b/a$ with $a \neq 0$ and $b \neq 0$. Show that the slope of the lines bisecting the angle between $\ell_1$ and $\ell_2$ is $\frac{a \pm b}{a - b}$. (*Note:* this problem was suggested by Richard Liu, a student in Mrs. Pam Giles's 1988–89 $C^2$PC class in Sandy, Utah.)

## 5.5 Parametric Equations

In this section we introduce parametric equations for a curve. We show that the graph of a polar function $r = f(\theta)$ can be obtained parametrically. The graphs of many important relations can often be obtained as graphs of parametric equations for the curve defined by the relation. Most graphing utilities can easily produce the graph of curves defined by parametric equations.

- **DEFINITION** Let $f$ and $g$ be two functions defined on an interval $I$. The relation consisting of all ordered pairs $(f(t), g(t))$ for $t$ in $I$ is called a **plane curve** $C$, or more simply a **curve** $C$. The equations $x = f(t)$ and $y = g(t)$ for $t$ in $I$ are called **parametric equations** for $C$. The variable $t$ is called a **parameter**. The parametric equations $x = f(t)$ and $y = g(t)$, with $t$ in $I$, are said to give a **parametrization** of $C$ in terms of $t$.

Figure 5.5.1 shows several possibilities for plane curves.

If $I$ is the interval $[a, b]$, then the points $(x(a), y(a))$ and $(x(b), y(b))$ are called **end points** of the curve $C$. It is possible that a curve described parametrically intersects itself. This means that more than one value of $t$ produces the same point in the coordinate plane. Usually a curve is not the graph of a function. However, any function $y = F(x)$ can be defined parametrically by the equations $x = f(t)$, $y = g(t)$ where $f(t) = t$ and $g(t) = F(t)$.

- **EXAMPLE 1:** Describe the plane curve $C$ defined by $x = 1 - 2t$ and $y = 2 - t$ for $t$ in $[0, 1]$.

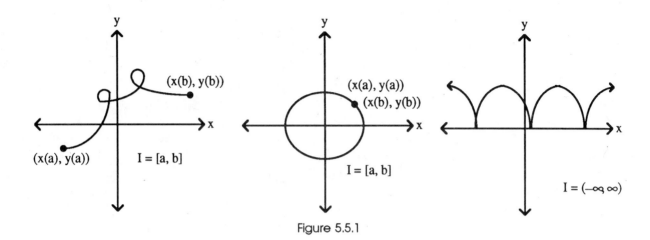

Figure 5.5.1

SOLUTION: The point corresponding to $t = 0$ is $(1, 2)$ and to $t = 1$ is $(-1, 1)$. Thus, the points $(1, 2)$ and $(-1, 1)$ are endpoints of $C$. Notice that $x$ varies from 1 to $-1$ and $y$ varies from 2 to 1 as $t$ varies from 0 to 1. In other words, as $t$ increases from 0 to 1, $x$ decreases from 1 to $-1$ and $y$ decreases from 2 to 1. We solve $y = 2 - t$ for $t$ ($t = 2 - y$) and substitute into $x = 1 - 2t$ obtaining $x = 1 - 2(2 - y)$ or $2y - x = 3$.

The graph of $2y - x = 3$ is a straight line. Thus, $C$ is the portion of the straight line between $(-1, 1)$ and $(1, 2)$ (Figure 5.5.2). •

In Example 1, we were able to remove the parameter $t$ and obtain an equation involving only the variables $x$ and $y$. Usually it is impossible to remove the parameter. Thus, in such cases a graph will need to be determined directly from the parametric equations.

The *Graphing Calculator and Computer Graphing Laboratory Manual* accompanying this textbook contains a simple program that permits graphs of curves defined by parametric equations to be obtained with a graphing calculator. The graphing utility *Master Grapher* used to draw graphs in this textbook has a parametric option. For example, the graph in Example 1 can be obtained by entering $x(t) = 1 - 2t$, $y(t) = 2 - t$, and choosing the range

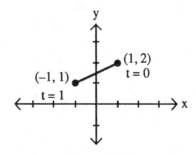

Figure 5.5.2

## 5.5 Parametric Equations

on $t$ to be $0 \leq t \leq 1$. We use the notation $t\min$ and $t\max$ for the smallest and largest values of $t$ entered into the graphing utility. The graph in Figure 5.5.2 can be confirmed using a parametric graphing utility.

- **EXAMPLE 2:** Find parametric equations for the circle with center at the origin and radius 2.

  **SOLUTION:** Let $P(x, y)$ be a point on the circle (Figure 5.5.3). Let $\theta$ be the angle $OP$ makes with the positive $x$-axis. We use $\theta$ as the parameter to obtain parametric equations for this circle. From the definition of the trig functions we have $x = 2\cos\theta$ and $y = 2\sin\theta$. Notice that we obtain a complete graph of the circle by letting $\theta$ vary through *any* interval of length $2\pi$. (Why?) In particular, $x = 2\cos\theta$, $y = 2\sin\theta$, for $t\min = 0$ and $t\max = 2\pi$ gives a parametrization of the circle centered at the origin with radius 2. •

  The domain of the relation graphed in Figure 5.5.3 is $-2 \leq x \leq 2$ and the range is $-2 \leq y \leq 2$. In general, the domain and range of a curve are not the same as the range of values of the parameter that produce the curve. For example, $0 \leq \theta \leq 2\pi$ will produce the graph in Figure 5.5.3.

- **EXAMPLE 3:** Draw the graph of the curve defined by $x = t\cos t$, $y = t\sin t$ for $0 \leq t \leq 4\pi$.

  **SOLUTION:** The graph of the curve is shown in Figure 5.5.4. •

  The graph in Figure 5.5.4 is exactly the same as one of the *polar* graphs of $r = \theta$ in Figure 5.3.16. This suggests that polar graphs and parametric equations are related. Indeed, this is true and the following theorem gives the connection.

- **THEOREM 1**    The polar graph $r = f(\theta)$ is the curve defined parametrically by $x(t) = f(t)\cos t$ and $y(t) = f(t)\sin t$.

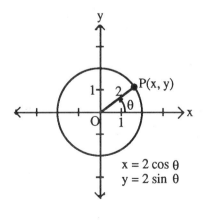

$x = 2\cos\theta$
$y = 2\sin\theta$

Figure 5.5.3

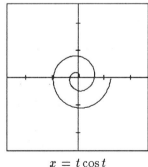

$x = t\cos t$
$y = t\sin t$
$[-27, 27]$ by $[-27, 27]$
$0 \leq \theta \leq 4\pi]$

Figure 5.5.4

**PROOF** Let $(x,y)$ be the rectangular coordinates of a point $P$ on the polar graph $r = f(t)$ (Figure 5.5.5). By definition of the trig functions we have

$$\sin t = \frac{y}{r}, \qquad \cos t = \frac{x}{r}.$$

Thus, $x = r\cos t = f(t)\cos t$ and $y = r\sin t = f(t)\sin t$. Notice that we have replaced $\theta$ by $t$ in $r = f(\theta)$.

Theorem 1 allows us to draw a graph of any polar equation using a parametric graphing utility. In fact, this theorem gives the basis for using a parametric graphing program to obtain polar graphs.

• **EXAMPLE 4:** Determine parametric equations for the four-leafed rose $r = 7\sin 2\theta$. Use a parametric graphing utility to draw a complete graph of the curve.

**SOLUTION:** Applying Theorem 1 we have $x(t) = 7\sin 2t \cos t$ and $y(t) = 7\sin 2t \sin t$ where we have replaced $\theta$ by $t$. A complete graph is produced for $0 \leq t \leq 2\pi$ as shown in Figure 5.5.6. •

• **EXAMPLE 5:** Draw the graph of the curve defined by $x(t) = 5\sin t$ and $y(t) = 5t\cos t$ for $0 \leq t \leq 20$.

**SOLUTION:** The curve is shown in Figure 5.5.7. The $x$ coordinates of the points are between $-5$ and $5$ regardless of the value of $t$. (Why?) The $y$ coordinates are between $-100$ and $100$ for all values of $t$ in $0 \leq t \leq 20$. (Why?) •

• **EXAMPLE 6:** Draw a complete graph of the curve defined by the parametric equations $x = 5t - 5\sin t$ and $y = 5 - 5\cos t$.

Figure 5.5.5

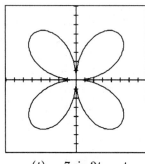

$x(t) = 7\sin 2t \cos t$
$y(t) = 7\sin 2t \sin t$
$0 \leq t \leq 2\pi$
$[-8, 8]$ by $[-8, 8]$
Figure 5.5.6

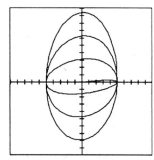

$x(t) = 5\sin t$
$y(t) = 5t\cos t$
$0 \leq t \leq 20$
$[-10, 10]$ by $[-100, 100]$
Figure 5.5.7

# 5.5 Parametric Equations

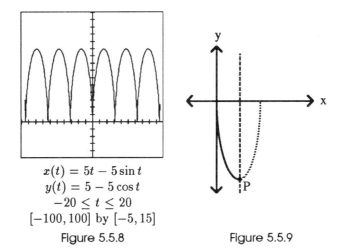

$x(t) = 5t - 5\sin t$
$y(t) = 5 - 5\cos t$
$-20 \leq t \leq 20$
$[-100, 100]$ by $[-5, 15]$
Figure 5.5.8

Figure 5.5.9

SOLUTION: The values of $\cos t$ vary between $-1$ and $1$. Thus, we can see from the equation for $y$ that the values of $y$ vary between $0$ and $10$. (Why?) Furthermore, $y$ is a periodic function of $t$ with period $2\pi$. The values of $\sin t$ vary between $-1$ and $1$. Thus, $-5\sin t$ varies from $-5$ to $5$. The values of $x$ vary from $-\infty$ to $\infty$ as the values of $t$ vary from $-\infty$ to $\infty$. (Why?) The graph in Figure 5.5.8 is a complete graph. You may need to view the graph in additional viewing rectangles to convince yourself that this is a complete graph of the curve. •

The graph in Figure 5.5.8 is called a cycloid. In fact, the graph of the parametric equations $x = a(t - \sin t)$ and $y = a(1 - \cos t)$ for any real number $a$ and all real $t$ is a **cycloid**. Graphs of cycloids are similar to the graph in Figure 5.5.8.

Cycloids are important in physics. They are sometimes called the *curve of quickest descent*. Suppose a cycloid is reflected through the $x$-axis. Then, out of all possible frictionless curves connecting the origin to the first local minimum (point $P$) to the right of $x = 0$, a cycloid is the one that an object will slide down in the *least* amount of time (Figure 5.5.9).

## • EXERCISES 5-5

Describe the curve defined parametrically. Sketch the graph without using a graphing utility. Check using a parametric graphing utility.

1. $x(t) = 1 + t$ and $y(t) = t$ with $t$ in $[0, 5]$
2. $x(t) = 1 + t$ and $y(t) = t$ with $t$ in $[-3, 2]$
3. $x(t) = 2 - 3t$ and $y(t) = 5 + t$ with $t$ in $[-1, 3]$
4. $x(t) = 2 - 3t$ and $y(t) = 5 + t$ with $t$ in $(-\infty, \infty)$
5. $x(t) = t$ and $y(t) = \frac{2}{t}$ with $t$ in $(0, 5]$
6. $x(t) = t + 2$ and $y(t) = \frac{2}{t}$ with $t$ in $[-3, 0) \cup (0, 3]$

7. $x(t) = 3\cos t$ and $y(t) = 3\sin t$ with $t$ in $[0, 4\pi]$
8. $x(t) = 4\sin t$ and $y(t) = 4\cos t$ with $t$ in $[-\frac{\pi}{2}, \frac{\pi}{2}]$
9. Which of the curves in Exercises 1–8 are graphs of functions?
10. Which of the curves in Exercises 1–8 are graphs of one-to-one functions?
11. What are the domain and range of the curve in Exercise 2?
12. What are the domain and range of the curve in Exercise 3?
13. What are the domain and range of the curve in Exercise 5?
14. What are the domain and range of the curve in Exercise 7?

Find parametric equations and draw the graph of each circle.

15. centered at the origin with radius 5
16. centered at $(10, 0)$ with radius 4
17. centered at $(0, 10)$ with radius 6
18. centered at $(a, b)$ with radius $r$

Draw a complete graph.

19. $x(t) = 4t - 2$ and $y = 8t^2$ with $t$ in $(-\infty, \infty)$
20. $x(t) = 4\cos t$ and $y = 8\sin t$ with $t$ in $(-2\pi, 2\pi)$
21. $x(t) = 1 + \frac{1}{t}$ and $y = t - \frac{1}{t}$ with $t$ in $(0, 20]$
22. $x(t) = 2t - 2\sin t$ and $y(t) = 2 - 2\cos t$ with $t$ in $[0, 40]$
23. $x(t) = 4t - 4\sin t$ and $y(t) = 4 - 4\cos t$ with $t$ in $[0, 40]$
24. $x(t) = 6\cos t - 4\cos(\frac{3}{2}t)$ and $y(t) = 6\sin t - 4\sin(\frac{3}{2}t)$ with $t$ in $[0, 2\pi]$. (This is an *epicycloid*, the path of a point on a circle rolling on another circle.)
25. $x(t) = 5\cos^3 t$ and $y(t) = 5\sin^3 t$ with $t$ in $[0, 2\pi]$. (This is a *hypocycloid* of four cusps.)
26. $x(t) = 4(\cos t + t\sin t)$ and $y(t) = 4(\sin t - t\cos t)$ with $t$ in $[0, 50]$. (This is the *involute* of a circle.) *Hint:* Use a large viewing rectangle.
27. Find the domain and range of the curve in Exercise 25.
28. Find the domain and range of the curve in Exercise 26.
29. Let $f$ be a function. Show that the curve defined parametrically by $x(t) = t$ and $y(t) = f(t)$ with $t$ in the interval $I$ is the *function $f$* with domain equal to the interval $I$.
30. Let $f(x) = 3 - x^2$. Show that the curve defined parametrically by $x(t) = at + b$ and $y(t) = f(t)$ with $t$ in the interval $I$ is a *function $y = g(x)$*. Find a rule for $g(x)$. What is the domain of $g$?
31. Find the maximum value (if any) of $y(t)$ for the curve defined parametrically by $x(t) = 2t - 1$ and $y(t) = 4 - t^2$.

Write the polar equation in parametric form. Graph *both* the parametric and polar form with an appropriate graphing utility.

32. $r = 3\theta$ for $\theta$ in $[0, 10\pi]$
33. $r = 5\sin\theta$ for $\theta$ in $[-\pi, \pi]$
34. $r = 5\sin 2\theta$ for $\theta$ in $[0, \pi]$
35. $r = 5\sin 3\theta$ for $\theta$ in $[0, 2\pi]$
36. Draw a complete graph of $x(t) = a(t - \sin t)$ and $y(t) = a(1 - \cos t)$ for $a = 1, 2, 3, 4,$ and $5$. Use zoom-in and determine the period of the curve and the maximum $y$-coordinate with least positive $x$-coordinate.

37. Based on the results of Exercise 36, make a conjecture for the *period* and the maximum $y$-value with least positive $x$-coordinate of the cycloid $x(t) = a(t - \sin t)$ and $y(t) = a(1 - \cos t)$ for an arbitrary value of $a$.

## 5.6 Motion Problems and Parametric Equations

In this section we continue the study of parametric equations. We see that the path of objects acted upon by certain forces can be described by parametric equations. We asume all motion occurs in a two dimensional plane.

In Section 5.4 we found that velocity can be described with a vector because it has both magnitude and direction. We know from Section 1.3 that if we assume that the only force acting on an object is due to gravity, then the distance above ground of an object thrown straight up with initial velocity $v_0$ from a point $s_0$ above ground is given by $s = -16t^2 + v_0 t + s_0$. Actually, the coefficient of $t^2$ is $-\frac{g}{2}$, where $g$ is an acceleration constant due to gravity. A usual approximation for $g$ on Earth at sea level is 32 ft/sec$^2$.

In Example 1 we consider an object that is propelled with an initial velocity that acts in a direction that makes an angle different from 90° with the horizontal. The velocity vector will have both a horizontal and vertical component in this case. We also neglect air resistance when we find parametric equations describing the path of the object.

• **EXAMPLE 1:** An object is launched with an initial velocity of 50 ft/sec at an angle of 60° with the positive $x$-axis. Assume that the only force acting on the object is due to gravity.

   (a) Find parametric equations that model the motion of the object.
   (b) When will the object hit the ground?
   (c) How far does the object travel in the horizontal direction?
   (d) Find the maximum height attained by the object, and when it reaches that height.
   (e) Show that the position at any time $t$ of the object written as a vector can be expressed as the sum of two vectors; one of the vectors depends only on the initial velocity and the other only on gravity.

   SOLUTION: Let $t$ be the time (in seconds) that the object is in flight. Assume the object is launched from the origin.

   (a) Let $v_0$ be the initial velocity vector. The horizontal component of the initial velocity vector is $50 \cos 60° = 25$ and the vertical component is $50 \sin 60° = 25\sqrt{3}$ (Figure 5.6.1). Thus, $v_0 = (25, 25\sqrt{3})$. Let $P(t) = P(x, y)$ be the position of the object at time $t$. Then, both $x$ and $y$ are functions of time. By assumption, the only force acting in the horizontal direction on the object is the horizontal component 25 of $v_0$. Using the standard distance formula $d = rt$ we have $x(t) = 25t$. In the vertical direction, gravity and the vertical component of $v_0$ are the only forces assumed to be acting on the object. We can use the formula of Section 1.3 to write $y(t) = -16at^2 + 25\sqrt{3}t$.

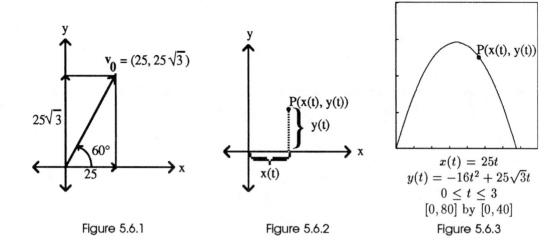

Figure 5.6.1    Figure 5.6.2    Figure 5.6.3

The position of the object at any time $t$ is given by $P(x(t), y(t))$ where $x(t) = 25t$ and $y(t) = -16t^2 + 25\sqrt{3}t$ (Figure 5.6.2).
(b) The object will hit the ground when $y(t) = 0$:

$$y(t) = -16t^2 + 25\sqrt{3}t = 0$$
$$t(25\sqrt{3} - 16t) = 0.$$

Now, $y(t) = 0$ when $t = 0$ or $t = \frac{25\sqrt{3}}{16}$. Thus, the object will hit the ground when $t = \frac{25\sqrt{3}}{16} = 2.7063294$, or approximately 2.71 seconds after launch.

(c) We need to determine the value of $x(t) = 25t$ when the object hits the ground. Using the value of $t$ from (b) we will find that $x(t) = 67.66$ when $t = 2.7063294$. Thus, the object travels 67.66 feet in the horizontal direction.

(d) The function $y(t) = -16t^2 + 25\sqrt{3}t$ is a parabola that opens downward and has zeros of 0 and $\frac{25\sqrt{3}}{16}$. Thus, the largest value of $y$ is attained when $t$ is half way between its two zeros, or when $t = \frac{25\sqrt{3}}{32}$. The value of $t$ is 1.3531647 and the corresponding values of $x$ and $y$ are 33.83 and 29.30, respectively. The object reaches its maximum height of 29.3 feet 1.35 seconds after launch. The graph of the parametric equations in Figure 5.6.3 can be used to confirm this information.

(e) The position at any time $t$ of the object in vector form is

$$p = (25t, 25\sqrt{3}t - 16t^2).$$

We can rewrite $p$ as follows:

$$p = (25t, 25\sqrt{3}t - 16t^2)$$
$$= (25t, 25\sqrt{3}t) + (0, -16t^2)$$
$$= t(25, 25\sqrt{3}) + (0, -16t^2).$$

Notice that the vector $v_1 = t(25, 25\sqrt{3}) = tv_0$ depends only on the initial velocity vector, and the vector $v_2 = (0, -16t^2)$ depends only on gravity (Figure 5.6.4). Thus, $p = tv_0 + v_2$.

A parametric graphing utility actually simulates the path (motion) of an object. That is, by observing a computer or graphing calculator draw the graph of $P(x(t), y(t))$ for $a \leq t \leq b$, we can see the path of the object from time $t = a$ to time $t = b$ as it actually evolves.

• **EXAMPLE 2:** A baseball is hit when it is 3 feet above the ground and leaves the bat with initial velocity of 150 feet per second and at an angle of elevation of 20°. A 6 mph wind is blowing in the horizontal direction against the batter. A 20 foot high fence is 400 feet from home plate. Will the hit be a home run?

**SOLUTION:** First we determine the parametric equations for the motion of the ball. Let $t$ be the time in seconds after the ball is hit and $P(t) = (x(t), y(t))$ be the coordinates of the position of the ball. Now $P(0) = (0, 3)$. We need to determine $x(t)$ and $y(t)$. The initial velocity that acts on the ball due to the bat is $v_0 = (150\cos 20°, 150\sin 20°)$. Thus, the effect of the bat on the ball at any time $t$ is given by the vector $v_1 = (150t\cos 20°, 150t\sin 20°)$. The effect on the ball due to gravity at any time $t$ is $v_2 = (0, 3 - 16t^2)$. It can be shown that 6 mph is the same as 8.8 feet per second. The effect on the ball of the wind is given by the vector $v_3 = (-8.8t, 0)$ because the wind is blowing in the horizontal direction against the batter (Figure 5.6.5). Therefore, the path of the ball is given by

$$P(t) = v_1 + v_2 + v_3$$
$$= (150t\cos 20°, 150t\sin 20°) + (0, 3 - 16t^2) + (-8.8t, 0)$$
$$= (150t\cos 20° - 8.8t, 150t\sin 20° + 3 - 16t^2).$$

Figure 5.6.4

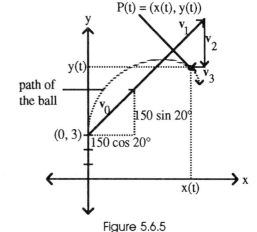

Figure 5.6.5

It now follows that

$$x(t) = 150t \cos 20° - 8.8t, \quad \text{and}$$
$$y(t) = 150t \sin 20° + 3 - 16t^2.$$

The ball will reach the fence when $x(t) = 400$. This occurs when $t$ is 3.03 seconds. At this value of $t$ we have $y(t) = 11.7$. Thus, the ball does not clear the fence. We need not do the above computations to determine if the hit is a home run. The graph of the curve defined by the above parametric equations together with the graph of the horizontal line $y = 20$ and the vertical line $x = 400$ in Figure 5.6.6 clearly shows that the hit is *not* a home run. Again, by watching a computer or graphing calculator draw a graph of the parametric equations we can see a simulation of the actual path of the ball. •

- **EXAMPLE 3:** Suppose the ball of Example 2 leaves the bat at an angle of elevation of 30° but all other conditions of the example remain the same. Will this hit be a home run?

    SOLUTION: The parametric equations for the path of the ball in this case are the following:

    $$x(t) = 150t \cos 30° - 8.8t$$
    $$y(t) = 150t \sin 30° + 3 - 16t^2.$$

    Notice only the angle value changes from the previous example. The graph in Figure 5.6.7 shows that this hit is a home run; and a big one at that! In the Exercises you will show that the ball reaches the fence in 3.30 seconds and the height of the ball at this time is 76.17 feet. •

- **EXAMPLE 4:** Let $C(0, 20)$ be the center of a circle with radius 20 and let $P(x, y)$ be any point on the circle. Find parametric equations for the circle in terms of the parameter $\theta$, where $\theta$ is the angle that $PC$ makes with the positive $x$-axis (Figure 5.6.8).

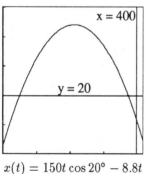

$x(t) = 150t \cos 20° - 8.8t$
$y(t) =$
$150t \sin 20° + 3 - 16t^2$
$0 \le t \le 5$
$[0, 420]$ by $[0, 50]$
Figure 5.6.6

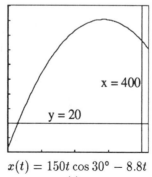

$x(t) = 150t \cos 30° - 8.8t$
$y(t) =$
$150t \sin 30° + 3 - 16t^2$
$0 \le t \le 5$
$[0, 420]$ by $[0, 100]$
Figure 5.6.7

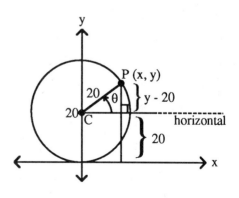

Figure 5.6.8

SOLUTION: We can think of $\theta$ as a central angle of the circle. Notice that $\cos\theta = \frac{x}{20}$ and $\sin\theta = \frac{y-20}{20}$. Thus, $x = 20\cos\theta$ and $y = 20 + 20\sin\theta$. A range on $\theta$ that produces the graph of the circle exactly once is given by any interval of length $2\pi$. The graph in Figure 5.6.8 can be confirmed by using a parametric graphing utility to graph $x(\theta) = 20\cos\theta$, $y(\theta) = 20 + 20\sin\theta$ with $\theta\min = 0$ and $\theta\max = 2\pi$. •

Now we use the information gained in Examples 1 and 4 to help solve the following problem posed by Neal Koblitz in the March, 1988 issue of *The American Mathematical Monthly*.

• **EXAMPLE 5:** You are standing on the ground at point $D$ (Figure 5.6.9), a distance of 75 feet from the bottom of a ferris wheel that is 20 feet in radius. Your arm is at the same level as the bottom of the ferris wheel. Your friend is on the ferris wheel, which makes one revolution (counterclockwise) every 12 seconds. At the instant when she is at point $A$ you throw a ball to her at 60 ft/sec at an angle of 60° above the horizontal. Take $g = 32$ ft/sec$^2$, and neglect air resistance. How close does the ball get to your friend?

SOLUTION: We assume that the ferris wheel is in a rectangular coordinate system with a diameter along the $y$-axis and the bottom at the origin. Then, the ferris wheel is a circle with center $C(0, 20)$ and radius 20 (Figure 5.6.10). We know from Example 4 that parametric equations for the position of the point $P = (x_A(\theta), y_A(\theta))$ on the ferris wheel are given by $x_A(\theta) = 20\cos\theta$ and $y_A(\theta) = 20 + 20\sin\theta$, where $\theta$ is the angle that the radius $CP$ makes with the positive $x$-axis.

Because the ferris wheel makes one complete revolution every 12 seconds we can write $\theta$ as a function of time $t$ (in seconds) as follows: $\theta = \frac{2\pi}{12}t = \frac{\pi}{6}t$. Thus, in terms of the parameter $t$ (time), we have the following parametric equations for the position $P$ of the

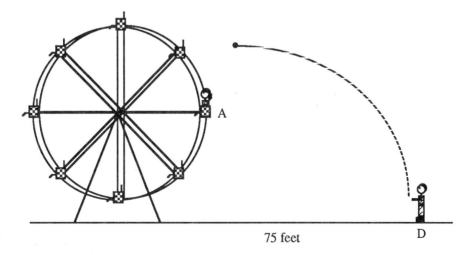

75 feet

D

Figure 5.6.9

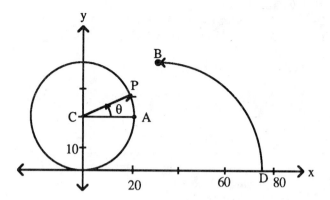

Figure 5.6.10

friend on the ferris wheel:

$$x_A(t) = 20\cos\left(\frac{\pi}{6}t\right)$$
$$y_A(t) = 20 + 20\sin\left(\frac{\pi}{6}t\right).$$

Next, we determine parametric equations for the path $B = (x_B(t), y_B(t))$ of the ball. The initial velocity at which the ball is thrown has horizontal and vertical components $60\cos 120°$ and $60\sin 120°$, respectively (Figure 5.6.11). Notice $\cos 120°$ is negative so that the horizontal component acts to the left.

Thus, we have the following parametric equations for the path of the ball:

$$x_B(t) = 75 + (60\cos 120°)t = 75 - 30t,$$
$$y_B(t) = -16t^2 + (60\sin 120°)t = -16t^2 + 30\sqrt{3}\, t.$$

Figure 5.6.11

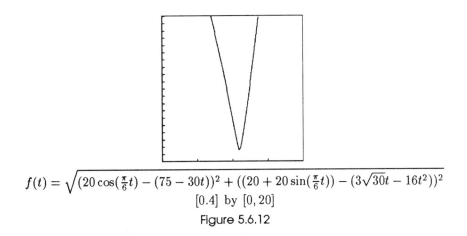

$$f(t) = \sqrt{(20\cos(\tfrac{\pi}{6}t) - (75 - 30t))^2 + ((20 + 20\sin(\tfrac{\pi}{6}t)) - (3\sqrt{30}t - 16t^2))^2}$$
$[0,4]$ by $[0,20]$
Figure 5.6.12

The parametric graphing utility that we use and graphing calculators permit us to draw the two curves simultaneously. That is, we can see the points being plotted on each curve for the same value of $t$. This simultaneous feature of the graphing utility permits us to actually simulate the real world problem situation. It can be shown that the ball comes close to the friend on the ferris wheel, but they do not meet. In fact, the distance between the ball and the friend steadily decreases to a minimum that is less than 2 and then increases again. The minimum occurs for $t$ between 2.1 and 2.3 seconds. •

• **EXAMPLE 6:** Find the minimum distance between the ball and the friend on the ferris wheel and the time at which the minimum occurs. (Figure 5.6.9).

SOLUTION: We have shown that $x_A(t) = 20\cos(\tfrac{\pi}{6}t)$ and $y_A(t) = 20 + 20\sin(\tfrac{\pi}{6}t)$ are parametric equations for the path of the friend on the ferris wheel, and that $x_B(t) = 75 - 30t$ and $y_B(t) = 30\sqrt{3}\,t - 16t^2$ are parametric equations for the path of the ball. For each $t$, the distance between the friend and the ball is given by $D(t) = \sqrt{(x_A(t) - x_B(t))^2 + (y_A(t) - y_B(t))^2}$. We can use $y_B(t)$ to show that the ball will hit the ground in a little more than 3 seconds. Thus, the graph of $D$ in the $[0,4]$ by $[0,20]$ viewing rectangle given in Figure 5.6.12 shows a complete graph of the problem situation. We can use zoom-in to show that $D$ has an absolute minimum of 1.59 and this value of $D$ occurs at $t = 2.19$. Therefore, the minimum distance between the ball and the friend on the ferris wheel is 1.59 feet and occurs 2.19 seconds after the ball is thrown. Furthermore, this answer has error at most 0.01. •

## EXERCISES 5-6

Draw a complete graph.
1. $x(t) = 2t - 6\sin t$ and $y(t) = 2 - 6\cos t$
2. $x(t) = \cos^2 t$ and $y(t) = (1 - \cos t)\sin t$
3. $x(t) = 2 - 5\cos t$ and $y(t) = -2 + 4\cos t$

4. $x(t) = \frac{4}{1+\sin t} \cos t$ and $y(t) = \frac{4}{1+\sin t} \sin t$

A dart is thrown upward with an initial velocity of 58 feet per second at an angle of elevation 41°. Consider the position of the dart at any time $t$ ($t = 0$ when the dart is thrown). Neglect air resistance.

5. Find parametric equations that model the problem situation.
6. Draw a complete graph of the model.
7. What portion of the graph represents the problem situation?
8. When will the dart hit the ground?
9. Find the maximum height of the dart. At what time will it reach maximum height?
10. How far does the dart travel in the horizontal direction? Neglect air resistance.
11. An arrow is shot with an initial velocity of 205 feet per second and at an angle of elevation of 48°. Find when and where the arrow will strike the ground.
12. A golfer hits a ball with an initial velocity of 133 feet per second and at an angle of 36° from the horizontal. Find when and where the ball will hit the ground.
13. Will the ball in Exercise 12 clear a fence 9 feet high that is at a distance of 275 feet from the golfer?
14. With what initial velocity must a ball be thrown from the ground at an angle of 35° from the horizontal if it is to travel a horizontal distance of 255 feet?
15. A batter hits a ball with an initial velocity of 92 feet per second and at an angle of 55° from the horizontal. Find the maximum height attained and the total horizontal distance traveled by the ball.

Chris and Linda are standing 78 feet apart. At the same time they each throw a softball toward each other. Linda throws her ball with an initial velocity of 45 feet per second with an angle of inclination of 44°. Chris throws her ball with an initial velocity of 41 feet per second with an angle of inclination of 39°.

16. Find two sets of parametric equations that represent a model of the problem situation.
17. Draw complete graphs of both sets of parametric equations in the same viewing rectangle.
18. What values of $t$ make sense in this problem situation?
19. Find the maximum height of each ball. How far does each ball travel in the horizontal direction? When does each ball hit the ground? Whose ball hits first?
20. By choosing $t$ max carefully (guess and check), estimate how close the two balls get, and the time when they are closest (minimum distance).
21. Use the distance formula and zoom-in to find the minimum distance (and the time at which it occurs) with error less than 0.01.
22. A river boat's paddle wheel has diameter of 26 feet and makes 1 revolution clockwise in 2 seconds at full speed. Write parametric equations describing the position of a point $A$ on the paddle. Assume at $t = 0$, $A$ is at the very top of the wheel as shown in Figure 5.6.13.
23. In Exercise 22, how far has point $A$, which is fixed on the wheel, moved in one minute?
24. Solve the ferris wheel and ball problem (Example 5) if the radius of the ferris wheel is 26 feet and the ball is thrown with initial velocity 76 feet per second at an angle of 52° with the horizontal

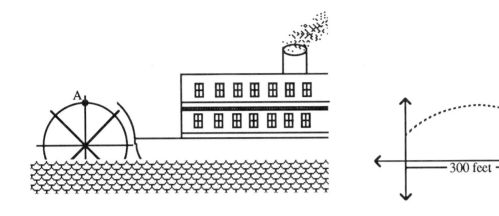

Figure 5.6.13              Figure 5.6.14

from a distance of 62 feet from the bottom of the ferris wheel. Use parametric simulation and estimate your answer. (Assume the same number of seconds per ferris wheel revolution.)

25. Solve Exercise 24 using the distance formula and zoom-in.

26. Use the parametric simulation and vary the position of the ball thrower (from 62 feet) and see how close you can get the ball to the friend on the ferris wheel. Use a guess and check method. (Keep all other values the same.)

27. Use the parametric simulation and vary the angle of elevation (from 52°) and see how close you can get the ball and the friend on the ferris wheel. Use a guess and check method. (Keep all other values the same.)

An NFL punter at the 15 yard line kicks a football downfield with initial velocity 85 feet per second at an angle of elevation of 56°.

28. Draw a complete graph of the problem situation.

29. How far downfield will the football first hit the field?

30. Determine the maximum height of the ball above the field.

31. What is the "hang time?" (That is, the total time the football is in the air.)

A Major League baseball player hits a fast ball with initial velocity 103 feet per second on a path 300 feet (measured horizontally) from the home run fence. The fence is 10 feet high. Draw a complete graph of the problem situation and determine if the hit is a home run (clears the fence). We assume (unless stated otherwise) that gravity is the only force affecting the path of the ball (Figure 5.6.14).

32. The hit is at an angle of elevation of 35°.

33. The hit is at an angle of elevation of 43°.

34. The hit is at an angle of elevation of 49°.

35. The hit is at an angle of elevation of 41° and there is a wind of 22 feet per second blowing in the same direction as the horizontal path of the ball.

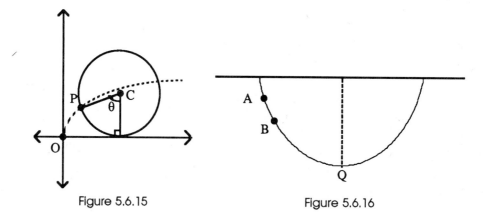

Figure 5.6.15        Figure 5.6.16

36. The hit is at an angle of elevation of 41° and there is a wind of 22 feet per second blowing in the opposite direction as the horizontal path of the ball.

37. The hit is at an angle of elevation of 41° and there is a wind of 22 feet per second blowing in the opposite direction as the horizontal path of the ball; the wind is blowing with angle of depression 12°.

38. Show algebraically that in Example 3, the ball reaches the fence in 3.30 seconds and is 76.17 feet above the ground at that time.

### The Cycloid Revisited

*39. Show that the path of a fixed point $P$ on a circle, with radius $a$ and center $C$, that rolls on a straight line is the **cycloid**
$$x(\theta) = a(\theta - \sin \theta),$$
$$y(\theta) = a(1 - \cos \theta),$$
where $\theta$ represents the angle the radius $CP$ has turned through. Assume the starting position is at the origin and $\theta$ is a positive angle, even though the rotation is clockwise (Figure 5.6.15).

*40. Design an experiment in physics to show the following remarkable property of cycloids. If $Q$ is the lowest point of an arch of an inverted cycloid, then the time it takes for a frictionless bead to slide down the curve to point $Q$ is *independent* of the starting point. That is, if $A$ and $B$ are dropped at the same time they will land at $Q$ at the same time (Figure 5.6.16).

## • KEY TERMS

Listed below are the key terms, vocabulary, and concepts in this chapter. You should be familiar with their definitions and meanings as well as be able to illustrate each of them.

Acceleration
Addition of vectors
Cardioid

Cis notation
Complex conjugate
Complex numbers

Complex plane
Cycloid
DeMoivre's theorem
Directed line segment
Gaussian plane
Imaginary axis
Imaginary part of a complex number
Initial point of vectors
Initial velocity
Linear combination
Magnitude of vectors
Modulus
Motion simulation
Parameters
Parametric equations
Parametrization
Plane curve

Polar coordinates
Polar graphs
Radial shrinking
Radial stretching
Real axis
Real part of a complex number
Rectangular coordinates
Regular $n$-gon
Roots of unity
Scalar multiplication of vectors
Scalars
Speed
Spiral of Archimedes
Terminal point of vectors
Trig form of complex numbers
Vector
Zero vector

## • REVIEW EXERCISES

This section contains representative problems from each of the previous six sections. You should use these problems to test your understanding of the material covered in this chapter.

Compute. Express the answers in the form $a + bi$ where $a$ and $b$ are real numbers.

1. $(4 + 3i) - (2 - i)$
2. $(1 - i)(3 + 2i)$
3. $\dfrac{-2 + 4i}{1 - i}$

Graph in the complex plane.

4. $-5 + 2i$
5. $2(\cos \frac{\pi}{3} + i \sin \frac{\pi}{3})$

Consider the complex number $z_1$ as represented in Figure R.1.

6. If $z_1 = a + bi$, determine $a$, $b$, and $|z_1|$.

Figure R.1

7. Determine the trig form of $z_1$.

Let $z_1 = -1 + 4i$, $z_2 = 3 + 4i$, and $z_3 = 5 - 2i$. Perform the indicated computation and write in the form $a + bi$.

8. $z_1 - z_2$      9. $z_2 z_3$      10. $|z_1 z_3|$

Write the complex number in the form $a + bi$.

11. $6(\cos 30° + i \sin 30°)$      12. $2.5(\cos 4 + i \sin 4)$

Write in trig form with $0 \leq \theta \leq 2\pi$, then give all possible trig forms.

13. $3 - 3i$      14. $-1 + \sqrt{2}i$      15. $3 - 5i$      16. $-2 - 2i$

Use Theorem 1 of Section 5.1 to compute. (Note: $\text{cis } \theta = \cos \theta + i \sin \theta$.)

17. $(3 \text{ cis } 30°)(4 \text{ cis } 60°)$      18. $\dfrac{3 \text{ cis } 30°}{4 \text{ cis } 60°}$

19. Write the two complex numbers in Exercise 17 in the form $a + bi$, perform the multiplication, and verify that the result is the same as that obtained using the trig form and Theorem 1 of Section 5.1.

20. Write the two complex numbers in Exercise 18 in the form $a + bi$, perform the division, and verify that the result is the same as that obtained using the trig form and Theorem 1 of Section 5.1.

Compute.

21. $(1 - i)^8$      22. $(3 \text{ cis } \frac{\pi}{4})^5$

Find the $n$th roots of each complex number for the specified value of $n$. Graph each $n$th root in the complex plane.

23. $n = 4$, $3 + 3i$      24. $n = 3$, $8$

25. Determine $z$ and the three cube roots of $z$ if one cube root of $z$ is $2 + 2i$.

26. Solve $z^4 = 4 - 4i$ and graph the solutions.

Compute a polar coordinate pair for the point with given rectangular coordinates.

27. $(-4, 5)$      28. $(8, -1)$

Compute *all* polar coordinate pairs for the point with given rectangular coordinates.

29. $(-2, 2)$      30. $(3, -1)$

Compute the rectangular coordinates for the point with given polar coordinates.

31. $(-4, 225°)$      32. $(3, \frac{4\pi}{3})$

33. Compute the distance between the points $(-2, 120°)$ and $(3, 60°)$.

Sketch a complete graph of each equation without using a graphing utility. Check your answer with a graphing utility.

34. $r = 5$      35. $r = 2 \sin \theta$      36. $r = 3 + 3 \cos \theta$

Use a graphing utility to draw the polar graph for $\theta$ in the given interval.

37. $r = 4\theta$; $0 \leq \theta \leq \frac{\pi}{2}$, $0 \leq \theta \leq \pi$, $0 \leq \theta \leq 2\pi$, $-\pi \leq \theta \leq \pi$, and $0 \leq \theta \leq 4\pi$.

38. $r = 5 \sin 3\theta$; $0 \leq \theta \leq \frac{\pi}{2}$, $0 \leq \theta \leq \pi$, $0 \leq \theta \leq 2\pi$, $-\pi \leq \theta \leq \pi$, and $0 \leq \theta \leq 4\pi$.

Use a graphing utility to draw a complete graph of the polar equation. Specify an interval $a \leq \theta \leq b$ of smallest length that gives a complete graph.

39. $r = 2 \cos 4\theta$      40. $r = 3\theta$

## Review Exercises

41. Determine the length of *one* rose "petal" for $r = 6\sin 2\theta$.

Solve the polar system by using a graphing utility.

42. $r = 3\cos\theta$
    $r = 3\sin\theta$

43. $r = 2 + \sin\theta$
    $r = 1 - \cos\theta$

Let $u = (2,-1)$, $v = (4,2)$, and $w = (1,-3)$ be vectors. Determine the following.

44. $u - v$
45. $2u - 3w$
46. $|u + v|$

47. Use the geometric definitions of addition and subtraction of vectors to find the sum $u+v$, the difference $u-v$, and then compare with the results obtained using the respective componentwise definitions for $u = (2,4)$ and $v = (-1,3)$.

Let $A = (2,-1)$, $B = (3,1)$, $C = (-4,2)$, and $D = (1,-5)$. Write each as a vector with *initial point at the origin*.

48. $3\overrightarrow{AB}$
49. $\overrightarrow{AB} + \overrightarrow{CD}$
50. Compute $|\overrightarrow{AB} + \overrightarrow{CD}|$.

Express the vector $\overrightarrow{AB}$ as a linear combination of $i = (1,0)$ and $j = (0,1)$.

51. $A = (4,0)$, $B = (2,1)$
52. $A = (3,1)$, $B = (5,1)$

Find the angle each vector makes with the positive $x$-axis. Then determine the angle between the vectors.

53. $u = (4,3)$, $v = (2,5)$
54. $u = (-2,4)$, $v = (6,4)$

55. Let $r = (1,2)$ and $s = (2,-1)$. Show that *any* vector $v = (x,y)$ can be expressed as a linear combination of $r$ and $s$.

Describe the curve defined parametrically. Sketch the graph without using a graphing utility. Check using a parametric graphing utility.

56. $x(t) = 2 + 3t$ and $y(t) = t$ with $t$ in $[0,5]$
57. $x(t) = 2t$ and $y(t) = \frac{3}{t}$ with $t$ in $[1,4]$
58. $x(t) = 4\cos\theta$ and $y(t) = 4\sin\theta$ with $\theta$ in $[-2\pi, 2\pi]$
59. Which of the curves in Exercises 56–58 are graphs of functions?
60–62. What are the domain and range of the curves in Exercises 56–58?
63. Which of the curves in Exercises 56–58 are graphs of one-to-one functions?
64. Find parametric equations and draw the graph of the circle centered at $(5,0)$ with radius 4.

Draw a complete graph.

65. $x(t) = 3t + 2$ and $y = 6t^2$ with $t$ in $(-\infty, \infty)$
66. $x(t) = 2 - \frac{1}{t}$ and $y = t + \frac{1}{t}$ with $t$ in $(0, 20]$
67. $x(t) = \sin^2 t$ and $y(t) = (1 - \sin t\cos t)$ with $t$ in $[0, 2\pi]$
68. $x(t) = t + \sin t$ and $y(t) = 2 + \cos t$ with $t$ in $[-10, 10]$
69. $x(t) = 12\cos t - 8\cos(\frac{3}{2}t)$ and $y(t) = 12\sin t - 8\sin(\frac{3}{2}t)$ with $t$ in $[0, 2\pi]$ (This is an *epicycloid*, the path of a point on a circle rolling on another circle.)
70. Find the domain and range of the curve in Exercise 69.

71. Write the polar equation $r = 2\theta$ for $\theta$ in $[0, 10\pi]$ in parametric form. Graph *both* the parametric and polar form with an appropriate graphing utility.
72. A plane is flying on a bearing of 10° east of south at 460 mph. A tail wind is blowing in the direction 20° west of south at 80 mph. Express the actual velocity of the plane as a vector. Determine the actual speed and direction of the plane.
73. A plane is flying on a bearing of 10° east of south at 460 mph. A 30 mph head wind is blowing in the direction of 20° east of north. Express the actual velocity of the plane as a vector. Determine the actual speed and direction of the plane.
74. A sailboat under auxiliary power is proceeding on a bearing of 25° north of west at 6.25 mph in still water. Then a tail wind blowing 15 mph in the direction 35° south of west alters the course of the sailboat. Express the actual velocity of the sailboat as a vector. Determine the actual speed and direction of the boat.

A dart is thrown upward with an initial velocity of 60 feet per second at an angle of elevation 35°. Consider the position of the dart at any time $t$ ($t = 0$ when the dart is thrown). Neglect air resistance.

75. Find parametric equations that model the problem situation.
76. Draw a complete graph of the model.
77. What portion of the graph represents the problem situation?
78. When will the dart hit the ground?
79. Find the maximum height of the dart. At what time will it reach maximum height?
80. A golfer hits a ball with an initial velocity of 125 feet per second and at an angle of 33° from the horizontal. Find when and where the ball will hit the ground.
81. A batter hits a ball with an initial velocity of 100 feet per second and at an angle of 50° from the horizontal. Find the maximum height attained and the distance traveled by the ball.

An NFL punter at the 20-yard line kicks a football downfield with initial velocity 80 feet per second at an angle of elevation of 55°.

82. Draw a complete graph of the problem situation.
83. How far downfield will the football first hit the field?
84. Determine the maximum height above the field of the ball.
85. What is the "hang time?" (That is, the total time the football is in the air.)

# 6  Conics

## • Introduction

In this chapter we study conic sections. A **conic section**, or **conic**, is any curve obtained by intersecting a double-napped right circular cone with a plane. The three basic conic sections are the **parabola**, the **ellipse**, and the **hyperbola**. A cone is the set of points that result when a line is rotated around another line, the axis of symmetry, that it intersects. A nappe is similar to an ice cream cone and is the portion of the cone formed by the rotation of one of the half lines. Figure 6.0.1 illustrates how the three conics are obtained as intersections of double-napped right circular cones and planes. Notice that the planes intersect one nappe to form a parabola or an ellipse and both nappes to form a hyperbola.

The position of the plane relative to the cone can sometimes result in **degenerate conics**. For example, if the plane contains the axis of symmetry of the cone, then the curve of intersection is a pair of intersecting lines. A pair of intersecting lines is a *degenerate hyperbola*.

Conics are important in applications. In this chapter we see that the path of any planet around the sun is an ellipse. The path of a comet can be hyperbolic. Conics have important *reflective properties*. Certain microphones and automotive headlamps are constructed using the reflective properties of parabolas. The reflective properties of parabolas and hyperbolas are used in optics. Ellipses are used in medicine. For example, the lithotripter is a device designed to break up kidney stones and remove the need for surgery. It uses shock waves and the reflective properties of an ellipse. Hyperbolas are used in navigation.

We will see that any conic is a graph of a quadratic equation in two variables of the form

$$Ax^2 + Bxy + Cy^2 + Dx + Ey + F = 0$$

• 413

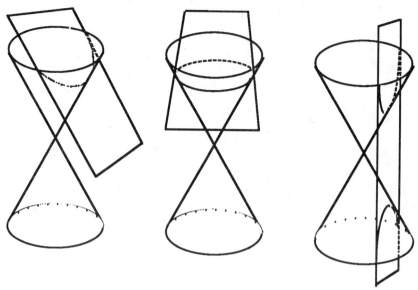

Figure 6.0.1

where at least one of $A$, $B$, or $C$ must be different from zero. When $B = 0$, the above equation can be rewritten in a form to help make a quick sketch of its graph. This can also be done when $B \neq 0$, but the algebraic procedure is very tedious. We treat the case $B \neq 0$ using the quadratic formula and a function graphing utility.

Systems of equations and inequalities that involve equations quadratic in two variables are solved. Polar equations for conics are obtained in this chapter using the focus-directrix definition of a conic.

## 6.1 Parabolas

In this section we give the focus-directrix definition of a parabola and obtain the standard form for the equation of a parabola with line of symmetry parallel to a coordinate axis. In Section 6.4 we consider parabolas with line of symmetry *not* parallel to a coordinate axis. An application that uses the reflective properties of a parabola is investigated.

The path of motion of an object thrown through the air is approximately a parabola. If air resistance is neglected, or if the object is thrown in a vacuum, then the path traveled by the object is a parabola. Parabolas have important reflective properties that are discussed in this section.

The graph of any quadratic function $y = ax^2 + bx + c$ can be obtained from the graph of $y = x^2$ by using vertical stretching or shrinking, reflection through the $x$-axis, vertical shifting, and horizontal shifting. The graph of $y = ax^2 + bx + c$ is a parabola that opens

## 6.1 Parabolas

up if $a > 0$ and opens down if $a < 0$. The line of symmetry of this parabola is a vertical line. If we rewrite

$$y = ax^2 + bx + c = a(x-h)^2 + k,$$

then the line of symmetry is the vertical line $x = h$ and the vertex is $(h, k)$ (Figure 6.1.1).

- **DEFINITION**  Let $L$ be any line and $F(x_0, y_0)$ any point not on $L$. The set of all points $P(x, y)$ in the plane determined by $L$ and $F$ that are equidistant from $L$ and $F$ is called a **parabola** (Figure 6.1.2). The point $F$ is called the **focus** of the parabola, and the line $L$ is called the **directrix** of the parabola.

Notice that the midpoint of the line segment from $F$ perpendicular to $L$ is on the graph of the parabola. This point $V$ is the **vertex** of the parabola. The vertex and focus lie on the same side of the directrix. The point $P'$ that is symmetric to any point $P$ on the parabola with respect to the line through $F$ and $V$ is a point of the parabola. (Why?) The line through $F$ and $V$ is a line of symmetry of the parabola.

We use the word *focus* in the definition of a parabola because of the *reflective properties* of a parabola. For example, light will be reflected by a parabola parallel to its line of symmetry if the source of the light is placed at the focus. Headlamps of motor vehicles use this principle. Similarly, sound waves traveling into the opening of a parabola parallel to its line of symmetry will be reflected through the focus of the parabola. Parabolic microphones use this principle.

In the first example we determine an equation of a parabola with given vertex, focus, and directrix.

Figure 6.1.1

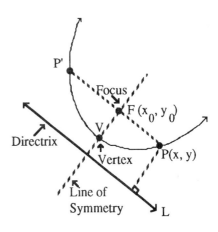

Figure 6.1.2

- **EXAMPLE 1:** Determine an equation for the parabola with focus $(0, a)$ and directrix $y = -a$ where $a > 0$ (Figure 6.1.3). Draw a complete graph of the parabola and give the coordinates of the vertex.

SOLUTION: The distance from $P(x, y)$ to the line $y = -a$ is $|y + a|$, and the distance from $P(x, y)$ to $(0, a)$ is $\sqrt{(x-0)^2 + (y-a)^2}$. (Why?) Applying the definition of a parabola, we obtain the following equation:

$$\sqrt{x^2 + (y-a)^2} = |y + a|. \tag{1}$$

Next, we square each side of equation (1) and simplify:

$$\sqrt{x^2 + (y-a)^2} = |y + a|$$
$$x^2 + (y-a)^2 = (y+a)^2$$
$$x^2 + y^2 - 2ay + a^2 = y^2 + 2ay + a^2$$
$$x^2 = 4ay. \tag{2}$$

Thus, every point on the parabola is a solution to equation (2). Because the steps used to derive equation (2) from equation (1) are reversible, every solution to equation (2) corresponds to a point on the parabola.

We can rewrite equation (2) in the form $y = \frac{1}{4a}x^2$ to see that the graph can be obtained by starting with $y = x^2$ and applying a vertical stretch or shrink by the factor $\frac{1}{4a}$. The coordinates of the vertex are $(0, 0)$. The computer drawn graph for $a = 1$ in Figure 6.1.4 is a complete graph and suggests what the graphs for $a > 0$ look like. •

An equation for the parabola with focus $(0, a)$ and directrix $y = -a$ with $a < 0$ is also $x^2 = 4ay$. However, the graph of this equation is the reflection with respect to the $x$-axis of the graph in Example 1 because $a < 0$ in this case. Furthermore, $|a|$ is equal to half the distance between the focus and the directrix in each case. If the parabola opens up (focus above directrix), then $a > 0$; if the parabola opens down (focus below directrix), then $a < 0$.

Figure 6.1.3

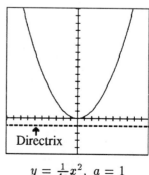

$y = \frac{1}{4a}x^2$, $a = 1$
$[-10, 10]$ by $[-5, 15]$

Figure 6.1.4

# 6.1 Parabolas

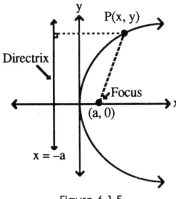

Figure 6.1.5

The $y$-axis is the line of symmetry of this parabola and is also the line through the focus perpendicular to the directrix. The vertex is $(0,0)$.

If the directrix of a parabola is the line $x = -a$ and the focus is the point $(a, 0)$ (Figure 6.1.5), then, as in Example 1, we would find that an equation of the parabola is $y^2 = 4ax$. That is, we only need to interchange $x$ and $y$ in the equation found in Example 1. The graph of $y^2 = 4ax$ is a parabola that opens to the right if $a > 0$, and a parabola that opens to the left if $a < 0$. In either case the line of symmetry of $y^2 = 4ax$ is the $x$-axis and the vertex is $(0,0)$ (Figure 6.1.6).

The next example illustrates that any parabola with directrix parallel to the $x$-axis can be obtained from $y = x^2$ by applying appropriate geometric transformations.

- **EXAMPLE 2:** Determine an equation for the parabola with focus $(3, 6)$ and directrix $y = -2$. Give the coordinates of the vertex and the line of symmetry, and sketch a complete graph. Give a sequence of geometric transformations that will produce the graph of this parabola from the graph of $y = x^2$.

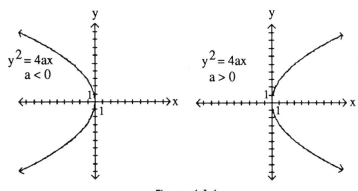

Figure 6.1.6

**SOLUTION:** The distance from $P(x,y)$ to $(3,6)$ is $\sqrt{(x-3)^2+(y-6)^2}$, and the distance from $P(x,y)$ to $y=-2$ is $|y+2|$ (Figure 6.1.7). We apply the definition of a parabola and then simplify the resulting equation:

$$\sqrt{(x-3)^2+(y-6)^2}=|y+2|$$
$$(x-3)^2+(y-6)^2=(y+2)^2$$
$$(x-3)^2+y^2-12y+36=y^2+4y+4$$
$$(x-3)^2=16y-32$$
$$(x-3)^2=16(y-2). \tag{3}$$

Notice that the vertex is $(3,2)$ and the line of symmetry is $x=3$.

We can rewrite equation (3) in the following way:

$$y=\frac{1}{16}(x-3)^2+2.$$

Now, we can see that a complete graph can be obtained from $y=x^2$ by applying a vertical shrink with factor $\frac{1}{16}$, followed by a horizontal shift right 3 units, and then a vertical shift up 2 units (Figure 6.1.8). •

Equation (3), $(x-3)^2=16(y-2)$, is often called the *standard form* for the equation of the parabola of Example 2. It follows from equation (2) that the coefficient 16 of $y-2$ is four times the distance between the focus and vertex, or, equivalently, four times the perpendicular distance between the vertex and directrix. (Why?) We can use this information to simplify writing equations for parabolas.

• **EXAMPLE 3:** Write an equation, determine the line of symmetry, the focus, and sketch a complete graph of the parabola with

(i) vertex $(-2,1)$ and directrix $y=-2$.
(ii) vertex $(3,-1)$ and directrix $y=1$.

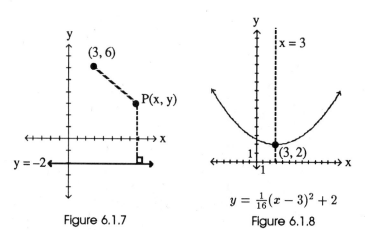

Figure 6.1.7         Figure 6.1.8

## 6.1 Parabolas

SOLUTION:

(i) This parabola opens upward because the vertex is above the directrix. The line of symmetry is $x = -2$, and an equation for the parabola is $4a(y-1) = (x+2)^2$ where $a > 0$. The perpendicular distance between the vertex and directrix is 3. Thus, $a = 3$ and an equation for the parabola is $12(y-1) = (x+2)^2$. The focus is 3 units above the vertex and is on the line $x = -2$; its coordinates are $(-2, 4)$. A complete graph of $12(y-1) = (x+2)^2$ or $y = \frac{1}{12}(x+2)^2 + 1$ can be obtained from the graph of $y = x^2$ by applying a vertical shrink with factor $\frac{1}{12}$, followed by a horizontal shift left 2 units, and then a vertical shift up 1 unit. (Why?) Alternatively, we can use the fact that the graph of $12(y-1) = (x+2)^2$ is a parabola that opens upward with vertex $(-2, 1)$ and compute a few more points to produce a quick sketch. The $x$-intercepts and $y$-intercepts are good points to include. However, you may only get one additional point this way. In this case, there are no $x$-intercepts and $(0, \frac{4}{3})$ is the $y$-intercept. Symmetry can be used to produce $(-4, \frac{4}{3})$ as a second point. (Why?) A complete graph is given in Figure 6.1.9. A function graphing utility could be used to confirm this graph.

(ii) This parabola opens downward because the vertex is below the directrix. The line of symmetry is $x = 3$, and an equation for the parabola is $4a(y+1) = (x-3)^2$ where $a < 0$. The perpendicular distance between the vertex and directrix is 2. Thus, $a = -2$, and an equation for the parabola is $-8(y+1) = (x-3)^2$. The focus is 2 units below the vertex and is on the line $x = 3$; its coordinates are $(3, -3)$. Again we can use geometric transformations or plot three points to sketch a complete graph. There are no $x$-intercepts and $(0, -\frac{17}{8})$ is the $y$-intercept. Thus, $(6, -\frac{17}{8})$ is another point on the graph. A complete graph is given in Figure 6.1.10. ●

We can extend Example 3 to any parabola with directrix parallel to the $y$-axis. We use the fact that an equation of such a parabola has the form $(y-k)^2 = 4a(x-h)$ where $(h, k)$ is the vertex of the parabola.

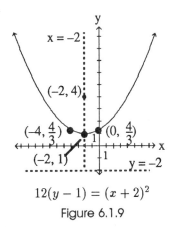

$12(y-1) = (x+2)^2$

Figure 6.1.9

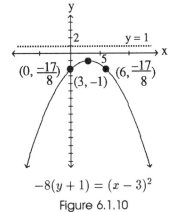

$-8(y+1) = (x-3)^2$

Figure 6.1.10

- **EXAMPLE 4:** Write an equation, determine the line of symmetry, the focus, and sketch a complete graph of the parabola with

  (i) vertex $(-2, -3)$ and directrix $x = -\frac{7}{2}$.
  (ii) vertex $(4, 1)$ and directrix $x = 6$.

Check each graph with a graphing utility.

SOLUTION:

(i) This parabola opens to the right because the vertex is to the right of the directrix. The line of symmetry is $y = -3$, and an equation for the parabola is $(y + 3)^2 = 4a(x+2)$ where $a > 0$. The perpendicular distance between the vertex and directrix is $\frac{3}{2}$. Thus, $a = \frac{3}{2}$, and an equation for the parabola is $(y + 3)^2 = 6(x + 2)$. The focus is $\frac{3}{2}$ units to the right of the vertex and is on the line $y = -3$; its coordinates are $(-\frac{1}{2}, -3)$. We can determine two points in addition to the vertex to sketch the complete graph given in Figure 6.1.11. The points $(0, 0.46), (0, -6.46)$ are the two $y$-intercepts.

If we solve for $y$ in terms of $x$ we obtain $y + 3 = \pm\sqrt{6(x+2)}$ or $y = -3 \pm \sqrt{6(x+2)}$. A complete graph of this parabola can be obtained using a function graphing utility to overlay the graphs of the two functions $y = -3 + \sqrt{6(x+2)}$ and $y = -3 - \sqrt{6(x+2)}$ in the same viewing rectangle (Figure 6.1.12). The graph of $y = -3 + \sqrt{6(x+2)}$ is the upper half of the parabola, and the graph of $y = -3 - \sqrt{6(x+2)}$ is the lower half of the parabola. Notice the gap between the two graphs near the vertex. This gap occurs because the graph of this parabola is very steep near the vertex. Gaps often occur in graphs of conics obtained with function graphing utilities, and even with special conic graphing utilities. These gaps can often be narrowed if high resolution graphs are drawn. Most graphs in this textbook will have the gaps removed or narrowed as much as possible.

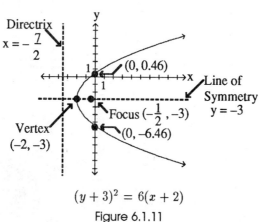

$(y + 3)^2 = 6(x + 2)$

Figure 6.1.11

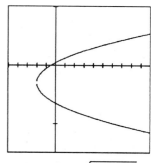

$y = -3 + \sqrt{6(x + 2)}$
$y = -3 - \sqrt{6(x + 2)}$
$[-5, 10]$ by $[-15, 10]$

Figure 6.1.12

(ii) This parabola opens to the left because the vertex is to the left of the directrix. The line of symmetry is $y = 1$, and an equation for the parabola is $(y-1)^2 = 4a(x-4)$ where $a < 0$. The perpendicular distance between the vertex and directrix is 2. Thus, $a = -2$, and an equation for the parabola is $(y-1)^2 = -8(x-4)$. The focus is 2 units to the left of the vertex and is on the line $y = 1$; its coordinates are $(2, 1)$. A complete graph is sketched in Figure 6.1.13. In the exercises you will be asked to confirm this graph with a function graphing utility. •

Geometric transformations can be used to produce the complete graphs in Example 4. However, usually it is much simpler to determine the vertex and two additional points to obtain a compete graph of any parabola. The equations found in Examples 3 and 4 are called the *standard form* for the equation of a parabola.

### Equations of Parabolas in Standard Form
If a parabola has a line of symmetry parallel to or coincident with a coordinate axis, then it is the graph of an equation of the type given in Theorem 1.

---

• **THEOREM 1**    (i)   The graph of the equation $(x-h)^2 = 4a(y-k)$ is a parabola with vertex $(h, k)$ and directrix $y = c$ where $c = k - a$. The line of symmetry is $x = h$, and the focus is $(h, k + a)$.

(ii)   The graph of the equation $(y-k)^2 = 4a(x-h)$ is a parabola with vertex $(h, k)$ and directrix $x = c$ where $c = h - a$. The line of symmetry is $y = k$, and the focus is $(h + a, k)$.

---

• **DEFINITION**    The equations of Theorem 1 are called **standard forms for the equation of a parabola**. •

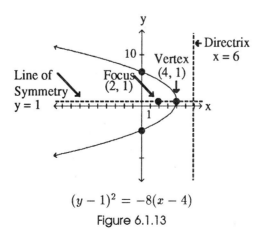

$(y-1)^2 = -8(x-4)$
Figure 6.1.13

The graphs of the equations of Theorem 1 can be obtained by applying horizontal and vertical shifts to the graphs of the equations $x^2 = 4ay$ and $y^2 = 4ax$.

It is not necessary to memorize the formulas in Theorem 1 for the directrix or the focus. In practice, you will use completing the square to put an equation of a parabola in standard form. Then, the values of $a$, $h$, and $k$ can be read and their meaning used to calculate the focus and directrix. We illustrate this use of the information about the standard form of the equation of a parabola in the next example.

• **EXAMPLE 5:** Find the vertex, line of symmetry, focus, and directrix of each parabola, and draw a complete graph.

(a) $(x + 1)^2 = -12(y - 3)$
(b) $(y - 2)^2 = -4(x + 3)$

SOLUTION:

(a) The graph of $(x + 1)^2 = -12(y - 3)$ is of the form $(x - h)^2 = 4a(y - k)$ with $h = -1$ and $k = 3$. This graph is a parabola that opens down with vertex $(-1, 3)$, line of symmetry $x = -1$, and directrix parallel to the $x$-axis. Because $4a = -12$, $a = -3$. The focus is $(-1, 0)$ because it is on the line $x = -1$ and 3 units below the vertex $(-1, 3)$. The directrix is $y = 6$ because it is a horizontal line 3 units above the vertex. In this case, $(0, \frac{35}{12})$ is the $y$-intercept. We find the $x$-intercepts by setting $y = 0$. Now, $(x + 1)^2 = 36$ so that $x + 1 = \pm 6$. Thus, $(-7, 0)$ and $(5, 0)$ are the two $x$-intercepts. A complete graph is sketched in Figure 6.1.14. Solving for $y$ in terms of $x$ we obtain $y = -\frac{1}{12}(x + 1)^2 + 3$. Now we can use a function graphing utility to confirm the graph in Figure 6.1.14.

(b) The graph of $(y-2)^2 = -4(x+3)$ is of the form $(y-k)^2 = 4a(x-h)$ with $k = 2$ and $h = -3$. This graph is a parabola that opens to the left with vertex $(-3, 2)$, line of symmetry $y = 2$, and directrix parallel to the $y$-axis. Because $4a = -4$, we see that $a = -1$. The focus is $(-4, 2)$ because it is on the line $y = 2$ and 1 unit to the left of the vertex $(-3, 2)$. The directrix is $x = -2$ because it is a vertical line 1 unit to the right of the vertex. There is no $y$-intercept and $(-4, 0)$ is the $x$-intercept.

$(x + 1)^2 = -12(y - 3)$

Figure 6.1.14

Thus, $(-4, 4)$ is also a point on the graph. (Why?) A complete graph is sketched in Figure 6.1.15. Again solving for $y$ in terms of $x$ we obtain $y = 2 \pm \sqrt{-4(x+3)}$. Check that overlaying the graphs of $y = 2 + \sqrt{-4(x+3)}$ and $y = 2 - \sqrt{-4(x+3)}$ in the appropriate viewing rectangle agrees with the graph in Figure 6.1.15. The graph of $y = 2 + \sqrt{-4(x+3)}$ gives the upper half of the parabola, and the graph of $y = 2 - \sqrt{-4(x+3)}$ gives the lower half of the parabola. Your graph may have a gap near the vertex. •

From the standard form we can see that the equation of any parabola with line of symmetry parallel to a coordinate axis is quadratic in one variable and linear in the other. Thus, parabolas with line of symmetry parallel to or coincident with a coordinate axis are graphs of quadratic equations in two variables. We can use completing the square to find the standard form of such a parabola.

• **EXAMPLE 6:** Find the standard form, vertex, focus, directrix, and line of symmetry of each parabola, and sketch a complete graph. Check using a graphing utility.

(a) $x^2 + 6x - 2y + 11 = 0$  (b) $y^2 - 6x + 2y + 13 = 0$

**SOLUTION:**

(a) First, we put the equation in standard form by completing the square on the terms involving the variable $x$. Note the equation is quadratic in the variable $x$:

$$x^2 + 6x - 2y + 11 = 0$$
$$x^2 + 6x = 2y - 11$$
$$x^2 + 6x + 9 = 2y - 11 + 9$$
$$(x + 3)^2 = 2y - 2$$
$$(x + 3)^2 = 2(y - 1).$$

From the standard form of the equation, we can see that this parabola opens up with vertex $(-3, 1)$ and has line of symmetry $x = -3$. Now, $4a = 2$ so that $a = 0.5$.

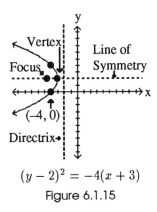

$(y - 2)^2 = -4(x + 3)$

Figure 6.1.15

Thus, the focus is $(-3, 1.5)$, and the directrix is $y = 0.5$. The $y$-intercept is $(0, \frac{11}{2})$. We can use this information to draw a rough sketch of the graph. However, we will use a graphing utility to determine the graph. Solving for $y$ in terms of $x$ we obtain $y = 0.5x^2 + 3x + 5.5$. The computer drawn graph given in Figure 6.1.16 is complete and confirms our algebraic analysis.

(b) This time we complete the square on the terms involving the variable $y$ because the equation is quadratic in $y$:

$$y^2 - 6x + 2y + 13 = 0$$
$$y^2 + 2y = 6x - 13$$
$$y^2 + 2y + 1 = 6x - 13 + 1$$
$$(y + 1)^2 = 6x - 12$$
$$(y + 1)^2 = 6(x - 2).$$

This parabola opens to the right, has vertex $(2, -1)$, and line of symmetry $y = -1$. Now $4a = 6$, so $a = 1.5$. The focus is $(3.5, -1)$ and the directrix is $x = 0.5$. We could, of course, use the above standard form to solve for $y$ in terms of $x$. Alternatively, we can rewrite the original equation in the form $y^2 + 2y + (13 - 6x) = 0$ and use the quadratic formula to solve for $y$:

$$y^2 + 2y + (13 - 6x) = 0$$
$$y = \frac{-2 \pm \sqrt{4 - 4(13 - 6x)}}{2}$$
$$= -1 \pm \sqrt{1 - (13 - 6x)} \quad \text{(Why?)}$$
$$= -1 \pm \sqrt{6x - 12}.$$

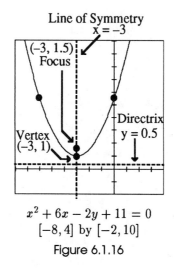

$x^2 + 6x - 2y + 11 = 0$
$[-8, 4]$ by $[-2, 10]$

Figure 6.1.16

# 6.1 Parabolas

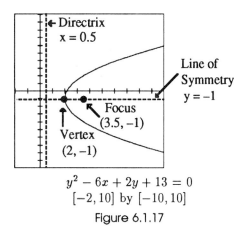

Figure 6.1.17

Thus, $y = -1 + \sqrt{6x - 12}$ and $y = -1 - \sqrt{6x - 12}$. The computer drawn graphs of these two functions given in Figure 6.1.17 give a complete graph of the parabola and confirm our algebraic analysis. •

- **EXAMPLE 7**: A parabolic microphone is formed by revolving the portion of the parabola $15y = x^2$ between $x = -9$ and $x = 9$ about its line of symmetry. Where should the sound receiver be placed?

    **SOLUTION**: The receiver should be placed at the focus of the parabola $15y = x^2$. Now, $4a = 15$ so that $a = 3.75$. The focus is at $(0, 3.75)$. Thus, the receiver should be placed 3.75 units from the vertex along its line of symmetry. •

## EXERCISES 6-1

Determine an equation, the vertex, and the line of symmetry of the parabola, and sketch a complete graph.

1. focus $(2, -1)$ and directrix $y = 3$
2. focus $(2, -1)$ and directrix $y = -3$
3. focus $(2, -1)$ and directrix $x = 3$
4. focus $(2, -1)$ and directrix $x = -3$

Determine an equation, the line of symmetry, and the focus of the parabola, and sketch a complete graph.

5. vertex $(3, 2)$ and directrix $x = 5$
6. vertex $(3, 2)$ and directrix $x = -5$
7. vertex $(3, 2)$ and directrix $y = 5$
8. vertex $(3, 2)$ and directrix $y = -5$

Determine an equation, the line of symmetry, and the directrix of the parabola, and sketch a complete graph.

9. focus $(-2, 0)$ and vertex $(-2, -6)$
10. focus $(-2, 0)$ and vertex $(-2, 4)$
11. focus $(3, 2)$ and vertex $(-1, 2)$
12. focus $(2, 4)$ and vertex $(0, 4)$

List a sequence of geometric transformations that when applied to the graph of $y = x^2$ will result in the graph of the given parabola.

13. $12(y+1) = (x-3)^2$   14. $y = 2x^2 - 6x + 9$

List a sequence of geometric transformations that when applied to the graph of $y^2 = x$ will result in the graph of the given parabola.

15. $(y+1)^2 = 3(x-2)$   16. $y^2 - 2y - 6 = 2x$

17. Consider the graph of $y = 4x^2$. Explain how a horizontal shrink of factor $\frac{1}{2}$ applied to the graph of $y = x^2$ produces the same graph as applying a vertical stretch of factor 4 to $y = x^2$. (*Hint*: What happens to the point $(1,1)$ on $y = x^2$?)

18. Consider the graph of $x = 4y^2$. Explain how a vertical shrink of factor $\frac{1}{2}$ applied to $x = y^2$ produces the same graph as applying a horizontal stretch of factor 4 to the graph of $x = y^2$.

19. Given that $y = Ax^2 + Bx + C$ is the equation of a parabola, show that the point $\left(-\frac{B}{2A}, \frac{4AC-B^2+1}{4A}\right)$ is the focus.

20. Draw a complete graph of the parabola $y^2 - 2y + 4 = x$ using only a *function* graphing utility.

Find the standard form, vertex, focus, directrix, and line of symmetry of each parabola, and sketch a complete graph without using a graphing utility. Check your graph with a graphing utility.

21. $x^2 + 2x - y + 3 = 0$   22. $3x^2 - 6x - 6y + 10 = 0$

23. $y^2 - 2y + 4x - 12 = 0$   24. $9y^2 - 9x - 6y - 5 = 0$

25. A parabolic microphone is formed by revolving the portion of the parabola $10y = x^2$ between $x = -7$ and $x = 7$ about its line of symmetry. Where should the sound receiver be placed for best reception?

26. A parabolic microphone is formed by revolving the portion of the parabola $18y = x^2$ between $x = -12$ and $x = 12$ about its line of symmetry. Where should the sound receiver be placed for best reception?

27. A parabolic headlight is formed by revolving the portion of the parabola $y^2 = 12x$ between the lines $y = -4$ and $y = 4$ about its line of symmetry. Where should the headlight bulb be placed for maximum illumination?

28. A parabolic headlight is formed by revolving the portion of the parabola $y^2 = 15x$ between the lines $y = -3$ and $y = 3$ about its line of symmetry. Where should the headlight bulb be placed for maximum illumination?

29. A large parabolic search light formed by revolving a parabola about its line of symmetry has its light source placed for maximum illumination at the focus of the parabola. The focus is 10 inches from the vertex of the search light. Determine an equation for the parabola. Explain what assumptions you make.

30. A large parabolic search light formed by revolving a parabola about its line of symmetry has its light source placed for maximum illumination at the focus of a parabola. The focus is 8 inches from the vertex of the search light. Determine an equation for the parabola. Explain what assumptions you make.

Find the points of intersection of the line and the parabola (if any).

31. $y = 2x^2 - 6x + 7$, $2x + 3y - 6 = 0$   32. $x = 3y^2 - 2y + 6$, $x - 4y - 10 = 0$

33. A parabola of the form $y = ax^2 + b$ passes through the points $(-1, 4)$ and $(2, 8)$. Determine an equation of the parabola.

34. A parabola of the form $y = ax^2 + b + c$ passes through the points $(0, 2)$, $(2, 5)$ and $(3, 8)$. Determine an equation of the parabola.

35. Find all points on the parabola $y^2 = -8x$ that are closest to the point $(-10, 0)$.

A paraboloid of revolution is determined by rotating the portion of the parabola $y = 6 - x^2$ above the $x$-axis about its line of symmetry. A cylinder with line of symmetry the $y$-axis and base on the $x$-axis is inscribed in the paraboloid of revolution. Let $(x, y)$ be a point on the parabola where $x$ represents the radius of the cylinder.

36. Determine an algebraic representation of the volume of the cylinder as a function of $x$.

37. Draw a complete graph of the algebraic representation.

38. Draw a complete graph of the problem situation. What values of $x$ make sense in the problem situation?

39. Determine the dimensions of the cylinder of maximum volume and the maximum volume.

40. Devise a real world problem scenario that fits Exercise 36.

41. The line $y = 3x - \frac{9}{8}$ is tangent to the curve $y = 2x^2$.

    (a) Find the point of tangency, $P$.
    (b) Find an equation of the line $\ell$ connecting $P$ and the focus of $y = 2x^2$.
    (c) Determine the angle between the line $\ell$ and the tangent line.
    (d) What is the angle between the tangent line and a line through $P$ that is parallel to the line of symmetry of $y = 2x^2$.
    (e) Do the angle sizes found in (c) and (d) help explain the reflective property of the parabola $y = 2x^2$?

42. Use a function graphing utility to confirm the graph in Example 4(ii).

## 6.2  Ellipses

In this section we give the two-foci definition of an ellipse and obtain the standard form for the equation of an ellipse with lines of symmetry parallel to the coordinate axes. We also consider applications that have an ellipse for an algebraic model.

- **DEFINITION**  Let $F_1$ and $F_2$ be any two points in the coordinate plane. An **ellipse** is the set of all points $P$ in the plane with the property that the sum of the distance from $P$ to $F_1$ and the distance from $P$ to $F_2$ is a constant (Figure 6.2.1). Each of the points $F_1$ and $F_2$ is called **a focus** of the ellipse.  •

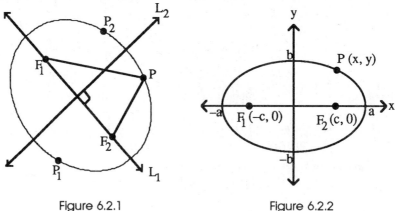

Figure 6.2.1          Figure 6.2.2

There are *no* points $P$ satisfying the condition that the sum of the distance from $P$ to $F_1$ and the distance from $P$ to $F_2$ is a constant unless that constant is greater than or equal to the distance between $F_1$ and $F_2$. (Why?) Let $L_1$ be the line determined by $F_1$ and $F_2$, and $L_2$ the perpendicular bisector of the line segment joining $F_1$ and $F_2$. The points $P_1$ and $P_2$ that are symmetric to any point $P$ on the ellipse with respect to $L_1$ and $L_2$, respectively, are also on the ellipse. (Why?) Thus, the ellipse is symmetric with respect to both $L_1$ and $L_2$. The point of intersection of $L_1$ and $L_2$ is called the **center** of the ellipse. An ellipse is symmetric with respect to its center. (Why?)

According to the laws of Newtonian mechanics the paths of objects due to forces such as gravitational attraction are conic sections. For example, if forces acting on the earth due to celestial bodies other than the sun are neglected, then the path of the earth around the sun is an ellipse with the sun as one focus. Orbits of the other planets and many comets around the sun are also ellipses with the sun as one focus.

Ellipses also have important reflective properties. If a source of light or sound is placed at one focus of an ellipse it will be reflected through the other focus. This principle was used to design a machine that is able to break up kidney stones and remove the need for surgery.

In this section we consider ellipses with lines of symmetry parallel to the coordinate axes. In Section 6.4 we remove this restriction. We begin by determining an equation for an ellipse with given foci. In this example the constant is $2a$.

- **EXAMPLE 1:** Let $F_1 = (-c, 0)$, $F_2 = (c, 0)$ be the two foci of the ellipse consisting of the points $P$ where $d(P, F_1) + d(P, F_2) = 2a$ with $a > c > 0$ (Figure 6.2.2). Determine an equation for the ellipse, the $x$-intercepts, the $y$-intercepts, and the center of the ellipse.

  **SOLUTION:** We can use the distance formula to obtain an equation for the ellipse:

  $$d(P, F_1) + d(P, F_2) = 2a$$
  $$\sqrt{(x+c)^2 + y^2} + \sqrt{(x-c)^2 + y^2} = 2a.$$

Next, we obtain an alternate form of this equation that does not involve radicals:

$$\sqrt{(x+c)^2+y^2}+\sqrt{(x-c)^2+y^2}=2a$$
$$\sqrt{(x+c)^2+y^2}=2a-\sqrt{(x-c)^2+y^2}$$
$$(x+c)^2+y^2=4a^2-4a\sqrt{(x-c)^2+y^2}+(x-c)^2+y^2$$
$$x^2+2cx+c^2+y^2=4a^2-4a\sqrt{(x-c)^2+y^2})+x^2-2cx+c^2+y^2$$
$$2cx=4a^2-4a\sqrt{(x-c)^2+y^2}-2cx$$
$$4a\sqrt{(x-c)^2+y^2}=4a^2-4cx$$
$$a\sqrt{(x-c)^2+y^2}=a^2-cx.$$

We continue by squaring each side of the above equation.

$$a\sqrt{(x-c)^2+y^2}=a^2-cx$$
$$a^2(x-c)^2+a^2y^2=a^4-2a^2cx+c^2x^2$$
$$a^2x^2-2a^2cx+a^2c^2+a^2y^2=a^4-2a^2cx+c^2x^2$$
$$a^2x^2-c^2x^2+a^2y^2=a^4-a^2c^2$$
$$(a^2-c^2)x^2+a^2y^2=a^2(a^2-c^2).$$

We divide each side of the above equation by $a^2(a^2-c^2)$ and obtain the following:

$$\frac{x^2}{a^2}+\frac{y^2}{a^2-c^2}=1.$$

Because $a>c$ we can define a positive number $b$ by $b^2=a^2-c^2$ and rewrite the above equation in the following form:

$$\frac{x^2}{a^2}+\frac{y^2}{b^2}=1. \qquad (1)$$

Thus, every point on the ellipse is a solution to equation (1). It can also be shown that every solution of (1) corresponds to a point on the ellipse. This form is called the *standard form* for the equation of an ellipse. The $x$-intercepts are $\pm a$, and the $y$-intercepts are $\pm b$. (Why?) The center of the ellipse is $(0,0)$. •

In Figure 6.2.2, the line segment from $(-a,0)$ to $(a,0)$ is called the **major axis** or **major diameter** of the ellipse, and the line segment from $(0,-b)$ to $(0,b)$ is the **minor axis** or **minor diameter**. We call $a$ the **semimajor axis length** and $b$ the **semiminor axis length**. Notice that the equation $\frac{x^2}{a^2}+\frac{y^2}{b^2}=1$ is not changed if $x$ is replaced by $-x$ or if $y$ is replaced by $-y$. This confirms algebraically that the graph in Figure 6.2.2 is symmetric with respect to both the $x$-axis and the $y$-axis. It follows that the graph is also symmetric with respect to the origin, the center of the ellipse.

It can be shown that the standard form for the equation of the ellipse with foci $F_1(0,-c)$ and $F_2(0,c)$ on the $y$-axis that satisfies $d(P,F_1)+d(P,F_2)=2a$ with $a>c>0$ is

$$\frac{x^2}{b^2}+\frac{y^2}{a^2}=1, \qquad (2)$$

where $b^2 = a^2 - c^2$. This equation is the same as equation (1) except that the number under $x^2$ is smaller than the number under $y^2$. A complete graph of this ellipse is given in Figure 6.2.3. The *major axis* is the line segment from $(0, -a)$ to $(0, a)$, and the *minor axis* is the line segment from $(-b, 0)$ to $(b, 0)$. The ellipse in Figure 6.2.3 is a complete graph of equation (2).

- **EXAMPLE 2**: Determine the endpoints of the major and minor axes, the standard form of the equation, and sketch a complete graph of each ellipse.

   (a) The sum of the distances from the foci is 10; the foci are $(-3, 0)$ and $(3, 0)$.
   (b) The sum of the distances from the foci is 10; the foci are $(0, -3)$ and $(0, 3)$.

SOLUTION:

   (a) The sum of the distances from any point on the ellipse to the two foci is $2a$. Thus, $2a = 10$ and $a = 5$. The foci are $(0, -3)$ and $(0, 3)$, so $c = 3$. Now, $b^2 = a^2 - c^2 = 5^2 - 3^2 = 16$. The endpoints of the major axis are $(-5, 0)$ and $(5, 0)$, and the endpoints of the minor axis are $(0, -4)$ and $(0, 4)$. The standard form for the equation is

$$\frac{x^2}{25} + \frac{y^2}{16} = 1.$$

   A complete graph is sketched in Figure 6.2.4.

   (b) Again, we find $a = 5$, $c = 3$, and $b = 4$. However, the standard form for the equation is

$$\frac{x^2}{16} + \frac{y^2}{25} = 1.$$

   The endpoints of the major axis are $(0, -5)$ and $(0, 5)$, and the endpoints of the minor axis are $(-4, 0)$ and $(4, 0)$. A complete graph is sketched in Figure 6.2.5. •

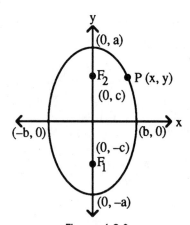

Figure 6.2.3

## 6.2 Ellipses

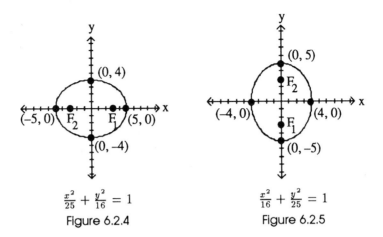

Figure 6.2.4: $\frac{x^2}{25} + \frac{y^2}{16} = 1$

Figure 6.2.5: $\frac{x^2}{16} + \frac{y^2}{25} = 1$

The graphs in Example 2 can be confirmed with a function graphing utility by solving for $y$ in terms of $x$. Note that two functions will have to be entered to graph each ellipse.

### Equations of Ellipses in Standard Form

If an ellipse has lines of symmetry parallel to or coincident with the coordinate axes, then it is the graph of an equation of the type given in Theorem 1.

• **THEOREM 1**   Let $a > b$. The graph of the equation $\frac{(x-h)^2}{a^2} + \frac{(y-k)^2}{b^2} = 1$ or the equation $\frac{(x-h)^2}{b^2} + \frac{(y-k)^2}{a^2} = 1$ is an ellipse with center at $(h, k)$, and lines of symmetry the lines $x = h$ and $y = k$. The foci of the first equation are $(h \pm c, k)$ where $c = \sqrt{a^2 - b^2}$ and lie on the line $y = k$. The foci of the second equation are $(h, k \pm c)$ where $c = \sqrt{a^2 - b^2}$ and lie on the line $x = h$ (Figure 6.2.6).

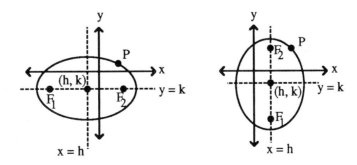

Figure 6.2.6

- **DEFINITION**  The equations of Theorem 1 are called **standard forms for the equation of an ellipse.**

The graphs of the equations of Theorem 1 can be obtained by applying horizontal and vertical shifts to the graphs of equations (1) and (2).

It is not necessary to memorize the formulas in Theorem 1 for the foci. In practice, you will use completing the square to put an equation of an ellipse in standard form. Then, the values of $a, b, h$ and $k$ can be read and used to compute $c$, the foci, and the endpoints of the major and minor axes.

- **EXAMPLE 3:** Give the endpoints of the major and minor axes, the coordinates of the center and foci, the lines of symmetry, and sketch a complete graph of the following ellipses.

(a) $\dfrac{(x+2)^2}{25} + \dfrac{(y-3)^2}{9} = 1$  (b) $\dfrac{(x-4)^2}{25} + \dfrac{(y+1)^2}{169} = 1$

SOLUTION:

(a) If we start with the graph of

$$\frac{x^2}{25} + \frac{y^2}{9} = 1$$

and shift horizontally left 2 units and vertically up 3 units we get the desired graph (Figure 6.2.7). Thus, the center of the ellipse is $(-2, 3)$, and the lines of symmetry are the lines $x = -2$ and $y = 3$. Now $a = 5$ and $b = 3$ so that $c^2 = a^2 - b^2 = 25 - 9 = 16$. The foci of $\frac{x^2}{25} + \frac{y^2}{9} = 1$ are $(-4, 0)$ and $(4, 0)$. The foci of $\frac{(x+2)^2}{25} + \frac{(y-3)^2}{9} = 1$ are $(-6, 3)$ and $(2, 3)$ because they are obtained by shifting the foci of $\frac{x^2}{25} + \frac{y^2}{9} = 1$ left 2 units and up 3 units. The endpoints of the major axis are $(-7, 3)$ and $(3, 3)$, and the endpoints of the minor axis are $(-2, 0)$ and $(-2, 6)$.

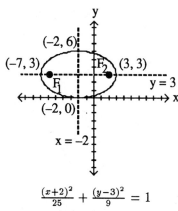

$\dfrac{(x+2)^2}{25} + \dfrac{(y-3)^2}{9} = 1$

Figure 6.2.7

(b) Shifting the graph of
$$\frac{x^2}{25} + \frac{y^2}{169} = 1$$
horizontally right 4 units and vertically down 1 unit produces the desired graph (Figure 6.2.8). The center of the ellipse is $(4, -1)$. The lines of symmetry are $x = 4$ and $y = -1$. Now $a = 13$, $b = 5$, so that $c = 12$. The foci are $(4, -13)$ and $(4, 11)$. The endpoints of the major axis are $(4, -14)$ and $(4, 12)$, and the endpoints of the minor axis are $(-1, -1)$ and $(9, -1)$. •

If $a^2 = b^2$ then the equations in Theorem 1 can be rewritten as follows.
$$\frac{(x-h)^2}{a^2} + \frac{(y-k)^2}{a^2} = 1$$
$$(x-h)^2 + (y-k)^2 = a^2$$

The graph of this equation is a circle with center $(h, k)$ and radius $a$. Thus, a circle is a degenerate case of an ellipse. In a circle, there is just one focus, and it is the center of the circle.

If the equations of Theorem 1 are cleared of fractions and parentheses are expanded, then the new equations are of the form $Ax^2 + Bxy + Cy^2 + Dx + Ey + F = 0$ where $B = 0$. Thus, ellipses with lines of symmetry parallel to or coincident with the coordinate axes are graphs of quadratic equations in two variables. The complete graphs obtained in Example 3 can be checked with a function graphing utility by first solving for $y$ in terms of $x$.

If we start with an equation in expanded form with $B = 0$, then we can use completing the square to write the equation in standard form and sketch its graph.

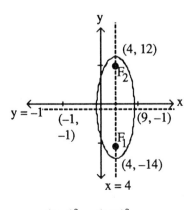

$$\frac{(x-4)^2}{25} + \frac{(y+1)^2}{169} = 1$$
Figure 6.2.8

- **EXAMPLE 4:** Use completing the square to write $9x^2 + 5y^2 - 54x + 40y + 116 = 0$ in standard form. Give the endpoints of the major and minor axes, the coordinates of the center and foci, the lines of symmetry, and sketch a complete graph.

SOLUTION: We group the terms involving $x$ and the terms involving $y$ and then complete the square:

$$9x^2 + 5y^2 - 54x + 40y + 116 = 0$$
$$9(x^2 - 6x) + 5(y^2 + 8y) = -116$$
$$9(x^2 - 6x + 9) + 5(y^2 + 8y + 16) = -116 + 9(9) + 5(16)$$
$$9(x-3)^2 + 5(y+4)^2 = 45$$
$$\frac{9(x-3)^2}{45} + \frac{5(y+4)^2}{45} = 1$$
$$\frac{(x-3)^2}{5} + \frac{(y+4)^2}{9} = 1.$$

The graph is an ellipse with center $(3, -4)$ and lines of symmetry $x = 3$ and $y = -4$. The endpoints of the major axis are $(3, -7)$ and $(3, -1)$, and the endpoints of the minor axis are $(3 - \sqrt{5}, -4) = (0.76, -4)$ and $(3 + \sqrt{5}, -4) = (5.24, -4)$. Notice that $a^2 = 9$, $b^2 = 5$ so that $c^2 = 4$. Thus, the foci are $(3, -6)$ and $(3, -2)$. Solving the above equation for $y$ in terms of $x$ we obtain

$$y = -4 \pm 3\sqrt{1 - \frac{(x-3)^2}{5}}.$$

Now, we can use a function graphing utility and enter

$$y = -4 + 3\sqrt{1 - \frac{(x-3)^2}{5}}$$

and

$$y = -4 - 3\sqrt{1 - \frac{(x-3)^2}{5}}$$

to obtain the complete graph shown in Figure 6.2.9. This graph confirms our algebraic computations. •

- **EXAMPLE 5:** Use the quadratic formula and a function graphing utility to obtain a complete graph of the ellipse $5x^2 + 3y^2 - 20x + 6y = -8$.

SOLUTION: First we rewrite the equation:

$$5x^2 + 3y^2 - 20x + 6y = -8$$
$$3y^2 + 6y + (5x^2 - 20x + 8) = 0.$$

Now, we use the quadratic formula to solve for $y$:

$$y = \frac{-6 \pm \sqrt{36 - 12(5x^2 - 20x + 8)}}{6}.$$

## 6.2 Ellipses

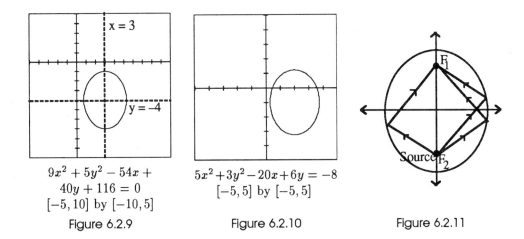

Figure 6.2.9: $9x^2 + 5y^2 - 54x + 40y + 116 = 0$, $[-5, 10]$ by $[-10, 5]$

Figure 6.2.10: $5x^2 + 3y^2 - 20x + 6y = -8$, $[-5, 5]$ by $[-5, 5]$

Figure 6.2.11

It is *not* necessary to simplify this equation before using a graphing utility. The graph of
$$y = \frac{-6 + \sqrt{36 - 12(5x^2 - 20x + 8)}}{6}$$
produces the upper half of the complete graph of the ellipse in Figure 6.2.10, and the graph of
$$y = \frac{-6 - \sqrt{36 - 12(5x^2 - 20x + 8)}}{6}$$
produces the lower half.

•

The next example is a real-world application that takes advantage of the reflective properties of an ellipse. Light or sound emitted at one focus of an ellipse is reflected by the ellipse through the other focus (Figure 6.2.11). A lithotripter is a special device that is used to break up kidney stones with shock waves through water. Roughly speaking a lithotripter is constructed as follows. An ellipse is rotated about its major axis and cut in half to form half of an *ellipsoid* (Figure 6.2.12). Careful measurements are made to put the patient's kidney stones at one focus and a source that produces ultra high frequency shock waves at the other focus. The shock waves are reflected by the ellipsoid to the other focus to break up the kidney stones.

Assume that the center of the ellipse that is rotated to construct a lithotripter is $(0, 0)$, the two ends of the major axis are $(-6, 0)$ and $(6, 0)$, and one end of the minor axis is $(0, -2.5)$. These are measurements in inches of an actual lithotripter.

- **EXAMPLE 6:** Determine the coordinates of the foci of the lithotripter described above.

**SOLUTION:** An equation for the ellipse that is rotated is $\frac{x^2}{6^2} + \frac{y^2}{2.5^2} = 1$. Thus, $a = 6$ and $b = 2.5$. Now, $c^2 = a^2 - b^2$ so that $c^2 = 29.75$ and $c = 5.45$. Therefore, the foci are at $(-5.45, 0)$ and $(5.45, 0)$. The shock source (an underwater spark discharge) is placed at the negative focus and the patient's kidney at the positive focus.

•

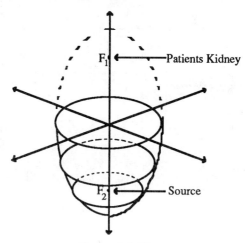

Figure 6.2.12

## EXERCISES 6-2

Determine the foci and standard form of the equation of the ellipse, and draw a complete graph. Do not use a graphing utility.

1. major axis of length 12 along the $x$-axis, minor axis of length 9 along the $y$-axis
2. major axis of length 6 along the $y$-axis, minor axis of length 5 along the $x$-axis

Determine the endpoints of the major and minor axes and standard form of the equation of the ellipse, and sketch a complete graph. Do not use a graphing utility.

3. foci at $(3,0)$ and $(-3,0)$, minor axis of length 14
4. foci at $(0,4)$ and $(0,-4)$, minor axis of length 9

Determine the endpoints of the major and minor axes and the standard form of the equation of the ellipse, and sketch a complete graph. Do not use a graphing utility.

5. The sum of the distances from the foci is 9; the foci are $(-4,0)$ and $(4,0)$.
6. The sum of the distances from the foci is 11; the foci are $(-0.5, 0)$ and $(0.5, 0)$.
7. foci at $(1,3)$ and $(1,9)$ and major axis of length 12
8. center at $(1,-4)$, foci at $(1,-2)$ and $(1,-6)$, and minor axis of length 10

Give the endpoints of the major and minor axes, the coordinates of the center and foci, and the lines of symmetry of the ellipse, and sketch a complete graph. Do not use a graphing utility.

9. $\dfrac{(x-1)^2}{2} + \dfrac{(y+3)^2}{4} = 1$
10. $(x+3)^2 + 4(y-1)^2 = 16$
11. $\dfrac{(x+2)^2}{5} + 2(y-1)^2 = 1$
12. $\dfrac{(x-4)^2}{16} + 16(y+4)^2 = 8$

## 6.2 Ellipses

Draw a complete graph using only a function graphing utility.

13. $2x^2 - 4x + y^2 - 6 = 0$
14. $3x^2 - 6x + 2y^2 + 8y + 5 = 0$

Write the equation of the ellipse in standard form, give the endpoints of the major and minor axes, the coordinates of the center and foci, and the lines of symmetry of the ellipse, and draw a complete graph. Check your answer using a graphing utility.

15. $9x^2 + 4y^2 - 18x + 8y - 23 = 0$
16. $3x^2 + 5y^2 - 12x + 30y + 42 = 0$
17. $4x^2 + y^2 - 32x + 16y + 124 = 0$
18. $2x^2 + 3y^2 + 12x - 24y + 60 = 0$

19. Explain why the graph of $5x^2 + y^2 + 5 = 0$ has no points.

20. A lithotripter is made by rotating the portion of an ellipse below its minor axis about its major axis. The major diameter (length of the major axis) is 26 inches, and the maximum depth from the major axis is 10 inches. Where should the shock source and patient be placed for maximum effect? Give the appropriate measurements.

21. There are real elliptical pool tables that have been constructed with a single pocket at one of the foci. Suppose such a table has major diameter 6 feet and minor diameter 4 feet.
    (a) Explain how a pool shark who knows conic geometry is at a great advantage over a "mark" who knows no conic geometry.
    (b) How should the ball be hit so it bounces off the cushion directly into the pocket? Give specific measurements.

The United States Capitol dome is in the shape of an ellipsoid of revolution. Assume the dome is formed by revolving the portion of an ellipse above its major axis about its minor axis. Further assume the diameter of the dome (major axis of the ellipse) is 400 feet and its maximum height is 60 feet (semiminor axis, one-half the minor axis length).

22. In 1849, a U.S. senator standing along the major axis of the ellipsoid under the capitol dome had his very private whispered conversation with a lady overheard by another senator standing quite a distance away (true story). How could this have happened?

23. Give the likely exact position(s) of the two senators in Exercise 22.

The **eccentricity** $e$ of an ellipse in standard form is defined by $e = c/a = \sqrt{a^2 - b^2}/a$. Compute the eccentricity of the ellipse.

24. $4x^2 + \frac{1}{8}y^2 = 16$
25. $2x^2 + 3y^2 + 6y - 4x - 1 = 0$

26. Show that $b = \sqrt{a^2(1 - e^2)}$ for an ellipse in standard form, where $e$ is its eccentricity.

Determine an equation and sketch a complete graph of the ellipse. Assume that the center is at the origin.

27. eccentricity 0.1 and major axis length 8
28. eccentricity 0.5 and major axis length 6

29. Show that $0 < e < 1$ for every ellipse.
30. What is the change in the nature of the ellipse as its eccentricity approaches 0?
31. What is the change in the nature of the ellipse as its eccentricity approaches 1?
32. Explain how eccentricity is a measure of the circularity of an ellipse.

| Planet | e | a (miles) |
|---|---|---|
| Mercury | 0.20 | $3.6 \times 10^7$ |
| Earth | 0.02 | $9.3 \times 10^7$ |
| Mars | 0.09 | $1.4 \times 10^8$ |
| Saturn | 0.50 | $8.9 \times 10^8$ |
| Neptune | 0.01 | $2.8 \times 10^9$ |
| Pluto | 0.25 | $3.7 \times 10^9$ |

Table 1

33. What should the eccentricity of a circle be?

A *Hohmann* ellipse is an ideal trajectory for space flight between two planets that requires a minimum of energy for the journey. Assume two planets 12,800,000 miles apart are located at the foci of a Hohmann ellipse. Determine the equation in standard form of the ellipse with the following eccentricity.

34. 0.2                                   35. 0.4

Table 1 gives the eccentricity $e$ of the elliptical orbits of selected planets and the length of their semimajor axes (one-half the length of the major axes). Determine an equation and draw the graph of each orbit in the same viewing rectangle. The sun is one of the foci. Assume the center of each orbit is at the origin, and each major axis is along the $x$-axis.

36. Earth and Mercury
37. Earth, Mars, and Saturn
38. Neptune and Pluto

## 6.3  Hyperbolas

In this section we give the two-foci definition of a hyperbola and obtain the standard form for the equation of a hyperbola with lines of symmetry parallel to the coordinate axes. The standard form is used to obtain a rough sketch of its graph. We use completing the square to put equations of hyperbolas into standard form. An application about lenses involving reflective properties of a hyperbola will be investigated.

In Chapter 1, we drew the graph of $xy = 1$, or $y = \frac{1}{x}$, which is an example of a *rectangular hyperbola*. A **rectangular hyperbola** is any hyperbola whose asymptotes are parallel to or coincident with the coordinate axes. Hyperbolas of this type will be investigated in the exercises of the next section.

The force of attraction between objects can cause the path of an object to be part of a hyperbola. For example, some comets travel through our solar system on hyperbolic paths. Applications involving navigation or locating sources of sound sometimes involve hyperbolas. We will study one such application in Section 6.5.

## 6.3 Hyperbolas

- **DEFINITION**  Let $F_1$ and $F_2$ be any two points in the coordinate plane. A **hyperbola** is the set of all points $P$ in the plane such that the absolute value of the difference of the distances from $P$ to $F_1$ and from $P$ to $F_2$ is a constant (Figure 6.3.1). Each of the points $F_1$ and $F_2$ is called a **focus** of the hyperbola.

There are *no* points $P$ satisfying the condition that the absolute value of the difference of the distances from $P$ to $F_1$ and from $P$ to $F_2$ is a constant unless that constant is less than the distance between the two foci. (Why?) Let $L_1$ be the line determined by $F_1$ and $F_2$, and $L_2$ the perpendicular bisector of the line segment joining $F_1$ and $F_2$. The points $P_1$ and $P_2$ that are symmetric to any point $P$ on the hyperbola with respect to $L_1$ and $L_2$, respectively, are on the hyperbola. (Why?) Thus, the hyperbola is symmetric with respect to both $L_1$ and $L_2$. The point of intersection of $L_1$ and $L_2$ is called the **center** of the hyperbola. A hyperbola is symmetric with respect to its center. The two points of the hyperbola that lie on $L_1$ are called **vertices** of the hyperbola. The line segment between the vertices is called the **transverse axis**. The line $L_1$ is called the **principal axis** of the hyperbola and the line $L_2$ the **conjugate axis**.

In this section we will consider hyperbolas with lines of symmetry parallel to the coordinate axes. This restriction will be removed in Section 6.4. We begin by determining an equation for a hyperbola with given foci.

It is convenient to let $2a$ be the difference of the distances from $P$ to $F_1$ and $F_2$.

- **EXAMPLE 1:** Let $F_1 = (-c, 0)$ and $F_2 = (c, 0)$ be the two foci of the hyperbola consisting of the points $P$ where $|d(P, F_1) - d(P, F_2)| = 2a$ with $0 < a < c$ (Figure 6.3.2). Determine an equation for the hyperbola, the $x$-intercepts, the $y$-intercepts, and the center of the hyperbola.

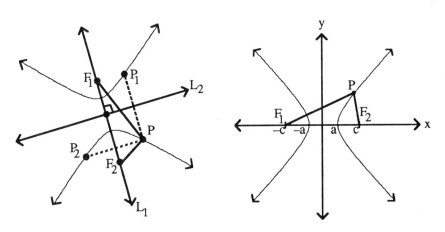

Figure 6.3.1  Figure 6.3.2

SOLUTION: We can use the distance formula to obtain an equation for the hyperbola:

$$|d(P, F_1) - d(P, F_2)| = 2a$$
$$|\sqrt{(x+c)^2 + y^2} - \sqrt{(x-c)^2 + y^2}| = 2a$$
$$\sqrt{(x+c)^2 + y^2} - \sqrt{(x-c)^2 + y^2} = \pm 2a$$
$$\sqrt{(x+c)^2 + y^2} = \sqrt{(x-c)^2 + y^2} \pm 2a.$$

As in the previous section we can obtain the following alternate form of this equation that does not involve radicals:

$$\frac{x^2}{a^2} - \frac{y^2}{c^2 - a^2} = 1.$$

Because $a < c$, we can define a positive number $b$ by $b^2 = c^2 - a^2$ and rewrite the above equation in the following form:

$$\frac{x^2}{a^2} - \frac{y^2}{b^2} = 1. \qquad (1)$$

Thus, every point on the hyperbola is a solution to equation (1). It can also be shown that every solution to (1) corresponds to a point on the hyperbola. This form is called a *standard form* for the equation of a hyperbola. Notice that the $x$-intercepts are $\pm a$ and are also the *vertices* of the hyperbola. The distance between the center $(0,0)$ and each vertex is one-half of the absolute value of the difference of the distances from $P$ to $F_1$ and from $P$ to $F_2$. There are *no* $y$-intercepts. •

Notice that the equation $\frac{x^2}{a^2} - \frac{y^2}{b^2} = 1$ of Example 1 is not changed if $x$ is replaced by $-x$ or if $y$ is replaced by $-y$. This confirms algebraically that the graph in Figure 6.3.2 is symmetric with respect to the $x$-axis, $y$-axis, and the origin.

It can be shown that a standard form for the equation of the hyperbola with foci $F_1(0, -c)$ and $F_2(0, c)$ on the $y$-axis that satisfies $|d(P, F_1) - d(P, F_2)| = 2a$ with $0 < a < c$ is

$$\frac{y^2}{a^2} - \frac{x^2}{b^2} = 1 \qquad (2)$$

where $b^2 = c^2 - a^2$ (Figure 6.3.3). Notice that this is the equation of Example 1 with $x$ and $y$ interchanged. The vertices $\pm a$ are the $y$-intercepts and there are *no* $x$-intercepts. The center is $(0,0)$.

### 6.3 Hyperbolas

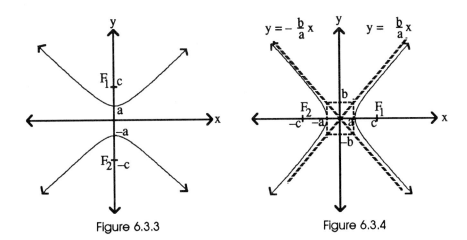

Figure 6.3.3  Figure 6.3.4

### Asymptotes

Equation (1) of the hyperbola in Example 1 can be rewritten as follows:

$$\frac{x^2}{a^2} - \frac{y^2}{b^2} = 1$$

$$\frac{x^2}{a^2} - 1 = \frac{y^2}{b^2}$$

$$y^2 = \frac{b^2 x^2}{a^2} - b^2$$

$$y^2 = \frac{b^2 x^2}{a^2}\left(1 - \frac{a^2}{x^2}\right). \tag{3}$$

We can see from equation (3) that as the absolute value of $x$ gets very large the equation of the hyperbola is approximately $y^2 = \frac{b^2 x^2}{a^2}$. Thus, an end behavior model of the hyperbola of Example 1 is $y^2 = \frac{b^2 x^2}{a^2}$. This will be illustrated in Example 3. Notice that the graph of $y^2 = \frac{b^2 x^2}{a^2}$ is the pair of intersecting lines $y = \frac{b}{a}x$ and $y = -\frac{b}{a}x$. These two lines are called the **asymptotes** of the hyperbola $\frac{x^2}{a^2} - \frac{y^2}{b^2} = 1$ and can be used to help sketch the graph. The asymptotes $y = \pm\frac{b}{a}x$ are the diagonals of the dashed rectangle of width 2a and length 2b in Figure 6.3.4. In the exercises you will show that the asymptotes of $\frac{y^2}{a^2} - \frac{x^2}{b^2} = 1$ are $y = \pm\frac{a}{b}x$.

- **EXAMPLE 2:** Determine the vertices, center, a standard form of the equation, and asymptotes of each hyperbola, and sketch a complete graph.

    (a) The absolute value of the difference of the distance from the foci is 8; the foci are $(-5, 0)$ and $(5, 0)$.
    (b) The absolute value of the difference of the distance from the foci is 8; the foci are $(0, -5)$ and $(0, 5)$.

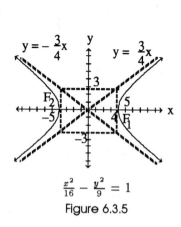

$\frac{x^2}{16} - \frac{y^2}{9} = 1$
Figure 6.3.5

$\frac{y^2}{16} - \frac{x^2}{9} = 1$
Figure 6.3.6

SOLUTION:

(a) The absolute value of the difference of the distances from any point on the hyperbola to the two foci is $2a$. Thus, $2a = 8$ and $a = 4$. The foci are $(-5, 0)$ and $(5, 0)$, so $c = 5$. Now, $b^2 = c^2 - a^2 = 25 - 16 = 9$. The standard form for the equation is $\frac{x^2}{16} - \frac{y^2}{9} = 1$. The vertices are $(\pm 4, 0)$, and the center is $(0, 0)$. The asymptotes are $y = \pm \frac{b}{a} x = \pm \frac{3}{4} x$. A complete graph of the hyperbola is sketched in Figure 6.3.5.

(b) Again, we find $a = 4$, $c = 5$, and $b = 3$. However, the standard form for the equation is $\frac{y^2}{16} - \frac{x^2}{9} = 1$. Notice this equation is the equation of (a) with $x$ and $y$ interchanged. The vertices are $(0, \pm 4)$, the center is $(0, 0)$, and the asymptotes are $y = \pm \frac{a}{b} x = \pm \frac{4}{3} x$. A complete graph of the hyperbola is sketched in Figure 6.3.6. •

• **EXAMPLE 3:** Show that the asymptotes $y = \pm \frac{3}{4} x$ of Example 2(a) correctly describe the end behavior of the hyperbola $\frac{x^2}{16} - \frac{y^2}{9} = 1$.

SOLUTION: If we use a sufficiently large viewing rectangle, the graphs of the hyperbola $\frac{x^2}{16} - \frac{y^2}{9} = 1$ and the pair of lines $y = \pm \frac{3}{4} x$ appear to be identical (Figure 6.3.7). This means that an end behavior model of the hyperbola is the pair of lines $y = \pm \frac{3}{4} x$. •

## Equation of Hyperbolas in Standard Form

If a hyperbola has lines of symmetry parallel to or coincident with the coordinate axes, then it is the graph of an equation of the type given in Theorem 1.

• **THEOREM 1** (i) The graph of the equation $\frac{(x-h)^2}{a^2} - \frac{(y-k)^2}{b^2} = 1$ is a hyperbola with center $(h, k)$ and lines of symmetry $x = h$ and $y = k$. The vertices are $(h \pm a, k)$, the foci are $(h \pm c, k)$ where $c = \sqrt{a^2 + b^2}$, and the asymptotes are $y - k =$

### 6.3 Hyperbolas

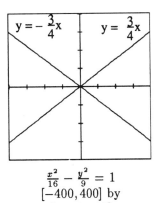

$\frac{x^2}{16} - \frac{y^2}{9} = 1$
$[-400, 400]$ by
$[-400, 400]$
Figure 6.3.7

$\pm \frac{b}{a}(x-h)$ (Figure 6.3.8). Notice that the branches of the hyperbola open left and right.

(ii) The graph of the equation $\frac{(y-k)^2}{a^2} - \frac{(x-h)^2}{b^2} = 1$ is a hyperbola with center $(h, k)$ and lines of symmetry $x = h$ and $y = k$. The vertices are $(h, k \pm a)$, the foci are $(h, k \pm c)$ where $c = \sqrt{a^2 + b^2}$, and the asymptotes are $y - k = \pm \frac{a}{b}(x - h)$ (Figure 6.3.9). Notice that the branches of the hyperbola open up and down.

- **DEFINITION** The equations of Theorem 1 are called **standard forms for the equation of a hyperbola.**

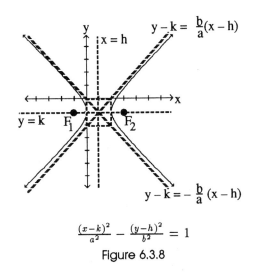

$\frac{(x-k)^2}{a^2} - \frac{(y-h)^2}{b^2} = 1$
Figure 6.3.8

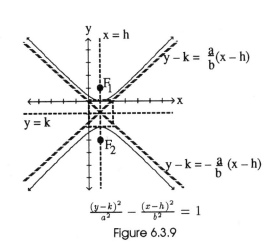

$\frac{(y-k)^2}{a^2} - \frac{(x-h)^2}{b^2} = 1$
Figure 6.3.9

The equations of Theorem 1 can be obtained by applying horizontal and vertical shifts to equations (1) and (2).

It is not necessary to memorize the formulas in Theorem 1 for the foci and the asymptotes. In practice, you will use completing the square to put an equation of a hyperbola in standard form. Then, the values of $a, b, h$ and $k$ can be read and used to compute $c$, the foci, and the asymptotes.

• **EXAMPLE 4:** Determine the coordinates of the center, foci, vertices, and the asymptotes of each hyperbola, and sketch a complete graph.

(a) $\dfrac{(x+2)^2}{9} - \dfrac{(y-1)^2}{16} = 1$  (b) $\dfrac{(y+3)^2}{4} - \dfrac{(x-2)^2}{25} = 1$

**SOLUTION:**

(a) The center is $(-2, 1)$. Because $a = 3$ and $b = 4$, $c^2 = a^2 + b^2 = 9 + 16 = 25$. Thus, $c = 5$. Because the branches of the hyperbola open left and right, the vertices are on the line $y = 1$ and are 3 units on either side of the center. Thus, the vertices are $(-5, 1)$ and $(1, 1)$. The foci are also on the line $y = 1$, but are 5 units on either side of the center. (Why?) The foci are $(-7, 1)$ and $(3, 1)$. The asymptotes are $y - 1 = \pm \frac{4}{3}(x + 2)$, and the graph sketched in Figure 6.3.10 is complete.

(b) The center is $(2, -3)$. Because $a = 2$ and $b = 5$, $c^2 = a^2 + b^2 = 4 + 25 = 29$. Thus, $c = \sqrt{29}$. The branches of this hyperbola open up and down, so the vertices are on the line $x = 2$ and are 2 units on either side of the center. Thus, the vertices are $(2, -5)$ and $(2, -1)$. The foci are also on the line $x = 2$, but are $\sqrt{29}$ units on either side of the center. The foci are $(2, -3 - \sqrt{29}) = (2, -8.39)$ and $(2, -3 + \sqrt{29}) = (2, 2.39)$. The asymptotes are $y + 3 = \pm \frac{2}{5}(x - 2)$, and the graph sketched in Figure 6.3.11 is complete. •

Figure 6.3.10

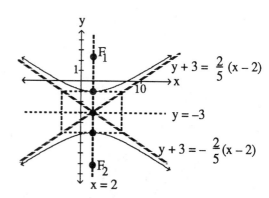

Figure 6.3.11

## 6.3 Hyperbolas

If the equations in Theorem 1 are cleared of fractions and if parentheses are expanded, then the resulting equations are of the form $Ax^2 + Bxy + Cy^2 + Dx + Ey + F = 0$ where $B = 0$. Thus, hyperbolas with lines of symmetry parallel to or coincident with the coordinate axes are graphs of quadratic equations in two variables. The coefficients $A$ and $C$ will have unlike signs in this case. However, when the equations of Theorem 1 of Section 6.2 are cleared of fractions, the coefficients $A$ and $C$ have like signs. The complete graphs obtained in Example 4 can be checked with a function graphing utility after first solving for $y$ in terms of $x$.

As in the previous section, if we have an equation in expanded form with $B = 0$, then we can use completing the square to write the equation in standard form and sketch a complete graph, or we can use the quadratic formula and a function graphing utility to draw a complete graph.

• **EXAMPLE 5:** Write each of the following equations in standard form. Determine the center, foci, vertices, lines of symmetry, asymptotes, and sketch a complete graph. Use the quadratic formula and a function graphing utility to check the graphs.

(a) $x^2 - 4y^2 + 2x - 24y = 39$

(b) $9y^2 - 5x^2 + 30x - 36y = 54$

**SOLUTION:**

(a) We group the terms involving $x$ and the terms involving $y$ and then complete the square:

$$x^2 - 4y^2 + 2x - 24y = 39$$
$$(x^2 + 2x) - 4(y^2 + 6y) = 39$$
$$(x^2 + 2x + 1) - 4(y^2 + 6y + 9) = 39 + 1 - 4(9)$$
$$(x + 1)^2 - 4(y + 3)^2 = 4$$
$$\frac{(x + 1)^2}{4} - \frac{(y + 3)^2}{1} = 1.$$

Now, we can see that the graph of this equation is a hyperbola with center $(-1, -3)$, and lines of symmetry $x = -1$ and $y = -3$. Notice that $a = 2$ and $b = 1$ so that $c^2 = 5$. The vertices are $(-3, -3)$ and $(1, -3)$, and the foci are $(-1 - \sqrt{5}, -3) = (-3.24, -3)$ and $(-1 + \sqrt{5}, -3) = (1.24, -3)$. The asymptotes are $y + 3 = \pm\frac{1}{2}(x + 1)$. A complete graph is sketched in Figure 6.3.12.

To check this graph with a function graphing utility we solve the original equation for $y$ using the quadratic formula:

$$x^2 - 4y^2 + 2x - 24y = 39$$
$$-4y^2 - 24y + (x^2 + 2x - 39) = 0$$
$$4y^2 + 24y - (x^2 + 2x - 39) = 0$$
$$y = \frac{-24 \pm \sqrt{576 + 16(x^2 + 2x - 39)}}{8}.$$

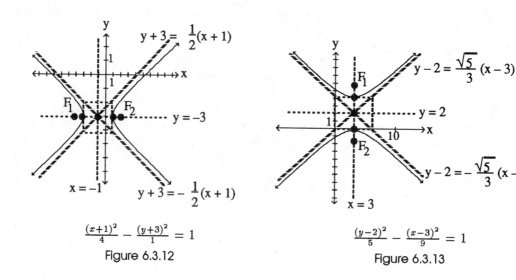

Figure 6.3.12

Figure 6.3.13

The graph of

$$y = \frac{-24 + \sqrt{576 + 16(x^2 + 2x - 39)}}{8}$$

produces the portion of the hyperbola in Figure 6.3.12 that is above the line $y = -3$, and the graph of

$$y = \frac{-24 - \sqrt{576 + 16(x^2 + 2x - 39)}}{8}$$

produces the portion below the line $y = -3$ confirming the above algebraic computations.

(b) Again, we group terms and complete the square:

$$9y^2 - 5x^2 + 30x - 36y = 54$$
$$9(y^2 - 4y) - 5(x^2 - 6x) = 54$$
$$9(y^2 - 4y + 4) - 5(x^2 - 6x + 9) = 54 + 9(4) - 5(9)$$
$$9(y - 2)^2 - 5(x - 3)^2 = 45$$
$$\frac{(y-2)^2}{5} - \frac{(x-3)^2}{9} = 1.$$

The graph of this equation is a hyperbola with center at $(3, 2)$ and lines of symmetry $x = 3$ and $y = 2$. Notice $a = \sqrt{5}$ and $b = 3$ so that $c^2 = 14$. The vertices are $(3, 2 - \sqrt{5}) = (3, -0.24)$ and $(3, 2 + \sqrt{5}) = (3, 4.24)$, and the foci are $(3, 2 - \sqrt{14}) = (3, -1.74)$ and $(3, 2 + \sqrt{14}) = (3, 5.74)$. The asymptotes are $y - 2 = \pm\frac{\sqrt{5}}{3}(x - 3)$. The complete graph in Figure 6.3.13 can be confirmed with a graphing utility as in (a). •

## 6.3 Hyperbolas

### Reflective Property of Hyperbolas

It can be proved that hyperbolas have a reflective property that is important in the construction of reflecting telescopes. The right branch of the hyperbola with foci $F_1$ and $F_2$ in Figure 6.3.14 is coated with a substance that causes it to reflect light. A light ray aimed toward the focus at $F_1$ is reflected by the surface through the focus at $F_2$ as illustrated in Figure 6.3.14. This fact is used to make a telescope using a parabolic lens and a hyperbolic lens. The main lens is parabolic with focus $F_1$ and vertex $F_2$, and the secondary lens is hyperbolic with foci $F_1$ and $F_2$ (Figure 6.3.15). The eye is positioned at the point $F_2$. The key to this lens arrangement is that one focus of the hyperbola *is* the focus of the parabola.

- **EXAMPLE 6:** The parabolic lens in Figure 6.3.15 is determined by a portion of the parabola $y^2 = 20x$ with $F_2 = (0,0)$. The vertex of the right branch of the hyperbola is at $(3,0)$. Determine

    (a) the focus of the parabolic lens.
    (b) the foci of the hyperbolic lens.
    (c) an equation for the hyperbola.

**SOLUTION:**

(a) The focus $F_1$ of the parabolic lens is on the positive $x$-axis $a$ units to the right of the origin, where $4a = 20$ (Figure 6.3.16). Thus, the focus of the parabola is $F_1 = (5,0)$.

(b) The foci of the hyperbolic lens are the vertex and focus of the parabolic lens. Thus, $(0,0)$ and $(5,0)$ are the foci of the hyperbolic lens.

(c) The foci of the hyperbola are $(0,0)$ and $(5,0)$. Thus, the center $H$ of the hyperbola is $(2.5, 0)$ and $a = 3 - 2.5 = 0.5$. Because $c = 5 - 2.5 = 2.5$, $b^2 = c^2 - a^2 = (2.5)^2 - (0.5)^2 = 6$. The equation of the hyperbola in standard form is

$$\frac{(x-2.5)^2}{0.25} - \frac{y^2}{6} = 1.$$

Figure 6.3.14

Figure 6.3.15

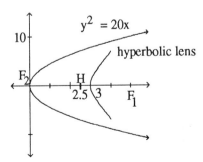

Figure 6.3.16

## EXERCISES 6-3

Determine the standard form of the equation, and sketch a complete graph of the hyperbola. Do not use a graphing utility.

1. The difference of the distances from any point on the hyperbola to the foci is 4; the foci are $(-3, 0)$ and $(3, 0)$.
2. The difference of the distances from any point on the hyperbola to the foci is 8; the foci are $(0, -8)$ and $(0, 8)$.
3. The difference of the distances from any point on the hyperbola to the foci is 3; the foci are $(-3, 2)$ and $(3, 2)$.
4. The difference of the distances from any point on the hyperbola to the foci is 6; the foci are $(1, -2)$ and $(9, -2)$.

Determine the coordinates of the center, foci, vertices, the lines of symmetry, and the asymptotes of the hyperbola, and sketch a complete graph. Do not use a graphing utility.

5. $\dfrac{x^2}{4} - \dfrac{(y-3)^2}{5} = 1$
6. $\dfrac{(y-3)^2}{9} - \dfrac{(x+2)^2}{4} = 1$
7. $4(y-1)^2 - 9(x-3)^2 = 36$
8. $4(x-2)^2 - 9(y+4)^2 = 1$

Use the quadratic formula and a function graphing utility to draw a complete graph.

9. $2x^2 - y^2 + 4x + 6 = 0$
10. $3y^2 - 5x^2 + 2x - 6y - 9 = 0$

Identify the conic. Write an equation of the conic in standard form; give the foci, vertices, center, the lines of symmetry, and asymptotes (if any), and sketch a complete graph. Check your answer using a graphing utility.

11. $9x^2 - 4y^2 - 36x + 8y - 4 = 0$
12. $y^2 - 4y - 8x + 20 = 0$
13. $25y^2 - 9x^2 - 50y - 54x - 281 = 0$
14. $9x^2 + 16y^2 + 54x - 32y - 47 = 0$
15. $4y^2 - 9x^2 - 18x - 8y - 41 = 0$
16. $16x^2 - y^2 - 32x - 6y - 57 = 0$
17. $2x^2 - 6x + 5y - 13 = 0$
18. $9x^2 + 4y^2 - 18x + 16y - 11 = 0$

The *eccentricity* $e$ of a hyperbola is defined by $c/a = \sqrt{a^2 + b^2}/a$. Note $a$ and $b$ are both positive so eccentricity is positive. Compute the eccentricity of the hyperbola.

19. $x^2 - 4y^2 + 4 = 0$
20. $4x^2 - 9y^2 - 24x - 18y - 9 = 0$

Write an equation for the conic in standard form and sketch a complete graph.

21. Eccentricity 2 and transverse axis length 8.
22. Eccentricity 5 and transverse axis length 8.
23. Eccentricity 100 and transverse axis length 8.
24. Show that the eccentricity of a hyperbola is greater than one.
25. What does the eccentricity measure in a hyperbola?
26. What is the change in the nature of the hyperbola as the eccentricity gets very large?

A comet with hyperbolic path will pass between the sun and the earth. One focus of the hyperbola is the sun. Assume that the line through the earth and sun is a line of symmetry of the hyperbola

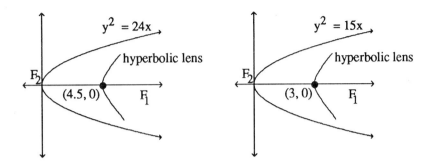

Figure 6.3.17    Figure 6.3.18

and that the distance between the earth and the sun is 93,000,000 miles. Determine the equation in standard form of the hyperbola representing the path of the comet satisfying the following condition.

27. Eccentricity 20        28. Eccentricity 400

Consider a parabolic and hyperbolic lens arrangement as illustrated in Figure 6.3.17. Assume that the generating equation of the parabolic lens is $y^2 = 24x$ and that the vertex of the right branch of the generating hyperbola is $(4.5, 0)$.

29. What are the coordinates of the foci of the hyperbola?
30. Determine the standard form of the equation of the generating hyperbola.
31. Draw complete graphs of both conics in the same viewing rectangle. Explain how this lens arrangement works.

Consider a parabolic and hyperbolic lens arrangement as illustrated in Figure 6.3.18. Assume that the generating equation of the parabolic lens is $y^2 = 15x$ and that the vertex of the right branch of the generating hyperbola is $(3, 0)$.

32. What are the coordinates of the foci of the hyperbola?
33. Determine the standard form of the equation of the generating hyperbola.
34. Draw complete graphs of both conics in the same viewing rectangle. Explain how this lens arrangement works.

A certain comet's path is known to be hyperbolic with the sun as one focus. Assume a space station is located at the center of the hyperbola. Further assume it has been determined that the distance between the comet and the space station is 5,000,000 miles at the closest point and that the sun and the space station are 12,000,000 miles apart.

35. Determine the standard form of the equation of the hyperbola representing the comet's path.
36. Draw a complete graph of the algebraic representation of the path of the comet. What portion of the graph represents the problem situation?
37. Describe the set of all points $P$ satisfying the condition that $d(P, F_1) - d(P, F_2) = k$ when $F_1$ and $F_2$ are distinct fixed points and $k$ is a positive real number.

## 6.4 Quadratic Forms and Conics

In this section we use a function graphing utility to determine complete graphs of equations of the form

$$Ax^2 + Bxy + Cy^2 + Dx + Ey + F = 0 \tag{1}$$

where not all of $A$, $B$, and $C$ are zero. Such an expression $Ax^2 + Bxy + Cy^2 + Dx + Ey + F$ is called a **quadratic form** in the variables $x$ and $y$. A complete graph of equation (1) is either an ordinary conic section or a degenerate conic. For example, degenerate forms are circles, parallel lines, intersecting lines, one line, a single point, or no graph.

In the previous sections of this chapter we found that conic sections with lines of symmetry parallel to the coordinate axes are the graphs of equations of the type given in (1) with $B = 0$. If the lines of symmetry are not parallel to the coordinate axes then we obtain an equation with $B \neq 0$. Example 1 illustrates this situation.

- **EXAMPLE 1:** Determine an equation for the parabola with focus $F(3,9)$ and directrix $y = 2x$, and determine the coordinates of the vertex.

  **SOLUTION:** Let $P(a,b)$ be any point on the parabola (Figure 6.4.1). Then choose the point $A$ on $y = 2x$ so that $PA$ is perpendicular to the line $y = 2x$. The parabola consists of all points $(a,b)$ satisfying $d(P,F) = d(P,A)$. Now, $d(P,F) = \sqrt{(a-3)^2 + (b-9)^2}$. To determine $d(P,A)$ we must find the coordinates of the point $A$.

  The slope of any line perpendicular to $y = 2x$ is $-\frac{1}{2}$. (Why?) An equation of the line through $P$ perpendicular to $y = 2x$ is $y - b = -\frac{1}{2}(x - a)$. $A$ is the point of intersection of the lines $y = 2x$ and $y - b = -\frac{1}{2}(x - a)$. The $x$-coordinate of the point of intersection can be found by solving each equation for $y$ and setting these two expressions equal to

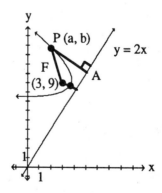

Figure 6.4.1

## 6.4 Quadratic Forms and Conics

each other:

$$2x = -\frac{1}{2}(x-a) + b$$
$$\frac{5x}{2} = \frac{a}{2} + b$$
$$x = \frac{a+2b}{5}.$$

The $y$-coordinate of $A$ is $2(\frac{a+2b}{5})$. Thus, $A$ has coordinates $(\frac{a+2b}{5}, \frac{2a+4b}{5})$. Next we find $d(P, A)$:

$$d(P, A) = \sqrt{(a - \frac{a+2b}{5})^2 + (b - \frac{2a+4b}{5})^2}$$
$$= \sqrt{(\frac{4a-2b}{5})^2 + (\frac{-2a+b}{5})^2}$$
$$= \sqrt{(\frac{2}{5})^2(2a-b)^2 + (-\frac{1}{5})^2(2a-b)^2}$$
$$= \sqrt{\frac{1}{5}(2a-b)^2}.$$

We use the definition of a parabola to determine an equation:

$$d(P, F) = d(P, A)$$
$$\sqrt{(a-3)^2 + (b-9)^2} = \sqrt{\frac{1}{5}(2a-b)^2}$$
$$(a-3)^2 + (b-9)^2 = \frac{1}{5}(2a-b)^2$$
$$5a^2 - 30a + 45 + 5b^2 - 90b + 405 = 4a^2 - 4ab + b^2$$
$$a^2 + 4ab + 4b^2 - 30a - 90b + 450 = 0.$$

The above equation holds for every point $(a, b)$ on the parabola. Thus, we have the following equation for the parabola:

$$x^2 + 4xy + 4y^2 - 30x - 90y + 450 = 0.$$

Next, we find the coordinates of the vertex. An equation for the line through $F$ perpendicular to $y = 2x$ is $y - 9 = -\frac{1}{2}(x - 3)$. It can be shown that the lines $y = 2x$ and $y - 9 = -\frac{1}{2}(x - 3)$ intersect at $(\frac{21}{5}, \frac{42}{5})$. The vertex $V$ of the parabola is the midpoint of the line segment joining $F(3, 9)$ and $(\frac{21}{5}, \frac{42}{5})$. Thus, $V = (\frac{18}{5}, \frac{87}{10}) = (3.6, 8.7)$.

## Graphing Conics with a Function Grapher

If $C \neq 0$, then equation (1) is quadratic in $y$ and can be rewritten and the quadratic formula used to solve for $y$:

$$Ax^2 + Bxy + Cy^2 + Dx + Ey + F = 0$$
$$Cy^2 + (Bx + E)y + (Ax^2 + Dx + F) = 0.$$
$$y = \frac{-(Bx + E) \pm \sqrt{(Bx + E)^2 - 4C(Ax^2 + Dx + F)}}{2C}.$$

A complete graph of equation (1) can be obtained by drawing the graphs of

$$f(x) = \frac{-(Bx + E) + \sqrt{(Bx + E)^2 - 4C(Ax^2 + Dx + F)}}{2C}$$

and

$$g(x) = \frac{-(Bx + E) - \sqrt{(Bx + E)^2 - 4C(Ax^2 + Dx + F)}}{2C}$$

in an appropriate viewing rectangle.

- **EXAMPLE 2:** Use the quadratic formula and a function graphing utility to determine a complete graph of the parabola $x^2 + 4xy + 4y^2 - 30x - 90y + 450 = 0$ of Example 1.

**SOLUTION:** The equation in Example 1 can be rewritten and the quadratic formula used to solve for $y$ in terms of $x$:

$$x^2 + 4xy + 4y^2 - 30x - 90y + 450 = 0$$
$$4y^2 + (4x - 90)y + (x^2 - 30x + 450) = 0$$
$$y = \frac{-(4x - 90) \pm \sqrt{(4x - 90)^2 - 16(x^2 - 30x + 450)}}{8}.$$

It is not necessary to simplify the above expression before using a function graphing utility. However, simplifying sometimes leads to functions that are easier to enter into a graphing utility. It can be shown that the above expression for $y$ simplifies to

$$y = \frac{45 - 2x \pm \sqrt{2225 - 60x}}{4}.$$

The graph of

$$f(x) = \frac{45 - 2x + \sqrt{225 - 60x}}{4} \quad \text{and} \quad g(x) = \frac{45 - 2x - \sqrt{225 - 60x}}{4}$$

in the $[-5, 5]$ by $[-5, 20]$ viewing rectangle gives a complete graph of the parabola (Figure 6.4.2). The graph of $f$ gives the top portion of the parabola and the graph of $g$ the lower portion. Depending on the resolution of your graph, you may see a gap in your graph. This gap occurs where the graph of $f$ and graph of $g$ meet. This is *not* at the vertex. (Why?) Generally, we remove or reduce gaps as much as possible in this textbook.

•

## 6.4 Quadratic Forms and Conics

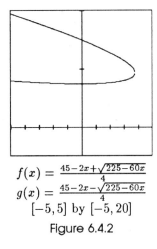

$f(x) = \frac{45 - 2x + \sqrt{225 - 60x}}{4}$
$g(x) = \frac{45 - 2x - \sqrt{225 - 60x}}{4}$
$[-5, 5]$ by $[-5, 20]$
Figure 6.4.2

Next we consider the case $C = 0$ in equation (1). If $C = 0$, then equation (1) becomes

$$(Bx + E)y + (Ax^2 + Dx + F) = 0. \tag{2}$$

This equation is linear in $y$. We can solve for $y$ unless the coefficient of $y$ is 0. This happens if $B = E = 0$. If $B = E = 0$, then the graph of equation (2) consists of a pair of vertical lines, one vertical line, or no graph depending on whether $Ax^2 + Dx + F$ has two real zeros, one real zero, or no real zeros. (Why?) In this case we cannot use a function graphing utility to obtain a graph of equation (2).

Now we determine the graph if one or both of $B$ and $E$ are different from zero. If $B = 0$ and $E \neq 0$, then we can solve equation (2) for $y$ as follows:

$$y = -\frac{Ax^2 + Dx + F}{E}.$$

The graph is a parabola in this case because $A \neq 0$. This is true because not all of $A$, $B$, and $C$ can be zero and we are assuming that $B = C = 0$. Notice that a function graphing utility can be used to graph equation (2). Finally, suppose that $B \neq 0$. We can have $E = 0$ or $E \neq 0$. If $-\frac{E}{B}$ is *not* a zero of $Ax^2 + Dx + F$, then $-\frac{E}{B}$ is not in the domain of the function defined by $(Bx + E)y + (Ax^2 + Dx + F) = 0$ so that

$$y = \frac{Ax^2 + Dx + F}{Bx + E}.$$

A function graphing utility can be used to graph equation (2) in this case. If $-\frac{E}{B}$ is a zero of $Ax^2 + Dx + F$, then $Bx + E$ is a factor of $Ax^2 + Dx + F$. Thus $Ax^2 + Dx + F = (Bx + E)(Gx + H)$ for some $G$ and $H$. Then, we can rewrite equation (2) as follows:

$$(Bx + E)y + (Ax^2 + Dx + F) = 0$$
$$(Bx + E)y + (Bx + E)(Gx + H) = 0$$
$$(Bx + E)(y + (Gx + H)) = 0.$$

Thus, in this case the graph of equation (2) is a pair of intersecting lines given by $Bx + E = 0$ and $y + Gx + H = 0$. Only one of these graph can be obtained with a function graphing utility. Therefore, except for the case $Bx + E = 0$ and the case $C = B = E = 0$, we can obtain the graph of equation (1) with a function graphing utility.

- **EXAMPLE 3:** Identify and draw a complete graph of $3x^2 + 4xy + 3y^2 - 12x + 2y + 7 = 0$.

  **SOLUTION:** We use the quadratic formula to solve for $y$ in terms of $x$ and determine the following pair of functions:

  $$f(x) = \frac{-(4x+2) + \sqrt{(4x+2)^2 - 12(3x^2 - 12x + 7)}}{6},$$

  $$g(x) = \frac{-(4x+2) - \sqrt{(4x+2)^2 - 12(3x^2 - 12x + 7)}}{6}.$$

  The graphs of $f$ and $g$ in Figure 6.4.3 give a complete graph of the conic. From this figure we can identify this conic as an ellipse. •

- **EXAMPLE 4:** Estimate the coordinates of the center, the endpoints of the major and minor axes, and the lines of symmetry of the ellipse of Example 3.

  **SOLUTION:** We magnify the view in Figure 6.4.3 to obtain the graph in Figure 6.4.4, from which we can *estimate* that the endpoints of the major axis have coordinates $(0.7, 0)$ and $(7.3, -6)$. The center is the midpoint of the major axis, and its coordinates are $(\frac{0.7+7.3}{2}, \frac{0+(-6)}{2}) = (4, -3)$. It turns out that the coordinates of the center are exactly $(4, -3)$. The slope of the major axis is $\frac{-6-0}{7.3-0.7} = -0.9$. Thus, the line of symmetry containing the major axis has equation $y = -0.9(x - 0.7)$. The line through the center $(4, -3)$ perpendicu-

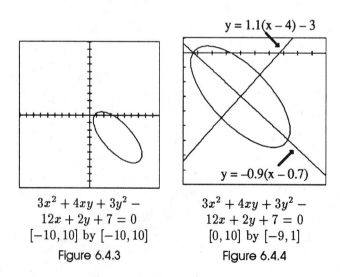

$3x^2 + 4xy + 3y^2 -$
$12x + 2y + 7 = 0$
$[-10, 10]$ by $[-10, 10]$

Figure 6.4.3

$3x^2 + 4xy + 3y^2 -$
$12x + 2y + 7 = 0$
$[0, 10]$ by $[-9, 1]$

Figure 6.4.4

lar to the major axis is the other line of symmetry. This line has equation $y+3 = 1.1(x-4)$. (Why?) We can estimate the endpoints of the minor axis to be $(2.7, -4.5)$ and $(5.3, -1.5)$. •

We can use the distance formula and the endpoints found in Example 3 to show that the lengths of the major and minor axes are 8.9 and 4.0, respectively. Thus, if the ellipse of Example 3 were in standard form with center at the origin and major axis on the $x$-axis, we would have $a = 4.5$ and $b = 2.0$. We can use $c^2 = a^2 - b^2$ to determine that $c = 4.0$. Therefore, the two foci are 4 units on either side of the center $(4, -3)$ on the line $y = -0.9(x - 0.7)$. The line $y = -0.9(x - 0.7)$ makes an angle of $-42.0°$ with the positive $x$-axis because $\tan^{-1}(-0.9) = -42.0°$. Thus, it appears that the ellipse in standard form with center at $(4, -3)$, $a = 4.5$, and $b = 2.0$ has been rotated $42.0°$ clockwise to produce the ellipse in Figure 6.4.4. It turns out that the exact value of the angle is $45°$.

We can use the above information to obtain the graph of the ellipse $3x^2 + 4xy + 3y^2 - 12x + 2y + 7 = 0$ in the following way from the graph of

$$\frac{x^2}{(4.5)^2} + \frac{y^2}{(2.0)^2} = 1.$$

We apply, in order, the following transformations:

1. A horizontal shift 4 units right to obtain $\frac{(x-4)^2}{(4.5)^2} + \frac{y^2}{(2.0)^2} = 1$.
2. A vertical shift 3 units down to obtain $\frac{(x-4)^2}{(4.5)^2} + \frac{(y+3)^2}{(2.0)^2} = 1$.
3. A $42.0°$ rotation clockwise of the conic of 2 about the point $(4, -3)$ to obtain $3x^2 + 4xy + 3y^2 - 12x + 2y + 7 = 0$.

The **translation** (horizontal and vertical shift) factors and the angle of rotation can be computed exactly, but the computation is tedious. It is much faster to approximate these values graphically. It is easy to check these approximations with some graphing utilities, such as *Master Grapher*, which include a special conic grapher.

It turns out that every conic in the plane can be obtained by applying geometric transformations similar to the ones above to an appropriate conic in standard form with center at the origin. The angle of rotation needed can be assumed to be of measure between $-45°$ and $45°$. If the conic is rotated through such an angle then its new equation contains an $xy$ term with nonzero coefficient.

Alternatively, if we start with a given conic in general form $Ax^2 + Bxy + Cy^2 + Dx + Ey + F = 0$, then it is possible to position the coordinate axes $x'$ and $y'$ so that an equation of the conic is one of the following (Figure 6.4.5):

**Parabola**

$(y')^2 = 4ax'$

$(x')^2 = 4ay'$

**Ellipse**

$\dfrac{(x')^2}{a^2} + \dfrac{(y')^2}{b^2} = 1$

$\dfrac{(x')^2}{b^2} + \dfrac{(y')^2}{a^2} = 1$

**Hyperbola**

$\dfrac{(x')^2}{a^2} - \dfrac{(y')^2}{b^2} = 1$

$\dfrac{(y')^2}{b^2} - \dfrac{(x')^2}{a^2} = 1$

The availability of graphing utilities reduces the need for determining the geometric transformations needed to obtain the graph of a given general conic from the graph of a conic in standard form centered at the origin. Now that we can quickly obtain a complete

456 • Chapter 6                                                                                                 Conics

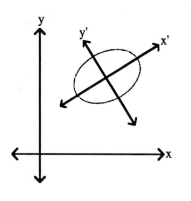

Figure 6.4.5

$x^2 + 2xy + y^2 - 4x - 16y + 19 = 0$
$[-10, 10]$ by $[-10, 10]$
Figure 6.4.6

graph of a conic we rarely need to know the precise rotation or shifting factors. Historically, one major use of these factors was to obtain a quick sketch of a graph.

- **Understanding** From now on in this textbook we will assume that either a special conic graphing utility or the quadratic formula and a function graphing utility are used to draw the graphs of conics.

- **EXAMPLE 5:** Identify and draw a complete graph of each conic.
  (a) $x^2 + 2xy + y^2 - 4x - 16y + 19 = 0$
  (b) $4x^2 - 6xy + 2y^2 - 3x + 10y - 6 = 0$

SOLUTION:

(a) The graph in Figure 6.4.6 is a complete graph. Thus, the conic in (a) is a parabola. However, we must be careful to zoom out in order to be sure that the conic is a parabola. A given viewing rectangle may only display one branch of a hyperbola or one end of an ellipse. Such a figure may suggest that the graph is a parabola when it is not. This point is illustrated in (b).

(b) The graph of this equation in the standard viewing rectangle suggests this conic is a parabola (Figure 6.4.7). Notice that the graph in Figure 6.4.8 clearly shows that this conic is a hyperbola. •

We give the following theorem without proof to be used as you work the exercises.

- **THEOREM 1**  Consider the equation $Ax^2 + Bxy + Cy^2 + Dx + Ey + F = 0$.

  (a) The graph of this equation is (possibly degenerate)
    1. a parabola if $B^2 - 4AC = 0$,
    2. an ellipse if $B^2 - 4AC < 0$, or
    3. a hyperbola if $B^2 - 4AC > 0$.

## 6.4 Quadratic Forms and Conics

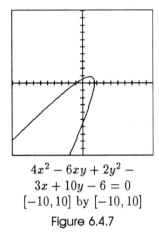

$4x^2 - 6xy + 2y^2 - 3x + 10y - 6 = 0$
$[-10, 10]$ by $[-10, 10]$
Figure 6.4.7

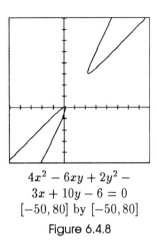

$4x^2 - 6xy + 2y^2 - 3x + 10y - 6 = 0$
$[-50, 80]$ by $[-50, 80]$
Figure 6.4.8

(b) The angle $\theta$ at which a conic in standard form with lines of symmetry parallel to the coordinate axes needs to be rotated to coincide (after appropriate horizontal and vertical shifts) with the given conic is $\frac{1}{2}\tan^{-1}(\frac{B}{A-C})$ if $A \neq C$, and $45°$ if $A = C$.

The angle of rotation in Theorem 1 could be clockwise (negative) or counterclockwise (positive) depending on the particular standard form we start with. Notice that $A = C = 3$ for the conic of Example 4. Thus, a $45°$ rotation clockwise of $\frac{(x-4)^2}{(4.5)^2} + \frac{(y+3)^2}{(2.0)^2} = 1$ produces $3x^2 + 4xy + 3y^2 - 12x + 2y + 7 = 0$. It can be shown that a $45°$ rotation counterclockwise of $\frac{(x-4)^2}{(2.0)^2} + \frac{(y+3)^2}{(4.5)^2} = 1$ produces the same ellipse.

- **EXAMPLE 6:** Identify the conic $2x^2 - 2xy + y^2 - 4x - 21 = 0$ and determine an angle $\theta$ that a conic in standard form needs to be rotated so that its lines of symmetry coincide with the lines of symmetry of the given conic.

**SOLUTION:** Now $A = 2$, $B = -2$, and $C = 1$, so $B^2 - 4AC = -4 < 0$ and the conic is an ellipse. Next $\theta = \frac{1}{2}\tan^{-1}\left(\frac{B}{A-C}\right) = \frac{1}{2}\tan^{-1}(-2) = -31.7°$. Check that a rotation of the given ellipse clockwise $31.7°$ about its center makes the lines of symmetry of the ellipse parallel to the coordinate axis, that is, makes the lines of symmetry horizontal and vertical. •

Some conic graphing utilities have features that allow a given conic to be rotated, translated (shifted), or graphs of functions to be overlaid with graphs of conics. *Master Grapher* is such a utility. The type of conic graphing utility you are using will influence how you do the exercises in this section.

## EXERCISES 6-4

Draw a complete graph and identify the conic.

1. $2x^2 - xy + 3y^2 - 3x + 4y - 6 = 0$
2. $-x^2 + 3xy + 4y^2 - 5x - 10y - 20 = 0$
3. $2x^2 - 4xy + 8y^2 - 10x + 4y - 13 = 0$
4. $2x^2 - 4xy + 2y^2 - 5x + 6y - 15 = 0$
5. $10x^2 - 8xy + 6y^2 + x - 5y + 20 = 0$
6. $10x^2 - 8xy + 6y^2 + x - 5y - 30 = 0$
7. $3x^2 - 2xy - 5x + 6y - 10 = 0$
8. $5xy - 6y^2 + 10x - 17y + 20 = 0$
9. $-3x^2 + 7xy - 2y^2 - x + 20y - 15 = 0$
10. $-3x^2 + 7xy - 2y^2 - 2x + 3y - 10 = 0$

Estimate the coordinates of the center, the endpoints of the major and minor axes, and the equations of the lines of symmetry of the ellipse.

11. $2x^2 - xy + 3y^2 + 2x + 6y - 10 = 0$
12. $2x^2 - 2xy + y^2 + 2x - 3y - 2 = 0$

Estimate the coordinates of the center and the equations of the asymptotes of the hyperbola.

13. $2x^2 - 4xy + y^2 + x + y - 2 = 0$
14. $2x^2 - 3xy + y^2 + 2x - 3y - 2 = 0$

Determine a pair of equations that model the end behavior of the hyperbola. Draw the two equations and the hyperbola in the same large viewing rectangle to check your answer.

15. $2x^2 + 5xy - y^2 + 2x - 8y + 10 = 0$
16. $-3x^2 + 8xy + 4y^2 - 10x + 5y - 30 = 0$
17. $20x^2 - 120xy - 15y^2 + 10x - 80y + 30 = 0$
18. $-5x^2 + 25xy + 6y^2 - 100x + 200y + 36 = 0$

Notice that $4x^2 + 9y^2 = 36$ can be expressed as $\left(\frac{x}{3}\right)^2 + \left(\frac{y}{2}\right)^2 = 1$.

19. Graph $4x^2 + 9y^2 = 36$ and $x^2 + y^2 = 1$ in the same viewing rectangle.

Give a sequence of transformations with order specified that when applied to a conic in standard form centered at the origin results in the given conic.

20. $4x^2 + 9y^2 = 36$
21. $9x^2 + 4y^2 - 18x + 16y - 11 = 0$
22. $9x^2 + 4y^2 + 36x - 18y + 4 = 0$
23. $xy - 2 = 0$
24. $xy + 2y - 1 = 0$

25. Determine an equation for the ellipse with foci $(-1, 1)$ and $(1, 2)$ such that the sum of the distances of its points from the foci is the constant 3.

26. Determine an equation for the hyperbola with foci $(-1, 1)$ and $(1, 2)$ such that the absolute value of the difference of the distances of its points from the foci is the constant 1.

27. Construct a new coordinate system for the conic in Exercise 25 such that the origin is at its center and the $x$-axis is along the major axis. Determine an equation in standard form of the conic relative to this new coordinate system.

28. Construct a new coordinate system for the conic in Exercise 26 such that the origin is at its center and the $x$-axis is along the major axis. Determine an equation in standard form of the conic relative to this new coordinate system.

Use the formula for the rotation angle $\theta$ given in Theorem 1 to determine the angle $\theta$ at which a conic in standard form needs to be rotated so the lines of symmetry coincide with the lines of symmetry of the given conic.

29. $2x^2 - 3xy + 2y^2 - 6x + 4y - 10 = 0$
30. the conic of Exercise 1

Carefully approximate the center and the lengths of the major and minor axes of the ellipse. Then list the transformations with order specified that when applied to an ellipse in standard form centered at the origin produce the given ellipse.

31. the ellipse of Exercise 3
32. the ellipse of Exercise 6

Consider 74 oz of a 56% acid solution. Let $x$ oz of pure acid be added to the solution. How much pure acid must be added to obtain a mixture that is 81% acid?

33. Determine an algebraic representation of the final acidity percentage as a function of the amount of pure acid added.
34. Draw a complete graph and identify the type of graph.
35. What are the domain and range of the algebraic representation?
36. What values of $x$ make sense in this problem situation?
37. Show how the intersection of two graphs can be used to solve the problem. Find the solution.

Consider 74 oz of a 56% acid solution. Let $x$ oz of pure water be added to the solution. How much pure water must be added to obtain a mixture that is 35% acid?

38. Determine an algebraic representation of the final acidity percentage as a function of the amount of pure water added.
39. Draw a complete graph and identify the type of graph.
40. What are the domain and range of the algebraic representation?
41. What values of $x$ make sense in this problem situation?
42. Show how the intersection of two graphs can be used to solve the problem. Find the solution.
43. **A rectangular hyperbola** has asymptotes coincident with or parallel to the coordinate axes. What conditions on $A$, $B$, $C$, $D$, $E$, and $F$ in the general form of a conic will result in a rectangular hyperbola? What angle of rotation applied to a rectangular hyperbola will result in a conic of the form $Ax^2 + Cy^2 + Dx + Ey + F = 0$?
44. Graph the following degenerate conics.
    (a) $x^2 = -1$
    (b) $x^2 = 0$
    (c) $x^2 = 1$
    (d) $x^2 - xy = 0$
    (e) $x^2 + y^2 = 0$
    (f) $x^2 - y^2 = 0$
    (g) $x^2 + y^2 + 1 = 0$

## 6.5 Nonlinear Systems of Equations and Inequalities

Some of the systems of equations solved in Section 1.5 were nonlinear. However, except for a few special exercises, each equation of the systems represented a function. In this section we solve systems of equations and inequalities where an equation or inequality need not represent a function. For example, equations of conics are involved in some of the systems. Two applications, one with a nonlinear system of equations for its model, and the other with a nonlinear system of inequalities, are investigated.

Three observers are strategically located with synchronized watches. They record the precise time that they hear sound being emitted from a certain source. For example, the source of sound may be a gun being fired. Let $d_1$, $d_2$, and $d_3$ be the distances of the observers $A, B,$ and $C$, respectively, from the source of sound. We assume sound travels at the rate of 1100 feet per second. Suppose observer $B$ hears the sound two seconds before observer $A$ does. Then the sound source must be 2200 feet further from observer $A$ than from observer $B$, that is, $d_1 - d_2 = 2200$. We can consider the positions of observers $A$ and $B$ to be foci of a hyperbola with $2a = 2200$. It follows that the sound source is a point of the hyperbola. Recall that a hyperbola is defined as the set of all points such that the absolute value of the *difference* of the distances from two fixed points (foci) is a constant denoted by $2a$. We must also have $a < c$, where $c$ is half the distance between the foci. In this case we must be sure that the distance between observers $A$ and $B$ is greater than 2200 feet. Suppose observer $C$ hears the sound four seconds before observer $A$ does. Then the sound source must be 4400 feet further from observer $A$ than from observer $C$, that is, $d_1 - d_3 = 4400$. We can consider the positions of observers $A$ and $C$ to be the foci of a hyperbola with $2a = 4400$. The location of the gun is at one point of intersection of the two hyperbolas. Thus, the location of the gun is a simultaneous solution to a system of equations consisting of the equations of the two hyperbolas.

- **EXAMPLE 1:** Observer $A$ is located at the origin, observer $B$ is located at $(0, 4000)$, and observer $C$ is located at $(7000, 0)$. Observer $B$ hears the sound of a gun 2 seconds before observer $A$, and observer $C$ hears the sound 4 seconds before observer $A$. Determine a model whose solution locates the position of the gun when it was fired.

SOLUTION: Let $P$ be the position of the gun (Figure 6.5.1). The hyperbola determined by $A$ and $B$ is defined by $|d(P, A) - d(P, B)| = 2200$, and the hyperbola determined by $A$ and $C$ is defined by $|d(P, A) - d(P, C)| = 4400$. Let $P$ have coordinates $(x, y)$. Next we determine an equation for each hyperbola. The center of the hyperbola determined by $A, C$, and the gun is $(3500, 0)$. Also $2a = 4400$, $c = 3500$, so that $b^2 = c^2 - a^2 = 7,410,000$. Thus, an equation is

$$\frac{(x-3500)^2}{(2200)^2} - \frac{y^2}{7,410,000} = 1.$$

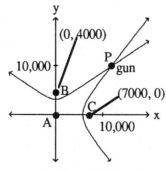

Figure 6.5.1

## 6.5 Nonlinear Systems of Equations and Inequalities

Similarly, an equation determined by $A, B$, and the gun is

$$\frac{(y-2000)^2}{(1100)^2} - \frac{x^2}{2{,}790{,}000} = 1.$$

The location of the gun is at one point of intersection of the two hyperbolas above. In other words, the location of the gun is a simultaneous solution to the following system of equations:

$$\begin{cases} \dfrac{(x-3500)^2}{(2200)^2} - \dfrac{y^2}{7{,}410{,}000} = 1 \\ \dfrac{(y-2000)^2}{(1100)^2} - \dfrac{x^2}{2{,}790{,}000} = 1. \end{cases}$$

- **EXAMPLE 2:** Find the simultaneous solutions to

$$\begin{cases} \dfrac{(x-3500)^2}{(2200)^2} - \dfrac{y^2}{7{,}410{,}000} = 1 \\ \dfrac{(y-2000)^2}{(1100)^2} - \dfrac{x^2}{2{,}790{,}000} = 1. \end{cases}$$

SOLUTION: Solving the first equation for $y$ we obtain

$$y = \pm\sqrt{7{,}410{,}000\left[\frac{(x-3500)^2}{(2200)^2} - 1\right]},$$

and solving the second equation for $y$ we obtain

$$y = 2000 \pm 1100\sqrt{\frac{x^2}{2{,}790{,}000} + 1}.$$

The graph of the resulting four functions in the $[-10{,}000, 20{,}000]$ by $[-10{,}000, 20{,}000]$ viewing rectangle given in Figure 6.5.2 gives a complete graph of the system of equations and shows that there are four simultaneous solutions. We can use zoom-in to determine that the four solutions are $(162.63, 3105.20)$, $(1241.99, 629.24)$, $(6414.37, -2365.08)$, and $(11{,}714.31, 9792.52)$. ●

Now we are in a position to locate the position of the gun in Example 1. We see in the next example that only one of the four possible positions found in Example 2 is the actual location of the gun.

- **EXAMPLE 3:** Determine the position of the gun of Example 1.

SOLUTION: Let $P$ be the position of the gun. From Example 2 we know there are four possibilities for $P$. From the statement of the problem we know that $d(P, A) - d(P, B) = 2200$ because $P$ must be further from $A$ than $B$. This means $P$ is on the upper half of the hyperbola determined by $A$ and $B$. (Why?) Similarly, $d(P, A) - d(P, C) = 4400$ because $P$ must be further from $A$ than $C$. This means that $P$ is on the right branch of the hyperbola

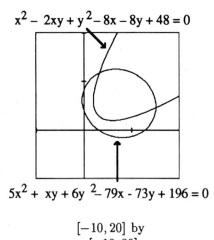

$[-10{,}000, 20{,}000]$ by $[-10{,}000, 20{,}000]$
Figure 6.5.2

$[-10, 20]$ by $[-10, 20]$
Figure 6.5.3

determined by $A$ and $C$. (Why?) Therefore, the gun must be located at $(11{,}714.31, 9792.52)$. Alternatively, $P$ cannot be $(162.63, 3105.20)$ or $(1241.99, 629.24)$ because $d(P,A)-d(P,C) = -4400$ in both of these cases. Similarly, $P$ cannot be $(6414.37, -2365.08)$ because $d(P,A) - d(P,B) = -2200$ in this case. It can be shown directly that $d(P, A) - d(P, C) = 4400$ and $d(P, A) - d(P, B) = 2200$ if $P = (11{,}714.31, 9792.52)$. •

The same principle used to locate the gun is used in navigation; radio waves are used in this case instead of sound.

• **EXAMPLE 4:** Find the simultaneous solutions to

$$\begin{cases} x^2 - 2xy + y^2 - 8x - 8y + 48 = 0 \\ 5x^2 + xy + 6y^2 - 79x - 73y + 196 = 0 \end{cases}$$

**SOLUTION:** We can use Theorem 1 of the previous section to see that the first conic is a parabola and the second an ellipse. Solving the two equations for $y$ gives the following:

$$y = \frac{2x + 8 \pm \sqrt{(2x+8)^2 - 4(x^2 - 8x + 48)}}{2}$$

$$y = \frac{73 - x \pm \sqrt{(73-x)^2 - 24(5x^2 - 79x + 196)}}{12}.$$

Now we can use a function grapher to obtain a complete graph of each equation in the system. Figure 6.5.3 gives a complete graph of each equation and shows that there are two points of intersection. We can use zoom-in to determine that the solutions are $(3.26, 11.74)$ and $(15.05, 4.60)$.  •

It is interesting to try to solve the system of equations in Example 4 algebraically. For example, the two expressions for $y$ can be equated and an algebraic process used to eliminate the radicals. This will lead to a polynomial equation of degree 4 and is an extremely tedious procedure. A graphing approach is much easier and faster.

Suppose that we replace the equal sign in $Ax^2 + Bxy + Cy^2 + Dx + Ey + F = 0$ by an inequality sign. If the equation represents a parabola or an ellipse, then the graph of a corresponding inequality will consist of the points on one side or the other of the conic. The conic itself will be included in the graph if $\geq$ or $\leq$ are involved. If the conic is a hyperbola, then the graph of a corresponding inequality consists of the points between the two branches or the points outside of the two branches. Again, the hyperbola may be part of the graph of an inequality. The definitions of the conics can be used to help establish these facts.

- **EXAMPLE 5:** Solve
$$\begin{cases} y > x^2, \\ 2x + 3y < 4. \end{cases}$$

SOLUTION: We know that the graph of $y > x^2$ consists of the portion on one side or the other of $y = x^2$. The parabola $y = x^2$ is the boundary of the region. A point can be checked to determine which portion is the solution of the inequality. The point $(0, 2)$ is *inside* of the parabola and is a solution to the inequality because $2 > 0$. Thus, the graph of $y > x^2$ consists of the portion *inside* of the parabola $y = x^2$. This is the shaded portion in Figure 6.5.4. The graph of the parabola $y = x^2$ is called the **boundary** of the shaded region. Notice that the parabola is dashed to indicate it is not part of the solution. The solution to $2x + 3y < 4$ consists of the points on one side or the other of the line $2x + 3y = 4$. Again, checking a single point will determine which side of the line is the solution to the inequality. You can verify that the solution is the portion below the line (Figure 6.5.5). Therefore, the solution to the system of inequalities consists of the intersection of the shaded portions of Figures 6.5.4 and 6.5.5. The solution to the system is the shaded portion of Figure 6.5.6. The parabola and line are dashed to indicate they are *not* part of the solution.  •

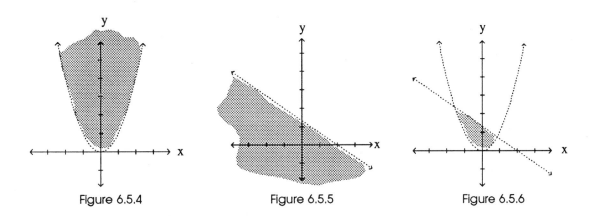

Figure 6.5.4     Figure 6.5.5     Figure 6.5.6

- **EXAMPLE 6:** Solve

$$\begin{cases} x^2 - 2xy + y^2 - 8x - 8y + 48 \leq 0, \\ 5x^2 + xy + 6y^2 - 79x - 73y + 196 > 0. \end{cases}$$

SOLUTION: Complete graphs of $x^2 - 2xy + y^2 - 8x - 8y + 48 = 0$ and $5x^2 + xy + 6y^2 - 79x - 73y + 196 = 0$ are given in Figure 6.5.3 of Example 4. We can use Figure 6.5.3 to convince ourselves that $(4, 4)$ is a point *interior* to both the parabola and the ellipse. Notice that $4^2 - 2(4)(4) + 4^2 - 8(4) - 8(4) + 48 = -16$ *is* less than 0 and $5(4^2) + (4)(4) + 6(4^2) - 79(4) - 73(4) + 196 = -220$ *is not* greater than 0. Thus, $(4, 4)$ is a solution to the first inequality but *not* to the second inequality. Therefore, the solution to this system of inequalities consists of the points *inside* the parabola and *outside* of the ellipse (Figure 6.5.7). The portion of the ellipse inside the parabola is not a solution to the second inequality. Thus, this portion of the boundary of the solution region is dashed to indicate it is not part of the solution to the system. Similarly, the portion of the parabola outside of the ellipse is a solution to both inequalities. Therefore, this portion of the boundary of the solution region is drawn with a solid curve to indicate it is part of the solution to the system. •

Let $x$ and $y$ represent the number of cc of red and white paint, respectively, that are mixed together to produce a new color. The new color depends on the amounts of each paint used.

- **EXAMPLE 7:** Red paint at $1.00 per cc is mixed with white paint at $0.20 per cc in a container with capacity 1050 cc. Color mixologists have determined that the product of $x$ and $y$ must be greater than 100,000 to produce *eye pleasing* colors. What are the possible mixtures that produce eye pleasing colors with total cost at most $500?

   (a) Determine an algebraic representation for the problem situation.
   (b) Draw a complete graph of the algebraic representation in (a).
   (c) Which portions of the graph in (b) represent the solutions to the problem situation?

Figure 6.5.7

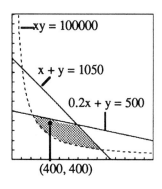

[0, 1500] by [0, 1500]
Figure 6.5.8

SOLUTION: Let $x$ be the amount (in cc) of red paint used, and $y$ the amount (in cc) of white paint used. There are many possible solutions to this problem. For example, we can show that $x = 400$ and $y = 400$ is a solution. First, the mixing container will hold 800 cc of paint. Second, the cost of this mixture is $400(0.2) + 400(1) = \$480$, and \$480 is less than \$500. Finally, $(400)(400) = 160{,}000$ is greater than 100,000 so the resulting color will be *eye pleasing*.

(a) We must have $x + y \leq 1050$ because the container holds at most 1050 cc. Next, $0.2x + y \leq 500$ because the total cost of the mixture is to be at most \$500. Finally, $xy > 100{,}000$ because the color of the mixture is to be eye pleasing. An algebraic representation of the problem situation is given by the following system of inequalities:

$$x + y \leq 1050,$$
$$0.2x + y \leq 500,$$
$$xy > 100{,}000.$$

Of course, we must also have $x \geq 0$ and $y \geq 0$. (Why?)

(b) Figure 6.5.8 shows the graphs of $x + y = 1050$, $0.2x + y = 500$, and $xy = 100{,}000$ in [0, 1500] by [0, 1500]. The complete graph of the algebraic representation in (a) consists of the region in the first quadrant that is below each of the lines $x + y = 1050$ and $0.2x + y = 500$, and above the branch of the hyperbola $xy = 100{,}000$. This is the shaded region in Figure 6.5.8. The straight line portions of the boundary are included, but the curved portion of the boundary given by the hyperbola is *not*.

(c) All pairs $(x, y)$ given by the shaded portion of Figure 6.5.8, including the straight line boundary segments and excluding the curved boundary segment, represent solutions to the problem situation. Notice that $(400, 400)$ is a point in the shaded region that corresponds to the solution found earlier. ●

## EXERCISES 6-5

Find the simultaneous solutions to each system.
1. $4x^2 + 9y^2 = 36$, $x + 2y = 2$
2. $4x^2 + 9y^2 = 36$, $x - 2y = 2$
3. $9x^2 - 4y^2 = 36$, $x + 2y = 4$
4. $9x^2 - 4y^2 = 36$, $x - 2y = 4$
5. $9x^2 - 4y^2 = 36$, $4x^2 + 9y^2 = 36$
6. $9x^2 - 4y^2 = 36$, $9x^2 + 4y^2 = 36$
7. $x^2 + 4y^2 = 4$, $y = 2x^2 - 3$
8. $x^2 - 4y^2 = 4$, $y = 2x^2 - 3$
9. $x^2 - 4y^2 = 4$, $x = 2y^2 - 3$
10. $x^2 + 4y^2 = 4$, $x = 2y^2 - 3$

Draw a complete graph of each equation and find the simultaneous solutions to each system.

11. $x^2 + xy + y^2 + x + y - 6 = 0$, $2x^2 - 3xy + 2y^2 + x + y - 8 = 0$.
12. $5x^2 - 40xy + 20y^2 - 17x + 25y + 50 = 0$, $xy - 3x = 0$.
13. $2x^2 - 3xy + 2y^2 + x + y - 8 = 0$, $2x^2 - 8xy + 3y^2 + x + y - 10 = 0$.
14. $3x^2 - 5xy + 6y^2 - 7x + 5y - 9 = 0$, $x^2 + y^2 - 2x - 6 = 0$.

Draw a graph representing the solutions to the system of inequalities.

15. $y \geq x^2$, $x^2 + y^2 \leq 4$
16. $y \leq x^2 + 4$, $x^2 + y^2 \geq 4$
17. $x^2 - y^2 < 4$, $x^2 + y^2 < 4$
18. $x^2 + 4y^2 - 2x + 4y - 6 > 0$, $y < \frac{1}{2}x + 1$
19. $x^2 + 2xy - 5y^2 + 2x + 4y - 10 > 0$, $4x^2 + xy + 4y^2 - 5x + 5y - 15 < 0$
20. $x^2 + 2xy + 5y^2 + 2x + 4y - 10 > 0$, $5y^2 + 2x + 3y - 6 < 0$

Consider an acid solution of 84 oz that is 58% pure acid. How many ounces of pure acid ($x$) must be added to obtain a solution that is 70% to 80% acid? Let $y$ be the concentration of acid in the final solution.

21. Determine a model consisting of a pair of inequalities and an equation whose simultaneous solution can be used to determine the solution to the problem situation.
22. Draw complete graphs of each inequality and the equation of the model of Exercise 21 and a graph of the solution to the model.
23. What values of $x$ and $y$ make sense in this problem situation?
24. Solve the problem.

Suppose three observers $A$, $B$, and $C$, listening for illegal dynamite explosions are positioned so the angle of the line between $A$ and $B$ and the line between $B$ and $C$ is 90°. Further suppose $A$ and $B$ are 6000 feet apart and $B$ and $C$ are 2000 feet apart. $A$ hears the explosion 4.06 seconds before $B$, and $C$ hears the explosion 0.63 seconds before $B$.

25. Determine a model whose solution locates the position of the explosion.
26. Solve the simultaneous system of equations in the model of Exercise 25.
27. Determine the position of the explosion.

"Hush Dog" shoes are made from two types of leather, $A$ and $B$. Type $A$ costs $0.25 per square inch, and type $B$ cost $0.65 per square inch. The "Hush Dog" Company must buy at least 3000 square inches of $A$ per day and 8600 square inches of $B$ per day (or the supplier will not do business). However, the total square inches of material provided by the supplier per day must be less than

30,000 square inches. Experience has shown the president of the company that the total cost of the material must remain less than $10,000 per day or no one will buy "Hush Dogs" (because they would need to be priced too high).

28. Determine a system of inequalities whose solution provides a model for all possible combinations of the amounts of leather $A$ and leather $B$ that could be used in this problem situation.

29. Graph the region that represents all possible combinations of the amounts of leather $A$ and leather $B$ that could be used in this problem situation.

## 6.6 Polar Equations of Conics

In Section 6.1 we defined a parabola as the locus of all points $P$ such that the perpendicular distance from $P$ to a fixed line (directrix) equals the distance from $P$ to a fixed point (focus) not on the directrix. In this section we give focus-directrix definitions for nondegenerate hyperbolas and ellipses. A circle is a degenerate ellipse. Then, we obtain polar equations for nondegenerate conics using their focus-directrix definitions.

First, we will extend the notion of eccentricity discussed in earlier exercises.

### Eccentricity

- **DEFINITION**    The **eccentricity $e$ of an ellipse or a hyperbola** is defined by $e = \frac{c}{a}$, where $2c$ is the distance between foci and $2a$ is the length of the major axis of an ellipse or the transverse axis of a hyperbola. The eccentricity of a parabola is defined to be 1. The eccentricity of a circle is defined to be 0. •

For an ellipse we have $0 < c < a$ so $e = \frac{c}{a} < 1$. In the exercises you will see that the closer $e$ is to 0, the more circular the ellipse is. Similarly, for a hyperbola $c > a > 0$ so $e = \frac{c}{a} > 1$. As $e \to \infty$, the graph of the hyperbola becomes more like a pair of parallel lines.

- **EXAMPLE 1:** Determine the eccentricity of each conic.

    (a) $\dfrac{x^2}{16} + \dfrac{y^2}{9} = 1$    (b) $\dfrac{x^2}{4} + \dfrac{y^2}{25} = 1$    (c) $\dfrac{x^2}{9} - \dfrac{y^2}{16} = 1$    (d) $\dfrac{y^2}{16} - \dfrac{x^2}{9} = 1$

    SOLUTION:

    (a) Here $a = 4$ and $b = 3$ because we must have $a > b$. Now, $c^2 = a^2 - b^2 = 16 - 9 = 7$. Thus, $e = \frac{c}{a} = \frac{\sqrt{7}}{4}$. Notice $e < 1$.
    (b) Here $a = 5$ and $b = 2$. Now, $c^2 = a^2 - b^2 = 25 - 4 = 21$. Thus, $e = \frac{c}{a} = \frac{\sqrt{21}}{5}$ and is again less than 1.
    (c) Here $a^2 = 9$, $b^2 = 16$, and $c^2 = a^2 + b^2 = 25$. Thus, $e = \frac{c}{a} = \frac{5}{3}$. Notice $e > 1$.
    (d) Here $a^2 = 16$, $b^2 = 9$, and $c^2 = a^2 + b^2 = 25$. Thus, $e = \frac{c}{a} = \frac{5}{4}$, and again $e > 1$. •

- **THEOREM 1**  Let $\frac{x^2}{a^2}+\frac{y^2}{b^2}=1$ be an ellipse with foci on the $x$-axis and let $l$ be the vertical line $x=\frac{a}{e}$ where $e$ is the eccentricity of the ellipse. (Figure 6.6.1). Let $P$ be any point on the ellipse and $Q$ the point on $l$ such that $PQ$ is perpendicular to $l$. Then $d(P,F)/d(P,Q) = e$ where $F = (c,0)$ and $c > 0$.

**PROOF**  If we solve $\frac{x^2}{a^2}+\frac{y^2}{b^2}=1$ for $y^2$ we obtain $y^2 = b^2 - \frac{b^2 x^2}{a^2}$. Now we compute $d(P,F)$ where $P=(x,y)$ is on the ellipse:

$$d(P,F) = \sqrt{(x-c)^2 + y^2}$$

$$= \sqrt{(x-c)^2 + b^2 - \frac{b^2 x^2}{a^2}}$$

$$= \sqrt{x^2 - 2cx + c^2 + b^2 - \frac{b^2 x^2}{a^2}}$$

$$= \sqrt{x^2 - 2cx + a^2 - b^2 + b^2 - \frac{b^2 x^2}{a^2}}$$

$$= \sqrt{x^2\left(1 - \frac{b^2}{a^2}\right) - 2cx + a^2}$$

$$= \sqrt{x^2\left(\frac{a^2 - b^2}{a^2}\right) - 2cx + a^2}$$

$$= \sqrt{\frac{c^2}{a^2}x^2 - 2cx + a^2}$$

$$= \sqrt{\frac{c^2}{a^2}\left(x^2 - \frac{2a^2}{c}x + \frac{a^4}{c^2}\right)}$$

$$= \sqrt{\frac{c^2}{a^2}\left(x - \frac{a^2}{c}\right)^2}$$

$$= \sqrt{\frac{c^2}{a^2}\left(x - \frac{a}{e}\right)^2}$$

$$= \frac{c}{a}\left|x - \frac{a}{e}\right|$$

$$= e\left|x - \frac{a}{e}\right|$$

$$= e \cdot d(P,Q).$$

Theorem 1 means that the vertical line $x = \frac{a}{e}$ can be taken as the directrix and the point $(c,0)$ as the focus for an ellipse in a focus-directrix definition. Later we will use the fact that the distance between the directrix and focus is $\frac{a}{e} - c$. It can be shown that the ellipse $\frac{x^2}{a^2}+\frac{y^2}{b^2}=1$ consists of precisely those points $P$ for which $d(P,F)/d(P,Q) = e$. The vertical line $x = -\frac{a}{e}$ and the point $(-c,0)$ can also be chosen as a focus and directrix of

## 6.6 Polar Equations of Conics

Figure 6.6.1

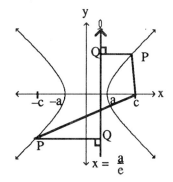

Figure 6.6.2

the ellipse of Theorem 1. In fact, Theorem 1 is also true for this pair. Again, the distance between the directrix and focus is $\frac{a}{e} - c$.

Consider an ellipse $\frac{x^2}{b^2} + \frac{y^2}{a^2} = 1$ with foci $(0, -c)$ and $(0, c)$ on the $y$-axis. Again, $e = \frac{c}{a}$, and we can choose the horizontal line $y = \frac{a}{e}$ as the directrix to pair with the focus $(0, c)$, or the horizontal line $y = -\frac{a}{e}$ as the directrix to pair with the focus $(0, -c)$, where $c > 0$. Again, the distance between the focus and directrix is $\frac{a}{e} - c$.

We state a theorem without proof for hyperbolas similar to Theorem 1.

- **THEOREM 2**  Let $\frac{x^2}{a^2} - \frac{y^2}{b^2} = 1$ be a hyperbola with foci on the $x$-axis and let $l$ be the vertical line $x = \frac{a}{e}$ where $e$ is the eccentricity of the hyperbola (Figure 6.6.2). Let $P$ be any point on the hyperbola and $Q$ the point on $l$ such that $PQ$ is perpendicular to $l$. Then $d(P,F)/d(P,Q) = e$ where $F = (c, 0)$ and $c > 0$.

Notice that $0 < \frac{a}{e} < a$ because $e > 1$. We can also use the vertical line $x = -\frac{a}{e}$ and the point $(-c, 0)$ as directrix and focus for the hyperbola. For the hyperbola $\frac{y^2}{a^2} - \frac{x^2}{b^2} = 1$ the horizontal line $y = \frac{a}{e}$ and the point $(0, c)$ with $c^2 = a^2 + b^2$ can be chosen as the directrix and focus, or the horizontal line $y = -\frac{a}{e}$ and the point $(0, -c)$ can be chosen as the directrix and focus. In all cases, the distance between the focus and directrix is $c - \frac{a}{e}$, the *opposite* of the distance for an ellipse.

Now we can give the following alternative definition of a conic.

- **DEFINITION**  Let $F$ be a fixed point and $l$ a line *not* containing $F$. A **conic** is the locus of all points $P$ with $d(P,F)/d(P,Q) = e$ where $e$ is a positive constant called the eccentricity and $Q$ is the point on $l$ with $PQ$ perpendicular to $l$. $F$ is called a **focus** and $l$ a **directrix** for the conic. •

It turns out that the conic in the above definition is an ellipse if $e < 1$, a parabola if $e = 1$, and a hyperbola if $e > 1$.

Next we obtain polar equations for nondengenerate conics. We choose our coordinate system so that a focus of the conic is at the *origin* and a directrix is parallel to the $x$- or $y$-axis.

## Polar Equations

We consider a new representation of a nondegenerate conic. Let $l$ be the line $x = -h$ with $h > 0$. Let $F = (0, 0)$ be a focus and $l$ a directrix of the conic. Let $P(r, \theta)$ be any point on the conic in polar form and $Q$ the point on $l$ with $PQ$ perpendicular to $l$ (Figure 6.6.3). First we assume that $P$ is on the same side of $l$ as $F$ and that we have chosen the polar coordinates of $P$ so that $r > 0$. Then, $d(P, F) = r$ and $D(P, Q) = r \cos \theta - (-h) = h + r \cos \theta$. The following equation is true for $P$:

$$d(P, F) = e \cdot d(P, Q)$$
$$r = e(h + r \cos \theta)$$
$$r - er \cos \theta = eh$$
$$r = \frac{eh}{1 - e \cos \theta}.$$

Now, suppose that $P$ is on the opposite side of $l$ (Figure 6.6.4). Notice if there is such a point, then we must have $e > 1$. We show that the alternate coordinates of $P$ given by $(r, \theta)$, where $r < 0$ as suggested by Figure 6.6.4 satisfy the same equation obtained above. Notice that $\cos \theta > 0$ so that $r \cos \theta < 0$:

$$d(P, F) = e \cdot d(P, Q)$$
$$-r = e(-h - r \cos \theta)$$
$$r = eh + er \cos \theta$$
$$r = \frac{eh}{1 - e \cos \theta}.$$

Figure 6.6.3

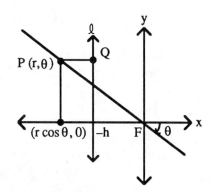

Figure 6.6.4

## 6.6 Polar Equations of Conics

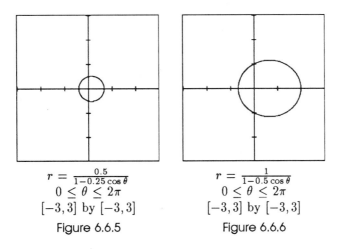

$r = \frac{0.5}{1 - 0.25 \cos \theta}$
$0 \leq \theta \leq 2\pi$
$[-3, 3]$ by $[-3, 3]$
Figure 6.6.5

$r = \frac{1}{1 - 0.5 \cos \theta}$
$0 \leq \theta \leq 2\pi$
$[-3, 3]$ by $[-3, 3]$
Figure 6.6.6

The *Graphing Calculator and Computer Graphing Laboratory Manual* accompanying this textbook contains a simple program that permits graphs of curves defined by polar equations to be obtained with a graphing calculator. The graphing utility *Master Grapher* used to draw graphs in this textbook, also has a polar equation option.

- **EXAMPLE 2:** Draw complete graphs of $r = \frac{2e}{1 - e \cos \theta}$ for $e$ equal to 0.25, 0.5, 1.5, and 4.

**SOLUTION:** Notice that $x = -2$ is a directrix for each of these polar equations, and the origin is a focus for each equation. Complete graphs are given in Figures 6.6.5–6.6.8. Notice that the first two graphs are ellipses with the first one more circular. The last two graphs are hyperbolas with the branches of the last one appearing least parabolic. •

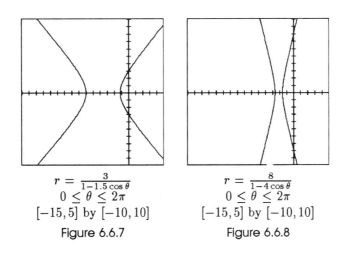

$r = \frac{3}{1 - 1.5 \cos \theta}$
$0 \leq \theta \leq 2\pi$
$[-15, 5]$ by $[-10, 10]$
Figure 6.6.7

$r = \frac{8}{1 - 4 \cos \theta}$
$0 \leq \theta \leq 2\pi$
$[-15, 5]$ by $[-10, 10]$
Figure 6.6.8

The next situation we consider is where the directrix is to the right of the $y$-axis (Figure 6.6.9). Let the directrix be the line $l$ defined by $x = h$, $h > 0$. First, we assume that $P$ is on the same side of $l$ as $F$ and $(r, \theta)$ is chosen so that $r > 0$:

$$d(P, F) = e \cdot d(P, Q)$$
$$r = e(h - r \cos \theta)$$
$$r = \frac{eh}{1 + e \cos \theta}.$$

Now, suppose $P$ is on the other side of $l$ (Figure 6.6.10). We show that the coordinates $(r, \theta)$ where $r < 0$ as suggested by Figure 6.6.10 satisfy the same equation obtained above. Notice that $\cos \theta < 0$ so that $r \cos \theta > 0$:

$$d(P, F) = e \cdot d(P, Q)$$
$$-r = e(r \cos \theta - h)$$
$$r = \frac{eh}{1 + e \cos \theta}.$$

Notice $h$ is the distance between a focus and directrix. For an ellipse, $h = \frac{a}{e} - c$, and for a hyperbola, $h = c - \frac{a}{e}$. Thus, $eh$ is $a - ce$ for an ellipse and $ce - a$ for a hyperbola. It can be shown that $h = 2a$ for a parabola. The numerators of the above equation can be replaced by $a - ce$, $ce - a$, or $2ae$ to give alternate polar forms for conics.

It can be shown that if the focus is $(0, 0)$ then an equation of a conic is also given by

1. $r = \frac{eh}{1 + e \sin \theta}$, if the directrix is $y = h$, with $h > 0$.
2. $r = \frac{eh}{1 - e \sin \theta}$, if the directrix is $y = -h$, with $h > 0$.

Again, $h$ can be replaced by $\frac{a}{e} - c$, $c - \frac{a}{e}$, or $2a$ to give alternative forms.

We summarize the above discussion in Theorem 3. However, we do not need to restrict $h$ to be positive because the theorem is true for $h < 0$.

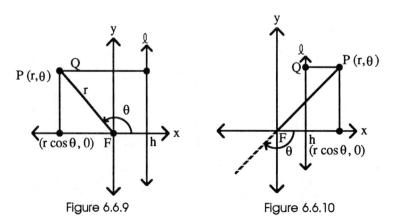

Figure 6.6.9          Figure 6.6.10

## 6.6 Polar Equations of Conics

- **THEOREM 3** A nondegenerate conic has one of the following for a polar equation where $e$ is the eccentricity of the conic:

$$r = \frac{he}{1 \pm e \cos \theta}, \qquad r = \frac{he}{1 \pm e \sin \theta}.$$

For the first equation, $x = \pm h$ are directrices, respectively; for the second equation, $y = \pm h$ are directrices, respectively. Furthermore, $he$ can be replaced by $\pm(ce - a)$ or $\pm 2ae$. Moreover, the graph of any such equation is a conic. The conic is an ellipse if $0 < e < 1$, a parabola if $e = 1$, and a hyperbola if $e > 1$.

- **EXAMPLE 3:** Identify and draw a complete graph of $r = \frac{6}{4 - 3 \cos \theta}$. Specify a directrix and a range for $\theta$ that produces a complete graph. Determine the standard form for the equation of the conic.

SOLUTION: We divide numerator and denominator of this equation by 4 to put it into the form given in Theorem 3. (Why?)

$$r = \frac{6}{4 - 3 \cos \theta},$$
$$r = \frac{1.5}{1 - 0.75 \cos \theta}.$$

Now, we can see that $e = 0.75$ (why?) and that $he = 1.5$, or $h = \frac{1.5}{e} = 2$. Thus, this conic is an ellipse with directrix $x = -2$. (Why?) A complete graph is given in Figure 6.6.11. We obtain a complete graph by letting $\theta$ vary from 0 to $2\pi$. In fact, any interval of length $2\pi$ for $\theta$ gives a complete graph. (Why?)

One focus of the ellipse is the origin. The ends of the major axis occur when $\theta = 0$ and $\theta = \pi$. (Why?) Thus, the endpoints of the major axis are $(-0.86, 0)$ and $(6, 0)$ in rectangular form. The center of the ellipse is halfway between the endpoints of the major axis. Thus, the coordinates of the center are $(2.57, 0)$. This also means $c = 2.57$ because $c$ is the distance

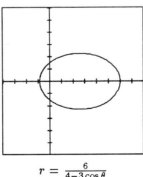

$r = \frac{6}{4 - 3 \cos \theta}$
$0 \leq \theta \leq 2\pi$
$[-4, 8]$ by $[-6, 6]$
Figure 6.6.11

from the center to a focus. Now, the length of the major axis is $2a = 6 - (-0.86) = 6.86$, so $a = 3.43$. We use the equation $b^2 = a^2 - c^2$ to determine that $b = 2.27$. The standard form for the equation of the conic is

$$\frac{(x - 2.57)^2}{(3.43)^2} + \frac{y^2}{(2.27)^2} = 1.$$

We must be careful before concluding that a given graph is an ellipse. We mentioned earlier in the section that the shape of an ellipse is more circular as $e$ gets closer to 0. However, to distinguish visually between circles and ellipses it is necessary to have the physical length representing one unit on each axis be the same. (Why?)

- **EXAMPLE 4:** Identify and draw a complete graph of $r = \frac{3}{1+\cos\theta}$. Specify a directrix and a range for $\theta$ that produces a complete graph.

  **SOLUTION:** We can see that $e = 1$, $h = 3$, and that this conic is a parabola with directrix $x = 3$. A complete graph is given in Figure 6.6.12 and can be obtained by letting $\theta$ vary from 0 to $2\pi$.

- **EXAMPLE 5:** Identify and draw a complete graph of $r = \frac{4}{3-4\sin\theta}$. Specify a directrix and a range for $\theta$ that produces a complete graph.

  **SOLUTION:** One focus of this conic is at the origin. First we divide numerator and denominator by 3.

  $$r = \frac{\frac{4}{3}}{1 - \frac{4}{3}\sin\theta}$$

  Thus, $e = \frac{4}{3}$, and this conic is a hyperbola. Also $h = \frac{4}{3e} = 1$ so that $y = -1$ is a directrix. We obtain a complete graph by allowing $\theta$ to vary from 0 to $2\pi$ (Figure 6.6.13).

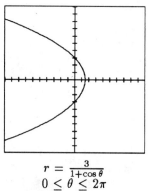

$r = \frac{3}{1+\cos\theta}$
$0 \leq \theta \leq 2\pi$
$[-10, 10]$ by $[-10, 10]$
Figure 6.6.12

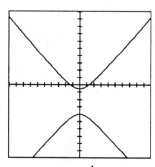

$r = \frac{4}{3-\sin\theta}$
$0 \leq \theta \leq 2\pi$
$[-10, 10]$ by $[-10, 10]$
Figure 6.6.13

## 6.6 Polar Equations of Conics

It can be shown that any interval of length $2\pi$ produces a complete graph of a conic with equation given by Theorem 3. (Why?) In the Exercises you will see that the graphs of the equations given in Theorem 3 are conics even if the numerators are negative. The graphs are also conics if the denominators are replaced by $-1 \pm e\cos\theta$ and $-1 \pm e\sin\theta$.

We assume that the orbit of the Earth around the sun is an ellipse with the sun as focus. The length of the semi-major axis is $9.3 \times 10^7$ miles, and the eccentricity of the ellipse is 0.02. Similarly, the orbit of the planet Mars is an ellipse with the sun as a focus, semi-major axis length of $1.4 \times 10^8$ miles, and eccentricity 0.1. We assume the orbits are in the same plane and have major axes along the $x$-axis.

• **EXAMPLE 6:** Draw complete graphs of the elliptical orbits of the Earth and Mars in the same viewing rectangle.

**SOLUTION:** We use the polar equation $r = \frac{a-ce}{1-e\cos\theta}$ of Theorem 3 where $e$ is the eccentricity, $a$ the length of the semi-major axis, and $c$ half the distance between the foci. We also know that $e = \frac{c}{a}$ so that $c = ea$. Thus, we can write the polar equation in the form $r = \frac{a-e^2 a}{1-e\cos\theta}$.

An equation for the orbit of the Earth is

$$r_1 = \frac{93,000,000 - (0.02)^2(93,000,000)}{1 - 0.02\cos\theta} = \frac{9.3 \times 10^7 - (0.02)^2(9.3 \times 10^7)}{1 - 0.02\cos\theta},$$

and for the orbit of Mars is

$$r_2 = \frac{140,000,000 - (0.1)^2(140,000,000)}{1 - 0.1\cos\theta} = \frac{1.40 \times 10^8 - (0.1)^2(1.40 \times 10^8)}{1 - 0.1\cos\theta}.$$

Figure 6.6.14 gives a complete graph of each orbit.

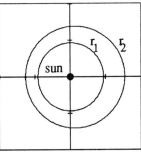

$0 \leq \theta \leq 2\pi$
$[-200,000,000, 200,000,000]$
by
$[-200,000,000, 200,000,000]$

Figure 6.6.14

## EXERCISES 6-6

Determine the eccentricity of the conic and draw a complete graph. Check using a graphing utility.

1. $4x^2 + y^2 = 4$
2. $4x^2 + 25y^2 = 100$
3. $9x^2 - 16y^2 = 144$
4. $4x^2 - y^2 = 4$
5. $2x^2 - 3y^2 - 4x + 6y - 10 = 0$
6. $-6x^2 + 8y^2 - 6x + 9y + 10 = 0$

Determine the center, length of the transverse axis, and the equations of the asymptotes.

7. the conic of Exercise 5
8. the conic of Exercise 6

Estimate the eccentricity of the ellipse. (*Hint*: First estimate $a$ and $b$. Then compute $c$ and $e$.) Draw a complete graph.

9. $2x^2 - 3xy + 6y^2 - 8x + 10y - 5 = 0$
10. $10x^2 + 52xy + 300y^2 - 100x + 250y - 500 = 0$
11. Draw a complete graph of $r = \frac{2e}{1 - e\cos\theta}$ for $e = 0.8, 0.9, 1, 1.1, 1.2, 1.3$ in the same viewing rectangle.
12. Draw a complete graph of $r = \frac{2e}{1 + e\cos\theta}$ for $e = 0.8, 0.9, 1, 1.1, 1.2$ in the same viewing rectangle.
13. Draw a complete graph of $r = \frac{2e}{1 - e\sin\theta}$ for $e = 0.8, 0.9, 1, 1.1, 1.2$ in the same viewing rectangle.
14. Draw a complete graph of $r = \frac{2e}{1 + e\sin\theta}$ for $e = 0.8, 0.9, 1, 1.1, 1.2$ in the same viewing rectangle.

Identify and draw a complete graph of the conic in polar form. Specify a directrix and a range for $\theta$ that produces a complete graph.

15. $r = \dfrac{5}{1 - \cos\theta}$
16. $r = \dfrac{4}{2 - 4\sin\theta}$
17. $r = \dfrac{3}{4 - 2\cos\theta}$
18. $r = \dfrac{4}{3 + 4\sin\theta}$
19. $r = \dfrac{-8}{6 + 3\cos\theta}$
20. $r = \dfrac{-4}{4 - 2\sin\theta}$

Draw a complete graph. Identify the conic. Determine the standard form of the conic.

21. $r = \dfrac{16}{12 + 9\cos\theta}$
22. $r = \dfrac{-4}{2 + 6\sin\theta}$

23. Consider an ellipse with major axis length $2a$ equal to 8 and length $c$ of a focus from the center equal to 2. Compute the eccentricity and draw the graph of $r_1$ and $r_2$ in the same viewing rectangle. What conclusions can you draw?

$$r_1 = \frac{a - ce}{1 + e\cos\theta} \qquad r_2 = \frac{ce - a}{1 - e\cos\theta}$$

24. Consider an ellipse with major axis length $2a$ equal to 12 and length $c$ of a focus from the center equal to 3.5. Compute the eccentricity and draw the graph of $r_1$ and $r_2$ in the same viewing rectangle. What conclusions can you draw?

$$r_1 = \frac{a - ce}{1 + e\cos\theta} \qquad r_2 = \frac{ce - a}{1 - e\cos\theta}$$

25. Consider the graph $r = \dfrac{eh}{-1 - e\cos\theta}$ for $e = 7$ and $h = 2$. Is the graph a conic? What conic? Why?

## 6.6 Polar Equations of Conics

26. Show that if an ellipse has major axis length $2a$ and minor axis length $2b$ then a polar equation of the ellipse is

$$r = \frac{b^2}{a + \sqrt{a^2 - b^2} \cos \theta}$$

where a focus is at the origin and the directrix is $x = h > 0$.

Let $a = 4$ and $b = \sqrt{12}$ in an ellipse. Compute the eccentricity and draw the polar graphs of $r_1$, $r_2$, $r_3$, and $r_4$ in the same viewing rectangle. What conclusions can you draw?

27. $r_1 = \dfrac{b^2}{a + \sqrt{a^2 - b^2} \cos \theta}$

28. $r_2 = \dfrac{-b^2}{a + \sqrt{a^2 - b^2} \cos \theta}$

29. $r_3 = \dfrac{-b^2}{a - \sqrt{a^2 - b^2} \cos \theta}$

30. $r_4 = \dfrac{b^2}{a - \sqrt{a^2 - b^2} \cos \theta}$

Draw the graph of the elliptical orbits of the indicated planets in the same viewing rectangle. Table 1 gives the eccentricity and the length of the semi-major axis. Assume the major axes are along the $x$-axis.

31. Mercury and Earth
32. Earth and Saturn
33. Earth, Saturn, and Pluto
34. All of the planets listed in the table.
35. Sketch the elliptical orbit of the Earth and the orbit of the asteroid Icarus in the same viewing rectangle. The eccentricity of the asteroid Icarus' orbit about the sun is 0.80 and its semi-major axis length is about 100,000,000 miles.
36. The elliptical orbit of Halley's comet has eccentricity about 0.97. Its major axis length is 3,348,000,000. What is its minor axis length? Draw a complete graph of its orbit.
37. How are $r_1 = \frac{h}{2 - \cos t}$, $h > 0$ and $r_2 = \frac{h}{2 - \cos t}$, $h < 0$ related?
38. (a) In the same viewing rectangle, draw complete graphs of

$$r = \frac{\frac{1}{2}h}{1 - \frac{1}{2}\cos t}$$

| Planet | $e$ | $a$ (miles) |
|---|---|---|
| Mercury | 0.2 | $3.6 \times 10^7$ |
| Earth | 0.02 | $9.3 \times 10^7$ |
| Mars | 0.09 | $1.4 \times 10^8$ |
| Saturn | 0.5 | $8.9 \times 10^8$ |
| Neptune | 0.01 | $2.8 \times 10^9$ |
| Pluto | 0.25 | $3.7 \times 10^9$ |

Table 1

for $h = \frac{1}{4}, \frac{1}{2}, 1, 2, 5, 10,$ and 20.

(b) Notice the eccentricity in (a) is constant. Describe the effects of changing the value of $h$. What is the line $y = h$?

(c) Repeat (a) for $h = -\frac{1}{4}, -\frac{1}{2}, -1, -2, -5, -10,$ and 20.

39. What is the graph of $r = \frac{eh}{-1 + e\cos\theta}$? Why?

40. What is the graph of $r = \frac{eh}{-1 + e\sin\theta}$? Why?

# • KEY TERMS

Listed below are the key terms, vocabulary, and concepts in this chapter. You should be familiar with their definitions and meanings as well as be able to illustrate each of them.

| | |
|---|---|
| Asymptotes | Lines of symmetry |
| Center | Major axis |
| Circle | Major diameter |
| Conic sections | Minor axis |
| Degenerate conics | Minor diameter |
| Directrix | Parabola |
| Eccentricity | Paraboloid |
| Ellipse | Polar form of conic equations |
| Ellipsoid | Quadratic form |
| End behavior | Reflective properties of conics |
| Endpoints | Semimajor axis |
| Expanded form of conic equations | Semiminor axis |
| Focus | Standard form of conic equations |
| Hyperbola | Transverse axis |
| Hyperboloid | Vertex |

# • REVIEW EXERCISES

This section contains representative problems from each of the previous six sections. You should use these problems to test your understanding of the material covered in this chapter.

Determine an equation for the parabola. Give the coordinates of the vertex and the line of symmetry, and draw a complete graph.

1. focus $(-2, 1)$ and directrix $y = -3$
2. focus $(-2, 1)$ and directrix $x = -3$
3. Determine an equation, the line of symmetry, and the focus of the parabola with vertex $(1, 4)$ and directrix $x = 7$, and draw a complete graph.

## Review Exercises

4. Determine an equation, the line of symmetry, and the directrix of the parabola with focus $(0, 2)$ and vertex $(0, -4)$, and draw a complete graph.

List a sequence of geometric transformations that when applied to the graph of $y = x^2$ will result in the graph of the given parabola.

5. $6(y - 1) = (x - 3)^2$
6. $x^2 - 2x - 6 = 2y$

7. The parabola $y^2 - 4y + 2 = x$ does not define $y$ as a function of $x$. How can it be graphed using only a function graphing utility?

Find the standard form, vertex, focus, directrix, line of symmetry, and draw a complete graph of each parabola without using a graphing utility. Check your graph with a graphing utility.

8. $x^2 + 4x - y + 6 = 0$
9. $6y^2 - 12y - 12x + 20 = 0$

10. Find the points of intersection of the line and the parabola (if any): $y = x^2 - 4x + 5$, $2x + y - 7 = 0$.

11. A parabola of the form $y = ax^2 + b$ passes through the points $(2, 4)$ and $(5, 8)$. Determine an equation of the parabola.

12. Determine the foci, standard form of the equation, and draw a complete graph of the ellipse without using a graphing utility: Major axis along the $x$-axis of length 10, minor axis along the $y$-axis of length 4.

Determine the endpoints of the major and minor axes, the standard form of the equation, and draw a complete graph of the ellipse. Do not use a graphing utility.

13. The sum of the distances from the foci is 16; the foci are $(-5, 0)$ and $(5, 0)$.
14. foci at $(2, 0)$ and $(-2, 0)$; minor axis of length 10
15. foci at $(2, 3)$ and $(2, 9)$; major axis of length 12

Give the endpoints of the major and minor axes, the coordinates of the center and foci, the lines of symmetry, and draw a complete graph of the ellipse. Do not use a graphing utility.

16. $\dfrac{(x+3)^2}{4} + \dfrac{(y-1)^2}{6} = 1$
17. $(x-1)^2 + 8(y+2)^2 = 16$

Write the equation of the ellipse in standard form, give the endpoints of the major and minor axes, the coordinates of the center and foci, the lines of symmetry, and draw a complete graph. Check your answer using a graphing utility.

18. $4x^2 + 9y^2 + 8x - 36y + 4 = 0$
19. $4x^2 + 5y^2 - 24x + 10y + 21 = 0$

Determine the standard form of the equation, and draw a complete graph of the hyperbola. Do not use a graphing utility.

20. The absolute value of the difference of the distances from any point on the hyperbola to the foci is 2; the foci are $(-2, 0)$ and $(2, 0)$.

21. The absolute value of the difference of the distances from any point on the hyperbola to the foci is 10; the foci are $(0, -2)$ and $(0, 10)$.

Determine the coordinates of the center, foci, vertices, the lines of symmetry, and the asymptotes of the hyperbola, and draw a complete graph. Do not use a graphing utility.

22. $\dfrac{x^2}{9} - \dfrac{(y+2)^2}{6} = 1$
23. $3(y-3)^2 - 5(x+1)^2 = 15$

Identify the conic. Write the equation of the conic in standard form, give the foci, vertices, center, the axes of symmetry, asymptotes (if any) and draw a complete graph. Check your answer using a graphing utility.

24. $25x^2 - 9y^2 - 50x - 36y - 236 = 0$

25. $y^2 - x + 6y + 9 = 0$

26. $5x^2 + 4y^2 + 30x - 8y + 29 = 0$

27. $4y^2 - 9x^2 + 24y + 35 = 0$

28. Explain how a conic of the form $Ax^2 + By^2 + Cx + Dy + E = 0$ can be graphed using a *function* graphing utility, not a special conic graphing utility.

Draw a complete graph and identify the conic.

29. $4x^2 + 3xy - y^2 - 10x - 5y - 20 = 0$

30. $2x^2 - 4xy + 2y^2 + 6x - 5y - 15 = 0$

31. $6x^2 - 8xy + 10y^2 - 5x + y - 30 = 0$

32. $-2x^2 + 7xy - 3y^2 + 3x - 2y - 10 = 0$

33. Estimate the coordinates of the center and the equations of the asymptotes of the hyperbola $2x^2 - 3xy + y^2 + 2x - 3y - 2 = 0$.

34. Estimate the coordinates of the center, the endpoints of the major and minor axes, and the equations of the lines of symmetry of the ellipse $3x^2 - xy + 2y^2 + 6x + 2y - 10 = 0$.

35. Determine a pair of equations that give an end behavior model of the hyperbola $4x^2 + 8xy - 3y^2 + 5x - 10y - 30 = 0$. Draw the two equations and the hyperbola in the same large viewing rectangle to check your answer.

36. Explain what transformations applied first to a circle in standard form will result in the ellipse $9x^2 + 16y^2 = 144$.

37. What transformations applied first to a conic in standard form will result in the conic $4x^2 + 9y^2 + 16x - 18y - 11 = 0$?

Find the simultaneous solutions to each system.

38. $9x^2 + 25y^2 = 225$, $x + 2y = 2$

39. $9x^2 - 25y^2 = 225$, $x + 2y = 4$

40. $9x^2 - 25y^2 = 225$, $9x^2 + 25y^2 = 225$

41. $x^2 + 9y^2 = 9$, $y = 2x^2 + 3$

Find the number of simultaneous solutions to each system.

42. $2x^2 - 3xy + 2y^2 + x + y - 8 = 0$, $3x^2 - 8xy + 2y^2 + x + y - 10 = 0$

43. $3x^2 - 5xy + 6y^2 - 7x + 5y - 9 = 0$, $x^2 + y^2 - 2y - 6 = 0$

44. Find the simultaneous solutions to the system in Exercise 42.

45. Find the simultaneous solutions to the system in Exercise 43.

Draw a graph representing the solution to the system of inequalities.

46. $x^2 + y^2 < 9$, $y > x^2$

47. $x^2 - y^2 > 9$, $x^2 + y^2 < 9$

Determine the eccentricity and draw a complete graph of the conic. Check using a graphing utility.

48. $x^2 - 9y^2 = 9$

49. $4x^2 + 9y^2 + 54y + 45 = 0$

50. Determine the center, length of the transverse axis, and the equations of the asymptotes for the conic in Exercise 48.

51. Draw a complete graph of $r = \frac{2e}{1 - e\cos\theta}$ for $e = 0.75$, 1, and 1.25 in the same viewing rectangle.

52. Draw a complete graph of $r = \frac{2e}{1-e\sin\theta}$ for $e = 0.75$, 1, and 1.25 in the same viewing rectangle. Identify and draw a complete graph of the conic in polar form. Specify a directrix and a range for $\theta$ that produces a complete graph.

53. $r = \dfrac{4}{1 - 2\cos\theta}$

54. $r = \dfrac{2}{4 + \sin\theta}$

55. $r = \dfrac{8}{8 - 4\sin\theta}$

56. A parabolic microphone is formed by revolving the portion of the parabola $12y = x^2$ between $x = -6$ and $x = 6$ about its axis of symmetry. Where should the sound receiver be placed for best reception?

57. A parabolic headlight is formed by revolving the portion of the parabola $y^2 = 10x$ between the lines $y = -3$ and $y = 3$ about its axis of symmetry. Where should the headlight bulb be placed for maximum lumination?

58. There are real elliptical pool tables that have been constructed with a single pocket at one of the foci. Suppose such a table has major diameter 8 feet and minor diameter 6 feet. How should the ball be hit so it bounces off the cushion directly into the pocket? Give specific measurements.

A certain comet's path is known to be hyperbolic with one focus at the sun. Assume a space station is located at the center of the hyperbola. Further assume it has been determined that the distance between the comet and the space station is 6,000,000 miles at the closest point, and that the sun and the space station are 12,000,000 miles apart.

59. Determine the equation in standard form of the hyperbola representing the comet's path.

60. Draw a complete graph of the algebraic representation of the path of the comet. What portion of the graph represents the problem situation?

Consider an acid solution of 80 oz that is 65% acid. Let $x$ oz of pure acid be added to the solution. How much pure acid must be added to obtain a mixture that is 81% acid?

61. Determine an algebraic representation of the final acidity percentage as a function of the amount of pure acid added.

62. Draw a complete graph and identify the type of graph.

63. What is the domain and range?

64. What values of $x$ make sense in this problem situation?

65. Show how the intersection of two graphs can be used to solve the problem. Find the solution.

Suppose three observers $A$, $B$, and $C$, listening for illegal dynamite explosions, are positioned so the angle of the line between $A$ and $B$ and the line between $B$ and $C$ is $90°$. Further, suppose $A$ and $B$ are 4000 feet apart and $B$ and $C$ are 1000 feet apart. $A$ hears the explosion 1 second before $B$, and $C$ hears the explosion 2 seconds before $B$.

66. Determine a model whose solution will locate the position of the explosion.

67. Solve the simultaneous system of equations in the model of Exercise 66.

68. Determine the position of the explosion.

"Hush Dog" shoes are made from two types of leather, $A$ and $B$. Type $A$ costs $0.25 per square inch while type $B$ costs $0.65 per square inch. The "Hush Dog" Company must buy at least 4000 square inches of $A$ per day and 8000 square inches of $B$ per day (or the supplier will not do business).

However, the total square inches of material provided by the supplier per day must be less than 30,000 square inches. Experience has shown the president that the total cost of the material must remain less than $10,000 per day or no one will buy "Hush Dogs" (because they would need to be priced very high).

69. Determine a system of inequalities whose solution provides a model for all possible combinations of the amounts of leather $A$ and leather $B$ that could be used in this problem situation.

70. Graph the region that represents all possible combinations of the amounts of leather $A$ and leather $B$ that could be used in this problem situation.

The ellipse near the Washington Monument in Washington, D.C. has eccentricity of about 0.52. Its major axis length is 458 meters.

71. What is its minor axis length?

72. Determine an equation for the boundary of the ellipse, assuming its center is at the origin.

# Answers

## • CHAPTER 1

### Exercises 1.1

1.  A: $(1, 3)$
    B: $(4, 2)$
    C: $(1, -2)$
    D: $(-4, -3)$
    E: $(-2, -1)$
    F: $(-4, 0)$
    G: $(-3, 1)$

3.

5. (a) Triangle

    (b) Trapezoid

    (c) Parallelogram

    (d) Pentagon

7. $T = -13$

9. Answers will vary

11. (a)

| Height (ft) | Weight (lbs) |
|---|---|
| 2 | 140 |
| 1 | 100 |
| 4 | 160 |
| 6 | 200 |

   (b) Domain: All nonnegative real numbers less than or equal to 6.

   Range: All nonnegative real numbers less than or equal to 200.

13.

• A1

**15.**

**17.**

**19.** (a)

| Input No. | Output No. (square root) |
|---|---|
| -3 | - |
| -2 | - |
| 0 | 0 |
| 1 | 1 |
| 2 | 1.41 |
| 3 | 1.73 |
| 4 | 2 |
| 5 | 2.24 |
| 6 | 2.45 |

(b) and (c)

(d) Domain and Range: All nonnegative real numbers

**21.** (a)

| Input | Output (sin) |
|---|---|
| 0 | 0 |
| 0.52 | 0.50 |
| 1.05 | 0.87 |
| 1.57 | 1.00 |
| 2.09 | 0.87 |
| 2.62 | 0.50 |
| 3.14 | 0 |
| 3.67 | -0.50 |
| 4.19 | -0.87 |
| 4.71 | -1.00 |
| 5.24 | -0.87 |
| 5.76 | -0.50 |
| 6.28 | 0 |
| 6.81 | 0.50 |
| 7.33 | 0.87 |
| 7.85 | 1.00 |
| 8.38 | 0.87 |
| 8.90 | 0.50 |
| 9.42 | 0 |
| 9.95 | -0.50 |

(b) and (c)

(d) Domain: All real numbers.

Range: All real numbers $\geq -1$ and $\leq 1$.

**23.** $ab = 100$

**25.** $w = 0.25z - 35.75$

**\*27.** (a) $y \leq x$ where $x$ represents the first number & $y$ the second.

(b) See graph.

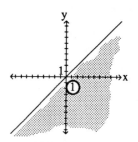

(c) Domain and Range: All Real Numbers

# Exercises 1.2

1. (a) and (c).
3. yes
5. yes
7. no
9. yes
11. $g(0) = 1$
    $g(1) = \frac{1}{2}$
    $g(3) = \frac{1}{4}$
    $g(-5) = -\frac{1}{4}$
    $g(t) = \frac{1}{t+1}$
    $g(-t) = \frac{1}{-t+1}$
    $g(\frac{2}{t}) = \frac{t}{2+t}$
    $g(a+h) = \frac{1}{a+h+1}$
    $\frac{g(a+h)-g(a)}{h} = \frac{-1}{(a+h+1)(a+1)}$
13. (a) $V(0) = 0$;
    $V(-5) = -10,000$;

$V(4) = 300$;
$V(11) = 0$;
(b) $x = 0, 6.5$ or $11$
(c) $x = 0.5, 6$ or $11.5$
(d) $x = 13.5$
(e) $x = -0.75, 7$ or $10.5$

15.

17. b, d, e
19. (d)
21. A function.

23. A function.

25.

27.

29.

31. Dom: All real numbers.
    Range: All real numbers.
33. Dom: All real numbers.
    Range: All real numbers $\geq -27$.
35. (a) $A(x) = 4x^2 + 21x$
    (b)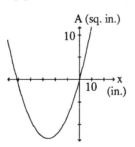
    (c) Dom: All real numbers.
        Range: All real numbers $\geq -27.56$.
    (d) All positive real numbers.
    (e) $x = 0.84$ inches.

37.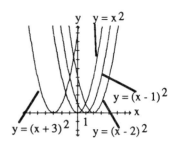

39. $y = x^2$
    $y^2 = x$
41. If $(a, b)$ and $(a, c)$ are ordered pairs of $y = 3x - 1$, then $b = 3a - 1$ and $c = 3a - 1$. Thus, $b = c$ and $y = 3x - 1$ is a function.

# Exercises 1.3

1.

3.

5.

7.

9.

11.

13. Dom: All real numbers $\geq -1$.

Range: All nonnegative real numbers.

15. Dom: All real numbers.
17. Dom: All real numbers.
19. Dom: All real numbers $\geq 5$.
21. The graphs are identical.
23. (e).
25. (c).
27. Dom: All real numbers.
Range: $[0, \infty)$

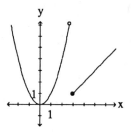

29. Dom: All real numbers.
Range: All real numbers $\leq -4$ as well as 6

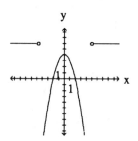

31. Dom: All real numbers
Range: All real numbers $\geq \frac{15}{4}$

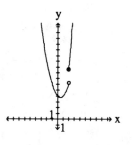

33. (a) $F = 3t + 5$

(b)

(c) Dom: All real numbers.

Range: All real numbers.

(d) All values of $t$ greater than 0 represent the problem situation.

(e) $t = 3.2$ hours

35. (a) $P = 1.50x - [1.25x + 20,000] = .25x - 20,000$

(b)

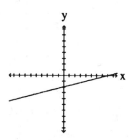

The scale is $10^4$ units per tick mark.

# 1 Answers to Odd-Numbered Exercises

(c) 80,000 lbs.

(d) 120,000 lbs.

(e) 80,120 lbs.

*37. (a) $C = .48 + .28(\text{INT}(t))$ if $t$ is *not* a whole number.

$C = .48 + .28(\text{INT}(t-1))$
or simply
$C = .48 + .28(t-1)$
if $t$ is a whole number.

(b)

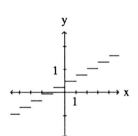

(c) Dom: All real numbers.

Range: $\{\ldots, -0.08, 0.20, 0.48, \ldots\}$

(d) All values of $t$ greater than 0

represent the problem situation.

(e) Bill can talk to his girlfriend up to 6 minutes.

39. (a) $D = \sqrt{(3-x)^2 + (5-(1-x))^2}$

(b) Dom: All real numbers.

(c)

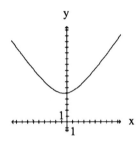

(d) $P : (6, -5)$ or $(-7, 8)$ (these are approximations).

41. a, c, f

43. a, d, e

45.

Dom: All real numbers
Range: All real numbers $\geq 0$

47.

Dom: All real numbers
Range: All real numbers $\geq 0$
Line of sym: $x = 0$

49.

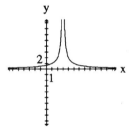

Dom: All real numbers except 3
Range: All real numbers $> 0$
Line of sym: $x = 3$

# Exercises 1.4

1. $x = -2, 5$

3. $x = -12, 4$

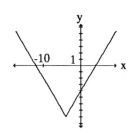

5. $x = \pm\sqrt{14}$
7. no real solutions
9. $x = -4, 8$
11. $x = 8, -3$
13. No; $-1$ and $2$ are solutions. Another solution is $x = -3$. The $x$-intercepts of the graph are the solutions to the equation.
15. 3
17. $x = \frac{2}{3}$
19. Answers will vary. One possible solution is: $[-2, -1]$ by $[-1, 1]$, $[-1.9, -1.8]$ by $[-0.1, 0.1]$, $[-1.9, -1.89]$ by $[-0.01, 0.01]$, $[-1.894, -1.893]$ by $[-0.001, 0.001]$
21. 1.52 with error at most 0.06 for $f(x)$
23. 10.19 with error at most 0.09 for $f(x)$
25. $x = 0.61, 4.39$
27. $x = 5, 15, 40$
29. 0.70, 3.39
31. 1.86, 5, 8.14
33. Change $b_1 x^3 + b_2 x^2 + b_3 x + b_4 = 0$, where $b_1 \neq 0$, to $x^3 + \frac{b_2}{b_1} x^2 + \frac{b_3}{b_1} x + \frac{b_4}{b_1} = 0$.

Use the formula for the solution to $x^3 + a_1 x^2 + a_2 x + a_3 = 0$ substituting $\frac{b_2}{b_1}$ for $a_1$, $\frac{b_3}{b_1}$ for $a_2$, and $\frac{b_4}{b_1}$ for $a_3$.

35. (a) $y = \frac{216}{x}$

(b)

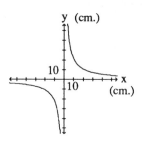

(c) The part of the graph in the first quadrant represents the problem situation.

(d) Width $= y = 9$ centimeters.

37. $C = \frac{5(F-32)}{9}$

39. (a) $n = \frac{1}{r}\left(\frac{S}{p} - 1\right)$

(b) 25 years

41. Side length $= 11.98$ units

43. (a)

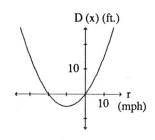

(b) Dom: All real numbers.

Range: All real numbers $\geq -4.96$.

All positive values of $r$ make sense in this problem, although, realistically, one is confined to a speed limit.

(c) 67.88 mph.

45. (a)

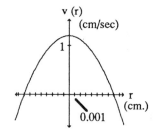

(b) Dom: All real numbers.

Range: All real numbers $\leq 1.19$.

All positive values of $r < 0.008$ make sense in this problem situation.

(c) $r = 0.003$ centimeters.

47. $p = \$12.24$; $x = 553$ units (Integer answers

given for problem situation.)

49.

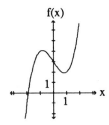

(a) 1
(b) 3
(c) $[-0.82, 0.82]$

51. $x = 3$

## Exercises 1.5

1. $(-1, 7)$
3. $(8, -2)$
5. (a) There is not a simultaneous solution for $E_1$ & $E_2$ because the lines are parallel.
   (b) There is an infinite number of simultaneous solutions because $E_1$ & $E_3$ are the same line.

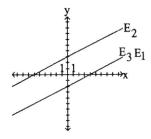

7. $(x, y) = (9, -5)$
9. $(x, y) = (0, 0); (\sqrt{2}, \sqrt{2}); (-\sqrt{2}, -\sqrt{2})$
11. $(-2.89, -2.36), (2.49, -0.20)$
13. $(1.32, 2.27)$
15. $(-3.84, -4.76), (0.62, 9.62), (4.23, -7.86)$
17. $(-1.73, -0.99), (1.06, 0.87)$
19. $(-1.32, -0.97), (0, 0), (1.32, 0.97)$
21. White sugar: 7.08 ounces
    Brown sugar: 15.92 ounces
23. Rowing speed: 3.56 mph
    Current speed: 1.06 mph
25. $a = \frac{1}{3}, b = \frac{-11}{3}$
27. 5%: $5494.50
    6%: $1098.90
    10%: $6593.40

29. 79
31. (a) $V(x) = x(40 - 2x)(30 - 2x)$
    (b) $V(x) = 4x^3 - 140x^2 + 1200x$
    $V(x) = 1200$
    (c)

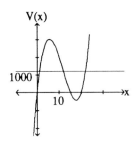

   (d) 3; the 2 solutions less than 20 are also solutions to the problem situation. The side lengths of the cut out squares can be no longer than 15 since one side of the box is 30 inches.
   (e) $x = 1.15$, maximum error in $x$ is 0.0001.
   $x = 11.89$, maximum error in $x$ is 0.0001.

33.

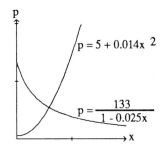

$53.74$ ($x = 59$ units)

35. a

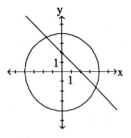

(b) 2
(c) $(-1.65, 3.65)$ & $(3.65, -1.65)$
(d) see (c)
(e)

zero points of intersection

1 point of intersection

2 points of intersection

*37. $b = \pm\sqrt{32}$

## Exercises 1.6

1. $(-1, 1) \cup (2, \infty)$
3. $[-1, 0] \cup (1, \infty)$
5.

7.

9. $(-\infty, -3) \cup (5, \infty)$; $x > 5$ or $x < -3$; The set of all real numbers less than $-3$ or greater than 5.
11. $(-3, 2) \cup (6, \infty)$; $-3 < x < 2$ or $x > 6$; The set of all real numbers between $-3$ and 2 or greater than 6.
13. $(-1, 7)$; $|x - 3| < 4$; The set of all real numbers less than 4 units from the point 3.
15. $(-\infty, 1] \cup [6, \infty)$; $|x - 3.5| \geq 2.5$; The set of all real numbers at least 2.5 units from 3.5.
16. $(-2, 6)$

17. $(-\infty, -1), (7, \infty)$

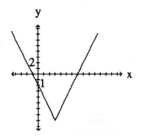

19. $(-\infty, -2] \cup [8, \infty)$
21. $(-\infty, -2) \cup (-2, 3)$
23. (a) $f(-3) > 0$
    (b) yes
    (c) $(-\infty, -2) \cup (8, \infty)$
25. $[0, 3]$
27. $(-3, 7)$
29. $(0, 1)$
31. $(-3, 1)$
33. $[-2, 0] \cup [4, \infty)$
35. $(2, \infty)$
37. $[-9, -4]$
39. The solution set $(-2.26, 0.76) \cup (3.51, \infty)$ is found where $f(x)$ lies above $g(x)$ instead of along the $x$-axis. The endpoints of the intervals are determined by the intersection of $f(x)$ and $g(x)$.
41. $[0.28, 2.39]$
43. $(-\infty, -1.98) \cup (3.94, 7.04)$
45. $(-\infty, -3.66)$
47. $[0, 0.41) \cup (2.73, 6.69) \cup (9.01, 10]$
49. (a) $s = -16t^2 + 275t + 155$

(b)

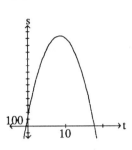

(c) Domain: $(-\infty, \infty)$

Range: $(-\infty, 1336.64]$

(d) The portion of the graph in the first quadrant.

(e) For $0 \le t < 17.73$ seconds

(f) $t = 17.73$ seconds

(g) $s = 1336.64$ ft.

51. i)

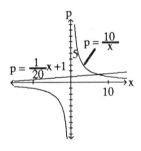

ii) The portion of the graphs in the first quadrant represent the problem situation.

iii) $1.37

iv) 8

53. (a) $30000 \le 5000(1+18r)$

(b)

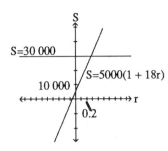

(c) $r \ge 0.28$ or $28\%$

55. (a)

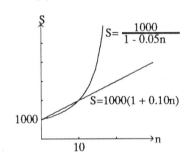

(b) yes, 10 years.

(c) yes, for $n > 10$ years.

## Exercises 1.7

1. $[-5, 5]$ by $[-2, 10]$

3. $[-100, 700]$ by $[-5000, 20000]$

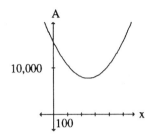

5. $[-5, 30]$ by $[-60000, 100000]$

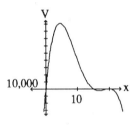

7. $[-1000, 3000]$ by $[-15000000, 2000000]$

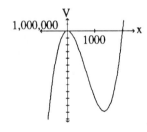

9. local minimum at $(\frac{3}{14}, \frac{831}{28})$

11. local min at $(1, 0)$

13.

15. zeros: $x = \pm\sqrt{10}$
local maximum at $(0, 10)$

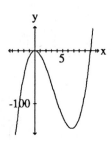

zeros: $x = 0, x = 10$
local max at $(0, 0)$
local min at $(6.67, -148.15)$

17.

no real zeros
local max at $(-0.41, 12.59)$
local min at $(-1.16, 11.91)$
and $(1.57, 3.51)$

19.

zeros: $x = 3.81$
Local maximum $(0.33, -29.85)$
Local minimum $(1, -30)$

21.

zeros: $x = -.93$
Local maximum $(-0.44, 8.08)$
Local minimum $(0.38, 2.59)$ and $(-0.93, 0)$

23. zeros: $x = 2.36$.
Local maximum $(0.79, 12.10)$.
Local minimum $(0.21, 11.90)$.

25. (a) $s(t) = -16t^2 + 40t + 300$
(b) $t = 5.76$ sec.
(c) maximum height of 325 ft. at 1.25 seconds.

27. 262.5 ft. by 525 sq. ft. with maximum area 137812.5 sq.ft.

29. (a) $R = x(2 + .002x - .0001x^2)$
(b)

(c) See (b).

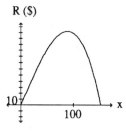

(d) $x = 89$, $R = \$123.35$

31. Length of sides are 10.93 in., 33.1 in., and 64.1 in. which gives maximum volume of 23231.91 cu. in.

# Exercises 1.8

1. Vertical shift down 4 units.

3. Vertical stretch by a factor of 3, horizontal shift left 4 units then vertical shift down 5 units.

5. Reflect through the x-axis, horizontal shift right 2 units, then vertical shift down 2 units.

7. Increasing on $[1.5, \infty)$ and decreasing on $(-\infty, 1.5]$.
   Minimum at $x = 1.5$

9. Increasing on $(-\infty, -1.79] \cup [1.12, \infty)$ and decreasing on $[-1.79, 1.12]$
   Maximum at $x = -1.79$
   Minimum at $x = 1.12$

11. Increasing on $(-\infty, 0.33] \cup [1, \infty)$ and decreasing on $[0.33, 1]$
    Maximum at $x = 0.33$
    Minimum at $x = 1$

13. $g$ is not an increasing function nor a decreasing function on any interval.

15.

17.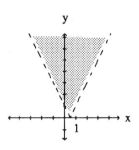

19. (a) $f(x) \to \infty$ as $x \to \infty$, $f(x) \to -\infty$ as $x \to -\infty$
    (b) $f(x) \to \infty$ as $|x| \to \infty$,
    (c) $f(x) \to -\infty$ as $x \to \infty$ $f(x) \to \infty$ as $x \to -\infty$
    (d) $f(x) \to -\infty$ as $|x| \to \infty$,

21.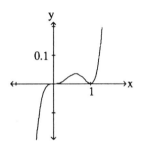
    local maximum $(0.60, 0.03)$
    local minimum $(1.00, 0)$
    real zero: $x = 0, 1$
    increasing on $(-\infty, 0.60] \cup [1.00, \infty)$
    decreasing on $[0.60, 1.00]$

23.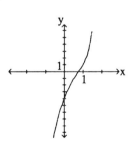
    no local minimum or maximum
    real zero: $x = 0.77$

increasing on $(-\infty, \infty)$

25. $[0.99, 1.03]$ by $[38.9997, 39.0001]$

27. (a) $y = x(335 - 2x)$
    (b)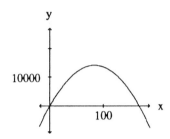
    (c) The portion of the graph in the first quadrant represents the problem situation.
    (d) Increasing on $(-\infty, 83.75]$ and decreasing on $[83.75, \infty)$.

29. (a) $y = -16t^2 + 80t + 186$
    (b)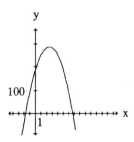
    (c) The portion of the graph in the first quadrant represents the problem situation.
    (d) Increasing on $(-\infty, 2.5]$ and

decreasing on [2.5, ∞).

31. (a) $y = x(20 - 2x)$

(b)

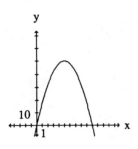

(c) The portion of the graph for which $0 < x < 10$ represents the problem situation.

(d) Increasing on $(-\infty, 5]$ and decreasing on $[5, \infty)$.

---

## Exercises 1.9

1. Dom: $(-\infty, \infty)$
   Range: $(-\infty, \infty)$
   no discontinuities

3. Dom: $(-\infty, \infty)$
   Range: $(2, \infty)$
   no discontinuities

5. Dom: $[-4, 2) \cup (2, \infty)$
   Range: $(-\infty, \infty)$
   discontinuous at $x = -4$

7. Dom: $(-\infty, 4) \cup (4, \infty)$
   Range: $[0, \infty)$
   discontinuous at $x = 2, x = 4$

9. Dom: $(-\infty, \infty)$
   Range: All integers.
   discontinuous at every integer value

11. Dom: $(-\infty, \infty)$
    Range: $(-\infty, \infty)$
    no discontinuities

13. Dom: $(-\infty, \infty)$
    Range: $(0, \infty)$
    discontinuous at $x = 0$

15. (e)

17. Answers will vary.

19. Answers will vary.

21. (a) $f(x) = -5x^4[1 - \frac{2}{5x} + \frac{1}{5x^3} - \frac{3}{5x^4}]$

    (b) Since each $r_i(x)$ approaches 0 as $|x| \to \infty$, $f(x)$ approaches $-5x^4$.

23. $f(1) = -3$
    $f(2) = 1$
    Since $f$ is continuous, it assumes every value between $-3$ and $1$.

    $c = \frac{7}{4}$

25. $c = \frac{L-b}{a}$

27. $f(x) \to \infty$

29. $f(x) \to 0$

31. $f(x) \to -\infty$

33. (a) $[0, 676.52]$

(b)

(c) The portion of the graph in the first quadrant.

(d) max revenue: $1615.01
    pounds of cookies: 676.52

*35. (a)

(b) $L(-1) = -20$
    $L(0) = 5$
    $L(1) = 8$
    $L(2) = 1$

(c) $L(1.5) = 5$

37. Triangles $DEF$ and $BDF$ are congruent by SSS.

From Pythagorean Theorem, $y = \frac{9-x^2}{6}$.

$A = \frac{1}{2}xy = \frac{1}{2}x\left(\frac{9-x^2}{6}\right) = \frac{9x-x^3}{12} = \frac{3}{4}x - \frac{1}{12}x^3$

# Review Exercises

1.
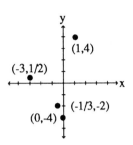

3. $a = 5$ or $-5$

5.

7.

9.
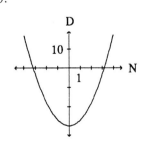

11. Dom: All real numbers.
    Range: All real numbers.

13. Dom: All real numbers.
    Range: All real numbers $\leq 5$.
    Line of Symmetry: $x = 0$.
    $y$ defined as a function of $x$.

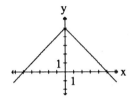

15. (a) (c)

17. Yes, it is a function.

19. (a) $f(0) = 0$
    $f(-1) = -2$
    $f(2) = 4$
    (b) $x = 0, 3$
    (c) $x = 0.25, 2.75$

21. d

In problems 23-28 we know these are complete graphs because all critical portions are present and the end behavior is suggested.

23.

25.

27.

29. Dom: All real numbers.
    Range: All real numbers $\geq -40$.

31. Dom: All real numbers $\geq 3$.
    Range: All real numbers $\geq 0$.

33. Dom: All real numbers.
    Range: All real numbers.

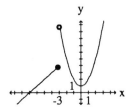

35. (a) $t = \frac{450}{R}$

(b)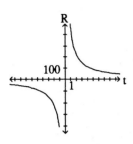

(c) The portion of the graph in the first quadrant represents the problem situation.

(d) $t = 11.25$ hours

37. (a) $P = 0.50x - 18.25$

(b)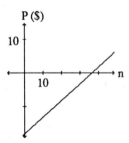

(c) Dom: All real numbers.
Range: All real numbers.

(d) The points $(x, P)$, for nonnegative integers $x$, represent the problem situation.

(e) 37 tickets

39. (a) $C = 85x + 75000$

(b)

(c) Dom: All real numbers.
Range: All real numbers.

(d) The portion of the graph in the first quadrant represents possible total annual costs of producing bikes.

(e) 800 bikes.

(f) $A$ represents the number of bikes made.

41. (a) $A = x\left(\frac{72-x}{2}\right)$

(b) Dom: All real numbers.
Range: All real numbers $\leq 648$.

(c)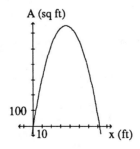

(d) All positive values of $x$ less than 72 may be used in this problem situation.

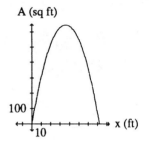

(e) $x = 65.94$ feet or $x = 6.06$ feet.

(f) Yes; 648 square feet.

43. (a) $A = 54x + 4x^2$

(b)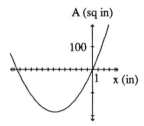

(c) Domain: All real numbers.

(d) All positive values of $x$ may be used in this problem situation.

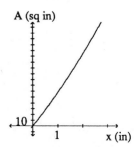

45. $x = \frac{27}{40}$
47. $x = 7, -3$
49. $x = \pm\sqrt{3}$
51. $x = -1.95$
53. 1 solution
55.

57. $-4 \le x < 2$; The set of all real numbers less than 2 but greater than or equal to $-4$.
59. $(-4, 1) \cup (7, \infty)$; $x > 7$ or $-4 < x < 1$
61. $(-\infty, -\frac{9}{2}]$
63. $[1, \frac{23}{4})$
65. No real solutions.
67. $(-\infty, 2]$
69. $(-5, 2)$
71. $(-\infty, 0.23] \cup [1.43, \infty)$
73. $[-0.91, \infty)$
75. (a) $y = \frac{420}{x}$
   (b) 35 cm.
77. $0 < w < 72.5$ inches
79. (a) $s = -16t^2 + 75t + 165$
   (b)

(c) Dom: All real numbers.
   Range: All real numbers $\le 252.89$.
(d) The portion of the graph in the first quadrant.
(e) $0 \le t < 6.32$
(f) $t = 6.32$ seconds
(g) $s = 252.89$ feet
(h) $t = 2.34$ seconds

81. $(-\infty, 3)$
83. Increasing on $[3.5, \infty)$
   Decreasing on $(-\infty, 3.5]$.
   Real zeros: $x = 2, 5$
85. Increasing on $(-\infty, \infty)$.
   Real zeros: $x = 1.64$
87.

89.

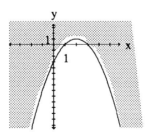

91. $(-\infty, -3] \cup [0, 3]$
93. $x \ge 1$

95.

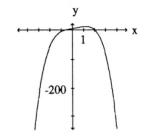

V.R. $[-5, 5]$ by $[-20, 20]$
Real zeros:
$x = -0.67, 2.03$

97. Increasing on $(-\infty, 0.21] \cup [3.12, \infty)$
   Decreasing on $[0.21, 3.12]$
99. Local maximum at $(-1.33, -13.63)$
   Local minimum at $(0, -16)$

101.

local maximum at $(0.06, -2.97)$

(e) $x = 0.87$ inches.

local minimum at
$(-1.44, -12.43)$ and
$(1.38, -9.60)$

103. $W = 14$ ft., $L = 16$ ft.

105. (a) $s = -16t^2 + 70t + 200$

(b)

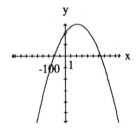

(c) The portion of the graph in the first quadrant represents the problem situation.

(d) $t = 0..39$ and $t = 3.98$

(e) $0.39 < t < 3.98$

107. (a) $R = x(1.5x - 0.03x^2)$

(b)

(c)

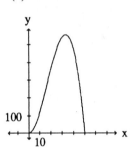

(d) $x = 33$ produces maximum daily revenue $R = \$555.56$ maximum daily revenue

109. Dom: $(-\infty, -\frac{1}{2}) \cup (-\frac{1}{2}, \infty)$
Range: $(-\infty, \infty)$
Discontinuous at $x$: $-\frac{1}{2}$

111. Dom: $(-\infty, \infty)$
Range: $[-1, \infty)$
Points of Discontinuity: none

113. Dom: $(-\infty, -5) \cup (-5, \infty)$
Range: $(-\infty, 0) \cup (0, \infty)$
Discontinuous at $x = -5$

115. $f(x) \to \infty$

117. V.R.: $[-12, 12]$ by $[-700, 700]$

## • CHAPTER 2

### Exercises 2.1

1.

3.

5.

7. III Quadrant

9. II Quadrant

11. I Quadrant

13. 408°

15. 345°

17. 415°, 775°, −305°, −665° are possible answers.

19. 50°, 770°, −310°, −670° are possible answers.

21. 180°

23. 45°

25. 315°

27. $2\sqrt{2}$

29. $\sqrt{13}$

31. $2\sqrt{17}$

33. $37° + n \cdot 360°$, $n$ any integer

35. $-72° + n \cdot 360°$, $n$ any integer

37. $\frac{\pi}{3}$

39. 15.58

41. All four ratios are the same: $S = \frac{5}{18}\pi$

43. About 6107 feet

45.

47.

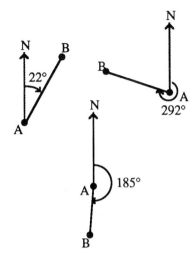

49. If a rational number has the form $\frac{a \cdot n}{b \cdot n}$ where $a$ and $b$ have no common factors, we divide out the common factor $n$ to get $\frac{a}{b}$ in lowest terms. If an angle has the form $\theta + n \cdot 360°$, with $0° \leq \theta \leq 360°$, we subtract $n \cdot 360°$ to get $\theta$ in "lowest terms."

## Exercises 2.2

1. $\sin 23° = 0.39$
   $\cos 23° = 0.92$
   $\tan 23° = 0.42$
   $\csc 23° = 2.56$
   $\sec 23° = 1.09$
   $\cot 23° = 2.36$

3. $\sin 38° = 0.62$
   $\cos 38° = 0.79$
   $\tan 38° = 0.78$
   $\csc 38° = 1.62$
   $\sec 38° = 1.27$
   $\cot 38° = 1.28$

5. $\tan 30° = \frac{1}{\sqrt{3}}$
   $\cos 60° = \frac{1}{2}$
   $\sec 30° = \frac{2}{\sqrt{3}}$
   $\csc 60° = \frac{2}{\sqrt{3}}$
   $\cot 60° = \frac{1}{\sqrt{3}}$

7. For $\theta = 56.44269024$
   $\sin \theta = 0.8\overline{3}$
   $\cos \theta = 0.5528$
   $\tan \theta = 1.5076$
   $\csc \theta = 1.2$
   $\sec \theta = 1.809$
   $\cot \theta = 0.6633$
   These agree with the values in Example 1, to 4 decimal places.

9. $\sin \angle B = \frac{7}{\sqrt{113}}$
   $\cos \angle B = \frac{8}{\sqrt{113}}$
   $\tan \angle B = \frac{7}{8}$
   $\csc \angle B = \frac{\sqrt{113}}{7}$
   $\sec \angle B = \frac{\sqrt{113}}{8}$
   $\cot \angle B = \frac{8}{7}$

11. $\sin \angle B = \frac{2\sqrt{26}}{15}$
    $\cos \angle B = \frac{11}{15}$
    $\tan \angle B = \frac{2\sqrt{26}}{11}$
    $\csc \angle B = \frac{15}{2\sqrt{26}}$
    $\sec \angle B = \frac{15}{11}$
    $\cot \angle B = \frac{11}{2\sqrt{26}}$

13. $\sin \theta = \frac{1}{\sqrt{10}}$
    $\cos \theta = \frac{3}{\sqrt{10}}$
    $\tan \theta = \frac{1}{3}$

$\csc\theta = \sqrt{10}$
$\sec\theta = \frac{\sqrt{10}}{3}$
$\cot\theta = 3$

15. $55.54°$
17. $82.82°$
19. $71.57°$
21. $70.35°$
23. $\frac{1}{\cot\theta} = \frac{1}{b/a} = \frac{a}{b} = \tan\theta$. Taking reciprocals, we get $\frac{1}{1/\cot\theta} = \frac{1}{\tan\theta}$ or $\cot\theta = \frac{1}{\tan\theta}$.
25. In a right triangle with a very small acute angle, the lengths $a$ and $b$ could differ by a great deal so that the ratio $\frac{a}{b} = \tan\theta$ is very small (for $a < b$) or very large (for $a > b$). For example, if $a = 1$ and $b = 1000$, $\tan\theta = \frac{1}{1000} = 0.0001$.
27. $74.78°$
29. $28.07°$

Note: For 30 – 40, $\angle C = 90°$

31. $\angle A = 36.87°$
    $\angle B = 53.13°$
    $c = 5$
33. $\angle B = 35°$
    $a = 22.25$
    $c = 27.16$
35. $\angle A = 23.46°$
    $\angle B = 66.54°$
    $b = 46.56$
37. $\angle A = 31°$
    $b = 8.32$
    $c = 9.71$
39. $\angle B = 79.8°$
    $a = 2.57$

$b = 14.27$
41. $\sin^2\theta + \cos^2\theta = 1$ for each $\theta$ given.

## Exercises 2.3

1. $D = 5$
3. $D = 9$
5. $D = 5\sqrt{2}$
7. 100 feet
9. 680.55 feet
11. 106.12 feet
13. 101.57 feet
15. 52.35 feet
17. 84.85 miles; bearing of 140°
19. 19°
21. 637.85 feet
23. 9.78 feet
25. 2931.09 feet
27. $\sin\theta = b$
    $\cos\theta = a$
    $\tan\theta = c$
    $\csc\theta = f$
    $\sec\theta = e$
    $\cot\theta = d$
29. From the Pythagorean Theorem, we know the third side of the right triangle is $\sqrt{1-a^2}$. Therefore, $\sin\theta = \sqrt{1-a^2}$ and $\tan\theta = \frac{\sqrt{1-a^2}}{a}$.
31. $\sin\theta = \sqrt{1-a^2}$.

## Exercises 2.4

1. $\sin(83°) = 0.99$
   $\cos(83°) = 0.12$
   $\tan(83°) = 8.14$
   $\csc(83°) = 1.01$
   $\sec(83°) = 8.21$
   $\cot(83°) = 0.12$
3. $\sin(-400°) = -0.64$
   $\cos(-400°) = 0.77$
   $\tan(-400°) = -0.84$
   $\csc(-400°) = -1.56$
   $\sec(-400°) = 1.31$
   $\cot(-400°) = -1.19$
5. (b) $P(0,1)$ $r = 1$
   $\sin\theta = \frac{1}{1} = 1$
   $\cos\theta = \frac{0}{1} = 0$
   $\tan\theta$ (no value)
   $\csc\theta = \frac{1}{1} = 1$
   $\sec\theta$ (no value)
   $\cot\theta = \frac{0}{1} = 0$
   (c) $P(-1,0)$ $r = 1$
   $\sin\theta = \frac{0}{1} = 0$
   $\cos\theta = \frac{-1}{1} = -1$
   $\tan\theta = \frac{0}{-1} = 0$
   $\csc\theta$ (no value)
   $\sec\theta = \frac{1}{-1} = -1$
   $\cot\theta$ (no value)
7. In Quadrant I both $x$ and $y$ values of a point on the terminal side of an angle are positive. The radius is always positive. Thus, the ratios formed are positive for all six trig functions.
9. In Quadrant III both the x value and the y value

## 2 Answers to Odd-Numbered Exercises • A19

are negative. Thus the only positive functions are those using both x and y (tan and cot). The rest are negative.

11. $\sin \theta = 0$
    $\cos \theta = -1$
    $\tan \theta = 0$
    $\csc \theta$ (no value)
    $\sec \theta = -1$
    $\cot \theta$ (no value)
    $\sin \theta' = 0$
    $\cos \theta' = 1$
    $\tan \theta' = 0$
    $\csc \theta'$ (no value)
    $\sec \theta' = 1$
    $\cot \theta'$ (no value)

(-3,0)   $\theta = \theta'$

13. $\sin \theta = \frac{3}{5}$
    $\cos \theta = \frac{4}{5}$
    $\tan \theta = \frac{3}{4}$
    $\csc \theta = \frac{5}{3}$
    $\sec \theta = \frac{5}{4}$
    $\cot \theta = \frac{4}{3}$
    $\sin \theta' = \frac{3}{5}$
    $\cos \theta' = \frac{4}{5}$
    $\tan \theta' = \frac{3}{4}$
    $\csc \theta' = \frac{5}{3}$
    $\sec \theta' = \frac{5}{4}$

$\cot \theta' = \frac{4}{3}$

(4,3)  $\theta' = \theta$

15. $\sin \theta = -\frac{3}{\sqrt{13}}$
    $\cos \theta = -\frac{2}{\sqrt{13}}$
    $\tan \theta = \frac{3}{2}$
    $\csc \theta = -\frac{\sqrt{13}}{3}$
    $\sec \theta = -\frac{\sqrt{13}}{2}$
    $\cot \theta = \frac{2}{3}$
    $\sin \theta' = \frac{3}{\sqrt{13}}$
    $\cos \theta' = \frac{2}{\sqrt{13}}$
    $\tan \theta' = \frac{3}{2}$
    $\csc \theta' = \frac{\sqrt{13}}{3}$
    $\sec \theta' = \frac{\sqrt{13}}{2}$
    $\cot \theta' = \frac{2}{3}$

(-4, -6)

17. $\sin \theta = -\frac{1}{\sqrt{2}}$
    $\cos \theta = \frac{1}{\sqrt{2}}$
    $\tan \theta = -1$

$\csc \theta = -\sqrt{2}$
$\sec \theta = \sqrt{2}$
$\cot \theta = -1$
$\sin \theta' = \frac{1}{\sqrt{2}}$
$\cos \theta' = \frac{1}{\sqrt{2}}$
$\tan \theta' = 1$
$\csc \theta' = \sqrt{2}$
$\sec \theta' = \sqrt{2}$
$\cot \theta' = 1$

(22, -22)

19.

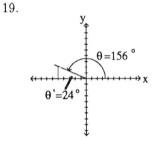
$\theta = 156°$
$\theta' = 24°$

21.

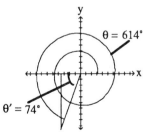
$\theta = 614°$
$\theta' = 74°$

23. $\sin \theta = 0.41$
    $\cos \theta = 0.91$
    $\tan \theta = 0.45$
    $\csc \theta = 2.46$

$\sec\theta = 1.09$
$\cot\theta = 2.25$
$\sin\theta' = 0.41$
$\cos\theta' = -0.91$
$\tan\theta' = -0.45$
$\csc\theta' = 2.46$
$\sec\theta' = -1.09$
$\cot\theta' = -2.25$

25. $\sin\theta = 0.96$
$\cos\theta = 0.28$
$\tan\theta = 3.49$
$\csc\theta = 1.04$
$\sec\theta = 3.63$
$\cot\theta = 2.90$
$\sin\theta' = -0.96$
$\cos\theta' = -0.28$
$\tan\theta' = 3.49$
$\csc\theta' = -1.04$
$\sec\theta' = -3.63$
$\cot\theta' = 2.90$

27. $\sin\theta = -\frac{1}{\sqrt{2}}$
$\cos\theta = -\frac{1}{\sqrt{2}}$
$\tan\theta = 1$
$\csc\theta = -\sqrt{2}$
$\sec\theta = -\sqrt{2}$
$\cot\theta = 1$

29. $\sin\theta = \frac{3}{5}$
$\cos\theta = \frac{4}{5}$
$\tan\theta = \frac{3}{4}$
$\csc\theta = \frac{5}{3}$
$\sec\theta = \frac{5}{4}$
$\cot\theta = \frac{4}{3}$

31. $\sin\theta = -\frac{2}{\sqrt{13}}$
$\cos\theta = -\frac{3}{\sqrt{13}}$
$\tan\theta = \frac{2}{3}$

$\csc\theta = -\frac{\sqrt{13}}{2}$

$\sec\theta = -\frac{\sqrt{13}}{3}$

$\cot\theta = \frac{3}{2}$

33. $74°$

35. $23°$

37. $34°$

39. $63.43°$

41. $-71.57°$

43. III quadrant

45. III quadrant

47. $\cot\theta\tan\theta =$
$\left(\frac{1}{\tan\theta}\right)\tan\theta = 1$

49. $\sec^2\theta - \sin^2\theta =$
$\frac{1}{\cos^2\theta} - (1 - \cos^2\theta) =$
$\frac{1-\cos^2\theta}{\cos^2\theta} + \cos^2\theta = \frac{\sin^2\theta}{\cos^2\theta} +$
$\cos^2\theta = \tan^2\theta + \cos^2\theta$

51. Height of pole: 30.09 feet

Length of wire: 34.08 feet

53. 3290.53 feet

55. (a) $\sin 0° = 0$
$\cos 0° = 1$
$\tan 0° = 0$
$\cot 0° =$ no value
$\sec 0° = 1$
$\csc 0° =$ no value

(b) $\sin 30° = \frac{1}{2}$
$\cos 30° = \frac{\sqrt{3}}{2}$
$\tan 30° = \frac{1}{\sqrt{3}}$
$\cot 30° = \sqrt{3}$
$\sec 30° = \frac{2}{\sqrt{3}}$
$\csc 30° = 2$

(c) $\sin 45° = \frac{1}{\sqrt{2}}$
$\cos 45° = \frac{1}{\sqrt{2}}$
$\tan 45° = 1$
$\cot 45° = 1$
$\sec 45° = \sqrt{2}$
$\csc 45° = \sqrt{2}$

(d) $\sin 60° = \frac{\sqrt{3}}{2}$
$\cos 60° = \frac{1}{2}$
$\tan 60° = \sqrt{3}$
$\cot 60° = \frac{1}{\sqrt{3}}$
$\sec 60° = 2$
$\csc 60° = \frac{2}{\sqrt{3}}$

(e) $\sin 90° = 1$
$\cos 90° = 0$
$\tan 90° =$ no value
$\cot 90° = 0$
$\sec 90° =$ no value
$\csc 90° = 1$

(f) $\sin 180° = 0$
$\cos 180° - 1$
$\tan 180° = 0$
$\cot 180° =$ no value
$\sec 180° = -1$
$\csc 180° =$ no value

(g) $\sin 270° = -1$
$\cos 270° = 0$
$\tan 270° = $ no value
$\cot 270° = 0$
$\sec 270° = $ no value
$\csc 270° = -1$

(h) $\sin 360° = 0$
$\cos 360° = 1$
$\tan 360° = 0$
$\cot 360° = $ no value
$\sec 360° = 1$
$\csc 360° = $ no value

Period: $\frac{2\pi}{3}$

19. Domain: $(-\infty, \infty)$

Range: $[1, 3]$

Period: $2\pi$

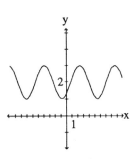

21. Domain: $(-\infty, \infty)$

Range: $[-4, 2]$

Period: $2\pi$

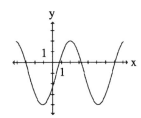

23. Domain: $(-\infty, \infty)$

Range: $[-4, 2]$

Period: $2\pi$

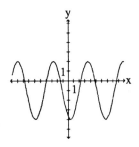

25. Horizontally shift right $180°$, then a vertical stretch of 2 followed by a reflection about the $x$-axis.

27. Let $-\theta$ be the angle determined by $OP$ and the $x$-axis and $\theta$ be the angle in standard position. Then $\cos(-\theta) = \frac{a}{c}$ and $\cos\theta = \frac{a}{c}$ where $c = \sqrt{a^2 + b^2}$. Since $\cos(-\theta) = \cos\theta$, cosine is an even function.

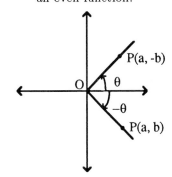

29. $\theta_1$ measures 2 radians
$\sin 2 = \sin(\pi - 2)$

$\theta_2$ measures 3 radians
$\sin 3 = \sin(\pi - 3)$

## Exercises 2.5

1. $\frac{\pi}{180} = 0.02$
3. $0.52$
5. $-2.09$
7. $57.30°$
9. $135°$
11. $-131.78°$
13. $\sin 200° = -0.34$
$\cos 200° = -0.94$
$\tan 200° = 0.36$
$\csc 200° = -2.92$
$\sec 200° = -1.06$
$\cot 200° = 2.75$
15. $\sin -2 = -0.91$
$\cos -2 = -0.42$
$\tan -2 = 2.19$
$\csc -2 = -1.10$
$\sec -2 = -2.40$
$\cot -2 = 0.46$
17. Domain: $(-\infty, \infty)$
Range: $[-1, 1]$

sin(π − 2) > sin(π − 3) since an angle with measure (π − 3) is smaller than angle with measure (π − 2) and for angles less than 90° the sine values increase with increased angle measure. Therefore sin 2 > sin 3.

31. sin 10 = sin(10 − 2π) = sin(3.72). The angle with measure 3.72 lies in quadrant III, so the sin(3.72) would be less than zero.
33. $x = 0.52$, $x = 2.62$
35. $x = 0.93$, $x = 2.21$
37. $x = 0.93$, $x = 5.36$
39. $0 \le \theta \le 36.87°$ and $143.13° \le \theta \le 360°$
41. (a) True; the graphs of each side of the equation overlay each other.
 (b) True; the graphs of each side of the equation overlay each other.
 (c) False; the graphs of each side of the equation do **not** overlay each other.
 (d) True; the graphs of each side of the equation overlay each other.
43. 5.76 inches
45. 50.27 inches
47. 7539.82 radians per minute
49. 3769.91 radians per minute
51. Angular speed of wheel: 177,408 rads/hr.
 Angular speed of pedal sprocket: 55,440 rads/hr.
 Angular speed of wheel sprocket: 177,408 rads/hr.(same as wheel)
53. 18.71 inches
55.

y = 2(sin x + 3)

y = 2 sin x + 3

## Exercises 2.6

1(a)

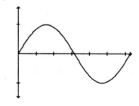

(b) [0, −2π] by [−2, 2]

(c) [−2π, 0] by [−2, 2]

(d) [−8π, −6π] by [−2, 2]

[10π, 12π] by [−2, 2]

3. Domain: All reals except $x = \frac{\pi}{2} + k\pi$, $k$ any integer
 Range: All reals
 Period: $\pi$
 Asymptotes: $x = \frac{\pi}{2} + k\pi$, $k$ any integer

5. Domain: All reals except $x = \frac{\pi}{2} + k\pi$, $k$ any integer
 Range: $(-\infty, -\frac{1}{2}] \cup [\frac{1}{2}, \infty)$
 Period: $2\pi$

Asymptotes: $x = \frac{\pi}{2} + k\pi$, $k$ any integer

7. Domain: All reals except $x = k\pi$, $k$ any integer
Range: $(-\infty, -3] \cup [3, \infty)$
Period: $2\pi$
Asymptotes: $x = k\pi$, $k$ any integer

9. Domain: All reals
Range: $[-3, 3]$
Period: $2\pi$
Asymptotes: none

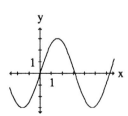

11. Domain: All reals except $x = \pi + 2k\pi$, $k$ any integer
Range: All reals
Period: $2\pi$

Asymptotes: $x = \pi + 2k\pi$, $k$ any integer

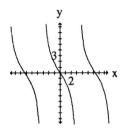

13. Domain: All reals except $x = k\pi$, $k$ any integer
Range: $(-\infty, -2] \cup [2, \infty)$
Period: $2\pi$
Asymptotes: $x = k\pi$, $k$ any integer

15. Domain: All reals
Range: $[-1, 1]$
Period: $\pi$
Asymptotes: none

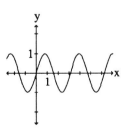

17. Domain: All reals except $x = \frac{\pi}{4} + k \cdot \frac{\pi}{2}$, $k$ any integer
Range: $(-\infty, -1] \cup [1, \infty)$
Period: $\pi$

Asymptotes: $x = \frac{\pi}{4} + k \cdot \frac{\pi}{2}$, $k$ any integer

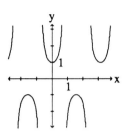

19. Domain: All reals
Range: $[-1, 1]$
Period: $4\pi$
Asymptotes: none

21. Domain: All reals
Range: $[-2, 2]$
Period: $\frac{2}{3}\pi$
Asymptotes: none

23. Domain: All reals
Range: $[-3, 3]$
Period: $4\pi$

Asymptotes: none

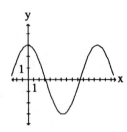

25. Domain: All reals
    Range: $[-2, 2]$
    Period: $2\pi$
    Asymptotes: none

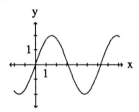

27. Domain: All reals except $x = \frac{1}{2} + k$, $k$ any integer
    Range: All reals
    Period: 1
    Asymptotes: $x = \frac{1}{2} + k$, $k$ any integer

29. Domain: All reals except $x = \frac{\pi}{4} + k \cdot \frac{\pi}{2}$, $k$ any integer
    Range: $(-\infty, -3] \cup [3, \infty)$
    Period: $\pi$

Asymptotes: $x = \frac{\pi}{4} + k \cdot \frac{\pi}{2}$, $k$ any integer

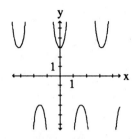

31. Horizontally shrink $y = \sin x$ by a factor of $\frac{1}{3}$ then apply a vertical stretch of 2.

33. Horizontally shrink $y = \cos x$ by a factor of $\frac{1}{\pi}$ then apply a vertical stretch of 3.

35. $x = 0.85, 2.29, -3.99, -5.43$

37. $x = 1.27, 4.41, -1.87, -5.01$

39. $(-5.82, -3.61) \cup (0.46, 2.68)$

41. $(-2\pi, -5.09) \cup (-\frac{3\pi}{2}, -1.95) \cup (-\frac{\pi}{2}, 1.19) \cup (\frac{\pi}{2}, 4.33) \cup (\frac{3\pi}{2}, 2\pi)$

43. $\cot(x + \pi) = \frac{1}{\tan(x+\pi)} = \frac{1}{\tan x} = \cot x$

45. $\csc(x + 2\pi) = \frac{1}{\sin(x+2\pi)} = \frac{1}{\sin x} = \csc x$

51. $x + 2k\pi$, $k$ any integer, where $x = 0.25$ or $2.89$

53. $x + 2k\pi$, $k$ any integer, where $x = 1.14$ or $5.15$

55. $x + 2k\pi$, $k$ any integer, where $-3.29 < x < 0.15$

57. $x + \frac{2}{3}k\pi$, $k$ any integer, where $x = 0.19$ or $0.85$

59. $x + 4k\pi$, $k$ any integer, where $0.51 \leq x \leq 5.78$ or $2\pi < x < 4\pi$

## Exercises 2.7

1.

3.

5.

7.

9.

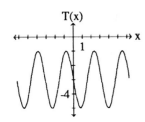

11. Dom: All real numbers.
Range: $[-4, 2]$
Period: $2\pi$
Asymptotes: none

13. Dom: All real numbers except $(2k+1)\pi$ where $k$ is any integer.
Range: $[-\infty, \infty]$
Period: $2\pi$
Asymptotes: $(2k+1)\pi$ where $k$ is any integer

15. Local maxima: $(0.79, 3.00), (3.93, 3.00)$
Local minima: $(2.36, -3.00), (5.50, -3.00)$
Increasing: $[0, 0.79] \cup [2.36, 3.93] \cup [5.50, 2\pi]$
Decreasing: $[0.79, 2.36] \cup [3.93, 5.50]$

17. No extrema
Increasing: $(0, \pi) \cup (\pi, 2\pi)$

Asymptotes: $x = 0, \pi$, or $2\pi$

19. Maxima at: $(\frac{\pi}{8}, 5.00)$, $(\frac{5\pi}{8}, 5.00)$, $(\frac{9\pi}{8}, 5.00)$, $(\frac{13\pi}{8}, 5.00)$

Minima at: $(\frac{3\pi}{8}, -1.00)$, $(\frac{7\pi}{8}, -1.00)$, $(\frac{11\pi}{8}, -1.00)$, $(\frac{15\pi}{8}, -1.00)$

Increasing in: $[0, \frac{\pi}{8}] \cup [\frac{3\pi}{8}, \frac{5\pi}{8}] \cup [\frac{7\pi}{8}, \frac{9\pi}{8}] \cup [\frac{11\pi}{8}, \frac{13\pi}{8}] \cup [\frac{15\pi}{8}, 2\pi]$

Decreasing in: $[\frac{\pi}{8}, \frac{3\pi}{8}] \cup [\frac{5\pi}{8}, \frac{7\pi}{8}] \cup [\frac{9\pi}{8}, \frac{11\pi}{8}] \cup [\frac{13\pi}{8}, \frac{15\pi}{8}]$

21. $\frac{\pi}{18} + \frac{2k\pi}{3}$ with $k$ any integer

$\frac{5\pi}{18} + \frac{2k\pi}{3}$ with $k$ any integer

23. $2.66 + 4k\pi$ with $k$ any integer

$-2.66 + 4k\pi$ with $k$ any integer

25. $x = 0, 2.13$

27. In the interval $(0.69, 3.83)$

29. $y = 3\sin(2x - \pi)$

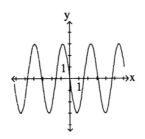

31. $y = 2\sin(\frac{1}{2}x + \frac{\pi}{8})$

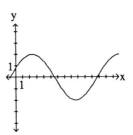

33. Vertical stretch by a factor of 4, a reflection through the $x$-axis, then horizontal shrink by a factor of 2, a horizontal shift right $\pi$ units, then a vertical shift up 3 units.

35. A reflection through the $x$-axis, followed by a horizontal shift right $\pi$ units, and a vertical shift up 2 units.

37. Period $= \frac{2\pi}{3}$
Frequency $= \frac{3}{2\pi}$

39. Period $= 4$
Frequency $= \frac{1}{4}$

41. Period $= 4$ seconds
$A = -15$
$B = \frac{\pi}{2}$

43. The part to the right of the $y$-axis

45. $\frac{1}{3}$ seconds
$A = -28$
$B = 6\pi$

47. for $x \geq 0$

49.

51. $d = 49.50$ inches
$\alpha = 1.43$ radians
belt $= 2d + (2\pi - 2\alpha)22$
$\qquad + (2\pi - 2\alpha')15$
$\qquad = 264.36$ inches
(where $\alpha'$ is the angle between 50 and the dashed line)

53. 2.09 radians/second
55. 11.42 miles/hour

## Review Exercises

1. $\sin 32° = 0.53$
$\cos 32° = 0.85$
$\tan 32° = 0.62$
$\csc 32° = 1.89$
$\sec 32° = 1.18$
$\cot 32° = 1.60$

3. $\sin 30° = \frac{1}{2}$
$\cos 270° = 0$
$\tan 135° = -1$
$\sin \frac{5\pi}{6} = \frac{1}{2}$
$\cos \frac{2\pi}{3} = -\frac{1}{2}$
$\cot 135° = -1$
$\csc \frac{\pi}{2} = 1$
$\sec 270° = $ (no value)

$\cot \frac{\pi}{4} = 1$
$\csc 210° = -2$
$\sec 330° = \frac{2}{\sqrt{3}}$
$\cot \frac{5\pi}{4} = 1$

5. $\sin -\frac{\pi}{6} = -\frac{1}{2}$
$\cos -\frac{\pi}{6} = \frac{\sqrt{3}}{2}$
$\tan -\frac{\pi}{6} = -\frac{1}{\sqrt{3}}$
$\cot -\frac{\pi}{6} = -\sqrt{3}$
$\sec -\frac{\pi}{6} = \frac{2}{\sqrt{3}}$
$\csc -\frac{\pi}{6} = -2$

7. $\sin 60° = \frac{\sqrt{3}}{2}$
$\cos 60° = \frac{1}{2}$
$\tan 60° = \sqrt{3}$
$\cot 60° = \frac{1}{\sqrt{3}}$
$\sec 60° = 2$
$\csc 60° = \frac{2}{\sqrt{3}}$

9. $\sin 300° = -\frac{3}{\sqrt{2}}$
$\cos 300° = \frac{1}{2}$
$\tan 300° = -\sqrt{3}$
$\cot 300° = -\frac{1}{\sqrt{3}}$
$\sec 300° = 2$
$\csc 300° = -\frac{2}{\sqrt{3}}$

11. $\sin \theta = \frac{2\sqrt{6}}{7}$; $\csc \theta = \frac{7}{2\sqrt{6}}$
$\cos \theta = \frac{5}{7}$; $\sec \theta = \frac{7}{5}$
$\tan \theta = \frac{2\sqrt{6}}{5}$; $\cot \theta = \frac{5}{2\sqrt{6}}$

13. $\sin \theta = \frac{a}{c}$
$\cos \theta = \frac{b}{c}$
$\cot \theta = \frac{b}{a}$
$\frac{\cos \theta}{\sin \theta} = \frac{\frac{b}{c}}{\frac{a}{c}} = \frac{b}{a} = \cot \theta$

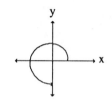

15. $\angle B = 55°$
$a = 8.60$
$b = 12.29$

17. 1.05 radians

19.
(a) 270°

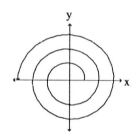

(b) 900°

21. $\theta = 202.62°$ and $\theta = 22.62°$

23. $\sin \theta = \frac{4}{5}$
$\cos \theta = \frac{3}{5}$
$\tan \theta = \frac{4}{3}$
$\csc \theta = \frac{5}{4}$
$\sec \theta = \frac{5}{3}$
$\cot \theta = \frac{3}{4}$

25. $\frac{\sin^2 \theta}{1 - \cos \theta} = \frac{1 - \cos^2 \theta}{1 - \cos \theta}$
$= \frac{(1 + \cos \theta)(1 - \cos \theta)}{1 - \cos \theta}$
$= 1 + \cos \theta$

27. 15.95°
29. II
31.
(a) 394°
(b) $\frac{5\pi}{4}$

33. $x = \frac{\pi}{6} + 2k\pi$, $k$ any integer; $x = \frac{11\pi}{6} + 2k\pi$

35.

37.

39. Domain: $(-\infty, \infty)$

Range: $[-3, 3]$

Period: $\pi$

41. Domain: all real numbers different from $\frac{\pi}{6} + \frac{k\pi}{3}$, $k$ any integer

Range: $(-\infty, -1] \cup [3, \infty)$

Period: $\frac{2\pi}{3}$

Asymptotes: $x = \frac{\pi}{6} + \frac{k\pi}{3}$, $k$ any integer

43. Apply a horizontal stretch of 2 and a vertical shift down 5 units to the graph of $y = \tan x$.

45. $x = 0$

47. False; the graphs do not overlay each other.

49. II

51. 23.78 feet

53. 1.62 miles

55. 4669.58 feet

57. $\sin \theta' = 0.41$
$\cos \theta' = 0.91$
$\tan \theta' = 0.45$
$\csc \theta' = 2.46$
$\sec \theta' = 1.09$
$\cot \theta' = 2.25$
$\sin \theta = 0.41$
$\cos \theta = -0.91$
$\tan \theta = -0.45$
$\csc \theta = 2.46$
$\sec \theta = -1.09$
$\cot \theta = -2.25$

# CHAPTER 3

## Exercises 3.1

1. $y = \sqrt{13}\sin(x + 0.59)$

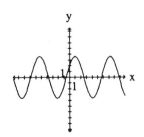

3. $y = \sqrt{2}\sin(x + \frac{\pi}{4})$

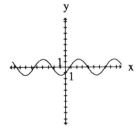

5. $A = 6.5$

7. Domain: $(-\infty, \infty)$

Range: $[-2, 1.12]$

Period: $2\pi$

absolute maxima: $(0.25, 1.12)$, $(2.89, 1.12)$

local minimum: $(1.57, 0)$

abs min: $(4.71, -2)$

9. Domain: $(-\infty, \infty)$
Range: $[-1.88, 1.88]$
Period: $2\pi$
local maxima:
$(2.56, 0.15)$, $(4.82, 1.06)$
absolute maximum:
$(0.47, 1.88)$
local minima:
$(1.68, -1.06)$,
$(5.70, -0.15)$
abs min: $(3.61, -1.88)$

15. End behavior is periodic with period $\pi$

Vertical asymptote at $x = k\pi$, $k$ any integer

17. Symmetric with respect to the $y$-axis

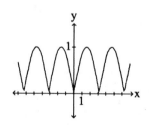

19. Symmetric with respect to the $y$-axis

Horizontal asymptote at $y = 1$

Undefined at $x = 0$

11. $y \to 0$ as $|x| \to \infty$
Undefined at $x = 0$

13. $y \to -1$ as $|x| \to \infty$
Undefined at $x = 0$

21. $V(\theta) = 16{,}000\pi \sin\theta \cos^2\theta$

23. $\theta = 0.62$ radians
$h = 40\sin\theta = 23.24$ units
$r = 20\cos\theta = 16.28$ units
$V = 19{,}347.18$ cubic units

25. $V(r) = \pi r^2 \sqrt{400 - 4x^2}$
Max vol. $2418.40$ cu. in.

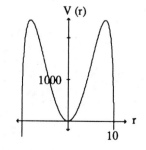

# 3 Answers to Odd-Numbered Exercises • A29

27.
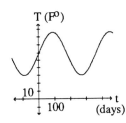

29. 90° on 86th day (June 26) 30° on 270th day (Dec. 26)

31. $f(x) \to 0.5$ as $x \to 0$

33. $f(x) \to 1$ as $x \to 0$

35. $\tan \theta = \frac{200t}{2055}$

## Exercises 3.2

1. $\cos \theta$
3. $-\sin x$
5. $\tan x$
7. $\frac{\sin^2 x + \cos^2 x}{\cos x \cdot \sin x} = \frac{1}{\cos x \cdot \sin x} = \sec x \cdot \csc x$
9. $\frac{\sin^2 x + \cos^2 x}{\cos x} = \frac{1}{\cos x} = \sec x$
11. $\frac{1 + \cos^2 x - \sin^2 x}{2 \sin x \cos x} = \frac{2 \cos^2 x}{2 \sin x \cos x} = \frac{2 \sin x \cos x}{2 \sin^2 x} = \frac{\sin 2x}{1 - \cos 2x}$

13. $\frac{\frac{\sin x + \cos x}{\cos x}}{\frac{\cos x + \sin x}{\sin x}} = \frac{\sin x}{\cos x} = \tan x$

15. true; graphs coincide

17. true; graphs coincide

19. $1 + \frac{\cos^2 \theta}{\sin^2 \theta} = \frac{\sin^2 \theta + \cos^2 \theta}{\sin^2 \theta} = \frac{1}{\sin^2 \theta} = \csc^2 \theta$

21. $\cos^2 \theta + 1 = \cos^2 \theta + \sin^2 \theta + \cos^2 \theta = 2\cos^2 \theta + \sin^2 \theta$

23. $\frac{1+\sqrt{3}}{2\sqrt{2}}; \frac{1+\sqrt{3}}{2\sqrt{2}}$

25. $\cos x = \frac{\sqrt{5}}{3}; \csc x = \frac{3}{2}$

27. $\sec x = -2$

29. $\tan 2x = -\sqrt{3}$

31. Domain: $x$ any real numbers different from $k \cdot (\sqrt{\frac{\pi}{2}})$ where $k$ is any integer.
Range: $(-\infty, -1] \cup [1, \infty)$

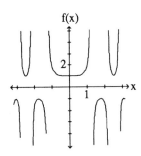

33. Domain: all real numbers
Range: $[0, 3]$

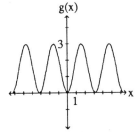

35. $\cos \theta (2 \sin \theta + 1)$

37. $3 \sin \theta \cos^2 \theta + \cos^2 \theta - \sin^3 \theta - \sin^2 \theta$

39. $\frac{\pi}{6}$ and $\frac{5\pi}{6}$

41. $\frac{\pi}{3}$ and $\frac{2\pi}{3}$

43. $\frac{\pi}{2}$ and $\frac{3\pi}{2}$

45. $0, \pi,$ and $2\pi$

47. IV

49.
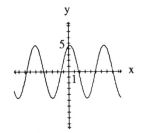

Period: $2\pi$
Phase shift: 1.03
$A = 5.83$

51. $A \sin(x + \alpha) = A(\sin x \cos \alpha + \cos x \sin \alpha) = A \sin x \cos \alpha + A \sin \alpha \cos x = (A \cos \alpha) \sin x + (A \sin \alpha) \cos x = a \sin x + b \cos x$

## Exercises 3.3

1. $6y + x = 16$
3. no
5. $(-4, -4)$

7.

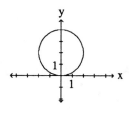

9. $f^{-1}(x) = x^2 + 2$ with $x \geq 0$, yes.

Domain of $f$: $[2, \infty)$

Range of $f$: $[0, \infty)$

Domain of $f^{-1}$: $(0, \infty)$

Range of $f^{-1}$: $[2, \infty)$

11.

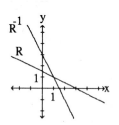

Inverse: $x = 3 - 2y$; function.

13.

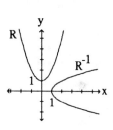

Inverse: $x = y^2 + 1$; not a function.

15.

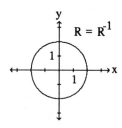

Inverse: $y^2 + x^2 = 4$; not a function.

17. $f$ is one-to-one

$x = 3y - 6$; function; $y = \frac{1}{3}x + 2$

19. $h$ is one-to-one

$x = \sqrt{y+2}$; function; $y = x^2 - 2$ with $x \geq 0$

21. $f$ is one-to-one

$x = \frac{y+3}{y-2}$; function;

$y = \frac{2x+3}{x-1}$

23. Domain of $g$: $(-\infty, \infty)$

Range of $g$: $[0, \infty)$

Domain of $g^{-1}$: $[0, \infty)$

Range of $g^{-1}$: $(-\infty, \infty)$

25. $f$ is not one-to-one

27. $f$ is not one-to-one

29. $f$ is one-to-one

31. Domain of f: $(-\infty, \infty)$
    Range of f: $(-\infty, \infty)$
    Domain of $f^{-1}$: $(-\infty, \infty)$
    Range of $f^{-1}$: $(-\infty, \infty)$

*33. Assume $f$ is not one-to-one, then we can find two distinct ordered pairs, $(a, f(a))$ and $(b, f(b))$, where $f(a) = f(b)$ but $a \neq b$. But this is a contradiction to the statement "$f(x) = f(y)$ implies $x = y$ for all $x$ and $y$ in the domain of $f$." Hence, $f$ is one-to-one.

35. $f(2\sqrt{2}) = 0$ and $f(-2\sqrt{2}) = 0$. Hence, $f$ is not one-to-one, since we have found two distinct ordered pairs, $(2\sqrt{2}, 0)$ and $(-2\sqrt{2}, 0)$, where $f(2\sqrt{2}) = f(-2\sqrt{2})$, but $2\sqrt{2} \neq -2\sqrt{2}$.

37. $f(x) = x$; $f(x) = c - x$, $c$ any constant; $f(x) = 1/x$

39. Domain $f$ : $(-\infty, \infty)$
    Range $f$ : $(-\infty, 1) \cup [2, \infty)$
    Domain $f^{-1}$ : $(-\infty, 1) \cup [2, \infty)$
    Domain $f^{-1}$ : $(-\infty, \infty)$
    $f^{-1}(x) = \begin{cases} \frac{x}{x-1} & \text{for } x < 1 \\ \sqrt{x-1} & \text{for } x \geq 2 \end{cases}$

## Exercises 3.4

1. 21.22°
3. 7.13°
5. 0.48
7. 1.17
9. $\frac{\pi}{2}$
11. $\frac{\pi}{4}$
13. $-\frac{\pi}{3}$
15. 0.36
17. 0.42
19. no value
21. no value
23. 0.74
25. 1.34
27. $\frac{\sqrt{3}}{2}$
29. $\frac{1}{2}$
31. 0.8
33. Domain: $[-1, 1]$
    Range: $[-0.57, 2.57]$

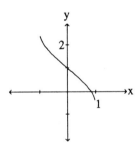

35. Domain: $(-\infty, \infty)$
    Range: $(-0.4, 0.4)$

37. Domain: $[-1, 1]$
    Range: $[-1, 1]$

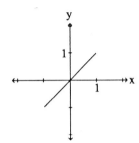

39. $[-1, 1]$
41. $\sin(.5)$
43. $\frac{x}{\sqrt{x^2+1}}$
45. $\frac{x}{\sqrt{1-x^2}}$
47. $1.16 + 2k\pi$ or $4.30 + 2k\pi$, where $k$ is any integer.
49. $1.23 + 2k\pi$ or $5.05 + 2k\pi$, where $k$ is any integer.
51. $f(x) \to 1$
53. $[-\pi, \pi]$
55. $\sin(-x) = -\sin x$ for $-\frac{\pi}{2} \leq x \leq \frac{\pi}{2}$, so for this interval $\sin^{-1}(-x) = -\sin^{-1}(x)$
57. Let $x = \arccos \theta$. Then $\cos x = \theta$, for $0° \leq x \leq 180°$. Since $\cos x = \sin(90° - x)$, $\theta = \sin(90° - x)$. So, $\arcsin \theta = 90° - x$. Since $x + (90° - x) = 90°$, $\arccos \theta + \arcsin \theta = 90°$.

59. $\theta = \tan^{-1}\left(\frac{50}{L}\right)$, $L \geq 0$

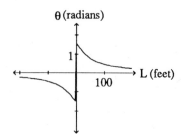

*61. $\sin\frac{\theta}{2} = \frac{c/2}{a} = \frac{c}{2a}$, then
$c = 2a\sin\frac{\theta}{2}$.
$\cos\frac{\theta}{2} = \frac{h}{a}$ then
$h = a\cos\frac{\theta}{2}$.
$A = \frac{1}{2}ch =$
$\frac{1}{2}\left(2a\sin\frac{\theta}{2}\right)\left(a\cos\frac{\theta}{2}\right)$
$= a^2\sqrt{\frac{1-\cos\theta}{2}} \cdot \sqrt{\frac{1+\cos\theta}{2}}$
$A = \frac{1}{2}a^2\sin\theta$

63.

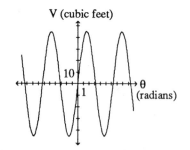

65. $V = 45\sin\theta$
67. $-1 \leq x \leq 1$
69. Compute $\tan^{-1}\frac{x}{\sqrt{1-x^2}}$.
71. $\sqrt{2}\cos\left(x - \frac{\pi}{4}\right)$
73. $5\cos(2x - 0.64)$

## Exercises 3.5

1. $0.775 + 2k\pi$, $2.366 + 2k\pi$

3. No solution.
5. $\frac{\pi}{4}, \frac{5\pi}{4}$
7. $\frac{\pi}{6}, \frac{5\pi}{6}$
9. $\frac{1}{2}\arcsin(\frac{1}{3})$:
   0.62, 2.53, 3.76, 5.66
11. $0, \pi, 2\pi$
13. $\frac{\pi}{2}, \frac{3\pi}{2}$
15. $k\pi$, where $k$ is any integer
17. $\frac{\pi}{6} + 2k\pi$ and $\frac{5\pi}{6} + 2k\pi$, where $k$ is any integer
19. $\frac{\pi}{6} + 2k\pi$ and $\frac{5\pi}{6} + 2k\pi$, where $k$ is any integer.
21. $2k\pi \pm \frac{\pi}{3}$ and $(2k-1)\pi$, where $k$ is any integer
23. $\frac{\pi}{2} + 2k\pi$, where $k$ is any integer
25. $0 \leq x \leq \frac{\pi}{6}$, $\frac{5\pi}{6} < x \leq 2\pi$
27. $-\frac{3\pi}{8} + k\pi < x < \frac{\pi}{8} + k\pi$, where $k$ is any integer.
29. $-2.28, 0, 2.28$
31. $0, 1.28$
33. $1.57 + 2k\pi < x < 2.84 + 2k\pi$ and $4.71 + 2k\pi < x < 6.59 + 2k\pi$, where $k$ is any integer.
35. $x = a + 2k\pi$ for $a = 1.18$, 2.36, 2.75, 4.32, 5.50, 5.89 where $k$ is any integer
37. Local max: $(0, 5)$, $(12.56, 5)$
    Local min: $(6.28, -1)$

39. Local max: $(0.64, 1)$,
    $(2.64, 1)$
    Local min: $(1.64, -3)$

41. Local max: $(3.62, 1.91)$,
    $(26.97, 1.22)$, $(16.79, 0.22)$
    Local min: $(10.72, -1.22)$,
    $(20.91, -0.22)$,
    $(34.08, -1.91)$

    These extrema are in $[0, 12\pi]$ and will repeat every $12\pi$ which is the period of the function.

43. $A = 2$, $B = -3$, $C = -2$, and $D = \frac{3\pi}{2}$

45. $y = \frac{27}{2} + \frac{27}{2}\sin\left(\frac{2\pi}{40}(t - \frac{40}{4})\right)$

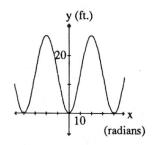

47. 0.15 radians ($\sim 9°$)

49. 45,000.00 feet
51. a)

b) $0 < \theta < 180°$
c) $\theta = 54.72°$ or 0.96 radians
d) 7.72 sq. in.

53. $x = \frac{\pi}{2}$, $\sin^{-1}(\frac{-1+\sqrt{5}}{2})$, $\pi - \sin^{-1}(\frac{-1+\sqrt{5}}{2})$

55. $2k\pi < x < \frac{\pi}{2} + 2k\pi$ (otherwise at least one term is undefined.)

## Exercises 3.6

1. Answers will vary, but once one side is chosen, the other 2 are determined. For example, $a = 11.2$cm then $b = 20.4$cm and $c = 20.2$cm.
3. $\gamma = 110°$, $a = 12.86$, $c = 18.79$
5. $\alpha = 90°$, $\gamma = 60°$, $c = 6\sqrt{3}$ or 10.39
7. $\gamma = 68°$, $a = 3.88$, $c = 6.61$
9. $\beta = 73.25°$, $\gamma = 56.75°$, $c = 4.37$
or
$\beta = 106.76°$, $\gamma = 23.24°$, $c = 2.06$
11. $\alpha = 55.17°$, $\gamma = 86.83°$, $c = 19.46$ or $\alpha = 124.83°$, $\gamma = 17.17°$, $c = 5.75$
13. no triangle possible.
15. $\alpha = 22.06°$, $\gamma = 5.94$, $c = 2.20$
17. (a) $6.69 < b < 10$
    (b) $b = 6.69$ or $b \geq 10$
    (c) $b < 6.69$
19. $a = 19.70$ miles
    $b = 15.05$ miles
    $h = 11.86$ miles
21. 0.72 miles
23. First, put in an altitude from $B$ to $AC$ to use in the solution of the triangle.
    $c = 3.85$, $\beta = 128.89°$, $\alpha = 29.11°$

## Exercises 3.7

1. no solution
3. $\alpha = 24.56°$, $\beta = 99.22°$, $\gamma = 56.22°$
5. $a = 9.83$, $\beta = 89.32°$, $\gamma = 35.68°$
7. no solution
9. no solution
11. (a) $6.78 < b < 8$
    (b) $b = 6.78$ or $b \geq 8$
    (c) $b < 6.78$
13. 17.55 square units
15. 110.34 square units

17. 5.56 square units
19. 841.22 feet
21. 42.50 feet
23. 41.56 feet
25. 893.64 feet, 1,123.56 feet
27. 61.73 feet

## Review Exercises

1. $\gamma = 68°$, $b = 3.88$, $c = 6.61$
3. $\beta = 113.50°$, $b = 27.55$, $c = 18.16$
5. $\gamma = 72°$, $a = 2.94$, $b = 5.05$
7. $\alpha = 44.4°$, $\beta = 78.4°$, $\gamma = 57.2°$
9. i.) $5.63 < b < 12$
   ii.) $b \geq 5.63$
   iii.) $0 < b < 5.63$
11. 22.98 square units
13. $y = -5\sin(x - 0.64)$
15. $x = 5 - 3y$; function: $y = \frac{5-x}{3}$

17. $f$ is not one-to-one

$x = (y+2)^2$; not a function

19.

| | f(x) | $f^{-1}$(x) |
|---|---|---|
| Domain | $(-\infty, \infty)$ | $[0, \infty)$ |
| Range | $[0, \infty)$ | $(-\infty, \infty)$ |

21. $f$ is not one-to-one

23. $\tan^2 x - \sin^2 x =$
$\frac{\sin^2 x - \sin^2 x \cos^2 x}{\cos^2 x} =$
$\frac{\sin^2 x (1 - \cos^2 x)}{\cos^2 x} =$
$\sin^2 x \tan^2 x$

25. Domain: $(-\infty, \infty)$
Range: $(-\infty, \infty)$

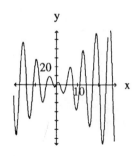

27. Domain: all real numbers except $\pi k$, where $k$ is any integer
Range: $(-\infty, \infty)$

29. $y \to 0$ as $|x| \to \infty$

31. $|y| \to \infty$ as $|x| \to 0$

33. Domain: $(-\infty, \infty)$
Range: $[-1.5, 3]$
Period: $2\pi$

35. Maxima: $(4.71, -1), (6.28, -1)$,
Absolute max at $(1.57, 3)$
Minima: $(0, -1)$

Absolute min at $(3.67, -1.50)$, $(5.76, -1.50)$

37. $\csc x - \cos x \cot x =$
$\frac{1}{\sin x} - \cos x \cdot \frac{\cos x}{\sin x} =$
$\frac{1 - \cos^2 x}{\sin x} = \sin x$

39. $\frac{1+\tan\theta}{1-\tan\theta} + \frac{1+\cot\theta}{1-\cot\theta} ==$
$\frac{2(1-\tan\theta\cot\theta)}{(1-\tan\theta)(1-\cot\theta)} = 0$

41. $x \neq \frac{\pi}{2}$

43. $\frac{\sqrt{2+\sqrt{2}}}{2}$

45. $\sqrt{\frac{3-\sqrt{5}}{6}}$

47. Domain: $(-\infty, \infty)$
Range: $[0, 5]$
Period: $\frac{\pi}{2}$

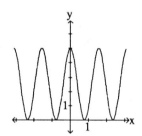

49. $\frac{\pi}{3}, \frac{5\pi}{3}, \pi$

51. $50°$

53. $\frac{\pi}{3}$

55. $0.25$

57. no value

59. $\frac{1}{\sqrt{3}}$

61. Domain: $[-1, 1]$
Range: $[0, \frac{\pi^2}{4}]$

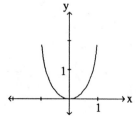

63. $\sqrt{1-x^2}$
65. $-4.40 \leq x \leq -1.88$, $0.63 \leq x \leq 1.05$, $5.24 \leq x \leq 5.65$
67. $\frac{\pi}{3}, \frac{5\pi}{3}$
69. $\frac{3\pi}{2} + 2k\pi$, where $k$ is any integer
71. no solution
73. $\frac{\pi}{2} < x < \frac{3\pi}{2}$
75. Maxima: $(-5.54, 16.04)$, $(-2.46, 9.81)$, $(0, 5)$, $(2.46, 9.81)$, $(5.54, 16.04)$
    Minima: $(-6.28, 5)$, $(-3.99, -2.92)$, $(-1.01, 3.18)$, $(3.99, -2.92)$, $(1.01, 3.18)$, $(6.28, 5)$

77. 0.61 miles

79. 1.25
81. a) $t = \sec\theta + 2 - \frac{1}{2}\tan\theta$
    b)

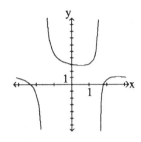

    c) $0 \leq \theta < 75.97°$
    d) $\theta = 30°$, $t = 2.87$ hours

# • CHAPTER 4

## Exercises 4.1

1. $(f+g)(x) = x^2 + 2x - 1$
   Dom: $(-\infty, \infty)$
   $(f-g)(x) = 2x - 1 - x^2$
   Dom: $(-\infty, \infty)$
   $(fg)(x) = 2x^3 - x^2$
   Dom: $(-\infty, \infty)$
   $(\frac{f}{g})(x) = \frac{2x-1}{x^2}$
   Dom: $(-\infty, 0) \cup (0, \infty)$

3. $(f+g)(x) = x^2 + 2x$
   Dom: $(-\infty, \infty)$
   $(f-g)(x) = x^2 - 2x$
   Dom: $(-\infty, \infty)$
   $(fg)(x) = 2x^3$
   Dom: $(-\infty, \infty)$
   $(\frac{f}{g})(x) = \frac{x}{2}$
   Dom: $(-\infty, 0) \cup (0, \infty)$

5. $(f+g)(x) = \frac{5x+8}{3}$
   Dom: $(-\infty, \infty)$
   $(f-g)(x) = \frac{x+10}{3}$
   Dom: $(-\infty, \infty)$
   $(fg)(x) = \frac{(x+3)(2x-1)}{3}$
   Dom: $(-\infty, \infty)$
   $(\frac{f}{g})(x) = \frac{3x+9}{2x-1}$
   Dom: $(-\infty, \frac{1}{2}) \cup (\frac{1}{2}, \infty)$

7. $(f+g)(x)$: Range: $(-\infty, \infty)$
   $(f-g)(x)$: Range: $(-\infty, \infty)$
   $(fg)(x)$: Range: $(-2.04, \infty)$
   $(\frac{f}{g})(x)$: Range: $(-\infty, 1.5) \cup (1.5, \infty)$

9. The product of functions is a different operation than the composition of functions. Let $f(x) = x + 3$ and $g(x) = x - 1$. Then $fg(x) = (x+3)(x-1)$ and $f \circ g(x) = x + 2$.

11. Multiplication of polynomial functions is commutative, associative, has an identity ($f(x) = 1$) and has an inverse $\left(\frac{1}{f(x)} \text{ where } f(x) \neq 0\right)$.

13.

15.

17. Domain: $(-\infty, \infty)$
Range: $(-\infty, -2] \cup (2, \infty)$

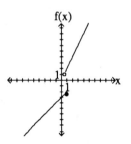

19. Domain: $(-\infty, \infty)$
Range: $(0, \infty)$

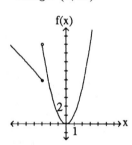

21. Domain: $(-\infty, \infty)$
Range: $(-\infty, 4]$

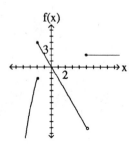

23. $f$ is discontinuous at $x = -2$ and $x = 5$

25. Domain: $(-\infty, \infty)$
Range: $\{-1, 0, 1\}$

27. Domain: $(-\infty, 5) \cup (5, \infty)$
Range: $\{-1, 1\}$

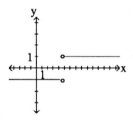

29. Domain: $(-\infty, \infty)$
Range: $\{0, \frac{1}{2}, 1\}$

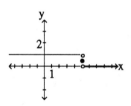

31. Domain: $(-\infty, 5) \cup (5, \infty)$
Range: $\{0, 1\}$

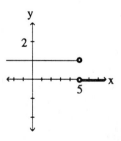

33. Domain: $(-\infty, \infty)$
Range: $(-0.19, \infty)$

Let $f_1(x) = x^3 + x + 1$ and $f_2(x) = x^4 - x^2$, then $g = C_{(-\infty, 2)} f_2 + C_{(2, \infty)} f_1$ agrees with $f$ except at $x = 2$.

35. $f$ is discontinuous at $x = 2$.

37. By definition the signum function has domain $(-\infty, \infty)$ and range $\{-1, 0, 1\}$. The graph of $\operatorname{sgn}(x - a)$ is a transformation of the graph of $\operatorname{sgn}(x)$ by a horizontal shift of $a$ units.

39. $\$2677.50 + 0.28 \times (\$43,150 - \$17,850) = \$9761.50$

41. Domain: $[0, \infty)$
Range: $[0, \infty)$

43. $C(x) = \begin{cases} 0.72x + 0.72 & 0 < x \le 5 \\ 0.63x + 1.08 & 5 < x \le 15 \\ 0.51x + 2.76 & 15 < x \end{cases}$

45.

47.

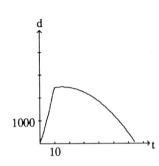

49. $d = 2475$ miles, $t = 15$ minutes

51. $5808.33

53. $B(10) = \$128,000$;
    $B(15) = \$153,000$;
    $B(25) = \$203,000$.

55.

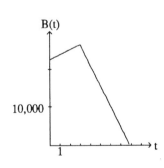

57. If $x < 0$ then
    $\text{sgn}(x)\text{abs}(x) = -1(-x) = x$

    If $x = 0$ then
    $\text{sgn}(x)\text{abs}(x) = 0 \cdot 0 = x$

If $x > 0$ then
$\text{sgn}(x)\text{abs}(x) = 1 \cdot x = x$

## Exercises 4.2

1. The graphs are the same.

3.

5.

7.

9. Domain: $(-\infty, \infty)$
   Range: $(0, \infty)$

11. $x = 4$

13. $x = -4$

15. $x = -79.80$

*17. This problem requires logarithms or a grapher to solve $e^{2x} = 3$ to get $x = 0.55$.

19. Domain: $(-\infty, \infty)$

21. Domain: $(-\infty, \infty)$

23. Increasing: $(-\infty, \infty)$
    Range: $(0, \infty)$

25. Increasing: $[-0.91, \infty)$
    Decreasing: $(-\infty, -0.91]$
    Minimum: $(-0.91, -0.33)$
    Range: $[-0.33, \infty)$

27. Increasing: $[-0.85, 0.85]$
    Decreasing: $(-\infty, -0.85] \cup [0.85, \infty)$
    Maximum: $(0.85, 0.52)$
    Minimum: $(-0.85, -0.52)$
    Range: $[-0.52, 0.52]$

29. From the horizontal line test, we see $f$ is one-to-one.

31.

33.

35.

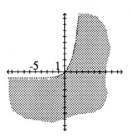

37. $x = -0.77, 2, 4$. For $x \in (-0.77, 2) \cup (4, \infty)$, $2^x > x^2$. For $x \in (-\infty, -0.77) \cup (2, 4)$, $x^2 > 2^x$.

39. (a) 10,000

(b)

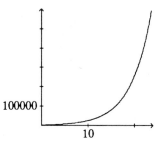

(c) 15.97 hours

41. (a) $P(t) = 475,000(1.0375)^t$

(b) The problem situation is represented by the portion of the graph in the first quadrant.

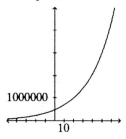

(c) after 20.22 years

43. $P(t) = 6250(1.0375)^t$

(b) The problem situation is represented by the portion of the graph in the first quadrant.

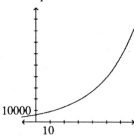

(c) in 1915–15,600 people

(d) in 1940–39,381 people

45. $A(t) = 3.5(0.5)^{\frac{t}{65}}$

(b)

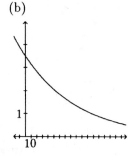

(c) $t = 117.49$

# Exercises 4.3

1.

3.
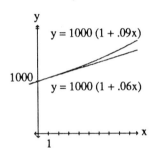

5. 1a: Amount in account after $x$ years at 5% simple interest.
   1b: Amount in account after $x$ years at 5% compounded annually.
   2a: Amount in account after $x$ years at 8% simple interest.
   2b: Amount in account after $x$ years at 8% compounded annually.

7. $r = 0.125$
9. $P = 1096.49$
11. $t = \frac{S-P}{rP}$
13. $r = 0.079$
15. $P = 729.73$
17. $P = \frac{S}{(1+r)^t}$
19. 11.55 years
21. 6.63 years
23. $r = 0.08$
25. 12.14 years
27. (a) $669.11
    (b) $673.43
    (c) $674.43
    (d) $674.91
    (e) $674.93

29.
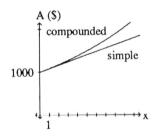

31. 6% compounded quarterly
33. 7.20% compounded daily

35.
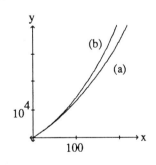

37. The values of $y$ in 35(a) represent the future value of the annuity after $x$ payments at 6% APR. The values of $y$ in 35(b) represent the future value of the annuity after $x$ payments at 8% APR.

39. $187.71
41. $36013.70
43. 14 years, 4 months
45. a) $884.61
    b) $905.77
    c) $946.93
    d) $1032.14

47. a)
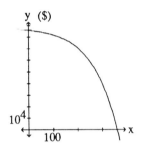

b) The $y$ value is the outstanding balance after $n$ of the 360 payments have been made.

49. a) $R(x) = \frac{860}{1-(1.01)^{-x}}$

b)
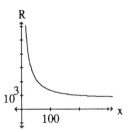

c) $R$ represents the monthly payment for a $86,000 mortgage at 12% APR to be paid in $x$ installments.

51. $i = 0.0125$; $i$ is the monthly interest rate.

*53. Each $R(1+i)^k$ is the future value of one payment after $k$ months. The number of months on deposit varies from $n-1$ for the first payment to 0 for the last payment. Summing the future values gives the desired result.

*55. From Exercise 52, we know $A(1+i)^n = S$. From Exercise 54 we know $S = R\left[\frac{(1+i)^n - 1}{i}\right]$. Therefore, $A = S(1+i)^{-n} = R\left[\frac{(1+i)^n - 1}{i}\right](1+i)^{-n} = R\left[\frac{1-(1+i)^{-n}}{i}\right]$

57.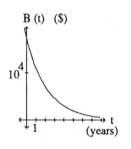

59. $B(4.25) = \$2599$

61. If $r < 0$, then $C(1-r)^n = C(1+|r|)^n$, where $|r| \geq 0$. But, this is equal to $C(1+r)^n = S$ when $r \geq 0$. $(1+r)^n$ will always increase as $n$ gets larger (i.e. the length of the term increases), will always be positive, and will be greater than 1. So $S = C(1+r)^n > C$.

63. $S(7) = \$94{,}260.33$
65. $\$0.46$

## Exercises 4.4

1. 2
3. $-4$
5. No solution. Not defined $x < 0$.
7. 81
9. 5
11. 1.63
13. $\pm 3$
15.

17.

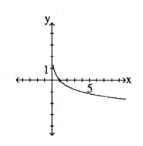

19. $\log_2(x+y) = 8$
21. $3^{-2} = \frac{x}{y}$
23. $\log 5000 + \log x + 360\log(1+r)$
25. The domain of $f(x)$ is $(-\infty, 0) \cup (0, \infty)$, while the domain of $g(x)$ is $(0, \infty)$.
27. $f(x) \to 0$
29. $f(x) \to \infty$
31. $P(t) = 123{,}000(0.97625)^t$ in 37.45 years
33. $A(t) = 6.58(\frac{1}{2})^{t/14}$ when $t > 38.05$ days
35. Table 2 does not have a power rule since the points $(\ln x, \ln y)$ are not linear.

Table 3 has a power rule since the points $(\ln x, \ln y)$ are linear.

Table 4 does not have a power rule since the points $(\ln x, \ln y)$ are not linear.

37. Table 4 has an exponential algebraic representation since the points $(x, \ln y)$ are linear.
39. $t = 11.91$
41. $t = 15.52$
43. $\frac{\ln S - \ln p}{\ln(1+r)} = t$
45. 15.75 years
47. 40.75
49. $x = 0.01$
51. $m = 1.51$

$b = -21.80$

|         | ln x    | ln P   |
|---------|---------|--------|
| Earth   | 18.344  | 5.900  |
| Mercury | 17.399  | 4.477  |
| Venus   | 18.022  | 5.416  |
| Mars    | 18.769  | 6.532  |
| Jupiter | 19.996  | 8.373  |
| Saturn  | 20.602  | 9.283  |

53. $43{,}504.80$ days

55. $p = 49.90 x^{.25}$

## Exercises 4.5

1.
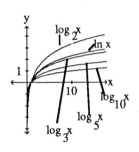

3. $(1, \infty)$
5. $(1, \infty)$
7.

9.

11. Domain: $(2, \infty)$
    Range: $(-\infty, \infty)$

13. Domain: $(-\infty, -3) \cup (-3, \infty)$
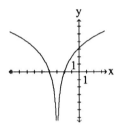

15. Domain: $(0, 1) \cup (1, \infty)$

17. Domain: $(-\infty, \infty)$

19. $f(x) \to \infty$ as $x \to \infty$
    $f(x)$ not defined
    $x \in (-\infty, 0)$

21. $f(x) \to 0$ as $x \to \infty$

    $f(x)$ not defined
    $x \in (-\infty, 0)$

23. local minimum:
    $(0.61, -0.18)$
    decreasing on $(0, 0.61]$
    increasing on $[0.61, \infty)$

25. local minimum:
    $(2.72, 0.37)$
    decreasing on $[2.72, \infty)$
    increasing on $(0, 2.72]$

27.
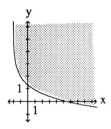

29. $(-2.02, 2.09)$ and $(2.23, 2.98)$

31. $f(x) > g(x)$ in $(-\infty, 2.48) \cup (3, \infty)$
    $f(x) = g(x)$ at $(2.48, 15.22)$ and $(3, 27)$
    $f(x) < g(x)$ in $(2.48, 3)$

33. $f(x) > g(x)$ in $(0, 10)$

$f(x) = g(x)$ at $(10, 10^{10})$ and $(0, 1)$

$f(x) < g(x)$ at $(10, \infty)$

$y = x^x$ is usually undefined for $x < 0$

35. $f(x) > g(x)$ in $(0, \infty)$

$g(x)$ is defined only in $(1, \infty)$

37. graph $y = \dfrac{\ln 4}{\ln x}$

39.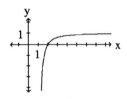

41. Domain: $(1, \infty)$

Range: $(-\infty, 1)$

43. Let $a_1$ be amplitude when 6 on Richter scale.

Let $a_2$ be amplitude when 4 on Richter scale.

Then $6 = \log(\frac{a_1}{T}) + B$ and $4 = \log(\frac{a_2}{T}) + B$ and

$2 = \log(\dfrac{a_1}{T}) - \log(\dfrac{a_2}{T})$

$2 = \log(\dfrac{a_1}{T} \cdot \dfrac{T}{a_2})$

$2 = \log(\dfrac{a_1}{a_2})$

$10^2 = \dfrac{a_1}{a_2}$ or $a_2 100 = a_1$

45. $I = 10^{(-0.00235x + \log 12)}$

$= \dfrac{12}{10^{(0.00235x)}}$

47. $I = 10.20$

49. $i(t) = \frac{10}{3}(1 - e^{(-\frac{3t}{0.03})})$

51. When $0 \le r \le 0.2$

53.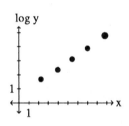

55. $f(x) = ab^x$

$\log(f(x)) = \log(ab^x)$

$\log(f(x)) = \log a + x \log b$

$\log(f(x)) = (\log b)x + \log a$

57. $\log(f(x)) = (\log b)x + \log a$

$\log f(x_1) = x_1 \log b + \log a$

$\log f(x_2) = x_2 \log b + \log a$

$\dfrac{\log f(x_2) - \log f(x_1)}{x_2 - x_1} = \log b$

59. (a)

(b) $923.17

(c) $\sim 252$ payments (21 years)

## Exercises 4.6

1. 10,000
3. $5\frac{1}{4}$
5. 3
7. 10.04
9. 1
11. $y = 2 \log x$

Domain: $(0, \infty)$

Range: $(-\infty, \infty)$

$y = \log x^2$

Domain: $(-\infty, 0) \cup (0, \infty)$

Range: $(-\infty, \infty)$

13. $y = \log_5 \sqrt{x+6} - \log_5 x$

Domain: $(0, \infty)$

Range: $(-\infty, \infty)$

$y = \log_5 \dfrac{\sqrt{x+6}}{x}$

Domain: $(-\infty, 0) \cup (0, \infty)$

Range: $(-\infty, \infty)$

15. $x = \dfrac{\log 4.1}{\log 1.06}$

17. $x = \dfrac{\log\left(\frac{12 + \sqrt{148}}{2}\right)}{\log 2}$

19. 24.22

21. 3.59

23. Domain: $(-\infty, \infty)$
    Range: $[-0.37, \infty)$
    Local minimum: $(-1.00, -0.37)$
    Increasing in: $[-1, \infty)$
    Decreasing in: $(-\infty, -1]$

25. Domain: $(-\infty, \infty)$
    Range: $(-\infty, \infty)$
    Increasing in: $(-\infty, \infty)$
    No local extrema

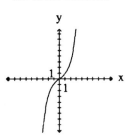

27. Domain: $(-\infty, -2) \cup (0, \infty)$
    Range: $(-\infty, \infty)$
    Increasing in: $(0, \infty)$
    Decreasing in: $(-\infty, -2)$
    No local extrema

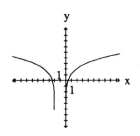

29. $(9, \infty)$

31. 1.31

33. $(0, 1.71)$

35. $1.08, 58.77$

37. $\frac{\ln x}{x} = \frac{1}{10}$ can be rewritten as $\ln x = \frac{1}{10} x \ln 2$ so $x = 2^{\frac{1}{10}x}$ or $x^{10} = 2^x$

39.

41.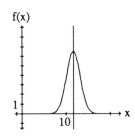

43. $f_2(x) = \frac{x^2}{2!} + x + 1 = \frac{x^2}{2!} + f_1(x)$

    $f_3(x) = \frac{x^3}{3!} + \frac{x^2}{2!} + x + 1$

    $= \frac{x^3}{3!} + f_2(x)$

    $\vdots$

    $f_n(x) = \frac{x^n}{n!} + f_{n-1}(x)$

45. $[-1.32, 1.17]$

47.

49. Relative maximum of 2.38 at $r = 1.73$

51. $B = \$6295.11$.

53. The straight line method allows for the greatest depreciation in the first two years.

55. $\approx 10.5\ \%$

## Review Exercises

1. $f + g$: $\frac{x^3 + 3x^2 - 3}{x - 1}$
   Dom: $(-\infty, 1) \cup (1, \infty)$
   $f - g$: $\frac{-x^3 - 3x^2 + 5}{x - 1}$
   Dom: $(-\infty, 1) \cup (1, \infty)$
   $fg$: $\frac{(x+2)^2}{x-1}$
   Dom: $(-\infty, 1) \cup (1, \infty)$
   $f/g$: $\frac{1}{(x-1)(x+2)^2}$
   Dom: $(-\infty, -2) \cup (-2, 1) \cup (1, \infty)$

3. Domain: $(-\infty, \infty)$

Range: $(0, \infty)$

5. Domain: $(-\infty, \infty)$

   Range: $\{-1, 0, 1\}$

7. Domain: $(-\infty, \infty)$

   Range: $(-\infty, \infty)$

   Let $g_1(x) = x^3 - x - 1$ and $g_2(x) = 2x + x^5$. Then $f = C_{(2,\infty)}g_1 + C_{(-\infty,2)}g_2$ agrees with $g$ except at $x = 2$.

9. Domain: $(-\infty, \infty)$

   Range: $(-\infty, 3)$

11.

13.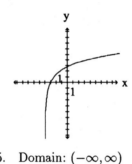

15. Domain: $(-\infty, \infty)$

    Range: $(-\infty, 4.246)$

17. Increasing: $(-\infty, 1.44]$

    Decreasing: $[1.44, \infty)$

    Maximum: $(1.44, 4.25)$

19. no solution

21. 64

23. $\dfrac{\log z}{\log(23+x)}$

25. $g(x) \to \infty$ as $x \to 2^+$

27. $(0, 1)$

29. Domain: $(-\infty, 5) \cup (5, \infty)$

Range: $(-\infty, \infty)$

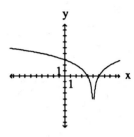

31. Decreasing: $(0, 0.72]$

    Increasing: $[0.72, \infty)$

    Local minimum: $(0.72, -0.12)$

33. 5

35. $\dfrac{\log 0.90}{\log 1.5}$

37. Domain: $(-\infty, 0) \cup (3, \infty)$

    Range: $(-\infty, 0) \cup (0, \infty)$

    Decreasing in: $(-\infty, 0)$ and $(3, \infty)$

    No local extrema

39. $(3.52, \infty)$

41. $(0.72, \infty)$

# 5 Answers to Odd-Numbered Exercises • A45

43. $(-\infty, 0.69]$

45. (a) $701.28
    (b) $707.39
    (c) $708.81
    (d) $709.51
    (e) $709.53

47. Approximately 11 years

49. (a) $P(t) = 625,000(1.0405)^t$

(b)

(c) In 11.84 years.

51. 6.29

# • CHAPTER 5

## Exercises 5.1

1. $2 + 7i$

3. $14 - 8i$

5. $i$

7.

9.

11.

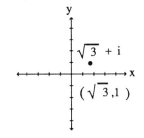

13. $a = 2, b = 2, |z_1| = 2\sqrt{2}$

15. $2\sqrt{2}(\cos\frac{\pi}{4} + i\sin\frac{\pi}{4})$

17. $1 + 7i$

19. $11 + 16i$

21. $5\sqrt{13}$

23. $5 + 7i$

25. $1 + \sqrt{3}i$

27. $0.96 - 3.26i$

29. $\sqrt{13}(\cos(0.98) + i\sin(0.98))$ and $\sqrt{13}(\cos(0.98 + 2k\pi) + i\sin(0.98 + 2k\pi))$, where $k$ is any integer.

31. $\sqrt{66}(\cos(4.89) + i\sin(4.89))$ and $\sqrt{66}(\cos(4.89 + 2k\pi) + i\sin(4.89 + 2k\pi))$, where $k$ is any integer.

33. If $u = r\cos\theta = a + bi$, then $b = 0$ and $\bar{u} = a + bi = r\cos\theta - 0 \cdot i = r\cos\theta$. But $\cos\theta = \cos(-\theta)$. So $\bar{u} = r\cos\theta = r\cos(-\theta)$.

35. $uv = 0$ if and only if $|u \cdot v| = 0$. By Exercise 34, $|u \cdot v| = |u| \cdot |v|$. So $|u \cdot v| = |u| \cdot |v|$ if and only if $|u| = 0$ or $|v| = 0$; that is, if and only if $u = 0$ or $v = 0$.

37. $\frac{\sqrt{3}}{3} - \frac{1}{3}i$

39. $5\sqrt{13}\ cis\ 7.51 = 6 + 17i$

41. $2\ cis\ 30° = 2(\frac{\sqrt{3}}{2} + \frac{1}{2}i) = \sqrt{3} + i$

$3 \operatorname{cis} 60° = 3(\frac{1}{2} + \frac{\sqrt{3}}{2}i) = \frac{3}{2} + \frac{3\sqrt{3}}{2}i$

$\frac{\sqrt{3}+i}{\frac{3}{2}+\frac{3\sqrt{3}}{2}i} \cdot \frac{(\frac{3}{2}-\frac{3\sqrt{3}}{2}i)}{(\frac{3}{2}-\frac{3\sqrt{3}}{2}i)} = = \frac{\sqrt{3}}{3} - \frac{i}{3}$

43. $6 + 17i$

# Exercises 5.2

1. $5(\cos 5.36 + i \sin 5.36)$
3. $\sqrt{34}(\cos 0.54 + i \sin 0.54)$
5. $2(\cos \frac{3\pi}{2} + i \sin \frac{3\pi}{2})$
7. $6 \operatorname{cis} 90° = 6i$
9. $(2 \operatorname{cis} 30°)(3 \operatorname{cis} 60°) = (\sqrt{3} + i)(\frac{3}{2} + \frac{3\sqrt{3}}{2}i) = 6i$
11. $\frac{-\sqrt{2}}{2} - \frac{\sqrt{2}}{2}i$
13. $64$
15. Roots: $1, i, -1, -i$

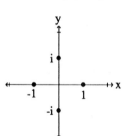

17. Roots: $\sqrt[8]{8} \operatorname{cis} \frac{\pi}{16}$, $\sqrt[8]{8} \operatorname{cis} \frac{9\pi}{16}$, $\sqrt[8]{8} \operatorname{cis} \frac{17\pi}{16}$, $\sqrt[8]{8} \operatorname{cis} \frac{25\pi}{16}$

19. Roots: $\sqrt[12]{8} \operatorname{cis} \frac{\pi}{8}$, $\sqrt[12]{8} \operatorname{cis} \frac{11\pi}{24}$, $\sqrt[12]{8} \operatorname{cis} \frac{9\pi}{8}$, $\sqrt[12]{8} \operatorname{cis} \frac{19\pi}{24}$, $\sqrt[12]{8} \operatorname{cis} \frac{35\pi}{24}$, $\sqrt[12]{8} \operatorname{cis} \frac{43\pi}{24}$

21. $-2, 1+\sqrt{3}i, 1-\sqrt{3}i$; $z = -8$

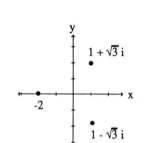

23. $\sqrt[8]{50} \operatorname{cis} \theta$ for $\theta = \frac{7\pi}{16}, \frac{15\pi}{16}, \frac{23\pi}{16}, \frac{31\pi}{16}$

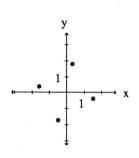

25. roots: $\sqrt{2} \operatorname{cis} \theta$ where $\theta = \frac{\pi}{12}, \frac{3\pi}{12}, \frac{5\pi}{12}, \frac{7\pi}{12}, \frac{9\pi}{12},$ $\frac{11\pi}{12}, \frac{13\pi}{12}, \frac{15\pi}{12}, \frac{17\pi}{12}, \frac{19\pi}{12},$ $\frac{21\pi}{12}, \frac{23\pi}{12}$

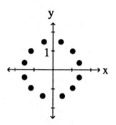

27. One possible pentagon has vertices $4 \operatorname{cis} \frac{2\pi}{5}$, $4 \operatorname{cis} \frac{4\pi}{5}$, $4 \operatorname{cis} \frac{6\pi}{5}$, $4 \operatorname{cis} \frac{8\pi}{5}$, $4 \operatorname{cis} 2\pi$

29. $(x+y)^6 = x^6 + 6x^5y + 15x^4y^2 + 20x^3y^3 + 15x^2y^4 + 6xy^5 + y^6$

31. $(x-y)^6 = x^6 - 6x^5y + 15x^4y^2 - 20x^3y^3 + 15x^2y^4 - 6xy^5 + y^6$

33. $56$

35. The coefficient of $a^{n-k}b^k$ is $\frac{n(n-1)\cdots(n-k+1)}{1 \cdot 2 \cdots k} = \frac{n(n-1)\cdots(n-k+1)}{k!} \cdot \frac{(n-k)\cdots 2 \cdot 1}{(n-k)\cdots 2 \cdot 1} = \frac{n!}{k!(n-k)!}$

37. (a) $\binom{n}{k} = \frac{n!}{(n-k)!k!}$ and $\binom{n}{n-k} = \frac{n!}{(n-(n-k))!(n-k)!} = \frac{n!}{k!(n-k)!}$

(b) $\binom{n}{1} = \frac{n!}{(n-1)!1!} = \frac{n(n-1)\cdots 2 \cdot 1}{(n-1)!1!} = \frac{n}{1!} = n$

# Exercises 5.3

1. $(\sqrt{2}, 45°)$
3. $(5, 126.87°)$

5 Answers to Odd-Numbered Exercises • A47

5. $(\sqrt{8}, \frac{\pi}{4} + 2k\pi)$, where $k$ is any integer
7. $(0,0)$
9. $(2.12, -2.12)$
11. $3.50$
13.
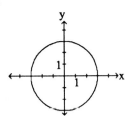
15. The graph is the y-axis.
17.

19.

21.
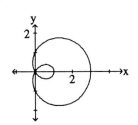

23. $0 \leq \theta \leq \frac{\pi}{2}$:

$0 \leq \theta \leq \pi$:

$0 \leq \theta \leq 2\pi$:
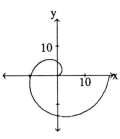

$-\pi \leq \theta \leq \pi$:
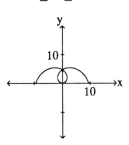

$0 \leq \theta \leq 4\pi$:
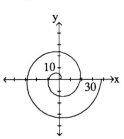

25. $0 \leq \theta \leq \frac{\pi}{2}$:
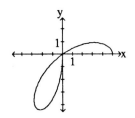

$0 \leq \theta \leq \pi$, $0 \leq \theta \leq 2\pi$,
$-\pi \leq \theta \leq \pi$, $0 \leq \theta \leq 4\pi$:
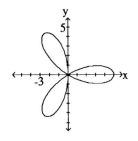

27. $-4\pi \leq \theta \leq 4\pi$:
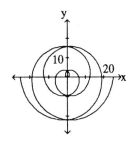

29. $0 \leq \theta \leq 2\pi$:
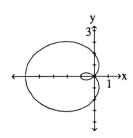

31. $0 \leq \theta \leq 2\pi$:

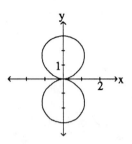

33. If $n$ is odd, there are $n$ petals.
    If $n$ is even, there are $2n$ petals.
    $r = 5\sin\theta$ is a circle.
    $r = 5\sin 2\theta$ is a four-petal rose.
    $r = 5\sin 3\theta$ is a three-petal rose.
    $r = 5\sin 4\theta$ is a eight-petal rose.
    $r = 5\sin 5\theta$ is a five-petal rose.
    $r = 5\sin 6\theta$ is a twelve-petal rose.

35. The length of the petal is determined by $a$. All graphs are three-petal roses.

37. 5
39. 8
41.

43.

45. $\left(\frac{a}{\sqrt{2}}, \theta\right)$ for $\theta = \frac{\pi}{4}, \frac{3\pi}{4}, \frac{5\pi}{4}, \frac{7\pi}{4}$

47.

49. $(x, y) = (0, 0), (2.5, 2.5),$ or $(-2.5, 2.5)$

51. $(x, y) = (0, 0),$
    $(0.96, 0.16), (0.96, -0.16),$
    $(2.29, 1.43), (2.29, -1.43),$
    $(0.41, 1.47), (0.41, -1.47),$
    or $(-0.9, 0.73)$

## Exercises 5.4

1. $(1, 7)$
3. $(-3, 8)$
5. $(4, -9)$
7. $\sqrt{50}$

9. $u + v = (4, 9)$
   $u - v = (-2, -3)$

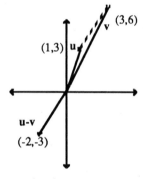

11. $(4, 2)$
13. $(12, 6)$
15. $(0, 15)$
17. $\sqrt{20}$
19. 0
21.

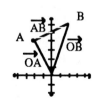

23. $3i - j$
25. $-i - 6j$
27. $u : 116.57°; v : 33.69°;$
    $\beta : 53.13°$
29. $u : 303.69°; v : 239.03°;$
    $\beta : 64.65°$

31. $(x, y) = (\frac{x+2y}{5})(1, 2) + (\frac{2x-y}{5})(2, -1)$

33. $v = (-223.99, 480.34)$

35. $v = (-136.87, 336.04)$;

    Speed = 362.84 mph;

    22.16° west of north

37. $v = (-17.95, -5.96)$;

    Speed = 18.92 mph;

    18.37° south of west

39. 758.91 mph

    in the direction

    1.05° east of north

41. $v = (0, -64)$;
    $B = (0, 156)$;

    $v = (0, -144)$;
    $B = (0, 76)$;

    $v = (0, -256)$; actually hits ground before $t = 4$.

43. $u + g = (36t \cos 70°, 36t \sin 70° - 16t^2)$

    $t = 1$: $(12.31, 17.83)$

    $t = 2.5$: already on ground

    $t = 4$: already on ground

45. 17.88 feet high, 25.98 feet in horizontal direction (use exercise 43).

## Exercises 5.5

1. Line segment from $(1, 0)$ to $(6, 5)$

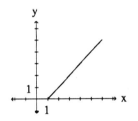

3. Line segment from $(-7, 8)$ to $(5, 4)$

5. Part of one arm of a hyperbola with asymptote at $x = 0$

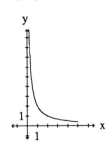

7. Circle with radius 3 centered at $(0, 0)$

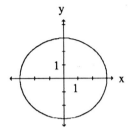

9. All except Exercises 7.

11. Domain: $[-2, 3]$;
    Range: $[-3, 2]$

13. Domain: $(0, 5]$;
    Range: $[\frac{2}{5}, \infty)$

15. $x(t) = 5 \cos t$ and $y(t) = 5 \sin t$ with $t$ in $[0, 2\pi]$.

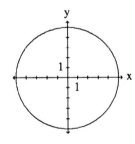

17. $x(t) = 6 \cos t$ and $y = 10 + 6 \sin t$ with $t$ in $[0, 2\pi]$.

19.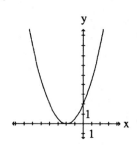

21.

23.

25.

27. Domain: $[-5, 5]$
    Range: $[-5, 5]$

29. The function $f$ is all ordered pairs $(x, f(x))$. The parametric curve defined by $x(t) = t$, $y(t) = f(t)$ is all ordered pairs $(x, y) = (t, f(t))$ for $t$ in $I$. So the curve is the function $f$ on $I$.

31. Maximum: $y(t) = 4$ when $t = 0$.

33. $x(t) = 5 \sin t \cos t$ and $y(t) = 5 \sin^2 t$ with $t$ in $[-\pi, \pi]$

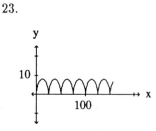

35. $x(t) = 5 \sin 3t \cos t$ and $y(t) = 5 \sin 3t \sin t$ with $t$ in $[0, 2\pi]$

37. period $= 2a\pi$; max $y = 2a$

## Exercises 5.6

1.

3.

5. $x(t) = 58t \cos 41°$ and $y(t) = -16t^2 + 58t \sin 41°$

7. The portion in the first quadrant

9. 22.62 feet at 1.19 seconds

11. 9.52 seconds, 1306.09 feet

13. yes

15. Max. height: 88.74 feet, Distance: 248.55 feet

17.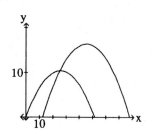

19. Chris: Max. height: 10.4 feet,
    Distance: 51.30 feet,
    Impact at 1.61 seconds
    Linda: Max. height: 15.27 feet,
    Distance: 63.12 feet,
    Impact at 1.95 seconds
    Chris' ball hits first.

21. 6.60 feet at 1.21 seconds

23. 2450.44 feet

25. 1.35 feet

27. about 51°

29. 209.34 feet, near the opponents' 15 yard line

31. 4.40 seconds

33. Home run:

35. Home run:

37. Not a home run:

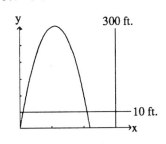

39. Point P is given by the coordinates $(a\theta - a\sin\theta, a - a\sin\theta)$. Thus, the parametric equations would be $x(\theta) = a(\theta - \sin\theta)$ and $y(\theta) = a(1 - \cos\theta)$

## Review Exercises

1. $2 + 4i$
3. $-3 + i$
5. 

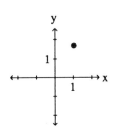

7. $5(\cos 5.35 + i\sin 5.35)$
9. $23 + 14i$
11. $3\sqrt{3} + 3i$
13. $3\sqrt{2}(\cos\frac{7\pi}{4} + i\sin\frac{7\pi}{4})$; $3\sqrt{2}(\cos\theta + i\sin\theta)$ for $\theta = \frac{7\pi}{4} + 2k\pi$, $k$ any integer
15. $\sqrt{34}(\cos 5.25 + i\sin 5.25)$; $\sqrt{34}(\cos\theta + i\sin\theta)$ for $\theta = 5.25 + 2k\pi$, $k$ any integer
17. $12\operatorname{cis} 90°$ or $12i$
19. $(3\operatorname{cis} 30°)(4\operatorname{cis} 60°) = 3(\frac{\sqrt{3}}{2} + \frac{1}{2}i) \cdot 4(\frac{1}{2} + \frac{\sqrt{3}}{2}i) = 3\sqrt{3} + 3i + 9i - 3\sqrt{3} = 12i$
21. $16\operatorname{cis} 14\pi = 16$
23. $\sqrt[8]{18}\operatorname{cis}\theta$ for $\theta = \frac{\pi}{16}, \frac{9\pi}{16}, \frac{17\pi}{16}, \frac{25\pi}{16}$

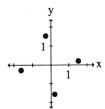

25. $z = -16 + 16i$; roots are $2\operatorname{cis}\theta$ for $\theta = \frac{\pi}{4}, \frac{11\pi}{12}, \frac{19\pi}{12}$
27. $(\sqrt{41}, 2.25)$
29. $(-2\sqrt{2}, \frac{7\pi}{4} + 2k\pi)$ and $(2\sqrt{2}, \frac{3\pi}{4} + 2k\pi)$ where $k$ is any integer
31. $(2\sqrt{2}, 2\sqrt{2})$
33. $d = 1$
35.

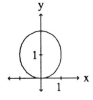

**A52** • Chapter 5

37.

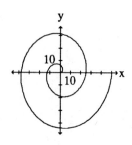

39. $0 \leq \theta \leq 2\pi$

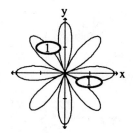

41. length $= 6$

43. $(x,y) = (-2,0)$; $(x,y) = (0,-1)$

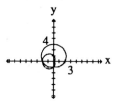

45. $(1,7)$

47. $u + v = (1,7)$

$u - v = (3,1)$

49. $(6,-5)$

51. $-2i + j$

53. $u: 36.87°$, $v: 68.20°$

angle between $u$ and $v$ is $31.33°$

55. Let $(x,y) = ar + bs$ where $a = \frac{x+2y}{5}$ and $b = \frac{2x-y}{5}$. Then $\frac{x+2y}{5}(1,2) + \frac{2x-y}{5}(2,-1) = (x,y)$

57. The curve is part of the hyperbola $y = \frac{6}{x}$.

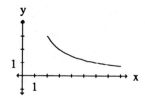

59. The curves in Exercises 56 and 57 are functions.

61. Domain: $[2,8]$; Range: $[\frac{3}{4}, 3]$

63. The curves in Exercises 56 and 57 are one-to-one.

65.

67.
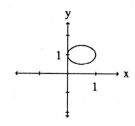

# 6 Answers to Odd-Numbered Exercises • A53

69.

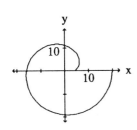

71. $x(t) = 2t\cos t$

$y(t) = 2t\sin t$

73. Velocity: $v = (90.145, -424.826)$

Speed: 434.28 mph

Direction: 11.98° east of south

75. $x(t) = (60\cos 35°)t = 49.15t$

$y(t) = -16t^2 + (60\sin 35°)t = -16t^2 + 34.41t$

77. The portion in the first quadrant.

79. Max. height = 18.5 feet when $t = 1.075$ seconds.

81. Max. height = 91.68 feet. The ball travels 307.73 feet.

83. 187.94 feet down field from the 20 yard line, around the opponents' 20 yard line. Note: this model ignores air resistance, so the actual distance would be less.

85. 2.05 seconds.

# • CHAPTER 6

## Exercises 6.1

1. Std. form:
   $(x-2)^2 = -8(y-1)$

   Vertex: $(2,1)$

   Line of Symmetry: $x = 2$

   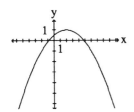

3. Std. form:
   $(y+1)^2 = -2(x - \frac{5}{2})$

   Vertex: $(\frac{5}{2}, -1)$

   Line of Symmetry: $y = -1$

5. Std. form:
   $(y-2)^2 = -8(x-3)$

   Line of Symmetry: $y = 2$

   Focus: $(1, 2)$

7. Std. form:
   $(x-3)^2 = -12(y-2)$

   Line of Symmetry: $x = 3$

   Focus: $(3, -1)$

   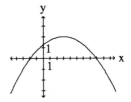

9. Std. form:
 $(x+2)^2 = 24(y+6)$

 Line of Symmetry:
 $x = -2$

 Directrix: $y = -12$

 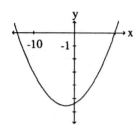

11. Std. form:
 $(y-2)^2 = 16(x+1)$

 Line of Symmetry: $y = 2$

 Directrix: $x = -5$

 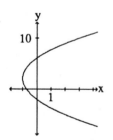

13. A horizontal stretch of factor 12, followed by a vertical shift down 1 unit, then a horizontal shift right 3 units.

15. A vertical shift down 1 unit, followed by a vertical stretch of factor 3, then a horizontal shift right 2 units.

17. Horizontal shrink by $\frac{1}{2}$ yields $y = (2x)^2 = 4x^2$.

Vertical stretch by 4 yields $y = 4x^2$.

19. Put $y = Ax^2 + Bx + C$ in standard form by completing the squares to get $y = A(x^2 + \frac{B}{A}x + \frac{B^2}{A^2}) + C - \frac{B^2}{4A}$. Simplify to get $\frac{1}{A}(y + \frac{B^2 - 4AC}{4A}) = (x + \frac{B}{2A})^2$. So $a = \frac{1}{4A}$, $h = \frac{-B}{2A}$, and $k = \frac{4AC - B^2 + 1}{4A}$. The focus is at $(h, k+a) = (\frac{-B}{2A}, \frac{4AC - B^2 + 1}{4A})$.

21. Std. form: $(x+1)^2 = y-2$
 Vertex: $(-1, 2)$
 Focus: $(-1, \frac{9}{4})$
 Directrix: $y = \frac{7}{4}$
 Line of Symmetry: $x = -1$

 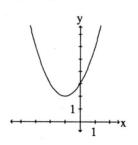

23. Std. form:
 $(y-1)^2 = -4(x - \frac{13}{4})$
 Vertex: $(\frac{13}{4}, 1)$
 Focus: $(\frac{9}{4}, 1)$
 Directrix: $x = \frac{17}{4}$

Line of Symmetry: $y = 1$

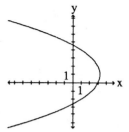

25. At the focus $(0, \frac{5}{2})$
27. At the focus $(3, 0)$
29. The equation is $x^2 = 40y$. Assume that the vertex is at $(0,0)$, the focus is $(0, 10)$, the directrix is $y = -10$ and the line of symmetry is $x = 0$.
31. No points of intersection.
33. $y = \frac{4}{3}x^2 + \frac{8}{3}$
35. $(-6, 4\sqrt{3})$ and $(-6, -4\sqrt{3})$
37.

39. Radius 1.73, height 3.01, volume 28.27 cubic units.
41. a) $(\frac{3}{4}, \frac{9}{8})$
 b) $y = \frac{4}{3}x + \frac{1}{8}$
 c) $18.43°$
 d) $18.43°$

# Exercises 6.2

1. Foci: $(\frac{-3\sqrt{7}}{2}, 0)$ and $(\frac{3\sqrt{7}}{2}, 0)$

   Std. form: $\frac{x^2}{36} + \frac{y^2}{9/4} = 1$

   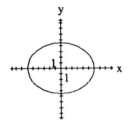

3. Endpts of major axis: $(-\sqrt{58}, 0)$ and $(\sqrt{58}, 0)$

   Endpts of minor axis $(0, -7)$ and $(0, 7)$

   Std. form: $\frac{x^2}{58} + \frac{y^2}{49} = 1$

   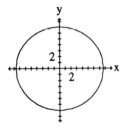

5. Endpts of major axis: $(-9/2, 0)$ and $(9/2, 0)$

   Endpoints of minor axis: $(0, -\frac{\sqrt{17}}{2})$ and $(0, \frac{\sqrt{17}}{2})$

   Std. form: $\frac{x^2}{81/4} + \frac{y^2}{17/4} = 1$

   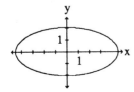

7. Endpts of major axis: $(1, 0)$ and $(1, 12)$

   Endpts of minor axis: $(1 - 3\sqrt{3}, 6)$ and $(1 + 3\sqrt{3}, 6)$

   Std. form: $\frac{(x-1)^2}{27} + \frac{(y-6)^2}{36} = 1$

   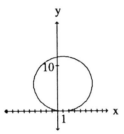

9. Endpts of major axis: $(1, -5)$ and $(1, -1)$

   Endpts of minor axis: $(1 - \sqrt{2}, -3)$ and $(1 + \sqrt{2}, -3)$

   Center: $(1, -3)$

   Foci: $(1, -3 - \sqrt{2})$ and $(1, -3 + \sqrt{2})$

   Lines of symmetry: $x = 1$ and $y = -3$

   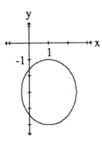

11. Endpts of major axis: $(-2 - \sqrt{5}, 1)$ and $(-2 + \sqrt{5}, 1)$

    Endpts of minor axis: $(-2, \frac{2-\sqrt{2}}{2})$ and $(-2, \frac{2+\sqrt{2}}{2})$

    Center: $(-2, 1)$

    Foci: $(\frac{-4-3\sqrt{2}}{2}, 1)$ and $(\frac{-4+3\sqrt{2}}{2}, 1)$

    Lines of symmetry: $x = -2$ and $y = 1$

13.

    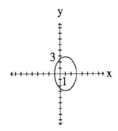

15. Std. form: $\frac{(x-1)^2}{4} + \frac{(y+1)^2}{9} = 1$

    Endpts of major axis: $(1, -4)$ and $(1, 2)$

    Endpts of minor axis: $(-1, -1)$ and $(3, -1)$

    Center: $(1, -1)$

    Foci: $(1, -1 - \sqrt{5})$ and $(1, -1 + \sqrt{5})$

    Lines of symmetry: $x = 1$ and $y = -1$

17. Std. form: $\frac{(x-4)^2}{1} + \frac{(y+8)^2}{4} = 1$

Endpts of major axis:
$(4, -10)$ and $(4, -6)$

Endpts of minor axis:
$(3, -8)$ and $(5, -8)$

Center: $(4, -8)$

Foci: $(4, -8 - \sqrt{3})$ and $(4, -8 + \sqrt{3})$

Lines of symmetry: $x = 4$ and $y = -8$

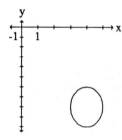

19. $5x^2 + y^2 + 5 = 0$ simplifies to $5x^2 + y^2 = -5$ which is impossible since $5x^2$ and $y^2$ are both positive. So this equation cannot be graphed. Also note that, if we convert to standard form, we get $\frac{x^2}{-1} + \frac{y^2}{-5} = 1$ but $a^2$ and $b^2$ must be positive so this is not an ellipse.

21. (a) The pool shark knows that any object passing through one focus of an ellipse will be reflected through the other focus (the pocket, in this case.)

(b) Any ball passing through the other focus will bounce off the cushion into the pocket, if another ball is not in the way, no matter how it is hit. If we denote the center of the table $(0, 0)$, the equation for the elliptical cushion is $\frac{x^2}{9} + \frac{y^2}{4} = 1$, the pocket is at $(-\sqrt{5}, 0)$, and the ball should pass through $(\sqrt{5}, 0)$.

23. If the center of the rotunda is $(0, 0)$, the foci are at $(-20\sqrt{91}, 0)$ and $(20\sqrt{91}, 0)$. The senators stood along the major axis about 190.8 feet on either side of the center.

25. $e = 0.58$

27. $\frac{x^2}{16} + \frac{y^2}{15.84} = 1$

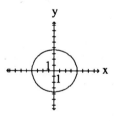

29. Since $a \geq b > 0$, we have $e = \frac{\sqrt{a^2 - b^2}}{a} \geq 0$ because $a > 0$ and $\sqrt{a^2 - b^2} > 0$. Also $b = \sqrt{a^2(1 - e^2)} > 0$ (from Exercise 10.2.26) and $a^2 > 0$, so $1 - e^2 > 0$. Thus, $e^2 < 1$ and $e < 1$. So we have $0 \leq e < 1$.

31. As $e$ approaches 1, the distance from the focus to the center approaches the length of the major axis so the ellipse becomes more elongated.

33. $e = \frac{c}{a} = \frac{0}{a} = 0$ for any circle.

35. $\frac{x^2}{2.56 \times 10^{14}} + \frac{y^2}{2.15 \times 10^{14}} = 1$

37. Earth: $\frac{x^2}{8.649 \times 10^{15}} + \frac{y^2}{8.646 \times 10^{15}} = 1$

Mars: $\frac{x^2}{1.96 \times 10^{16}} + \frac{y^2}{1.94 \times 10^{16}} = 1$

Saturn: $\frac{x^2}{7.92 \times 10^{17}} + \frac{y^2}{5.94 \times 10^{17}} = 1$

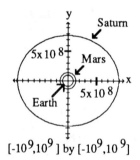

$[-10^9, 10^9]$ by $[-10^9, 10^9]$

# Exercises 6.3

1. Std. form: $\frac{x^2}{4} - \frac{y^2}{5} = 1$

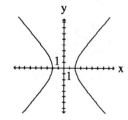

3. Std. form:
$\frac{4x^2}{9} - \frac{4(y-2)^2}{27} = 1$

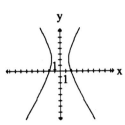

5. Center: $(0, 3)$
Foci: $(3, 3), (-3, 3)$
Vertices: $(2, 3), (-2, 3)$
Lines of symmetry: $x = 0$, $y = 3$
Asymptotes:
$y = \pm\frac{\sqrt{5}}{2}x + 3$

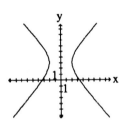

7. Center: $(3, 1)$
Foci: $(3, 1 + \sqrt{13}), (3, 1 - \sqrt{13})$
Vertices: $(3, 4), (3, -2)$
Lines of symmetry: $x = 3$, $y = 1$
Asymptotes:
$y - 1 = \pm\frac{3}{2}(x - 3)$

9.

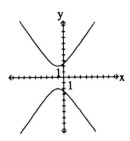

11. Hyperbola
Std. form:
$\frac{(x-2)^2}{4} - \frac{(y-1)^2}{9} = 1$
Center: $(2, 1)$
Foci: $(2 + \sqrt{13}, 1), (2 - \sqrt{13}, 1)$
Vertices: $(0, 1), (4, 1)$
Lines of symmetry: $x = 2$, $y = 1$
Asymptotes:
$y - 1 = \pm\frac{3}{2}(x - 2)$

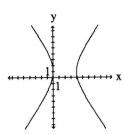

13. Hyperbola
Std. form:
$\frac{(y-1)^2}{9} - \frac{(x+3)^2}{25} = 1$
Center: $(-3, 1)$
Foci: $(-3, 1 + \sqrt{34}), (-3, 1 - \sqrt{34})$
Vertices: $(-3, 4), (-3, -2)$
Lines of symmetry:
$x = -3$, $y = 1$

Asymptotes:
$y - 1 = \pm\frac{3}{5}(x + 3)$

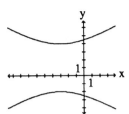

15. Hyperbola
Std. form:
$\frac{(y-1)^2}{9} - \frac{(x+1)^2}{4} = 1$
Center: $(-1, 1)$
Foci: $(-1, 1 + \sqrt{13}), (-1, 1 - \sqrt{13})$
Vertices: $(-1, 4), (-1, -2)$
Lines of symmetry:
$x = -1$, $y = 1$
Asymptotes:
$y - 1 = \pm\frac{3}{2}(x + 1)$

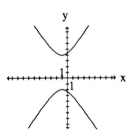

17. Parabola
Std. form:
$(x - \frac{3}{2})^2 = -\frac{5}{2}(y - \frac{7}{2})$
Center: $(\frac{3}{2}, \frac{7}{2})$
Focus: $(\frac{3}{2}, \frac{23}{8})$

Line of symmetry: $x = \frac{3}{2}$

19. $e = \sqrt{5}$
21. Std. form: $\frac{x^2}{16} - \frac{y^2}{48} = 1$

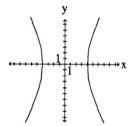

23. Std. form: $\frac{x^2}{16} - \frac{y^2}{159,984} = 1$

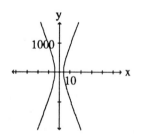

25. Eccentricity measures the ratio of the distance from the foci to the center versus the distance from the vertices to the center.
27. Std. form: $\frac{x^2}{8.649 \times 10^{15}} - \frac{y^2}{3.450951 \times 10^{18}} = 1$
29. Foci: $(6, 0), (0, 0)$
31. Light reflects from the parabolic surface toward the focus, strikes the hyperbolic surface, and reflects back to the eye of the viewer (focus).

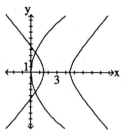

33. $\frac{(x-1.875)^2}{1.265625} - \frac{y^2}{2.25} = 1$
35. Std. form: $\frac{x^2}{2.5 \times 10^{13}} - \frac{y^2}{1.19 \times 10^{14}} = 1$
37. The graph would be one branch of a hyperbola.

## Exercises 6.4

1. Ellipse

3. Ellipse

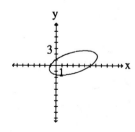

5. Degenerate conic; no graph
7. Hyperbola

9. Hyperbola

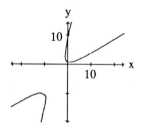

11. Major axis: $(1.87, -0.22)$
    $(-3.30, -2.37)$

    Minor axis: $(0.23, -2.98)$
    $(-1.55, 0.82)$

    Center: $(-0.59, -1.24)$

    Line of Symmetry:
    $y = 0.41x - 1$
    $y = -2.41x - 2.5$

13. Center: $(0.83, 1.06)$

Asymptotes:
$y = 3.3x - 1.5$
$y = 0.58x + 0.6$

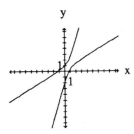

15. $y = 4.6x - 6$;
    $y = -0.36x - 1.1$

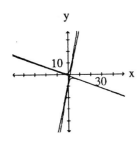

17. $y = 0.17x$;
    $y = -6.8x - 5$

19.

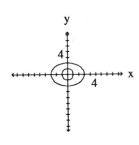

21. Vertical shift down 2 units, horizontal shift right 1 unit applied to $\frac{x^2}{9} + \frac{y^2}{4} = 1$

23. Rotation of 45° applied to $\frac{x^2}{4} - \frac{y^2}{4} = 1$

25. $20x^2 - 16xy + 32y^2 + 24x - 96y + 36 = 0$

27. $\frac{x^2}{8.978} + \frac{y^2}{3.9784} = 1$

29. $\theta = 45°$

31. Major axis: 8.78

    Minor axis: 3.49

    Center: $(2.92, 0.37)$

    Transformations applied to

    $\frac{x^2}{4.39^2} + \frac{y^2}{1.74^2} = 1$:

    Horizontal shift right 2.92 units

    Vertical shift up 0.37 units

    Rotation of 16.85°

33. Final acidity percentage
    $= \frac{0.56(74)+x}{74+x} \times 100$

35. Domain: $(-\infty, -74) \cup (-74, \infty)$

    Range: $(-\infty, 100) \cup (100, \infty)$

37. Intersection of
    $y = \frac{0.56(74)+x}{74+x}(100)$ and
    $y = 8$

    Solution: $x = 96.37$ oz. acid

39. Hyperbola

41. Positive values of $x$ make sense for this problem

43. $B \neq 0$ and $A = C$;
    45° rotation

## Exercises 6.5

1. $(-1.48, 1.74)$ & $(2.92, -0.46)$

3. $(-3.10, 3.55)$ & $(2.10, 0.95)$

5. $(-2.20, 1.36)$ & $(2.20, 1.36)$ & $(-2.20, -1.36)$ & $(2.20, -1.36)$

7. $(-1.36, 0.70), (1.36, 0.70), (-1.03, -0.89),$ & $(1.03, -0.89)$

9. $(-2.32, 0.58), (-2.32, -0.58), (4.32, 1.91), (4.32, -1.91)$

11. $(0.21, 1.87)$, $(1.87, 0.21)$,
    $(-0.53, -2.75)$,
    $(-2.75, 0.53)$

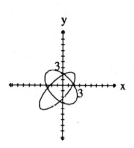

13. $(0.15, 1.84)$, $(1.63, -0.24)$,
    $(-2.08, 0.19)$,
    $(-0.37, -2.61)$

15.

17.

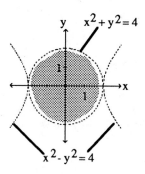

19. The shaded part of the graph is the area intersected by the right branch of the hyperbola and the ellipse.

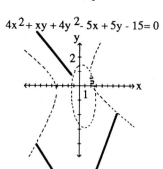

21. $f(x) = \frac{0.58(84)+x}{84+x}$
    $g(x) \geq 0.7$
    $h(x) \leq 0.8$

23. $x \geq 0$, $y \geq 0$

25. $\frac{(x-1000)^2}{346.5^2} - \frac{y^2}{879{,}937.75} = 1$

    $\frac{(y-3000)^2}{2233^2} - \frac{x^2}{4{,}013{,}711} = 1$

27. Closest solution is $(3953.20, 7939.73)$. Answers will vary widely due to roundoff error. The explosions are at about $(3950, 7950)$.

29. Vertices that define the region are $(3000, 14230.77)$, $(3000, 8600)$, and $(17640, 8600)$.

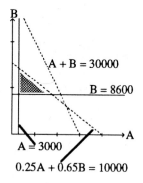

## Exercises 6.6

1. $e = \frac{\sqrt{3}}{2}$

3. $e = \frac{5}{4}$

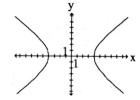

# 6 Answers to Odd-Numbered Exercises

5. $e = \frac{\sqrt{15}}{3}$

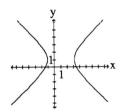

7. Center: $(1, 1)$
   Transverse length: $4.5$
   Asym: $y = 1 \pm \frac{\sqrt{6}}{3}(x - 1)$

9. $e = 1.15$ (Use estimates of $a = 5.72$, $b = 3.24$)

11.

13.

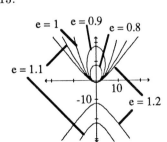

15. Parabola with $e = 1$,
    Directrix $x = -5$,
    $0 \leq \theta \leq 2\pi$

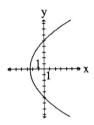

17. Ellipse with $e = \frac{1}{2}$,
    Directrix $x = -\frac{3}{2}$,
    $0 \leq \theta \leq 2\pi$

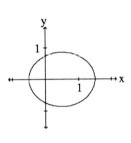

19. Ellipse with $e = \frac{1}{2}$,
    Directrix $x = \frac{8}{3}$,
    $0 \leq \theta \leq 2\pi$

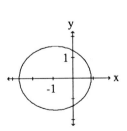

21. Ellipse with $e = \frac{3}{4}$

Std form: $\frac{(x+2.29)^2}{(6.09)^2} + \frac{y^2}{(5.64)^2} = 1$

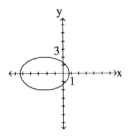

23. $e_1 = e_2 = \frac{1}{2}$. The graphs coincide.

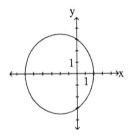

25. Yes, a hyperbola since $e > 1$

27. $e = \frac{1}{2}$. The equations in Exercises 27 and 29 yield the same graph.

29. see 27.

31.

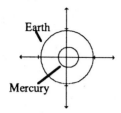

$[-2 \times 10^8, 2 \times 10^8]$ by $[-2 \times 10^8, 2 \times 10^8]$

33.

$[-4 \times 10^9, 6 \times 10^9]$ by $[-5 \times 10^9, 5 \times 10^9]$

35.

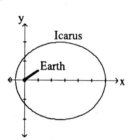

$[-1.25 \times 10^9, 7 \times 10^9]$ by $[-3 \times 10^9, 3 \times 10^9]$

37. They are similar ellipses.

39. $r = \frac{eh}{-1+e\cos\theta} = -\left(\frac{eh}{1-e\cos\theta}\right)$. The type of conic is determined by $e$; the sign of the directrix is reversed.

## Review Exercises

1. Std. form: $(x+2)^2 = 8(y+1)$

   Vertex: $(-2, -1)$

   Line of symmetry: $x = -2$

   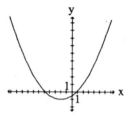

3. Std. form: $(y-4)^2 = -24(x-1)$

   Focus: $(-5, 4)$

   Line of symmetry: $y = 4$

   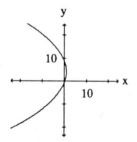

5. Horizontal shift right 3 units, followed by a vertical shrink by a factor of $\frac{1}{6}$, then a vertical shift up 1 unit.

7. Solve for $y$ as a function of $x$: $y = 2 + \sqrt{x+2}$ or $y = 2 - \sqrt{x+2}$. Then graph each function on the same coordinate axis.

9. Std. form: $(y-1)^2 = 2(x - \frac{7}{6})$

   Vertex: $(\frac{7}{6}, 1)$

   Focus: $(\frac{5}{3}, 1)$

   Directrix: $x = \frac{2}{3}$

   Line of symmetry: $y = 1$

   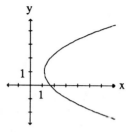

11. $y = \frac{4}{21}x^2 + \frac{68}{21}$

13. Std. form: $\frac{x^2}{64} + \frac{y^2}{39} = 1$

    Endpts. of major axis: $(-8, 0), (8, 0)$

    Endpts. of minor axis: $(0, -\sqrt{39}), (0, \sqrt{39})$

    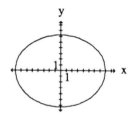

15. Std. form: $\frac{(x-2)^2}{27} + \frac{(y-6)^2}{36} = 1$

    Endpts. of major axis: $(2, 0), (2, 12)$

    Endpts. of minor axis: $(2 - 3\sqrt{3}, 6), (2 + 3\sqrt{3}, 6)$

    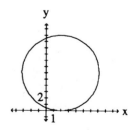

17. Endpts. of major axis: $(-3,-2), (5,-2)$

Endpts. of minor axis: $(1,-2-\sqrt{2})$, $(1,-2+\sqrt{2})$

Center: $(1,-2)$

Foci: $(1-\sqrt{14},-2)$, $(1+\sqrt{14},-2)$

Lines of symmetry: $x=1$, $y=-2$

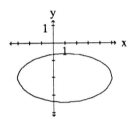

19. Std. form: $\frac{(x-3)^2}{5}+\frac{(y+1)^2}{4}=1$

Endpts. of major axis: $(3-\sqrt{5},-1)$, $(3+\sqrt{5},-1)$

Endpts. of minor axis: $(3,-3), (3,1)$

Center: $(3,-1)$

Foci: $(2,-1), (4,-1)$

Lines of symmetry: $x=3$, $y=-1$

21. Std. form: $\frac{(y-4)^2}{25}-\frac{x^2}{11}=1$

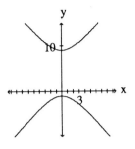

23. Center: $(-1,3)$

Foci: $(-1,3-\sqrt{8})$, $(-1,3+\sqrt{8})$

Vertices: $(-1,3-\sqrt{5})$, $(-1,3+\sqrt{5})$

Lines of symmetry: $x=-1$, $y=3$

Asymptotes: $y=3\pm\sqrt{\frac{5}{3}}(x+1)$

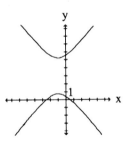

25. Parabola: $(y+3)^2=x$

Focus: $(\frac{1}{4},-3)$

Vertex: $(0,-3)$

Line of symmetry: $y=-3$

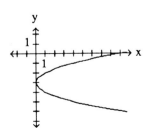

27. Hyperbola: $\frac{(y+3)^2}{\frac{1}{4}}-\frac{x^2}{\frac{1}{9}}=1$

Center: $(0,-3)$

Foci: $(0,-3-\frac{\sqrt{13}}{6})$, $(0,-3+\frac{\sqrt{13}}{6})$

Vertices: $(0,-3.5)$, $(0,-2.5)$

Lines of symmetry: $x=0$, $y=-3$

Asymptotes: $y=\pm\frac{3}{2}x+3$

29. Hyperbola.

31. Ellipse.

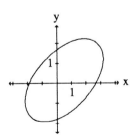

33. Asymptotes (estimated): $y=2x+4$ and $y=x-1$

Center (estimated): $(-4.7, -5.6)$

35. $y = 3.3x - 3$ and $y = -0.44x - 1$

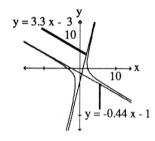

37. A horizontal shift left 2 units, a vertical shift up 1 unit, applied to $\frac{x^2}{9} + \frac{y^2}{4} = 1$

39. $(5.08, -0.54)$ and $(-23.26, 13.63)$

41. No solution.

43. 3 solutions.

45. $(-1.23, -1.34)$, $(2.48, 1.92)$

47. No solution

49. $e = 0.75$

51.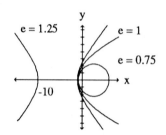

53. Hyperbola

Directrix: $x = -2$; $0 \leq \theta \leq 2\pi$

55. Directrix: $y = -2$; $0 \leq \theta \leq 2\pi$

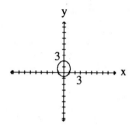

57. At the focus: $(\frac{5}{2}, 0)$

59. $\frac{x^2}{3.6 \times 10^{13}} - \frac{y^2}{1.08 \times 10^{14}} = 1$
    (Assume the space station is at $(0,0)$.)

61. $y = \frac{52+x}{80+x}(100)$

63. Domain: $(-\infty, -80) \cup (-80, \infty)$
    Range: $(\infty, 100) \cup (100, \infty)$

65. The graphs of $y = \frac{52+x}{80+x}(100)$ and $y = 81$ intersect at 67.37.

67. $(-1635.17, 2721.88)$, $(-401.66, 1438.25)$, $(3022.11, 975.52)$, $(5645.91, 3705.94)$
    Answers will vary due to round-off error

69. $A \geq 4000$
    $B \geq 8000$
    $A + B < 30,000$
    $0.25A + 0.65B < 10,000$

71. 391.20 meters

# Index

Absolute maximum, 186
Absolute minimum, 186
Absolute value function, 34
Accuracy Agreement, 52
Addition Formulas, 209
Algebraic representation, 8
Amplitude, 188
Angle Between Two Lines, 271
Angle between Vectors, 389
Angle between $\ell_1$ and $\ell_2$, 155
Angle of depression, 141
Angle of elevation, 141
Annual percentage rate, 303
Annuity, 308
Apple Macintosh, 23
Arcsin, 234
Area of Triangles, 268
Asymptotes, 441

Binomial coefficient, 369
Bisection method, 103
Boundary, 463
Built-In Functions, 32

Cardan cubic equation formulas, 64

Cardan formulas, 48, 51
Cardioid, 374
Cartesian coordinate system, 6
Center of the hyperbola, 439
Circular property of the trig functions, 243
Circular Property of Trig Functions, 149
Coefficients, 2
Common logarithmic function, 315
Complementary, 135
Complete, 23
Completing the square, 44
Compound Interest, 303
Compound Interest Formula, 303
Compounded continuously, 307
Conic, 469
Conjugate axis, 439
Constant term, 2
Continuous, 98, 99
Continuous on the interval $J$, 98
Coordinate, 6
Coordinate axes, 6
Coordinate plane, 6
Cycle, 193
Cycloid, 397

Damped Motion, 194
Data Analysis, 319
De Moivre's Theorem, 362
Decreasing on an interval $J$, 91
Degree, 2
Dependent variable, 8
Directrix, 469
Discontinuous, 100
Discriminant, 45
Distance Formula, 144
Domain of the equation, 8
Domain of the function, 19
Domain of the relation, 7
Double-Angle Formulas, 216

Eccentricity, 437
Eccentricity $e$ of an ellipse or a hyperbola, 467
End behavior, 103
End behavior model, 105, 105
End points, 393
Epicycloid, 398
Equation of Hyperbolas in Standard Form, 442
Equivalent, 42, 70
Error, 49
Exponential and logarithmic functions, 335
Exponential function, 298

Factorial, 369
First, 6
Focus, 469
Focus of the hyperbola, 439
Four-leafed rose, 377
Fourth quadrants, 6
Frequency, 193
Function, 19, 31
Function graphing utility, 23
Function that is not periodic, 249
Future value, 308

Geometric representation, 7
Graph, 186
Graph of a function, 22
Graph of an equation, 373

Graph of the equation, 8
Graph of the inverse of a relation, 225
Graph of the problem situation, 7
Graph of the relation, 7
Graphing Calculator and Computer Graphing Laboratory Manual, 471
Graphing Conics with a Function Grapher, 452
Greatest integer function, 33

Half-Angle Formulas, 217
Half-life, 298
Harmonic Motion, 193
Heron's Formula, 269
Hohmann ellipse, 438
Horizontal Line Test, 227
Horizontal shrinking, 184
Horizontal Stretch or Shrink, 182
Horizontal stretching, 184
Hyperbola, 439
Hypocycloid, 398

Identities for Complementary Angles, 135
Identity, 156
Increasing on an interval $J$, 91
Independent variable, 8
Inequalities involving trig functions, 248
Integers, 5
Interest, 302
Intermediate Value Property, 101
Inverse cosine function, 237
Inverse relation, 222
Inverse sine function, 234
Inverse tangent function, 237
Inverse trig functions, 232
Irrational number, 5

Law of sines, 254
Leading coefficient, 2
Linear combination, 387
Linear polynomial, 2
Local extreme value, 86
Local maximum value, 84
Local minimum value, 84
Logarithm of $x$ with base $a$, 317

Logarithmic function with base $a$, 317
Logarithmic functions, 316

Master Grapher, 455, 471
Modeling, 8
Multiplication by a Scalar, 386

Natural logarithmic function, 315
Negative integers, 5
Nondecreasing on an interval $J$, 92
Nonincreasing on an interval $J$, 92
Nonlinear system of equations, 459
Nonlinear system of inequalities, 459
Normal, 155

Oblique triangle, 254
One-to-one, 227
Ordered pair, 6
Ordered pairs, 6
Ordinary annuities, 308
Orthogonal, 155

Parameter, 393
Parametric equations, 393
Parametrization, 393
Path of objects, 399
Periodic, 246
Periodicity, 244
Phase shift, 189
Phase shift right, 189
Piecewise defined, 31
Piecewise-Defined Functions, 37
Plane curve, 393
Point of discontinuity, 100
Polar axis, 379
Polar Coordinate System, 379
Polar coordinates, 370
Polar equation, 473
Polar Graphs, 373
Pole, 379
Polynomial equation, 2
Polynomial equation of degree $n$, 2
Polynomial function, 2
Polynomial in $x$, 2
Polynomials of Degree 3, 85

Polynomials of Degree 4 or 5, 86
Positive integers, 5
Present value, 308
Principal, 302
Principal axis of the hyperbola, 439
Properties of Logarithms, 317

Quadrantal Angles, 151
Quadratic, 43
Quadratic form, 450
Quadratic formula, 44, 45
Quadratic polynomial, 2

Radioactive Decay, 298
Range of the equation, 8
Range of the function, 19
Range of the relation, 7
Rate of interest, 302
Rational number, 5
Real number line, 6
Real solutions, 42
Rectangular hyperbola, 438, 459
Reference angle, 152
Reference triangle, 152
Reflective Property of Hyperbolas, 447
Regular $n$-gon, 366
Relation, 7, 8
Root, 42
Roots of unity, 362
Rule for $f^{-1}$, 231

Scalar multiple, 386
Scale, 4
Second, 6
Semiminor axis length, 429
Shrink radially, 378
Sign chart, 74, 75
Sign chart method, 70
Simple Interest, 302, 303
Simultaneous solutions, 58
Sinusoid, 191
Solution, 8
Solution to a system of equations in two variables, 58
Solution to an equation, 41

Solution to an inequality, 70
Solution with error of at most $r$, 50
Solve equations and inequalities, 243
Solved the equation, 42
Solving a right triangle, 139
Speed, 390
Spiral of Archimedes, 378
Square root function, 32
Standard, 24
Standard form for the equation of a hyperbola, 440
Standard forms for the equation of an ellipse, 432
Step function, 34
Stretch radially, 377
Subscript, 2
Supplementary angles, 211

The Law of Cosines, 263
The Law of Sines, 255
Third, 6
Translation, 455
Transverse axis, 439
Trig identities, 209
Trigonometric Functions of Any Angle, 148
Trigonometric Graphs, 186
Trigonometric Identities, 157

Understood Domain of a Function, 31
Union, 71
Unit vectors, 387

Velocity, 390
Vertices of the hyperbola, 439
Viewing rectangle, 23, 24

Well-defined property of a function, 318
Whole numbers, 5

$x$-axis, 6
$x$-intercept, 42
$x$-intercepts, 42

$y$-axis, 6

Zero, 42
Zeros of Periodic Functions, 246
Zoom-in, 48, 51, 57, 63, 78
Zoom-out, 107